Current Topics in
Developmental Biology

Volume 49

The Centrosome in Cell Replication and Early Development

The Centrosome in Cell Replication and Early Development

Edited by

Robert E. Palazzo
Department of Molecular Biosciences
University of Kansas
Lawrence, Kansas

Gerald P. Schatten
Department of Obstetrics–Gynecology and Cell and Developmental Biology
Oregon Regional Primate Research Center
Oregon Health Sciences University
Beaverton, Oregon

ACADEMIC PRESS

A Harcourt Science and Technology Company

San Diego San Francisco New York Boston London Sydney Tokyo

Front cover photograph: Reprinted from *Molecular Biology of the Cell,* (1999, volume 10, 2955–2969), with permission by the American Society for Cell Biology.

This book is printed on acid-free paper. ♾

Academic Press
A Harcourt Science and Technology Company
525 B Street, Suite 1900, San Diego, California 92101-4495, USA
http://www.academicpress.com

Academic Press
Harcourt Place, 32 Jamestown Road, London NW1 7BY, UK
http://www.academicpress.com

International Standard Book Number: 0-12-153149-X

PRINTED IN THE UNITED STATES OF AMERICA
00 01 02 03 04 05 SB 9 8 7 6 5 4 3 2 1

In Memory
of
Daniel Mazia
Teacher, Scholar, Friend

Contents

4

γ-Tubulin of Budding Yeast

Jackie Vogel and Michael Snyder

5

The Spindle Pole Body of *Saccharomyces cerevisiae:* Architecture and Assembly of the Core Components

Susan E. Francis and Trisha N. Davis

6

The Microtubule Organizing Centers of *Schizosaccharomyces pombe*

Iain M. Hagan and Janni Petersen

7

Comparative Structural, Molecular, and Functional Aspects of the *Dictyostelium discoideum* Centrosome

Ralph Gräf, Nicole Brusis, Christine Daunderer, Ursula Euteneuer, Andrea Hestermann, Manfred Schliwa, and Masahiro Ueda

8

Are There Nucleic Acids in the Centrosome?

Wallace F. Marshall and Joel L. Rosenbaum

9

Basal Bodies and Centrioles: Their Function and Structure

Andrea M. Preble, Thomas M. Giddings, Jr., and Susan K. Dutcher

10

Centriole Duplication and Maturation in Animal Cells

B. M. H. Lange, A. J. Faragher, P. March, and K. Gull

11

Centrosome Replication in Somatic Cells: The Significance of the G_1 Phase

Ron Balczon

12

The Coordination of Centrosome Reproduction with Nuclear Events during the Cell Cycle

Greenfield Sluder and Edward H. Hinchcliffe

13

Regulating Centrosomes by Protein Phosphorylation

Andrew M. Fry, Thibault Mayor, and Erich A. Nigg

14

The Role of the Centrosome in the Development of Malignant Tumors

Wilma L. Lingle and Jeffrey L. Salisbury

15

The Centrosome-Associated Aurora/Ipl-like Kinase Family

T. M. Goepfert and B. R. Brinkley

16

Centrosome Reduction during Mammalian Spermiogenesis
G. Manandhar, C. Simerly, and G. Schatten

17

The Centrosome of the Early *C. elegans* Embryo: Inheritance, Assembly, Replication, and Developmental Roles
Kevin F. O'Connell

18

The Centrosome in *Drosophila* Oocyte Development
Timothy L. Megraw and Thomas C. Kaufman

19

The Centrosome in Early *Drosophila* Embryogenesis
W. F. Rothwell and W. Sullivan

Contributors

Numbers in parentheses indicate the pages on which authors' contributions begin.

Ron Balczon (251), Department of Cell Biology and Neuroscience, The University of South Alabama, Mobile, Alabama 36688

Michel Bornens (1), Institut Curie, Section Recherche, UMR 144 CNRS, 75248 Paris Cedex 05, France

B. R. Brinkley (331), Department of Molecular and Cellular Biology, Baylor College of Medicine, Houston, Texas 77030

Nicole Brusis (161), Adolf-Butenandt-Institut/Zellbiologie, Ludwig-Maximilians-Universität München, D-80336 Munich, Germany

Christine Daunderer (161), Adolf-Butenandt-Institut/Zellbiologie, Ludwig-Maximilians-Universität München, D-80336 Munich, Germany

Trisha N. Davis (105), Department of Biochemistry, University of Washington, Seattle, Washington 98195

Susan K. Dutcher (207), Department of Genetics, Washington University School of Medicine, St. Louis, Missouri 63110

Ursula Euteneuer (161), Adolf-Butenandt-Institut/Zellbiologie, Ludwig-Maximilians-Universität München, D-80336 Munich, Germany

A. J. Faragher (235), School of Biological Sciences, University of Manchester, Manchester M13 9PT, United Kingdom

Susan E. Francis (105), Department of Biochemistry, University of Washington, Seattle, Washington 98195

Andrew M. Fry (291), Department of Biochemistry, University of Leicester, Leicester LE1 7RH, United Kingdom

Thomas M. Giddings, Jr. (207), Molecular, Cellular, and Developmental Biology, University of Colorado, Boulder, Colorado 80309-0347

T. M. Goepfert (331), Department of Molecular and Cellular Biology, Baylor College of Medicine, Houston, Texas 77030

Ralph Gräf (161), Adolf-Butenandt-Institut/Zellbiologie, Ludwig-Maximilians-Universität München, D-80336 Munich, Germany

K. Gull (235), School of Biological Sciences, University of Manchester, Manchester M13 9PT, United Kingdom

Ruwanthi N. Gunawardane (55), Department of Embryology, Carnegie Institution of Washington, Baltimore, Maryland 21210; and Department of Biology, Johns Hopkins University, Baltimore, Maryland 21218

Iain M. Hagan (133), School of Biological Sciences, University of Manchester, Manchester M13 9PT, United Kingdom

Andrea Hestermann (161), Adolf-Butenandt-Institut/Zellbiologie, Ludwig-Maximilians-Universität München, D-80336 Munich, Germany

Edward H. Hinchcliffe (267), Department of Cell Biology, University of Massachusetts Medical School, Worcester, Massachusetts 01605

Dawn R. Hull (449), Department of Molecular Biosciences, University of Kansas, Lawrence, Kansas 66045

Thomas C. Kaufman (385), Department of Biology, Howard Hughes Medical Institute, Indiana University, Bloomington, Indiana 47405

Thomas Küntziger (1), Institut Curie, Section Recherche, UMR 144 CNRS, 75248 Paris Cedex 05, France

B. M. H. Lange (235), EMBL, D-69012 Heidelberg, Germany

Wilma L. Lingle (313), Tumor Biology Program, Division of Experimental Pathology, Mayo Clinic Foundation, Rochester, Minnesota 55905

Sofia B. Lizarraga (55), Department of Embryology, Carnegie Institution of Washington, Baltimore, Maryland 21210; and Department of Biology, Johns Hopkins University, Baltimore, Maryland 21218

G. Manandhar (343), Departments of Obstetrics–Gynecology and Cell and Developmental Biology, Oregon Regional Primate Research Center, Oregon Health Science University, Beaverton, Oregon 97006

P. March (235), School of Biological Sciences, University of Manchester, Manchester M13 9PT, United Kingdom

Wallace F. Marshall (187), Department of Molecular, Cellular, and Developmental Biology, Yale University, New Haven, Connecticut 06520

Thibault Mayor (291), Department of Molecular Biology, University of Geneva, CH-1211 Geneva 4, Switzerland

Timothy L. Megraw (385), Department of Biology, Indiana University, Bloomington, Indiana 47405

Erich A. Nigg (291), Department of Molecular Biology, University of Geneva, CH-1211 Geneva 4, Switzerland

Berl R. Oakley (27), Department of Molecular Genetics, The Ohio State University, Columbus, Ohio 43210

Kevin F. O'Connell (365), Laboratory of Molecular Biology, University of Wisconsin, Madison, Wisconsin 53706-1596

Robert E. Palazzo (449), Department of Molecular Biosciences, University of Kansas, Lawrence, Kansas 66045; and The Marine Biological Laboratory, Woods Hole, Massachusetts 02543

Janni Petersen (133), School of Biological Sciences, University of Manchester, Manchester M13 9PT, United Kingdom

Andrea M. Preble (207), Department of Biological Sciences, Stanford University, Stanford, California 94305-5020

Joel L. Rosenbaum (187), Department of Molecular, Cellular, and Developmental Biology, Yale University, New Haven, Connecticut 06520

W. F. Rothwell (409), Sinsheimer Laboratories, Department of Biology, University of California, Santa Cruz, California 95064

Jeffrey L. Salisbury (313), Tumor Biology Program, Mayo Clinic Foundation, Rochester, Minnesota 55905

G. Schatten (343), Departments of Obstetrics–Gynecology and Cell and Developmental Biology, Oregon Regional Primate Research Center, Oregon Health Science University, Beaverton, Oregon 97006

Manfred Schliwa (161), Adolf-Butenandt-Institut/Zellbiologie, Ludwig-Maximilians-Universität München, D-80336 Munich, Germany

Bradley J. Schnackenberg (449), Department of Molecular Biosciences, University of Kansas, Lawrence, Kansas 66045

C. Simerly (343), Departments of Obstetrics–Gynecology and Cell and Developmental Biology, Oregon Regional Primate Research Center, Oregon Health Science University, Beaverton, Oregon 97006

Greenfield Sluder (267), Department of Cell Biology, University of Massachusetts Medical School, Worcester, Massachusetts 01605

Michael Snyder (75), Department of Molecular, Cellular and Developmental Biology, Yale University, New Haven, Connecticut 06520-8103

W. Sullivan (409), Sinsheimer Laboratories, Department of Biology, University of California, Santa Cruz, California 95064

Masahiro Ueda (161), Adolf-Butenandt-Institut/Zellbiologie, Ludwig-Maximilians-Universität München, D-80336 Munich, Germany

Jacalyn M. Vogel (449), Department of Molecular Biosciences, University of Kansas, Lawrence, Kansas 66045

Jackie Vogel (75), Department of Molecular, Cellular and Developmental Biology, Yale University, New Haven, Connecticut 06520-8103

Christiane Wiese (55), Department of Embryology, Carnegie Institution of Washington, Baltimore, Maryland 21210

Andrew Wilde (55), Department of Embryology, Carnegie Institution of Washington, Baltimore, Maryland 21210

Xingyong Wu (449), Department of Molecular Biosciences, University of Kansas, Lawrence, Kansas 66045

Yixian Zheng (55), Department of Embryology, Carnegie Institution of Washington, Baltimore, Maryland 21210

Preface

After more than 100 years of research it now appears that an understanding of the molecular and structural base of the function and replication of centrosomes, spindle pole bodies, nuclear associated bodies, and related organelles may finally be attainable. For many years this field of research suffered from the inability to isolate sufficient quantities of these organelles for study and the lack of genetic approaches to the isolation of mutants that could bring important insights into the mechanisms controlling microtubule nucleation and replication of these organelles. However, as the authors of the chapters contained within this book will demonstrate, significant advances are now under way, and the identification and characterization of a number of important molecules that are broadly distributed among related organelles in various eukaryotic cells have already been achieved, even though these organelles are known to be diverse in ultrastructure. In part, these advances have resulted from a rapid and open exchange of information and reagents gained from studies on a wide variety of seemingly unrelated model organisms, leading to a convergence of thought that is essential for rapid progress in any scientific discipline. Thus, a sense of optimism is emerging that these crucial and mysterious cell organelles can finally be understood at a molecular level. Enthusiasm is further fueled by the extraordinary opportunities for biological research now emerging as a result of rapid progress in the complete sequencing of genomes from a variety of experimental model systems, advances in analytical biochemistry that allow identification and sequencing of minute quantities of proteins, and imaging technologies that allow detailed observations of the dynamic interactions of molecules in living cells. Given this background, this field of research appears to be at an important juncture where a rapid proliferation of relevant information is about to ensue.

With the hope of providing a useful compendium of current thought in this field, leading researchers were invited to contribute chapters to serve as reviews of their particular specialty areas and to provide some insights for future directions. This book is their achievement, and the editors thank all of the authors for the care and thought that have gone into each of their contributions.

R. E. Palazzo and Jerry Schatten

1

The Centrosome and Parthenogenesis

Thomas Küntziger and *Michel Bornens*
Institut Curie
Section Recherche, UMR 144 CNRS
75248 Paris Cedex 05
France

I. Introduction

We would like to use the title of this contribution, proposed by the editors of this book, to address some of the questions raised by the existence of the centrosome organelle. Besides its established role in microtubule nucleation, which remains to be understood in a comprehensive way (see other contributions in this volume; see also Knop *et al.,* 1999; Tassin and Bornens, 1999), the centrosome maintains a puzzling status as a cellular organelle. The most compelling feature is that the centrosome is not present in all multicellular organisms: Plants control their microtubule array during interphase as well as during mitosis, and divide without a centrosome. This original feature among eukaryotic cells is shared only with the chloroplast and the vacuole, which have a complementary situation. One is thus entitled to ask whether the centrosome is necessary for cell function, even though in both animal and yeast cells, the bipolarity of the cell division process rests on the reproductive capacity of the centrosome. On the other hand, a complete parthenogenetic development can take place in frog embryos when the sperm cell is replaced by the centrosome of a somatic cell, whether from the same or from divergent species.

This most challenging question should be asked in an evolutionary perspective: What does this organelle add to the cell economy that explains

Current Topics in Developmental Biology, Vol. 49

0070-2153/00 $35.00

its maintenance in so many organisms and its evolution among divergent species? Today, no one would question whether chloroplasts are necessary for cell function. Could the high evolutionary variability of the centrosome, which makes a structural definition difficult to achieve, find its natural explanation one day, when a comprehensive view of the centrosome–microtubule network is developed?

II. Toward a Functional Description of the Centrosome

A. The Centrioles/Matrix Debate

The structural evolution of the centrosome among divergent organisms and its complexity in animal cells, due to the conspicuous presence of a centriole pair not observed in other eukaryotic organisms, have raised conflicting views. A dominant view has long been that centrioles, as opposed to the centrosomal matrix in which they are "embedded," are not relevant for centrosome functions. In this view, the matrix, most often referred to as the pericentriolar material, is assumed to be the only centrosomal constituent harboring microtubule nucleating activity. Several results, ancient or more recent, call for a different view in which centrioles play an active role in the organization and the reproduction of the centrosome. A few are listed here (see also Marshall and Rosenbaum, 1999):

Isolating centrosomes has revealed that the centrosomal matrix binds the proximal wall and the proximal end of both centrioles tightly and links them together (Bornens *et al.,* 1987; Paintrand *et al.,* 1992). The manner in which the matrix interacts with the centriole walls clearly suggests that it is largely organized by the centrioles. The two sets of pericentriolar appendages associated with the mother centriole, which can be considered specialized parts of the matrix during interphase, have a distribution that respects the ninefold symmetry of the centriole.

Loading HeLa cells with a monoclonal antibody directed against polyglutamylated tubulin (Bobinnec *et al.,* 1998a) has led to the demonstration that the centriole pair is instrumental in maintaining centrosome structural stability and is essential for centrosome activity in animal cells. Polyglutamylation, a common modification of neuronal α- and β-tubulin (Wolff *et al.,* 1992), is minimally observed in nonneuronal cells. Centriolar tubulin is highly polyglutamylated, however, particularly on α-tubulin (Bobinnec *et al.,* 1998b). When loaded into cells by electropermeation or microinjection, anti-polyglutamylated tubulin mAbs are rapidly and specifically targeted to the centriole pair and induce a specific and transient disassembly of centrioles. In these cells, the centrosome is no longer detected after

36 to 48 h, either structurally using several markers, or functionally in microtubule regrowth experiments.

Centrioles have been shown to disappear during myogenesis, when myoblasts fuse into myotubes (Tassin *et al.*, 1985). This cellular reorganization is accompanied by the redistribution of the microtubule nucleating activity around the nucleus.

Experiments that aimed at identifying the centrosome-associated activity responsible for triggering parthenogenetic development in *Xenopus* have shown a correlation between the loss of activity and the disappearance of the centriole structure (Klotz *et al.*, 1990).

The reproductive capacity of centrosome in sea urchin eggs has been shown to depend upon the number of centrioles present (Sluder and Rieder, 1985), and centrosomes from blastomeres artificially deprived of centrioles cannot reproduce (Sluder *et al.*, 1989; see also Maniotis and Schliwa, 1991).

B. A Model for Centriole-Based Assembly of the Centrosome?

We do not yet understand how centrioles can organize the microtubule nucleating center in animal cells, but a possible clue lies in the fact that the centriole architecture itself is made of microtubules.

First, microtubule triplets provide a general polarity to centrioles: The minus ends of the wall microtubules define the proximal end of the centriole, their plus ends defining the distal end of the centriole from which a primary cilium can grow in some differentiated cells. Second, the nine triplet microtubules of each centriole are highly stable. As a consequence, they provide at the proximal end of centrioles a permanent set of microtubule minus ends with a fixed relative position that are able to anchor minus-end-binding proteins, which could include γ-tubulin-containing complexes. Because of their considerable stability with respect to that of the cytoplasmic microtubules, and despite their limited number (corresponding to 27 microtubule minus ends per centriole), they could successfully compete for minus-end-binding proteins, and thus concentrate these proteins in a limited space around centrioles. The triplet structure, which is unique to the proximal end of centrioles, as well as the close distance between the nine triplets of a centriole (between the A tubule of one triplet and the C tubule of the adjacent triplet), could compensate for the small number of centriole microtubules by providing additional properties that individual microtubules with 13 protofilaments cannot provide. Note, for example, that the centriole architecture results in a very high local density of protofilament ends (more than 300, concentrated at the periphery of a disk 0.2 mm in diameter), a density that should favor the anchoring and concentration of minus-end-binding proteins.

The stability of centriole microtubules apparently brings in an additional principle for centrosome assembly: It allows a heavy posttranslational polyglutamylation of centriole tubulin to take place (Bobinnec *et al.,* 1998b; Plessmann and Weber, 1997). The massive glutamylation of tubulin apparently confers long-term stability to centrioles (Bobinnec *et al.,* 1998a). How this is achieved is not known. One possibility is that polyglutamylation controls the association of specific centrosomal microtubule associated proteins (MAPs) (whose existence remain to be proven) to the centriole wall, in a manner similar to what has been shown for the binding of brain MAPs to tubulin (Boucher *et al.,* 1994; Larcher *et al.,* 1996).

Thus, centrioles could nucleate the assembly of the centrosomal matrix through two subsets of microtubule-binding proteins, one able to bind to their walls and another able to bind to their proximal ends. By the same token, one could imagine that a subset of plus-end-binding proteins could concentrate at the distal end of the centrioles. Their function could be, for example, to mediate the specific association of a set of appendages to the plus end of the parental centriole. An additional feature must be postulated in order for the centrosomal matrix to form and to act as a microtubule nucleating center: These proteins, whatever the subset they belong to must be able to bind in one way or another with each other in order to form oligomers and to assemble a matrix. This would create additional nucleating sites for microtubule assembly and could provide a mechanism for regulating their number (see Fig. 1, color insert).

This minimal model for centrosome assembly results in several predictions. One of them is that the centriole/matrix interaction should be weakened when a pool of stable cytoplasmic microtubule minus ends is formed, i.e., when the cellular tubulin pool is artificially polymerized into stable microtubules by a drug such as paclitaxel (Taxol), as these additional stable minus ends should be able to compete successfully for the binding of centrosomal proteins. This has been experimentally observed (see, for example, Gosti-Testu *et al.,* 1986; Infante *et al.,* 1999, and for a discussion, Bornens, 1992). A 1-h pulse treatment with a very low dose of docetaxel (Taxotere) during S phase is sufficient to displace the centrosomal matrix off the centrioles and to impair mitotic progression (Paoletti *et al.,* 1997). This could mean that the interaction between the centriole pair and the centrosomal matrix is dependent upon the dynamics of the cytoplasmic microtubules. If the centriole/matrix interaction controls the nucleating capacity of the centrosome, it would provide a feedback mechanism coupling microtubule nucleation by the centrosome and the global pool of microtubules detached from the centrosome.

This model is admittedly highly hypothetical. It basically proposes that the unique microtubular features of the centrioles are used to shape and control the size of the microtubule nucleating center. It does not address how the assembly of the centrioles themselves is controlled in the first place (see later discussion) and how the stability of centriole microtubules is brought about before polyglutamylation can take place and enforce it. Our knowledge of centriole composition is still scarce. The stability of centriole microtubules could be due to proteins such as tektins that are likely to play a role in the stability of the axoneme doublets (Steffen and Linck, 1988; Stephens and Lemieux, 1998). The C tubule, which is unique to the proximal side of centriole triplets, could also participate in this stability. Its presence has been linked to a novel, distinct tubulin protein, the δ-tubulin, first characterized as the product of the gene UN13 from *Chlamydomonas reinhardtii* and which appears necessary for the assembly of basal bodies (Dutcher and Trabuco, 1998). This gene is conserved from unicellular algae to mammalian organisms (Smrzka *et al.,* 2000), a result that could parallel the remarkable evolutionary conservation of the triplet organization of centriole microtubules. In agreement with this view, the yeast *Saccharomyces cerevisiae,* which has neither centrioles nor flagella, does not possess this gene.

We conclude this section by stressing that cells without centrioles and basal bodies can possess a bona fide centrosome, like the yeast SPB, indicating that the presence of centrioles is not a general criterion for identifying centrosomes. But when present, the centriole pair is a conspicuous component of the centrosome and plays a critical role in centrosome assembly and reproduction. An important step now would be to identify and characterize the proteins directly involved in the attachment of the matrix to the centrioles and find their direct partners. This could be done, for instance, by looking for proteins that interact with modified tubulin isoforms and are specifically found in the centrioles (e.g., polyglutamylated tubulin) and for proteins able to anchor the nucleating γ-tubulin complexes.

C. The Biochemical Nature and the Function of the Matrix

A comprehensive description of the centrosomal matrix, at the biochemical and structural levels, has yet to be done. There is increasing evidence that in addition to the nucleation of microtubules, the matrix can govern the release and capture of microtubules (Keating *et al.,* 1997; Mogensen, 1999). The complex organization of the animal centrosome contrasts with the simple molecular principles that seem to govern the assembly at the yeast spindle pole body SPB (Adams and Kilmartin, 1999; Knop *et al.,* 1999). Significant differences also exist between centrosomes from somatic cells and those from early segmenting embryos.

Observations of somatic centrosomes *in situ* using advanced immunofluorescence imaging by deconvolution have led some authors to emphasize the possible role of a "highly ordered" matrix, organized around pericentrin, as an architectural framework for centrosome activities (Dictenberg *et al.,* 1998; Zimmerman *et al.,* 1999). Three-dimensional electron microscopy of egg centrosomes has demonstrated a "complex lattice," built from 12- to 15-nm-thick elementary fibers (Schnackenberg *et al.,* 1998). Description of the matrix structure with a significant resolution will not be possible, however, without analysis of the biochemical complexity of this material and without molecular characterization of the major components. The most common view is that the centrosomal matrix is able to anchor γ-tubulin-containing complexes (Schnackenberg *et al.,* 1998, and references therein). In the model just outlined, which makes centrioles the cornerstone of centrosomal matrix assembly, γ-tubulin-containing complexes would directly participate in the organization of the matrix from the centrioles. Biochemical analysis of isolated centrosomes supports this view: After treatment with chaotropic agents, a significant part of the centrosomal γ-tubulin (Moudjou *et al.,* 1996) and of HsSpc98, a protein found in γ-tubulin complexes involved in microtubule nucleation (Knop and Schiebel, 1997; Martin *et al.,* 1998; Murphy *et al.,* 1998; Tassin *et al.,* 1998), is tightly associated with the centrosome remnants. The proportion of loosely bound γ-tubulin-containing complexes might be much higher in egg centrosomes, which usually display a huge pericentriolar matrix (Sluder and Rieder, 1985). Accumulation of such a material could possibly reflect an additional function of the centrosome during embryonic cell division cycles, namely, to distribute the centrosomal material, which has been stockpiled during oogenesis, in equal amounts to the rapidly dividing blastomeres (Piel and Bornens, 1998; Tassin and Bornens, 1999).

Most of the characterized components that participate in the centrosomal matrix (the other proteins participating either permanently or transiently in the centrosome edifice are not yet characterized; for a review, see Andersen, 1999) are large coiled-coil proteins, a feature that has been also noted for yeast SPB components. Most of them, however, are significantly larger than any known SPB component in yeast. Their precise functions are still unknown, but their size suggests than they can probably participate in several functions, including that of anchoring and clustering different enzymes in the same area (such proteins are often referred to as scaffold proteins). Some are encoded by genes displaying a great exon/intron complexity, which suggests that they can generate multiple splice variants (see later discussion).

If we attempt to compare them in terms of functions, the first characterized protein, pericentrin (Doxsey *et al.,* 1994), has been involved in the regulation of the γ-tubulin-containing complexes, but this aspect is not yet fully characterized. Interestingly, a human homolog sequence that is significantly larger than pericentrin has been established (and called kendrin; H. Joshi, unpublished). This protein possesses regions of significant homology with AKAP450/hyperion/CG-NAP, another, and otherwise unrelated, large protein participating in the centrosomal matrix (see later discussion). The Asp protein has also been reported as participating in the regulation of the γ-tubulin-containing complexes in *Drosophila* (Avides and Glover, 1999; Saunders *et al.,* 1997). Pericentrin/kendrin is probably an abundant protein within the matrix.

AKAP450/hyperion/CG-NAP, a large protein of more than 3900 aa, has been characterized by several laboratories as AKAP450 (Witczak *et al.,* 1999), as hyperion (W. Kemmner and U. Schwarz, personal communication), as AKAP350 (Schmidt *et al.,* 1999) in which the 374 N-terminal aa are missing, and as CG-NAP (Takahashi *et al.,* 1999). The sequence predicts numerous coiled-coil structures. This protein was identified long ago as an abundant protein of the centrosomal matrix (Gosti-Testu *et al.,* 1986) and was shown to redistribute with new microtubule nucleating sites during cell differentiation (Tassin *et al.,* 1985) together with γ-tubulin. One early identified function for this protein (called AKAP350 on the basis of its apparent MW) was to anchor protein kinase A-type II at the centrosome (Keryer *et al.,* 1993, 1998). Evidence demonstrates that this protein is also able to anchor regulatory or catalytic subunits of other enzymes, including PKN, a protein kinase that has a catalytic domain homologous to that of PKC, and the two protein phosphatases PP1 and PP2A (Takahashi *et al.,* 1999). We do not know yet whether this protein facilitates interactions among the bound enzymes in an active state or retains them in an inactive state until activation liberates them. In any case, such a protein has great structural and regulatory potential. Its possible interaction with γ-tubulin-containing complexes should be important for understanding the regulation of nucleation. Interestingly, in *S. cerevisiae,* a coiled-coil protein, Stu2p, interacts at the SPB with both microtubules and the γ-tubulin complex-anchoring protein Spc72p (Chen *et al.,* 1998). The gene encoding AKAP450/hyperion/CG-NAP protein contains 50 exons, and there is substantial evidence of isoforms with different sizes in the centrosome. This, together with the significant homology of some stretches of pericentrin/kendrin protein with AKAP450/hyperion/CG-NAP, suggests the intriguing possibility that the centrosomal matrix could be made of complexes between highly related but different isoforms. Another most intriguing feature of the AKAP450/hyperion/CG-NAP protein is that the N-terminal part of its open reading frame corresponds to a previously characterized cDNA encoding yotiao, an NMDA receptor subunit 1 binding protein (Lin *et al.,* 1998). As a matter of fact, hyperion was identified by a screen aiming at the identification of new membrane-associated proteins of retinal ganglion cells, and a number of isoforms of hyperion/AKAP450 were found in postsynaptic densities (W. Kemmner and U. Schwarz, personal communication). Thus, this large protein could govern the assembly of functional protein complexes not only in the centrosome but also in postsynaptic densities.

The two other characterized centrosomal matrix proteins, ninein (Bouckson-Castaing *et al.,* 1996a) and C-Nap1/Cep250 (Fry *et al.,* 1998a; Mack and Rattner, 1998), could have specific functions different from those of AKAP450/hyperion or pericentrin/kendrin. Ninein is a 235 to 240 kDa protein with three predicted distinct domains, two globular N- and C-terminal domains separated by a long coiled-coil domain (Bouckson-Castaing *et al.,* 1996b). When overexpressed, ninein accumulates massively at the centrosome, in agreement with the lack of a significant pool of noncentrosomal endogenous ninein. Transient expression of truncated forms of ninein has revealed that the N-terminal domain, which contains several predicted motives, including an EF-hand-like domain, a GTP-binding domain, a PEST sequence, and an NLS sequence, localizes within the nucleus, suggesting that the cytoplasmic localization probably depends on oligomerization by the central coiled-coil region. The coiled-coil domain alone is accumulated in perinuclear vesicle-like structures. The amount of endogenous ninein is apparently regulated in a cell-cycle-dependent manner: It steadily increases from G1 to M, where it abruptly decreases by an unknown mechanism. Deleting the PEST sequence did not increase the accumulation of the protein during mitosis.

A possible clue to the function of this protein comes from the study of ninein localization throughout the development of Kolliker epithelial cells, which clearly suggests its participation in the translocation of growing microtubules from the centrosome to the cell surface (Mogensen *et al.,* 1998, in preparation). Thus, in the centrosome of other cells, ninein may be a protein able to capture and stabilize microtubule minus ends close to their nucleation sites (for discussion, see Mogensen, 1999). Another candidate protein, R2, not yet fully characterized, seems, like ninein, concentrated at the subdistal appendages of the mother centriole. This class of centrosomal proteins might be important, because many microtubules, and in some cells the majority of them (up to 75%), can be free in the cytoplasm as a result of release from the centrosome, breaking, or spontaneous assembly (for a review, see Waterman-Storer and Salmon, 1997a). Notably, the anchorage of the minus end of microtubules at the centrosome can play a significant role in switching microtubule dynamics from dynamic instability to treadmilling (Keating and Borisy, 1999; Rodionov *et al.,* 1999).

C-Nap1/Cep250 (Fry *et al.,* 1998a; Mack and Rattner, 1998) is specifically present in the intercentriolar link and concentrated at the proximal end of each centriole. It seems to participate, in a phosphorylation-dependent manner, in the control of the distance between the two centrioles. The remarkable accumulation of this protein at the proximal end of each centriole is the expected distribution for a protein participating in the building and maintenance of the centrosomal matrix at the triplet minus ends of centrioles. A very similar distribution has been reported for Nek2, a NIMA-like protein kinase that apparently controls the organization of Nap1/Cep250 (Fry *et al.,* 1998b), suggesting that they both participate in the same complex at the proximal end of each centriole and within the intercentriolar link. This complex could be an important building block for the matrix. Interestingly, when overexpressed, an inactive Nek2 has apparently a dramatic effect on the centrosome, which is no longer identified as a functional microtubule nucleating center. If this reflects a structural disorganization of the centrosome, one could conclude that the structural stability of the centrosome, either of the matrix or of the centrioles themselves, depends upon an active Nek2, possibly to maintain C-Nap1/Cep250 in a phosphorylated state.

Less is known about the function of PCM-1, a 228-kDa large centrosomal protein, one of the earliest autoantigens characterized, which dissociates from the centrosome in G_2/M. Its role in centrosome biogenesis is, however, clearly supported by its cyclic expression during the cell cycle (Balczon *et al.,* 1994).

All these proteins are apparently genuine centrosomal proteins. Most of them are highly insoluble, a property that might be important for their role in the centrosomal matrix but that makes biochemical analysis, even with recombinant proteins, a difficult task. Probably because of their insolubility, most of them can elicit autoimmune responses (Mack *et al.,* 1998), and several were first identified and characterized using autoimmune sera (Balczon and West, 1991; Maunoury, 1978). How they are addressed and concentrated at the centrosome around the centriole pair is an important issue. The only way to dissociate these proteins from the centrioles *in vivo* is to significantly modify the dynamics of microtubules in the cell, using tubulin drugs. As already discussed (see also Bornens, 1992), this suggests that microtubule dynamics controls the localization of centrosomal matrix components and that one or all matrix components must interact with the microtubule minus ends. If the model briefly outlined in this contribution (Section I,B) is valid, it should predict that in control cells, one or all matrix components must interact with microtubule triplet minus ends of the centrioles (see Fig. 2, color insert). Each of them must have, however, a specific pattern of association and a specific turnover. This will be an important issue for understanding centrosome biogenesis. PCM-1 has been shown to associate to the centrosome in a cyclic way. Ninein accumulates also in a cyclic way, but with a different time course (Bouckson-Castaing *et al.,* in preparation).

Finally, the data reported by Meda *et al.,* (1997) that GDP-tubulin binds specifically to centrosome in permeabilized cells suggest that at least one tubulin-binding protein is concentrated at the centrosomal matrix. Is it γ-tubulin itself or one of the characterized aforementioned matrix proteins? This could favor the nucleation activity observed at the centrosome by locally increasing the tubulin concentration.

D. The Extension of the Centrosomal Matrix: The Golgi Apparatus Boundary

Defining the boundary of the centrosome organelle in animal cells has been difficult, as reflected by the old term "centrosphere," defined as a region of the cytoplasm that did not stain by classical May–Grünwald and Giemsa stainings like the rest of the cytoplasm because, as it was later discovered, it is devoid of ribosomes.

A pericentrosomal localization of the Golgi apparatus is observed in most cells and has obvious significance for membrane traffic (LippincottSchwartz, 1998). Recent results indeed suggest the existence of a pervasive network between the centriole pair and the Golgi apparatus and the existence of coordinated regulations between activities found in the centrosome and those found in the Golgi apparatus.

The AKAP450/Hyperion/CG-NAP protein can be observed, depending on the antibody used, either concentrated on the centriole pair or extending away from the centrioles up to the Golgi apparatus, along which it seems to line and form a detergent-insoluble scaffold (Keryer *et al.,* 1999; Takahashi *et al.,* 1999). Similar distribution has been observed for less characterized centrosomal antigens revealed by monoclonal antibodies (CTR210; Perret *et al.,* 1995).

A more compelling argument for an intimate association between the two organelles comes from the discovery that a cis-Golgi network membrane-associated protein (Rios *et al.,* 1994) is also able to bind minus end microtubules (Infante *et al.,* 1999). This 210-kDa protein, called GMAP-210 for golgi microtuble-associated protein, interacts with microtubules in a direct, tight, and nucleotide-independent manner. GMAP-210 associates with a subpopulation of microtubules enriched in detyrosinated α-tubulin. The full-length cDNA encoding GMAP-210 predicts a protein of 1979 aa with a very long central coiled-coil domain. Deletion analysis shows that the C terminus of GMAP-210 binds to microtubules, whereas the N terminus binds to Golgi membranes. Overexpression of GMAP-210-encoding cDNA induced a dramatic enlargement of the Golgi apparatus and perturbations in the microtubule network. These effects did not occur when a mutant lacking the C-terminal domain was expressed. When transfected in fusion with the green fluorescent protein, the N-terminal domain associated with the cis-Golgi network and induced the disorganization of the Golgi apparatus, whereas the C-terminal domain localized at the centrosome. These data suggest that GMAP-210 links CGN membranes to the minus ends of centrosome-nucleated microtubules. This interaction thus appears essential for ensuring the proper localization, morphology, and size of the Golgi apparatus.

Based on a release and capture model, such as that put forward by Mogensen *et al.,* (1997; see also Mogensen, 1999), one could propose that free microtubule ends could be captured by GMAP-210 via its C-terminal domain. Because of its ability to bind CGN membranes via its N-terminal domain, GMAP-210 would participate in addition in the anchoring of a whole organelle to the centrosome periphery. This might account for the existence of a subpopulation of short, stable microtubules that colocalize with the Golgi apparatus (Schulze and Kirschner, 1986). This subpopulation is rich in detyrosinated microtubules (Skoufias *et al.,* 1990), and GMAP-210 is specifically enriched in a microtubule fraction with similar characteristics. In this view, CGN membranes, rather than the centrosome itself, would be the anchoring sites for microtubules participating in Golgi dynamics and stability. Alternatively, this subset of Golgi microtubules could be nucleated in the direct vicinity of the Golgi membranes and stabilized by proteins such as GMAP-210. This possibility is suggested by data showing that γ-tubulin can be immunoprecipitated together with GMAP-210 (Rios and Bornens, unpublished).

Other proteins are suggestive of a direct cross-talk between centrosome and Golgi apparatus. This is the case of GAP-CenA, a G-protein activating protein for Rab6, a small G-protein of the Ras superfamily. This 150-kDa protein, found by a two-hybrid screen using Rab6 as a bait, has been shown to localize in part to the centrosome, and not to the Golgi apparatus where Rab6 is concentrated (Cuif *et al.,* 1999). The significance of that is unclear. Antibodies directed against this protein were shown to inhibit microtubule nucleation.

The way in which an intimate border between the centrosome and the Golgi apparatus is actively maintained will be an important aspect of the biology of both organelles. The results obtained with GMAP-210 indicate that, in addition to the activity of dynein for moving Golgi elements toward the centrosome, association of CGN membranes with the minus end of microtubules is necessary both for the proper localization and for the steady-state control of the morphology and size of the Golgi apparatus (Infante *et al.,* 1999). The size and shape of the centrosomal matrix, which can vary according to several parameters, including calcium (Baron *et al.,* 1994), and those of the Golgi apparatus, which depend on the membrane traffic, are likely to be permanently regulated in a coordinated manner.

III. The Centrosome Organelle versus Self-Assembled Centers

Although the centrosome has long been known to have a dominant role in microtubule nucleation and anchoring, a property that has been recently rediscovered (Heald *et al.,* 1997), there are many situations in which the cell microtubules are not attached to it. It is the case, for example, during the myoblast-to-myotube transition when centrosomes are eliminated and microtubules remain abundant in parallel arrays (Tassin *et al.,* 1985). In epithelial cells, many microtubules are not anchored in the centrosome but are found in bundles (Bacallao *et al.,* 1989) or associated with their minus ends to the cell surface (Mogensen *et al.,* 1989).

One obvious explanation for the existence of noncentrosomal microtubules is the observed severing of microtubules nucleated at the centrosome (Belmont *et al.,* 1990; Keating *et al.,* 1997; Rodionov *et al.,* 1999). Another source of free microtubules is due to their breakage within the cytoplasm (Waterman-Storer and Salmon, 1997b). The released microtubules move away from the centrosome by a combination of translocation and treadmilling (Keating *et al.,* 1997). The precise mechanisms by which microtubules are detached from the centrosome and are stabilized at their minus ends to prevent instant depolymerization remain undetermined, although katanin (McNally and Vale, 1993) and γ-tubulin ring complexes are candidates for the detachment step.

Cytoplasmic nucleation of microtubules independent of the centrosome represents another possibility to account for the existence of noncentrosomal microtubules. There are many examples of noncentrosomal nucleation in various experimental systems, such as *Xenopus* egg extracts (Gard and Kirschner, 1987), cytoplasts lacking centrosomes (Karsenti *et al.,* 1984), centrosome-deprived cytoplasm fragments (Maniotis and Schliwa, 1991; McNiven and Porter, 1988), or somatic cells (Vorobjev *et al.,* 1997; Yvon and Wadsworth, 1997). It would be important to decide in each case whether γ-tubulin is involved.

The existence of microtubule arrays that are not nucleated by a centrosome, and therefore not constitutively focused, raises the question of their organization. One interesting property of the microtubules is that even when dispersed all over the cytoplasm, they can organize into focused asters with the help of molecular motors. This is true both in interphase—for example, in specialized cells such as melanophores or erythrophores, in which self-centering properties of the cytoplasm have been demonstrated (McNiven and Porter, 1988; Rodionov and Borisy, 1997)—and in mitosis. Microtubules stabilized by heavy water, Taxol, or DMSO were shown to arrange themselves into asters during mitosis *in vitro* as well as *in vivo,* and this organization is dependent upon the activity of minus or plus-end motors (Heald *et al.,* 1996).

The assembly of the bipolar mitotic spindle has long been a paradigm to address the question of the autoorganization of microtubule arrays. The noncentrosomal microtubule assembly pathway has been illustrated in a spectacular way by the formation of mitotic spindles in the absence of centrosomes (Heald *et al.*, 1996). This situation can be naturally observed during formation of the female meiotic spindle in animal cells, and in all plant cells spindles. How mitotic chromatin can in some way trigger the nucleation of microtubules or can locally stabilize in its vicinity unstable, randomly growing microtubules has long remained unknown. It has been proposed that stathmin/Op18, which is known as a potent microtubule-destabilizing factor (Belmont and Mitchison, 1996), had its activity downregulated by mitotic chromatin, thus increasing microtubule growth around it (Andersen *et al.*, 1997). The demonstration by several laboratories of an involvement of Ran GTPase in microtubule nucleation during mitosis (for a review, see Kahana and Cleveland, 1999) has led to the discovery that mitotic chromatin-associated RCC1, which is the Ran exchange factor, actually triggers microtubule nucleation by regulating Ran (Kalab *et al.*, 1999; Carazo-Salas *et al.*, 1999). The observed inactivation of stathmin/Op18 through hyperphosphorylation has been shown not to be a direct effect of mitotic chromatin, but rather to result from changes in microtubule dynamics (Küntziger *et al.*, submitted). The microtubules assembled around chromatin lead to the activation of a kinase, or the inactivation of a phosphatase, resulting in the phosphorylation of one particular site (Ser 16) of stathmin/Op18. This results in downregulation of stathmin/Op18 microtubule destabilizing activity (Küntziger *et al.*, 1998), thus locally favoring the assembly of microtubules. The combination of these two mechanisms would explain how microtubules can first nucleate and then become stabilized by mitotic chromatin in the absence of centrosomes.

Why centrosomes have evolved if microtubules are able to self-organize and even to assemble bipolar spindles has been discussed elsewhere (Marshall and Rosenbaum, 1999; Paoletti and Bornens, 1997; Waters and Salmon, 1997). The kinetic advantage resulting from the capacity of the centrosome-associated components to catalyze the nucleation reaction explains the dominant role of the centrosome in the organization of the cell microtubules and probably its evolutionary success. Key reasons for the evolutive success of the centrosome is indeed that it is a structurally stable organelle able to nucleate microtubules throughout the cell division cycle and that it reproduces according to a precise duplicating process.

IV. Centrosome Reproduction and the Parthenogenetic Activity of the Centrosome

Centrosome reproduction is a most challenging question. Two main aspects have to be understood: the molecular mechanisms underlying centrosome biogenesis itself, and the coordination between the nuclear cycle and the centrosome cycle. The latter aspect has recently progressed (see the contribution of Sluder in this book; see also Hinchcliffe *et al.,* 1999; Lacey *et al.,* 1999; Matsumoto *et al.,* 1999; Meraldi *et al.,* 1999), although we do not know yet the centrosomal targets responding to cell-cycle regulators. An important notion, however, put forward by Sluder and co-workers, from the manipulation of sea urchin eggs is that the procentriole assembly that can be observed morphologically at the onset of S phase might be temporally distinct from the assembly of the real centriole precursor (Hinchcliffe *et al.,* 1998). This also has implications for the molecular mechanisms underlying centrosome biogenesis.

A similar proposal has been put forward from the comparison of experimental systems such as amphibian eggs, echinoderm eggs, or amoeboflagellates, in which centriole assembly takes place according to structurally different pathways (Paoletti and Bornens, 1997). They also involve a temporal disconnection between the assembly of the precursors and that of the centrioles. Although all the experimental facts are consistent with a procentriole orthogonal budding from each centriole being an early structural event in centrosome reproduction (Maniotis and Schliwa, 1991; Sluder *et al.,* 1989; Sluder and Rieder, 1985), we still do not know whether the microtubule-based centriole architecture itself is really necessary for the reproductive capacity of the centrosome. The procentriole assembly might be a late event in the pathway.

In the classical duplication pathway, the new organelle assembles in the vicinity of the old one, suggesting that some structure, or protein complex, is associated to the proximal wall of each centriole, where it is able upon a signal to promote the assembly of a new centriole. In yeast, such a complex would be the half-bridge associated to the edge of the SPB (see the detailed description of the process of SPB duplication; Adams and Kilmartin, 1999; Elliott *et al.,* 1999).

There are, however, many situations where the requirement for structural continuity is not observed, that is, when centriole assembly occurs in absence of preexisting centrioles. In several unicellular eukaryotes, the centriole structure totally disappears during some stage of the vegetative cycle and reforms upon specific signals (Levy *et al.,* 1996). In echinoderms, artificial activation of eggs by various treatments can trigger *"de novo"* assembly of centrioles and parthenogenetic development (Kuriyama and Borisy, 1983). In clams, activation-induced centriole assembly can even be reproduced in a cell-free system (Palazzo *et al.,* 1992). In mouse embryos, the early cell division cycles take place in the absence of centrioles. Centrioles reappear at the 16- to 32-cell stage (for a review, see Schatten, 1994). Finally, during terminal differentiation in vertebrates, ciliated cells assemble many centrioles according to a pathway distinct from classical duplication (see Tournier and Bornens, 1994, for a review): Centrioles emanate from a dense and uncharacterized body, in the vicinity of the centrosome and the Golgi apparatus, and migrate toward the apical pole to form the basal bodies necessary for ciliary growth.

One may assume that despite their apparent differences, all pathways involve some common principles as they all lead to centriole assembly. For example, while the essential precursor structure would be centriole-associated during the normal centriolar pathway of centrosome assembly in proliferating cells, it could accumulate in numerous copies in the cytoplasm, ahead of the centriole assembly step, in postmitoic cells that differentiate into ciliated cells. The mechanism of this accumulation, which is uncoupled from downstream events involved in centriole assembly itself, is not known. It has been postulated that it is due to an amplification by sequential rounds of duplication of the essential precursor structure for centriole assembly (Paoletti *et al.,* 1997). Such a possibility, which makes of the precursor structure the real "duplicating structure," would allow one to rationalize the apparently disparate set of data on centrosome biogenesis and structural continuity (for more details, see Paoletti *et al.,* 1997). The "duplication complex" would be the key precursor structure in the centrosome assembly pathway. Such a complex would possess the structural information essential for centriole assembly.

Potentially the most direct experimental approach to search for the essential centriole-associated precursor for centriole assembly is the parthenogenetic assay in the *Xenopus* system (Fig. 3). Like the eggs of echinoderms, *Xenopus* eggs have inactivated their centrosome during meiosis. But in contrast with echinoderm eggs, *de novo* assembly of centrioles and parthenogenetic development cannot be triggered artificially in *Xenopus* eggs. Only injection of a centrosome at the time of activation can do it (for a review see Tournier and Bornens, 1994). According to the hypothesis just outlined, *Xenopus* eggs do not contain preassembled precursor complexes, whereas echinoderm eggs do. In the latter case, artificial treatment would trigger catastrophic centriole assembly from each precursor complex at once, instead of the sequential assembly of two centrioles at each division cycle mediated by the controlled association of individual precursor complexes to the sperm centrioles. *Xenopus* eggs, despite the presence of a full set of centriole precursors, as judged by the ability of centrosome duplication to proceed in absence of protein synthesis (Gard *et al.,* 1990), would rely on centrosome-associated precursor complexes for centriole assembly. In that case, as well as in somatic cells, the association of the duplication complex with each centriole would allow strict control of the number of centrioles assembled and ensure the formation of a unique new copy of the centrosome at each cell cycle. Interestingly, in HeLa cells where centriole disassembly is achieved by injecting an anti-polyglutamylated tubulin mAb,

Figure 3 Parthenogenetic frogs. The centrosome, usually inherited from the sperm (the oocyte centrosome has been inactivated at the end of oogenesis), is necessary for sequential divisions after egg fertilization. Parthenogenetic development without fertilization is possible in the *Xenopus* frog by injection of a centrosome isolated from a somatic cell in a mature oocyte. The somatic cell can originate from widely divergent species. Metamorphosis and complete development can take place in few cases as for these two frogs, developed from eggs injected with a human centrosome and a rat centrosome, respectively. This experiment demonstrates the morphogenetic role of the centrosome and the conservation of centrosome properties among divergent species.

paired centriolar structures reappear during the subsequent cell cycles (Bobinnec *et al.*, 1998a), suggesting that the structural information required for centriole assembly has not been lost with the disassembly of the centrioles, and that a precursor complex distinct from centrioles themselves has been duplicated once to allow the reformation of two centrioles in each daughter cell.

All attempts to identify, using centrosome subfractionation, the component able to complement the unfertilized *Xenopus* egg upon the injection of a centrosome have failed to produce a soluble active fraction (for a review see Tournier and Bornens, 1994). The disruption of the centriole microtubule triplets was correlated with the loss of parthenogenetic activity, suggesting that the centriole structure was acting as a structural template, that is, was required to allow the assembly of the precursor complex in its immediate vicinity. Notably, the injected centrosome can come from divergent animal species, even from invertebrates such as echinoderms, a result that would fit with the great structural conservation of the centriole architecture. An additional argument for the injected centrosome acting as a structural template is the lack of parthenogenetic activity of bovine thymus centrosomes, the centriole pair of which display a specific structural arrangement (Tournier *et al.*, 1991b). To further investigate this aspect, the parthenogenetic activity of more evolutionarily distant centrosomes has been tested (Tournier *et al.*, 1999). Centrosomes devoid of centrioles, such as SPBs isolated from *Saccharomyces cerevisiae,* are inactive. However, *Drosophila* centrosomes, from embryonic or somatic cells, that possess a typical centriole architecture are also unable to trigger parthenogenesis, although they can nucleate actively the assembly of microtubule asters from the egg tubulin. Thus, the evolutionarily conserved ninefold symmetry cannot be sufficient for triggering procentriole assembly. *Drosophila* centrosomes contain, however, short-length centrioles compared to vertebrate ones (Callaini and Riparbelli, 1990), and the cartwheel structure still persists in the mature centriole, contrary to what is observed in vertebrate centrioles. The lack of parthenogenetic activity of the *Drosophila* centrosome could thus be the exception enforcing the rule: It would be the differences in the centriole structure between *Xenopus* and *Drosophila* centrosomes that would make them incompatible.

In proliferating cells, the precursor structure should be tightly associated with the centrioles and should be present at the earliest stages of duplicating centrosomes, namely, in the new procentriole buds. A candidate protein for participating in such a structure is centrin. Anti-centrin antibodies have revealed a very early accumulation of centrin centrioles, well before procentriole microtubules can be detected (Paoletti *et al.,* 1996). The use of cell lines with stable GFP-centrin expression could lead to a more direct demonstration of how this early accumulation of centrin is controlled (Piel *et al.,* 2000). Centrin is one of the rare proteins thus far demonstrated to be directly involved in the duplication of the centrosome in eukaryotic cells. The family of centrin, which belongs to the calmodulin superfamily, contains the *Saccharomyces cerevisiae CDC31* gene product, which is essential for SPB duplication and which has been localized to the half-bridge of the SPB (Spang *et al.,* 1993, 1995). Three distinct centrin genes have been identified in human and mouse cells. Two of them are highly related (Errabolu *et al.,* 1994; Lee and Huang, 1993) and are closer to the unicellular algae centrin than to the yeast *CDC31* gene, whereas the third one is more divergent (Middendorp *et al.,* 1997) and closer to the yeast gene. They actually define two centrin subfamilies that could participate in distinct animal centrosome-associated activities (Middendorp *et al.,* 1997; Paoletti *et al.,* 1996). The subfamily of *CDC31*-related genes may be more relevant to centrosome duplication than classical centrin genes. Data supporting the involvement of Cen3p in centrosome duplication have been obtained (Middendorp *et al.,* 2000).

γ-Tubulin could participate in late events of centrosome biogenesis: Inactivation of the γ-tubulin gene in *Paramecium* leads to absence of basal body duplication (Ruiz *et al.,* 1999). Surprisingly, however, these cells present an apparently normal microtubule network (J. Beisson, personal communication). This result contrasts with the data in *Saccharomyces cerevisiae,* where the different tub4 mutants do not lead to impairment of SPB duplication (Marschall *et al.,* 1996; Sobel and Synder, 1995; Spang *et al.,* 1996). Yeast SPB does not contain a centriole or any microtubule-based substructure. The mutant spc97-20 presents, however, a SPB duplication phenotype, whereas the spc97-14 mutant presents a normal SPB duplication (Knop *et al.,* 1997).

One must note that, although procentriole buds grow simultaneously from the two centrioles in the centrosome of animal cells, these two parental centrioles are not equivalent either structurally (Paintrand *et al.,* 1992) or biochemically (Lange and Gull, 1995) as long as they have not reached the G2-M border. In other words, duplication capacity precedes full maturation. There is also some evidence that centrioles acquire their capacity to support the growth of a bud in a sequential way: The manner in which the oocyte centrosome is inactivated during meiosis seems to clearly support this view (Saiki and Hamaguchi, 1998). A cell-free assay is now necessary to analyze the centrosome duplication pathway (Hinchcliffe *et al.,* 1999; Piel and Bornens, 1998; Tournier *et al.,* 1991a). The centrosome duplication pathway involves a series of discrete structural steps that are likely to be each under specific regulations. One thus needs to develop an assay in which the different steps of the duplication can be precisely monitored. Progress toward this goal has been achieved (Piel *et al.,* in preparation).

V. Conclusion

In recent years, discussions on the self-organizational properties of microtubules, particularly concerning the mitotic spindle assembly, have led many authors to question the role of the centrosome: "Why do cells need centrosomes at all if some cell types can make perfectly good spindles without them?" (Waters and Salmon, 1997). One can generalize the question to why the centrosomes have ever evolved if some cell types can live perfectly well without them, which is the case in plant cells, for example. Or, why do differentiated animal cells that no longer divide such as neurons or polymorphonuclear leukocytes, maintain a centrosome while others, such as myotubes, eliminate centrosomes?

Each individual cell system is placed under specific organizational constraints that are the evolutionary result of internal adaptation to many selective pressures. It is not possible to draw a comparison between two cell systems on a specific mechanism, such as the cell division process, without taking into account the whole survival strategy of each system. For example, controlling the position of the cleavage furrow is a general problem, common to all cell systems. But in all likelihood cytokinesis requires specific procedures that cannot be identical for cells possessing a constant shape imposed by a cell wall, as in plant cells, and for locomotory ameboid cells.

The role of the centrosome in the organization of the microtubule array, which itself is a major morphogenetic tool at the cell level, suggests that its structural evolution among divergent eukaryotic cells must reflect an adaptive answer at the cell level (Bornens, 1992). One cannot hope to get at a comprehensive understanding of the function of the centrosome among divergent species independent of a comparative analysis of the evolution of the cellular economy. This is why the study of this single-copy organelle is both important and attractive.

References

Adams, I. R., and Kilmartin, J. V. (1999). Localization of core spindle pole body (SPB) components during SPB duplication in *Saccharomyces cerevisiae*. *J. Cell Biol.* **145,** 809–823.

Andersen, S. S. (1999). Molecular characteristics of the centrosome. *Int. Rev. Cytol.* **187,** 51–109.

Andersen, S. S. L., Ashford, A. J., Tournebize, R., Gavet, O., Sobel, A., Hyman, A. A., and Karsenti, E. (1997). Mitotic chromatin regulates phosphorylation of Stathmin/Op18. *Nature* **389,** 640–643.

Avides, M. D., and Glover, D. M. (1999). Abnormal spindle protein, asp, and the integrity of mitotic centrosomal microtubule organizing centers. *Science* **283,** 1733–1735.

Bacallao, R., Antony, C., Dotti, C., Karsenti, E., Stelzer, E. H., and Simons, K. (1989). The subcellular organization of Madin–Darby canine kidney cells during the formation of a polarized epithelium. *J. Cell Biol.* **109,** 2817–2832.

Balczon, R., and West, K. (1991). The identification of mammalian centrosomal antigens using human autoimmune anticentrosome antisera. *Cell Motil. Cytoskel.* **20,** 121–135.

Balczon, R., Bao, L., and Zimmer, W. E. (1994). PCM-1, a 228-kD centrosome autoantigen with a distinct cell cycle distribution. *J. Cell Biol.* **124,** 783–793.

Baron, A. T., Suman, V. J., Nemeth, E., and Salisbury, J. L. (1994). The pericentriolar lattice of PtK2 cells exhibits temperature and calcium-modulated behavior. *J. Cell Sci.* **107,** 2993–3003.

Belmont, L. D., and Mitchison, T. J. (1996). Identification of a protein that interacts with tubulin dimers and increases the catastrophe rate of microtubules. *Cell* **84,** 623–631.

Belmont, L. D., Hyman, A. A., Sawin, K. E., and Mitchison, T. J. (1990). Real-time visualization of cell cycle-dependent changes in microtubule dynamics in cytoplasmic extracts. *Cell* **62,** 579–589.

Bobinnec, Y., Khodjakov, A., Mir, L. M., Rieder, C. L., Edde, B., and Bornens, M. (1998a). Centriole disassembly in vivo and its effect on centrosome structure and function in vertebrate cells. *J. Cell Biol.* **143,** 1575–1589.

Bobinnec, Y., Moudjou, M., Fouquet, J. P., Desbruyeres, E., Edde, B., and Bornens, M. (1998b). Glutamylation of centriole and cytoplasmic tubulin in proliferating non-neuronal cells. *Cell Motil. Cytoskel.* **39,** 223–232.

Bornens, M. (1992). Structure and function of isolated centrosomes. *In* "The Centrosome" (V. I. Kalnins, Ed.), pp. 1–43. Academic Press, San Diego.

Bornens, M., Paintrand, M., Berges, J., Marty, M. C., and Karsenti, E. (1987). Structural and chemical characterization of isolated centrosomes. *Cell Motil. Cytoskeleton* **8,** 238–249.

Boucher, D., Larcher, J. C., Gros, F., and Denoulet, P. (1994). Polyglutamylation of tubulin as a progressive regulator of in vitro interactions between the microtubule-associated protein Tau and tubulin. *Biochemistry* **33,** 12471–12477.

Bouckson-Castaing, V., Moudjou, M., Ferguson, D. J. P., Mucklow, S., Belkaid, Y., Milon, G., and Crocker, P. R. (1996a). Molecular characterisation of ninein, a new coiled-coil protein of the centrosome. *J. Cell Sci.* **109,** 179–190.

Bouckson-Castaing, V., Tassin, A. M., and Bornens, M. (1996b). Functional domains of a new centrosomal protein. *Mol. Biol. Cell* **7,** 206a.

Callaini, G., and Riparbelli, M. G. (1990). Centriole and centrosome cycle in the early Drosophila embryo. *J. Cell Sci.* **97,** 539–543.

Carazo-Salas, R. E., Guarguaglini, G., Gruss, O. J., Segref, A., Karsenti, E., and Mattaj, I. W. (1999). Generation of GTP-bound Ran by RCC1 is required for chromatin-induced mitotic spindle formation. *Nature* **400,** 178–181.

Chen, X. Y. P., Yin, H. W., and Huffaker, T. C. (1998). The yeast spindle pole body component Spc72p interacts with Stu2p and is required for proper microtubule assembly. *J. Cell Biol.* **141,** 1169–1179.

Cuif, M. H., Possmayer, F., Zander, H., Bordes, N., Jollivet, F., CouedelCourteille, A., JanoueixLerosey, I., Langsley, G., Bornens, M., and Goud, B. (1999). Characterization of GAPCenA, a GTPase activating protein for Rab6, part of which associates with the centrosome. *EMBO J.* **18,** 1772–1782.

Dictenberg, J. B., Zimmerman, W., Sparks, C. A., Young, A., Vidair, C., Zheng, Y. X., Carrington, W., Fay, F. S., and Doxsey, S. J. (1998). Pericentrin and gamma-tubulin form a protein complex and are organized into a novel lattice at the centrosome. *J. Cell Biol.* **141,** 163–174.

Doxsey, S. J., Stein, P., Evans, L., Calarco, P. D., and Kirschner, M. (1994). Pericentrin, a highly conserved centrosome protein involved in microtubule organization [see comments]. *Cell* **76,** 639–650.

Dutcher, S. K., and Trabuco, E. C. (1998). The UN13 gene is required for assembly of basal bodies of *Chlamydomonas* and encodes delta-tubulin, a new member of the tubulin superfamily. *Mol. Biol. Cell* **9,** 1293–1308.

Elliott, S., Knop, M., Schlenstedt, G., and Schiebel, E. (1999). Spc29p is a component of the Spc110p subcomplex and is essential for spindle pole body duplication. *Proc. Nat. Acad. Sci. USA* **96,** 6205–6210.

Errabolu, R., Sanders, M. A., and Salisbury, J. L. (1994). Cloning of a cDNA encoding human centrin, an EF-hand protein of centrosomes and mitotic spindle poles. *J. Cell Sci.* **107,** 9–16.

Fry, A. M., Mayor, T., Meraldi, P., Stierhof, Y. D., Tanaka, K., and Nigg, E. A. (1998a). C-Nap1, a novel centrosomal coiled-coil protein and candidate substrate of the cell cycle-regulated protein kinase Nek2. *J. Cell Biol.* **141,** 1563–1574.

Fry, A. M., Meraldi, P., and Nigg, E. A. (1998b). A centrosomal function for the human Nek2 protein kinase, a member of the NIMA family of cell cycle regulators. *EMBO J.* **17,** 470–481.

Gard, D. L., and Kirschner, M. W. (1987). Microtubule assembly in cytoplasmic extracts of Xenopus oocytes and eggs. *J. Cell Biol.* **105,** 2191–2201.

Gard, D. L., Hafezi, S., Zhang, T., and Doxsey, S. J. (1990). Centrosome duplication continues in cycloheximide-treated *Xenopus* blastulae in the absence of a detectable cell cycle. *J. Cell Biol.* **110,** 2033–2042.

Gosti-Testu, F., Marty, M. C., Berges, J., Maunoury, R., and Bornens, M. (1986). Identification of centrosomal proteins in a human lymphoblastic cell line. *EMBO J.* **5,** 2545–2550.

Heald, R., Tournebize, R., Blank, T., Sandaltzopoulos, R., Becker, P., Hyman, A., and Karsenti, E. (1996). Self-organization of microtubules into bipolar spindles around artificial chromosomes in *Xenopus* egg extracts. *Nature* **382,** 420–425.

Heald, R., Tournebize, R., Habermann, A., Karsenti, E., and Hyman, A. (1997). Spindle assembly in *Xenopus* egg extracts: Respective roles of centrosomes and microtubule self-organization. *J. Cell Biol.* **138,** 615–628.

Hinchcliffe, E. H., Cassels, G. O., Rieder, C. L., and Sluder, G. (1998). The coordination of centrosome reproduction with nuclear events of the cell cycle in the sea urchin zygote. *J. Cell Biol.* **140,** 1417–1426.

Hinchcliffe, E. H., Li, C., Thompson, E. A., Maller, J. L., and Sluder, G. (1999). Requirement of Cdk2-cyclin E activity for repeated centrosome reproduction in *Xenopus* egg extracts. *Science* **283,** 851–854.

Infante, C., RamosMorales, F., Fedriani, C., Bornens, M., and Rios, R. M. (1999). GMAP-210, a cis-Golgi network-associated protein, is a minus end microtubule-binding protein. *J. Cell Biol.* **145,** 83–98.

Kahana, J. A., and Cleveland, D. W. (1999). Beyond nuclear transport. Ran-GTP as a determinant of spindle assembly. *J. Cell Biol.* **146,** 1205–1210.

Kalab, P., Pu, R. T., and Dasso, M. (1999). The Ran GTPase regulates mitotic spindle assembly. *Curr. Biol.* **9,** 481–484.

Karsenti, E., Kobayashi, S., Mitchison, T., and Kirschner, M. (1984). Role of the centrosome in organizing the interphase microtubule array: properties of cytoplasts containing or lacking centrosomes. *J. Cell Biol.* **98,** 1763–1776.

Keating, T. J., and Borisy, G. G. (1999). Centrosomal and non-centrosomal microtubules. *Biol. Cell* **91,** 321–329.

Keating, T. J., Peloquin, J. G., Rodionov, V. I., Momcilovic, D., and Borisy, G. G. (1997). Microtubule release from the centrosome. *Proc. Natl. Acad. Sci. USA* **94,** 5078–5083.

Keryer, G., Rios, R. M., Landmark, B. F., Skalhegg, B., Lohmann, S. M., and Bornens, M. (1993). A high-affinity binding protein for the regulatory subunit of cAMP-dependent protein kinase II in the centrosome of human cells. *Exp. Cell Res.* **204,** 230–240.

Keryer, G., Yassenko, M., Labbe, J. C., Castro, A., Lohmann, S. M., Evain-Brion, D., and Tasken, K. (1998). Mitosis-specific phosphorylation and subcellular redistribution of the RIIalpha regulatory subunit of cAMP-dependent protein kinase. *J. Biol. Chem.* **273,** 34594–34602.

Keryer, G., Skalhegg, B. S., Landmark, B. F., Hansson, V., Jahnsen, T., and Tasken, K. (1999). Differential localization of protein kinase a type II isozymes in the Golgi-centrosomal area. *Exp. Cell Res.* **249,** 131–146.

Klotz, C., Dabauvalle, M. C., Paintrand, M., Weber, T., Bornens, M., and Karsenti, E. (1990). Parthenogenesis in *Xenopus* eggs requires centrosomal integrity. *J. Cell Biol.* **110,** 405–415.

Knop, M., and Schiebel, E. (1997). Spc98p and Spc97p of the yeast gamma-tubulin complex mediate binding to the spindle pole body via their interaction with Spc110p. *EMBO J.* **16,** 6985–6995.

Knop, M., Pereira, G., Geissler, S., Grein, K., and Schiebel, E. (1997). The spindle pole body component Spc97p interacts with the gamma-tubulin of *Saccharmyces cerevisiae* and functions in microtubule organization and spindle pole body duplication. *EMBO J.* **16,** 1550–1564.

Knop, M., Pereira, G., and Schiebel, E. (1999). Microtubule organization by the budding yeast spindle pole body [In Process Citation]. *Biol. Cell* **91,** 291–304.

Küntziger, T., Gavet, O., Manceau, V., Sobel, A., and Bornens, M. (1998). Phosphorylation of the tubulin-sequestering protein stathmin is regulated by microtubule assembly. *MCB* **9 suppl,** 257a.

Kuriyama, R., and Borisy, G. G. (1983). Cytasters induced within unfertilized sea-urchin eggs. *J. Cell Sci.* **61,** 175–189.

Lacey, K. R., Jackson, P. K., and Stearns, T. (1999). Cyclin-dependent kinase control of centrosome duplication. *Proc. Natl. Acad. Sci. USA* **96,** 2817–2822.

Lange, B. M., and Gull, K. (1995). A molecular marker for centriole maturation in the mammalian cell cycle. *J. Cell Biol.* **130,** 919–927.

Larcher, J. C., Boucher, D., Lazereg, S., Gros, F., and Denoulet, P. (1996). Interaction of kinesin motor domains with alpha- and beta-tubulin subunits at a tau-independent binding site. Regulation by polyglutamylation. *J. Biol. Chem.* **271,** 22117–22124.

Lee, V. D., and Huang, B. (1993). Molecular cloning and centrosomal localization of human caltractin. *Proc. Natl. Acad. Sci. USA* **90,** 11039–11043.

Levy, Y. Y., Lai, E. Y., Remillard, S. P., Heinzelman, M. B., and Fulton, C. (1996). Centrin is a conserved protein that forms diverse associations with centrioles and MTOCs in *Naegleria* and other organisms. *Cell Motil. Cytoskel.* **33,** 298–323.

Lin, J. W., Wyszynski, M., Madhavan, R., Sealock, R., Kim, J. U., and Sheng, M. (1998). Yotiao, a novel protein of neuromuscular junction and brain that interacts with specific splice variants of NMDA receptor subunit NR1. *J. Neurosci.* **18,** 2017–2027.

LippincottSchwartz, J. (1998). Cytoskeletal proteins and Golgi dynamics. *Curr. Opin. Cell Biol.* **10,** 52–59.

Mack, G. J., and Rattner, J. B. (1998). Identification of two novel coiled coil proteins of the centriole using human autoimmune sera. *MCB* **9 suppl,** 9a.

Mack, G. J., Rees, J., Sandblom, O., Balczon, R., Fritzler, M. J., and Rattner, J. B. (1998). Autoantibodies to a group of centrosomal proteins in human autoimmune sera reactive with the centrosome. *Arthritis Rheum.* **41,** 551–558.

Maniotis, A., and Schliwa, M. (1991). Microsurgical removal of centrosomes blocks cell reproduction and centriole generation in BSC-1 cells. *Cell* **67,** 495–504.

Marschall, L. G., Jeng, R. L., Mulholland, J., and Stearns, T. (1996). Analysis of Tub4p, a yeast gamma-tubulin-like protein: Implications for microtubule-organizing center function. *J. Cell Biol.* **134,** 443–454.

Marshall, W. F., and Rosenbaum, J. L. (1999). Cell division: The renaissance of the centriole. *Curr. Biol.* **9,** R218–R220.

Martin, O. C., Gunawardane, R. N., Iwamatsu, A., and Zheng, Y. X. (1998). Xgrip109: A gamma tubulin-associated protein with an essential role in gamma tubulin ring complex (gamma TuRC) assembly and centrosome function. *J. Cell Biol.* **141,** 675–687.

Matsumoto, Y., Hayashi, K., and Nishida, E. (1999). Cyclin-dependent kinase 2 (Cdk2) is required for centrosome duplication in mammalian cells. *Cur. Biol.* **9,** 429–432.

Maunoury, R. (1978). Localisation immunocytochimique de la centrosphère de cellules tumorales humaines par utilisation d'anticorps naturels de lapins. *C. R. Acad. Sci. Paris* **286,** 503–506.

McNally, F. J., and Vale, R. D. (1993). Identification of katanin, an ATPase that severs and disassembles stable microtubules. *Cell* **75,** 419–429.

McNiven, M. A., and Porter, K. R. (1988). Organization of microtubules in centrosome-free cytoplasm. *J. Cell Biol.* **106,** 1593–1605.

Meda, P., Chevrier, V., Edde, B., and Job, D. (1997). Demonstration and analysis of tubulin binding sites on centrosomes. *Biochemistry* **36,** 2550–2558.

Meraldi, P., Lukas, J., Fry, A. M., Bartek, J., and Nigg, E. A. (1999). Centrosome duplication in mammalian somatic cells requires E2F and Cdk2-Cyclin A. *Nature Cell Biol.* **1,** 88–93.

Middendorp, S., Küntziger, T., Abraham, Y., Holmes, S., Bondes, N., Paintrand, M., Paoletti, A., and Bornens M. (2000). A role for Centrin 3 in centrosome reproduction. *J. Cell Biol.* **748,** 405–415.

Middendorp, S., Paoletti, A., Schiebel, E., and Bornens, M. (1997). Identification of a new mammalian centrin gene, more closely related to *Saccharomyces cerevisiae CDC31* gene. *Proc. Natl. Acad. Sci. USA* **94,** 9141–9146.

Mogensen, M. M. (1999). Microtubule release and capture in epithelial cells. *Biol. Cell* **91,** 331–342.

Mogensen, M. M., Tucker, J. B., and Stebbings, H. (1989). Microtubule polarities indicate that nucleation and capture of microtubules occurs at cell surfaces in Drosophila. *J. Cell Biol.* **108,** 1445–1452.

Mogensen, M. M., Mackie, J. B., Doxsey, S. J., Stearns, T., and Tucker, J. B. (1997). Centrosomal deployment of gamma-tubulin and pericentrin: Evidence for a microtubule-nucleating domain and a minus- end docking domain in certain mouse epithelial cells. *Cell Motil. Cytoskel.* **36,** 276–290.

Mogensen, M. M., Henderson, C. G., Mackie, J. B., Lane, E. B., Garrod, D. R., and Tucker, J. B. (1998). Keratin filament deployment and cytoskeletal networking in a sensory epithelium that vibrates during hearing. *Cell Motil. Cytoskel.* **41,** 138–153.

Moudjou, M., Bordes, N., Paintrand, M., and Bornens, M. (1996). Gamma-tubulin in mammalian cells: The centrosomal and the cytosolic forms. *J. Cell Sci.* **109,** 875–887.

Murphy, S. M., Urbani, L., and Stearns, T. (1998). The mammalian gamma-tubulin complex contains homologues of the yeast spindle pole body components Spc97p and Spc98p. *J. Cell Biol.* **141,** 663–674.

Paintrand, M., Moudjou, M., Delacroix, H., and Bornens, M. (1992). Centrosome organization and centriole architecture: their sensibility to divalent cations. *J. Struct. Biol.* **108,** 107–128.

Palazzo, R. E., Vaisberg, E., Cole, R. W., and Rieder, C. L. (1992). Centriole duplication in lysates of *Spisula solidissima* oocytes [published erratum: (1992). *Science,* **256,** 1746]. *Science* **256,** 219–221.

Paoletti, A., and Bornens, M. (1997). Organisation and functional regulation of the centrosome in animal cells. *Prog. Cell Cycle Res.* **3,** 285–299.

Paoletti, A., Moudjou, M., Paintrand, M., Salisbury, J. L., and Bornens, M. (1996). Most of centrin in animal cells is not centrosome- associated and centrosomal centrin is confined to the distal lumen of centrioles. *J. Cell Sci.* **109,** 3089–3102.

Paoletti, A., Giocanti, N., Favaudon, V., and Bornens, M. (1997). Pulse treatment of interphasic HeLa cells with nanomolar doses of docetaxel affects centrosome organization and leads to catastrophic exit of mitosis. *J. Cell Sci.* **110,** 2403–2415.

Perret, E., Moudjou, M., Geraud, M. L., Derancourt, J., Soyer-Gobillard, M. O., and Bornens, M. (1995). Identification of an HSP70-related protein associated with the centrosome from dinoflagellates to human cells. *J. Cell Sci.* **108,** 711–725.

Piel, M., Meyer, P., Khodjakov, A., Rieder, C. L., and Bornens, M. (2000). The respective contributions of the mother and daughter centrioles to centrosome activity and behavior in vertebrate cells. *J. Cell Biol.* **149,** 317–330.

Piel, M., and Bornens, M. (1998). A cell free assay for centrosome duplication and the formation of mitotic spindle-like structures without chromosomes. *MCB* **9 suppl,** 12a.

Plessmann, U., and Weber, K. (1997). Mammalian sperm tubulin: an exceptionally large number of variants based on several posttranslational modifications. *J. Protein Chem.* **16,** 385–390.

Rios, R. M., Tassin, A. M., Celati, C., Antony, C., Boissier, M. C., Homberg, J. C., and Bornens, M. (1994). A peripheral protein associated with the cis-Golgi network redistributes in the intermediate compartment upon brefeldin A treatment. *J. Cell Biol.* **125,** 997–1013.

Rodionov, V. I., and Borisy, G. G. (1997). Self-centering activity of cytoplasm. *Nature* **386,** 170–173.

Rodionov, V., Nadezhdina, E., and Borisy, G. (1999). Centrosomal control of microtubule dynamics. *Proc. Natl. Acad. Sci. USA* **96,** 115–120.

Ruiz, F., Beisson, J., Rossier, J., and Dupuis Williams, P. (1999). Basal body duplication in *Paramecium* requires gamma-tubulin. *Curr. Biol.* **9,** 43–46.

Saiki, T., and Hamaguchi, Y. (1998). Aster-forming abilities of the egg, polar body, and sperm centrosomes in early starfish development. *Dev. Biol.* **203,** 62–74.

Saunders, R. D. C., Avides, M. D., Howard, T., Gonzalez, C., and Glover, D. M. (1997). The *Drosophila* gene abnormal spindle encodes a novel microtubule-associated protein that associates with the polar regions of the mitotic spindle. *J. Cell Biol.* **137,** 881–890.

Schatten, G. (1994). The centrosome and its mode of inheritance: the reduction of the centrosome during gametogenesis and its restoration during fertilization. *Dev. Biol.* **165,** 299–335.

Schmidt, P. H., Dransfield, D. T., Claudio, J. O., Hawley, R. G., Trotter, K. W., Milgram, S. L., and Goldenring, J. R. (1999). AKAP350, a multiply spliced protein kinase A-anchoring protein associated with centrosomes. *J. Biol. Chem.* **274,** 3055–3066.

Schnackenberg, B. J., Khodjakov, A., Rieder, C. L., and Palazzo, R. E. (1998). The disassembly and reassembly of functional centrosomes in vitro. *Proc. Natl. Acad. Sci. USA* **95,** 9295–9300.

Schulze, E., and Kirschner, M. (1986). Microtubule dynamics in interphase cells. *J. Cell Biol.* **102,** 1020–1031.

Skoufias, D. A., Burgess, T. L., and Wilson, L. (1990). Spatial and temporal colocalization of the Golgi apparatus and microtubules rich in detyrosinated tubulin. *J. Cell Biol.* **111,** 1929–1937.

Sluder, G., and Rieder, C. L. (1985). Centriole number and the reproductive capacity of spindle poles. *J. Cell Biol.* **100,** 887–896.

Sluder, G., Miller, F. J., and Rieder, C. L. (1989). Reproductive capacity of sea urchin centrosomes without centrioles. *Cell Motil. Cytoskel.* **13,** 264–273.

Smrzka, O. W., Delgehyr, N., and Bornens, M. (2000). Tissue-specific expression and subcellular localisation of mammalian δ-tubulin. *Curr. Biol.* **10,** 413–416.

Sobel, S. G., and Synder, M. (1995). A highly divergent gamma-tubulin gene is essential for cell growth and proper microtubule organization in Saccharomyces cerevisiae. *J. Cell. Biol.* **131,** 1775–1788.

Spang, A., Courtney, I., Fackler, U., Matzner, M., and Schiebel, E. (1993). The calcium-binding protein cell division cycle 31 of *Saccharomyces cerevisiae* is a component of the half bridge of the spindle pole body. *J. Cell Biol.* **123,** 405–416.

Spang, A., Courtney, I., Grein, K., Matzner, M., and Schiebel, E. (1995). The Cdc31p-binding protein Kar1p is a component of the half bridge of the yeast spindle pole body. *J. Cell Biol.* **128,** 863–877.

Spang, A., Geissler, S., Grein, K., and Schiebel, E. (1996). Gamma-tubulin-like Tub4p of *Saccharomyces cerevisiae* is associated with the spindle pole body substructures that organize microtubules and is required for mitotic spindle formation. *J. Cell Biol.* **134,** 429–441.

Steffen, W., and Linck, R. W. (1988). Evidence for tektins in centrioles and axonemal microtubules. *Proc. Natl. Acad. Sci. USA* **85,** 2643–2647.

Stephens, R. E., and Lemieux, N. A. (1998). Tektins as structural determinants in basal bodies. *Cell Motil. Cytoskel.* **40,** 379–392.

Takahashi, M., Shibata, H., Shimakawa, M., Miyamoto, M., Mukai, H., and Ono, Y. (1999). Characterization of a novel giant scaffolding protein, CG-NAP, that anchors multiple signaling enzymes to centrosome and the golgi apparatus. *J. Biol. Chem.* **274,** 17267–17274.

Tassin, A., and Bornens, M. (1999). Centrosome structure and microtubule nucleation in animal cells. *Biol. Cell* **91,** 343–354.

Tassin, A. M., Maro, B., and Bornens, M. (1985). Fate of microtubule-organizing centers during myogenesis in vitro. *J. Cell Biol.* **100,** 35–46.

Tassin, A. M., Celati, C., Moudjou, M., and Bornens, M. (1998). Characterization of the human homologue of the yeast Spc98p and its association with gamma-tubulin. *J. Cell Biol.* **141,** 689–701.

Tournier, F., and Bornens, M. (1994). Cell cycle regulation of centrosome function. *In* "Microtubules" (J. S. Hyams and C. W. Lloyd, Eds.), pp. 303–324. John Wiley & Sons, New York.

Tournier, F., Cyrklaff, M., Karsenti, E., and Bornens, M. (1991a). Centrosomes competent for parthenogenesis in *Xenopus* eggs support procentriole budding in cell-free extracts. *Proc. Natl. Acad. Sci. USA* **88,** 9929–9933.

Tournier, F., Komesli, S., Paintrand, M., Job, D., and Bornens, M. (1991b). The intercentriolar linkage is critical for the ability of heterologous centrosomes to induce parthenogenesis in *Xenopus*. *J. Cell Biol.* **113,** 1361–1369.

Tournier, F., Bobinnec, Y., Debec, A., Santamaria, P., and Bornens, M. (1999). *Drosophila* centrosomes are unable to trigger parthenogenetic development of *Xenopus* eggs. *Biol. Cell* **91,** 99–108.

Vorobjev, I. A., Svitkina, T. M., and Borisy, G. G. (1997). Cytoplasmic assembly of microtubules in cultured cells. *J. Cell Sci.* **110,** 2635–2645.

Waterman-Storer, C. M., and Salmon, E. D. (1997a). Microtubule dynamics: Treadmilling comes around again. *Curr. Biol.* **7,** R369–R372.

Waterman-Storer, C. M., and Salmon, E. D. (1997b). Actomyosin-based retrograde flow of microtubules in the lamella of migrating epithelial cells influences microtubule dynamic instability and turnover and is associated with microtubule breakage and treadmilling. *J. Cell Biol.* **139,** 417–434.

Waters, J. C., and Salmon, E. D. (1997). Pathways of spindle assembly. *Curr. Opin. Cell Biol.* **9,** 37–43.

Witczak, O., Skalhegg, B. S., Keryer, G., Bornens, M., Tasken, K., Jahnsen, T., and Orstavik, S. (1999). Cloning and characterization of a cDNA encoding an A-kinase anchoring protein located in the centrosome, AKAP450. *EMBO J.* **18,** 1858–1868.

Wolff, A., de Nechaud, B., Chillet, D., Mazarguil, H., Desbruyeres, E., Audebert, S., Edde, B., Gros, F., and Denoulet, P. (1992). Distribution of glutamylated alpha and beta-tubulin in mouse tissues using a specific monoclonal antibody, GT335. *Eur. J. Cell Biol.* **59,** 425–432.

Yvon, A. M. C., and Wadsworth, P. (1997). Non-centrosomal microtubule formation and measurement of minus end microtubule dynamics in A498 cells. *J. Cell Sci.* **110,** 2391–2401.

Zimmerman, W., Sparks, C. A., and Doxsey, S. J. (1999). Amorphous no longer: the centrosome comes into focus. *Curr. Opin. Cell Biol.* **11,** 122–128.

2

γ-Tubulin

Berl R. Oakley
Department of Molecular Genetics
The Ohio State University
Columbus, Ohio 43210

I. Centrosomes and Microtubule Nucleation

Centrosomes are, of course, the major microtubule organizing centers (MTOCs) in animal cells. They consist of two structural components, the centriole, which is composed of nine microtubule triplets linked into a cylindrical array, and the pericentriolar matrix (PCM). The PCM is amorphous and it is the component of the centrosome responsible for microtubule nucleation.

Microtubules are critically important for many essential cellular functions. They are the major functional components of the mitotic and meiotic apparatuses and they are essential for many other forms of directed organellar movement. They are composed of two related proteins, α- and β-tubulin, that form a heterodimer, tubulin, that is the building block of the microtu-

Current Topics in Developmental Biology, Vol. 49

0070-2153/00 $35.00

bule (Fig. 1). In microtubules, tubulin dimers associate longitudinally to form protofilaments and in most cells that have been studied, microtubules consist of 13 protofilaments. Adjacent protofilaments are slightly offset, creating a structure known as a B lattice (Fig. 1). Lateral interactions between protofilaments are mostly α-tubulin/α-tubulin and β-tubulin/β-tubulin interactions, but 13 protofilament B-lattice microtubules have a seam in which α-tubulin and β-tubulin interact laterally (Fig. 1).

Microtubules are inherently polar structures. The ends, which are designated plus and minus, have different assembly and disassembly characteristics. The motor molecules that generate force by interacting with microtubules are directional. They are plus-end-directed (i.e., they move toward the plus end of the microtubule) or they are minus-end-directed, but they rarely move in both directions. The polarity of micrtotubules in cells is almost never random (at least in the organisms that have been studied). In mitotic apparatuses, for example, the great majority of microtubules are arranged such that their minus ends are toward or at the poles and their plus ends are distal to the poles (McIntosh and Euteneuer, 1984). In interphase cells, minus ends are at the centrosomes and plus ends are distal.

The polarity of microtubules, in combination with the directionality of microtubule motors and the positions of microtubules within cells, is of paramount importance for directional movements within cells. In mitosis,

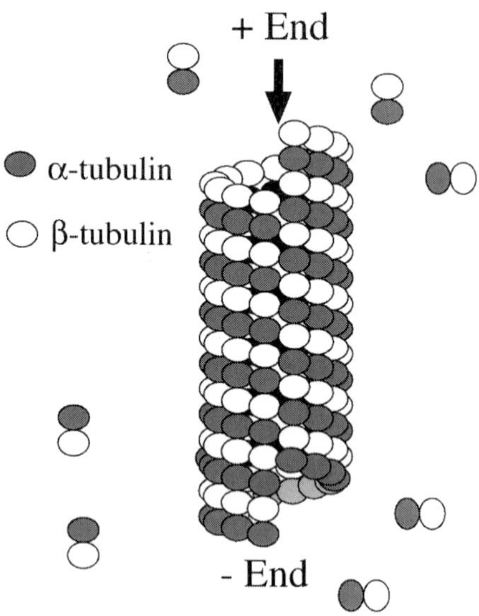

Figure 1 A schematic illustration of microtubule structure. The seam that is inherent in 13 protofilament B-lattice microtubules is shown by an arrow.

for example, a minus-end-directed motor at the kinetochore can, in principle, move chromosomes to the poles because the microtubules in the half spindle have a uniform polarity with the minus end at the pole. Likewise, movement of vesicles from the cell periphery to the vicinity of the nucleus could be carried out by a minus-end-directed motor because the plus ends of microtubules are at the cell periphery and the minus ends are at the centrosome adjacent to the nucleus. By the same token, movement of vesicles from the vicinity of the nucleus to the cell periphery could be carried out by a plus-end-directed motor because of the uniform polarity of microtubules.

These examples should suffice to demonstrate the importance of establishing microtubule polarity and position in cells. In many cell types microtubule polarity and positioning are largely determined by MTOCs. The major MTOC in animal cells is, of course, the centrosome, and centrosomes establish microtubule polarity and position by nucleating microtubule assembly with the minus end of the microtubule at the centrosome (Bergen *et al.*, 1980). Tubulin dimers have the ability to assemble into microtubules without centrosomes, but the initial steps of microtubule assembly (nucleation) are inefficient with pure tubulin *in vitro* (see Chapter 3, by Zheng and co-workers, in this volume for more discussion.) There must, thus, be one or more components of the centrosome that facilitate the nucleation of microtubule assembly such that microtubules assemble preferentially from the centrosome rather than assembling free in the cytoplasm of cells. It is also important to remember in this regard that the tubulin in living cells is in equilibrium with assembled microtubules such that the concentration of assembly-competent tubulin dimers is at or near the critical concentration for assembly. This means that spontaneous nucleation of microtubule assembly will be an infrequent event and assembly from nucleation sites at the centrosome or onto existing microtubules will be greatly favoured over spontaneous nucleation.

I hasten to point out that in many cells and organisms microtubule polarity and position can be established without centrosomes and, apparently, without other MTOCs. Many phyla (e.g., land plants) lack centrosomes and in some animal cells (notably meiotic cells) centrosomes are absent or unimportant for spindle formation. In these systems it appears that directional motors are responsible for bringing the ends of microtubules to a focus and for moving microtubules to create polar microtubule arrays (Matthies *et al.*, 1996; Heald *et al.*, 1996; and other data reviewed by Merdes and Cleveland, 1997). Nevertheless, centrosomes are clearly important for mitotic spindle formation in many types of cells, including most animal cells (Zhang and Nicklas, 1995 and earlier data reviewed by McIntosh, 1983, and Kalt and Schliwa, 1993).

The importance of centrosomes for organizing mitotic spindles and cytoplasmic microtubule arrays has naturally led to a great deal of interest in the components of centrosomes and other MTOCs that nucleate microtubule assembly. Biochemical approaches have led to the identification of many putative components of centrosomes (reviewed by Kimble and Kuriyama, 1992; Kalt and Schliwa, 1993), and there have been numerous attempts to determine the roles of these components through *in vivo* and *in vitro* antibody inhibition studies. A surprisingly large number of antibodies against putative centrosomal components inhibit microtubule assembly from centrosomes (reviewed by Kimble and Kuriyama, 1992; Kalt and Schliwa, 1993) and the number continues to grow each year. It is possible that many centrosomal components are actually involved in microtubule assembly and that the number of antibodies that inhibit microtubule assembly at centrosomes accurately reflects the number of proteins involved in this process. An alternative possibility, however, is that some antibodies may inhibit microtubule assembly by an indirect mechanism. A protein might be near an assembly site but might not be an active component of the site. An antibody against the protein might sterically hinder microtubule assembly even though the protein is not directly involved in microtubule assembly. It should also be remembered that antibodies have two antigen binding sites and, thus, have a tendency to bridge between antigen molecules. Antibodies against a protein distributed through the PCM would tend to bridge between the molecules and this would likely distort the PCM and perhaps hinder tubulin access to nucleation sites. Although the role of many centrosomal proteins in microtubule assembly remains open to debate and further study, many lines of evidence (discussed in Section IV and elsewhere in this volume) lead to the inescapable conclusion that γ-tubulin is important for assembly of microtubules from centrosomes and other MTOCs.

II. γ-Tubulin and the Tubulin Superfamily of Proteins

γ-Tubulin was discovered in a screen for suppressors of a β-tubulin mutation in the filamentous fungus *Aspergillus nidulans* (Oakley and Oakley, 1989). *A. nidulans* γ-tubulin is an acidic protein (predicted isoelectric pH of 5.55) with a predicted molecular mass of 50,825 (Oakley and Oakley, 1989). γ-Tubulin genes and/or cDNAs have subsequently been cloned from a wide variety of organisms, including members of all major phyla. It is, thus, safe to assume that γ-tubulin is ubiquitous in eukaryotes. γ-Tubulins share approximately 30% identity with α- and β-tubulins (Oakley and Oakley, 1989) and α-tubulins, in turn, share approximately 40% amino acid identity with β-tubulins (Krauhs *et al.,* 1981; Little and Seehaus, 1988). α-, β-, and γ-tubulins, thus, constitute families within the tubulin superfamily. An additional member of the tubulin superfamily, δ-tubulin, has been discovered in *Chlamydomonas,* and it is apparently involved in the function and maturation of basal bodies (Dutcher and Trabuco, 1998). The sequence similarities of the tubulins indicate clearly that they descend from a common ancestral protein.

A. Conservation of γ-Tubulin

α- and β-tubulin are highly conserved (Little and Seehaus, 1988). α-Tubulins share more than 60% amino acid identity among widely divergent organisms, as do β-tubulins (Little and Seehaus, 1988). γ-Tubulins are also conserved, but the situation is somewhat more complex than for α- and β-tubulins. Most γ-tubulin sequences that have been reported fall into a group that I will call "conventional." Conventional γ-tubulins share more than 60% amino acid identity and are quite conserved across great phylogenetic distances. γ-Tubulins of *Homo sapiens* and *A. nidulans* share more than 67% amino acid identity, for example (Zheng *et al.,* 1991), and *Xenopus laevis* (clawed frog) and *H. sapiens* γ-tubulins share 98% identity.

Conventional γ-tubulins also appear to be functionally conserved, although the data in this area are sparse. A *Schizosaccharomyces pombe* strain in which only a human γ-tubulin is expressed is viable and grows at about 90% of the rate of a control strain expressing *S. pombe* γ-tubulin (Horio and Oakley, 1994).

In addition to the many conventional γ-tubulins that have been identified, two divergent or "unconventional" γ-tubulins have been reported. They were revealed by the *Caenorhabditis elegans* and *Saccharomyces cerevisiae* genome projects. The product of the *TUB4* gene of *S. cerevisiae* (Tub4p) shares only a mean of 35.6% identity with 16 conventional γ-tubulins (Sobel and Snyder, 1995) and the *C. elegans* γ-tubulin homolog shares only 40–45% identity with γ-tubulins of other organisms and even less identity with Tub4p. These proteins are, thus, only slightly more similar to γ-tubulins than they are to α- and β-tubulins, and this led to the proposal that these tubulins be designated δ- and ε-tubulin (Burns, 1995). Although these designations might be justified on the basis of sequence, functional and localization data (see discussion later in this chapter and elsewhere in this volume) suggest that Tub4p is a γ-tubulin, albeit a highly divergent one. It is worth noting that the now-complete *S. cerevisiae* genome project has revealed no other γ-tubulin homologs, so *S. cerevisiae* does not possess a conventional γ-tubulin.

Few localization or functional data are available for the *C. elegans* γ-tubulin-like protein. The *C. elegans* genome project is essentially complete as of this writing, however, and has revealed no conventional γ-tubulin. It is, thus, likely that the unconventional γ-tubulin is the sole γ-tubulin in *C. elegans*.

Tub4p is not only divergent in sequence from other γ-tubulins, but it is functionally divergent as well. *TUB4* does not complement an *S. pombe* γ-tubulin null mutation (Marschall *et al.*, 1996) and a *TUB4* mutation was not complemented by *H. sapiens* or *X. laevis* γ-tubulin genes (Spang *et al.*, 1996; Marschall *et al.*, 1996).

Why the γ-tubulins of *S. cerevisiae* and *C. elegans* are so divergent from other γ-tubulins is not yet clear. Simple phylogenetic distance is not an adequate explanation. *A. nidulans* and *S. pombe* must be much more distant phylogenetically from humans than is *C. elegans*, and yet their γ-tubulins are much more similar to human γ-tubulins. It is more likely that the divergence of the *C. elegans* and *S. cerevisiae* γ-tubulins reflects a functional specialization of γ-tubulin in these organisms. If so, highly divergent, functionally specialized γ-tubulins may be found in other organisms as well.

B. γ-Tubulin Gene Families

Many organisms have multiple α- and β-tubulin genes, and the functions of these genes have been the subject of a great deal of study and speculation. It has, consequently, been of interest to determine if families of γ-tubulin genes also exist. Although the data are limited, examination of the databases reveals a clear pattern. Many organisms apparently have only a single γ-tubulin gene, whereas some organisms have two. To date, three γ-tubulin genes have been reported only in *Zea mays*. Families of γ-tubulin genes, thus, do exist, but they are typically smaller than the α- and β-tubulin families.

The presence of two γ-tubulin genes in some organisms raises the possibility that the genes have different expression patterns or that the γ-tubulins encoded by the two genes have different functions. The most extensive data on this subject are from *Drosophila melanogaster*. The two *D. melanogaster* γ-tubulins share 83% amino acid identity (Wilson *et al.*, 1997). This is a surprisingly great divergence if one considers that frog and human γ-tubulins, for example, share 98% amino acid identity. One of the two *D. melanogaster* genes is expressed zygotically and the other is expressed maternally (Wilson *et al.*, 1997; Tavosanis *et al.*, 1997). Although the expression patterns of the two γ-tubulins are different, both are present in cellularized embryos. There both localize to centrosomes so they share structural features necessary for binding to the centrosome (Wilson *et al.*, 1997). In addition, mutations in each of the γ-tubulin genes can cause mitotic defects (Sunkel *et al.*, 1995; Tavosanis *et al.*, 1997; Wilson and Borisy, 1998), albeit in different cells because of the different expression patterns of the two genes.

The situation in other organisms is even less clear. *Arabidopsis thaliana* contains two γ-tubulin genes that encode proteins with 98% amino acid identity (Liu *et al.,* 1994). Likewise, there are two γ-tubulin genes in *H. sapiens* that share 97% amino acid identity (Wise and Oakley, 1997). It would be surprising, in each case, if two such similar γ-tubulins had very different functions. In both cases the two genes are coexpressed in a number of tissues (Liu *et al.,* 1994; Wise and Oakley, 1997). It is possible that the presence of two γ-tubulin genes facilitates regulation of the levels of expression of γ-tubulin in different tissues, or perhaps γ-tubulin is sufficiently important that it is simply advantageous to have two γ-tubulin genes that function redundantly.

III. Distribution of γ-Tubulin in Cells

A. Localization of γ-Tubulin in Spindle Pole Bodies, Centrosomes, and Mitotic Spindles

γ-Tubulin was first localized by immunofluorescence microscopy in *A. nidulans* (Oakley *et al.,* 1990) where it was found to be located at spindle pole bodies (SPBs). SPBs are fungal MTOCs that share many of the functions of centrosomes. It was also localized to the SPBs of the fission yeast *S. pombe* (Horio *et al.,* 1991), and in this organism it was additionally located at cytoplasmic MTOCs that form at the end of mitosis and subsequently disappear. This result suggested that γ-tubulin could translocate in a cell-cycle-specific fashion.

γ-Tubulin was subsequently found to be located at other microtubule organizing centers including, most notably, centrosomes (Fig. 2) (Zheng *et al.,* 1991; Stearns *et al.,* 1991; Joshi *et al.,* 1992, and many subsequent studies). Although γ-tubulin is most obviously located at microtubule organizing centers, they are not the sole location. Some γ-tubulin is associated with the mitotic spindle in many cell types. The amount varies from cell type to cell type, but, in some cell types, the amount is apparently substantial (Lajoie-Mazenc *et al.,* 1994). There is also a particular accumulation of γ-tubulin at midbodies of some dividing cells (Julian *et al.,* 1993).

Interestingly, the amount of γ-tubulin staining at centrosomes changes as cells progress through the cell cycle (Zheng *et al.,* 1991; Stearns *et al.,* 1991; Lajoie-Mazenc *et al.,* 1994). Generally the amount of γ-tubulin staining at the centrosome (and the size of the stained region) is low in interphase, increases at the onset of prophase, remains high in prometaphase and metaphase, and decreases in anaphase and telophase (Zheng *et al.,* 1991; Lajoie-Mazenc *et al.,* 1994). As discussed in Section III,E and elsewhere

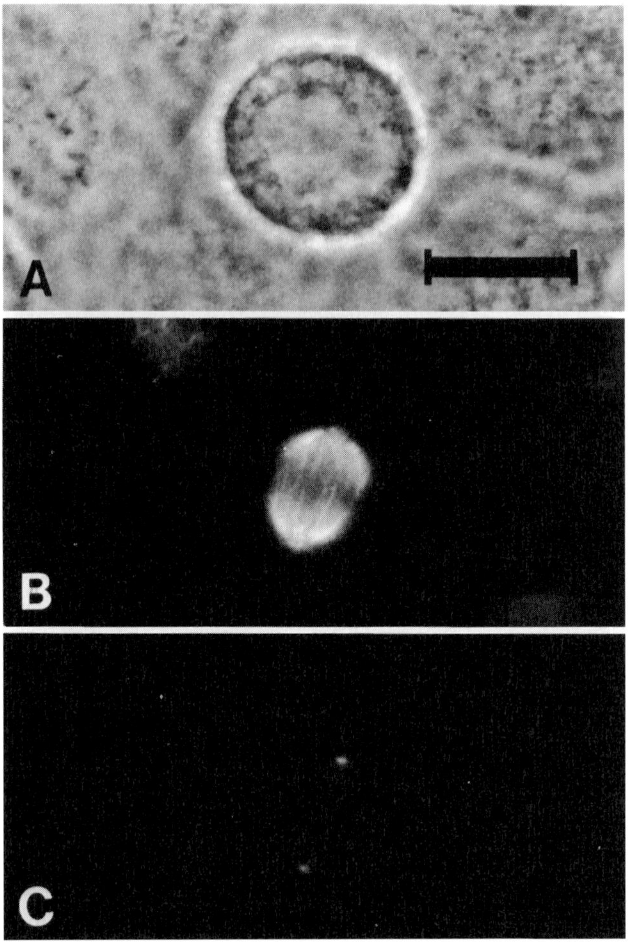

Figure 2 γ-Tubulin localization in human (HeLa) cells. The three panels are of the same field. A shows a phase contrast micrograph, B shows the mitotic spindle as revealed with an anti-β-tubulin antibody, and C shows the location of γ-tubulin as revealed with a specific γ-tubulin antibody. γ-Tubulin is most obviously located at the spindle poles, but there is very faint staining of the spindle as well. Scale in A = 20 γm. Reproduced from Zheng *et al.* (1991) by permission.

in this volume, a substantial amount of γ-tubulin is present as soluble complexes in the cytoplasm.

These immunofluorescence studies have been augmented by observations of a γ-tubulin green fluorescent protein fusion (γTGFP) in several mammalian cell lines (Khodjakov and Rieder, 1999). The γTGFP was not toxic to the cells and the patterns of localization are similar to those seen with immunofluorescence microscopy of γ-tubulin. γTGFP, thus, appears to be an accurate reporter of the location of γ-tubulin in living cells and, importantly, allows one to observe the behavior of γ-tubulin as cells progress through the cell cycle. Khodjakov and Rieder found that γTGFP levels at the centrosome were low in interphase, but it was rapidly recruited to the centrosome in prophase. Levels of centrosomal γTGFP remained high in metaphase but decreased in anaphase, reaching minimal levels after cytokinesis. As mitosis proceeded, γTGFP spread into the spindle, apparently from the centrosome, and at the end of mitosis some, but not all, midbodies showed γTGFP staining. Fluorescence recovery after photobleaching experiments revealed that γTGFP at the centrosome is exchangeable with soluble γTGFP in the cytoplasm and that there are two populations of γ-tubulin, a rapidly exchanging population and a more slowly exchanging population. Exchange does not require microtubules. Because γTGFP exchanges at both interphase and mitotic centrosomes, it is unlikely that the rapid accumulation of γTGFP at centrosomes in prophase is due to a change in the characteristics of γTGFP or complexes containing γTGFP. Rather, it appears that centrosomes suddenly gain the ability to bind more γTGFP.

As will be appreciated by readers of this volume, the centrosome is a complex structure, so it was of obvious interest to determine the location of γ-tubulin within the centrosome. The fact that the region of γ-tubulin staining changed in size as the cell entered and completed mitosis suggested that γ-tubulin is not limited to the centriole. Electron microscopic immunolocalization experiments revealed that γ-tubulin is, indeed, a component of pericentriolar matrix (PCM) (Stearns *et al.*, 1991). More detailed studies involving immunoelectron microscopic tomography have revealed that the γ-tubulin in the PCM is in ring-shaped complexes (Moritz *et al.*, 1995; Vogel *et al.*, 1997). These structures almost certainly correspond to the soluble γ-tubulin ring complexes (γ-TuRCs) that have been purified from *Xenopus* and *Drosophila* (Zheng *et al.*, 1995; Oegema *et al.*, 1999). γ-TuRCs are discussed in Section III,E and, in more detail, in Chapter 3. γ-Tubulin is apparently not limited to the PCM, however, because it has also been found in the core of the centriole (Fuller *et al.*, 1995).

Exactly what γ-TuRCs bind to in the PCM is still still to be determined, but one good candidate is pericentrin (Dictenberg *et al.*, 1998). Pericentrin is a 220 kDa coiled-coil protein that is a component of PCM (Doxsey *et al.*, 1994). Fluorescence energy transfer experiments indicate that pericentrin is associated with γ-tubulin in the PCM, and reconstructions of immunofluorescence images using a sophisticated deconvolution algorithm indicate that pericentrin and γ-tubulin are organized into a lattice structure in the PCM (Dictenberg *et al.*, 1998).

B. γ-Tubulin Location at MTOCs Other Than Centrosomes

Although centrosomes are the primary MTOC in most animal cells, other MTOCs are found in some cell types. It has, consequently, been of interest to determine if γ-tubulin is located at these MTOCs. Polarized epithelial cells of many types contain microtubules that are not associated with centrosomes but have their minus ends located at the apical surface (Meads and Schroer, 1995, and references therein). Data from microtubule regrowth experiments and other experiments suggest that the apical surface may function as an MTOC (Meads and Schroer, 1995). Perhaps surprisingly, γ-tubulin localization results vary among different polarized epithelial cell types. γ-Tubulin did localize to the apical surface in several cell lines (Meads and Schroer, 1995; Mogensen *et al.,* 1997) but not in others (Meads and Schroer, 1995). One interesting model that might explain some of the variation is that in some cells microtubules might be nucleated by the centrosome and then detach and dock at the apical surface (Mogensen *et al.,* 1997; see also Rizzolo and Joshi, 1993). The apical surface might not be a classical MTOC, but rather a microtubule docking site. Its ability to nucleate microtubule assembly could be due to small microtubule seeds that are resistant to depolymerization.

C. γ-Tubulin Location in Animal Cells without Centrioles

Mouse oocytes and early embryos lack centrioles. In these cells, γ-tubulin is located at the spindle poles of meiosis II oocytes and at mitotic poles in early cleavage stages (Gueth-Hallonet *et al.,* 1993). γ-Tubulin becomes focused into a dot in interphase at the 32-cell stage and centrioles appear at the 64-cell stage. Acentriolar cytoplasmic MTOCs are also present in these cells and γ-tubulin is located at these MTOCs (Gueth-Hallonet *et al.,* 1993).

The situation in *D. melanogaster* is quite different and the data are, in some instances, contradictory. Meiosis I spindles in *D. melanogaster* oocytes lack centrioles and γ-tubulin does not show any clear localization (Matthies *et al.,* 1996). In meiosis II, however, an unusual microtubule organizing center forms between the two spindles that are present in the same cytoplasmic compartment. This MTOC has been reported to stain for γ-tubulin, and the localization of γ-tubulin to this MTOC has been reported to depend on the kinesin-like motor protein Ncd (Endow and Komma, 1998). Another study failed to detect γ-tubulin at this MTOC, however (Wilson and Borisy, 1998).

Finally, γ-tubulin has been localized in a *D. melanogaster* cell line that lacks centrioles (Debec *et al.,* 1995). γ-Tubulin is not localized to any discrete structures in interphase but is located at the poles of mitotic spindles. The polar localization is dependent on microtubules, however, and it was hypothesized that the polar location might be due to the fact that γ-tubulin is located at the minus ends of microtubules that are located at the poles.

D. Localization in Plant Cells

Several immunofluorescence localization studies of γ-tubulin have now been carried out in plant cells (reviewed by Joshi and Palevitz, 1996). Plants have several microtubule arrays that reorganize as the cell approaches and passes through mitosis (reviewed by Wick, 1991; Goddard *et al.,* 1994; Vaughn and Harper, 1998). Punctate γ-tubulin staining is found in each of these microtubule arrays and is particularly prominent in kinetochore fibers. In prophase, two regions of the nuclear envelope are proposed to act as microtubule nucleation centers (Lambert, 1993), and these regions also stain prominently with anti-γ-tubulin antibodies. Interestingly, γ-tubulin is not distributed evenly throughout the microtubule arrays. Rather, it has been reported to avoid regions where there is a concentration of plus ends of microtubules (reviewed by Joshi and Palevitz, 1996). However, it has been reported that in three plant species an antibody against a γ-tubulin peptide stains kinetochore fibers near the kinetochore (a region where there should be no minus ends) and may even stain the kinetochore where plus ends of microtubules terminate (Binarová *et al.,* 1998). This is a very interesting result that is at odds with previous reports, and this area needs to be pursued with additional antibodies directed toward different regions of γ-tubulin.

The distribution of γ-tubulin in plant cells has led to two hypotheses, which are not mutually exclusive. The first is that γ-tubulin is really located at the minus ends of microtubules, but the minus ends of microtubules are distributed through microtubule arrays. This hypothesis is supported by studies in *Arabidopsis* with microtubule-active agents. Liu *et al.* (1995) treated *Arabidopsis* suspension cells with colchicine, an antimicrotubule agent, isopropyl *n*-3-chlorophenyl carbamate (IPC), a herbicide that appears to act on microtubules, and taxol, a microtubule hyperstabilizing compound. When microtubules were disassembled with colchicine, and the colchicine was then inactivated with ultraviolet light, microtubules reassembled and γ-tubulin was located at the sites of reassembly (e.g., the nuclear envelope). With taxol and IPC, which induce multipolar spindles, γ-tubulin was concentrated at foci from which microtubules extended.

The second hypothesis is that γ-tubulin associates with the sides of microtubules or is incorporated into microtubules. In support of this hypothesis, immunoelectron microscopy in the fern *Ceratopteris* appears to show that γ-tubulin is located along the length of mitotic spindle microtubules (Hoffman *et al.,* 1994).

Two things should be pointed out here. The first is that some animal cells show a substantial amount of γ-tubulin staining in the spindle, so γ-tubulin distribution may not be as different in plants and animals as was initially suspected. The second is that the distribution of a protein does not necessarily indicate its function. As Joshi and Palevitz (1996) point out, the γ-tubulin in the spindle could be functional in this location or it could be in an inactive or storage form.

E. Soluble Forms of γ-Tubulin

The changes in staining intensity of centrosomes as they pass through the cell cycle and the appearance and disappearance of γ-tubulin staining at cytoplasmic MTOCs in *S. pombe* suggested that γ-tubulin must exist in a soluble or, at least, transportable, form away from MTOCs. Indeed, γ-tubulin does exist in soluble forms in many, perhaps all, cells. γ-Tubulin is folded by the TCP-1 (also called c-cpn) chaperonin complex (Melki *et al.,* 1993) and part of the soluble γ-tubulin is probably in these complexes. The heterohexameric chaperone protein prefoldin binds to unfolded or improperly folded γ-tubulin (Geissler *et al.,* 1998; Vainberg *et al.,* 1998) and another fraction of the soluble γ-tubulin is probably in this complex. It is reasonable to presume that the amount of γ-tubulin in these two types of complexes depends on the rate of γ-tubulin synthesis and it could, thus, vary greatly among organisms and cell types.

In *S. cerevisiae,* γ-tubulin (Tub4p) forms a soluble 6S complex with two proteins Spc97p and Spc98p (discussed in detail elsewhere in this volume). The 6S complex apparently contains two or more molecules of γ-tubulin and a single molecule each of Spc97p and Spc98p (Knop *et al.,* 1997; Knop and Schiebel, 1997). The complex apparently binds to the SPB protein Spc110p on the nucleoplasmic face of the SPB (Knop and Schiebel, 1997; Sundberg and Davis, 1997) and another SPB protein Spc72 on the cytoplasmic face of the SPB (Knop and Schiebel, 1998).

Homologues of Spc97p and Spc98p have now been found in phylogenetically diverse organisms (Murphy *et al.,* 1998; Martin *et al.,* 1998; Tassin *et al.,* 1998; Oegema *et al.,* 1999). They cosediment and coimmunoprecipitate with γ-tubulin in extracts from mammalian cells and *X. laevis* (Murphy *et al.,* 1998; Martin *et al.,* 1998; Tassin *et al.,* 1998), and in *D. melanogaster* embryo extracts the Spc97p and Spc98p homologues, which are called Dgrip84 and Dgrip91, form a complex with γ-tubulin (Oegema *et al.,* 1999). This complex, the γ-tubulin small complex (γ-TuSC), appears to contain one molecule of Dgrip84, one molecule of Dgrip91, and two molecules of γ-tubulin (Oegema *et al.,* 1999). The γ-TuSC, thus, appears similar to the complex found in *S. cerevisiae.* Similar sized complexes have been found in other organisms (e.g., Akashi *et al.,* 1997), and it is likely that γ-TuSCs are universal.

Larger γ-tubulin-containing complexes (25S > 30S) have been identified in embryo extracts, eggs, animal cells and in *A. nidulans* (Stearns and Kirschner, 1994; Zheng *et al.,* 1995; Meads and Schroer, 1995; Akashi *et al.,* 1997; Murphy *et al.,* 1998; Martin *et al.,* 1998; Tassin *et al.,* 1998; Oegema *et al.,* 1999). Most, if not all, of these are the γ-TuRCs that are present in the pericentriolar material and have the ability to nucleate microtubule assembly (discussed in Section IV, C and elsewhere in this volume [Chapter 3]). These complexes are apparently made up of γ-TuSCs and approximately four other proteins (Zheng *et al.,* 1995; Murphy *et al.,* 1998; Oegema *et al.,* 1999). In *X. laevis,* α- and β-tubulin are apparently found in γ-TuRCs, but they have not been detected in mouse and *D. melanogaster* γ-TuRCs (Murphy *et al.,* 1998; Oegema *et al.,* 1999).

IV. Studies of γ-Tubulin Function

The function of γ-tubulin has now been studied using a variety of approaches in several organisms and experimental systems. All of the results point to an important role for γ-tubulin in microtubule nucleation and/or spindle formation, but there are variations in the results that raise a number of questions. I summarize the results and subsequently discuss some unresolved questions (Section V).

A. Genetic Studies

The first studies of γ-tubulin function were carried out in *A. nidulans* (Oakley *et al.,* 1990). The γ-tubulin gene of *A. nidulans* was disrupted, and the disruption proved to be lethal. Fortunately, in *A. nidulans* it is possible to maintain nuclei carrying recessive lethal mutations in heterokaryons, and one can determine the phenotype conferred by the lethal mutation in uninucleate spores (conidia) produced by the heterokaryon. The γ-tubulin disruption prevented mitotic spindle formation as judged by the fact that spindles were completely absent in germlings carrying the γ-tubulin disruption, even in nuclei that were in mitosis as judged by chromosomal condensation.

This work was extended by Martin *et al.* (1997), who disrupted the *A. nidulans* γ-tubulin gene in a strain that carried a temperature-sensitive mutation that blocked the cell cycle in G_2 at restrictive temperatures. (Actually, two different mutations that blocked the cell cycle in G_2 were used in duplicate experiments.) When conidia carrying the γ-tubulin disruption and the G_2-blocking mutation were germinated at a restrictive temperature, nuclei were blocked in G_2. A normal array of cytoplasmic microtubules was present at the block. When the block was released by shifting the germlings to a permissive temperature, the cytoplasmic microtubules disassembled normally and nuclei proceeded immediately into mitosis. Chromosomes condensed, but spindles did not form and nuclei were blocked with condensed chromosomes for nearly a cell cycle. The chromosomes eventually decondensed and nuclei reentered interphase but did not divide. Some cytoplasmic microtubules eventually reassembled, but they did not form normal interphase arrays. Instead, abnormally long microtubules or microtubule bundles formed and microtubules were often abnormally curved.

In *S. pombe,* the γ-tubulin gene is also essential for viability (Horio *et al.,* 1991; Stearns *et al.,* 1991). A strain carrying a disruption of the chromosomal γ-tubulin gene and a functional γ-tubulin gene on a plasmid is viable, however, and the phenotype caused by γ-tubulin depletion was observed in experiments in which the plasmid carrying the γ-tubulin gene was lost (Horio *et al.,* 1991). Depletion of γ-tubulin caused mitotic abnormalities of two sorts. In some cells, there was a tiny, barely visible spindle in the center of condensed chromosomes. In other cells there was an elongated spindle, but the chromosomes had not moved to the poles. These results were similar to those obtained in *A. nidulans* in that in both cases γ-tubulin appeared to be important for formation of a functional spindle. In *A. nidulans,* however, spindle formation was blocked completely whereas in *S. pombe* it was not. A reasonable explanation for the difference was offered by Horio *et al.,* (1991). It should be remembered that in the *S. pombe* plasmid loss experiments, γ-tubulin was depleted until it became growth limiting. It is possible that the cells containing tiny spindles might be almost completely depleted for γ-tubulin such that spindle assembly is nearly abolished. The cells with elongated spindles might have enough γ-tubulin to allow the formation of a spindle but not enough to allow the formation of one that is complete and functional. The incompletely formed spindle might then elongate but not complete mitosis successfully.

Mutations have been identified in each of the two *D. melanogaster* γ-tubulin genes. Sunkel *et al.* (1995) identified a P-element insertion into the zygotically expressed gene (*γtub23C*). The insertion resulted in reduction of transcription and translation of the gene to levels undetectable by Northern or Western analyses. Individuals homozygous for the insertion died during the late larval stage and had reduced brain and imaginal discs. Cytological analyses of cells in these larvae showed a variety of mitotic abnormalities, including polyploidy, alterations of the arrangement of chromosomes at the metaphase plate, and disorganization during anaphase. These data are consistent with the data from *A. nidulans* and *S. pombe* in that they indicate that γ-tubulin is essential for viability and normal mitosis. Mitotic spindles were present in many of the cells in larvae homozygous for the P-element insertion, however, and this may indicate that γ-tubulin is not an absolute requirement for spindle formation in *D. melanogaster.* It is difficult to be certain if this is the case, however, because it is possible that *γtub23C* is expressed at a low level in these cells and it is possible that some of the γ-tubulin expressed from the other, maternally expressed, γ-tubulin gene is carried over into the larvae. In addition to abnormalities in chromosome distribution and spindle structure, larvae homozygous for the P-element insertion displayed a variety of centrosomal abnormalities as judged by staining with an antibody for a *D. melanogaster* centrosomal-associated antigen. This result suggests that, in *D. melanogaster,* γ-tubulin may have a structural role in the PCM.

The function of the maternally expressed γ-tubulin gene of *D. melanogaster* has been investigated by two labs (Tavosanis *et al.,* 1997; Wilson and Borisy, 1998; Llamazares *et al.,* 1999). Studies from both labs indicate that null mutations or mutations that cause *γtub37C* to be dysfunctional are not lethal but cause female sterility. This is consistent with the fact that the γ-tubulin isoform encoded by *γtub37C* is only detected in ovaries and embryos. Studies from the two labs lead to quite different conclusions with respect to the effects of *γtub37C* on female meiosis, however. Tavosanis *et al.* (1997) concluded that the female meiotic spindle was "extremely disrupted" in mutant alleles, whereas Wilson and Borisy (1998) concluded that meiotic spindles were "close to normal." It is not completely clear how these apparently contradictory results can be resolved. One factor may be that the two groups have examined different mutant alleles. Another factor is that the extremely disrupted meiotic spindles were observed in stage 14 oocytes while the nearly normal meiotic spindles were observed in laid eggs and activated oocytes. The abnormal meiotic figures observed in stage 14 oocytes are apparently not seen in laid eggs or fertilized embryos (Llamazares *et al.,* 1999). They thus apparently exist only transiently. Some of the discrepancy may also have to do with the somewhat subjective scoring of spindles as normal or abnormal. It is important not to overlook, moreover, the important ways in which the studies agree. First, whether spindles are normal or abnormal, microtubules are generally present. Second, polar bodies are absent with stronger alleles. Third, mitotic spindles in developing embryos are abnormal and mitosis is severely compromised.

The function of Tub4p has also been studied extensively in *S. cerevisiae* (Sobel and Snyder, 1995; Spang *et al.,* 1996; Marschall *et al.,* 1996; see also Chapter 4 in this volume). TUB4 is essential for viability (Sobel and Snyder, 1995) and depletion of TUB4p or conditional mutations in Tub4 alter spindle structure but do not completely abolish spindle microtubule assembly (Sobel and Snyder, 1995; Spang *et al.*, 1996; Marschall *et al.*, 1996).

B. Antibody Inhibition Studies

The function of γ-tubulin has also been probed *in vivo* using antibodies. Joshi *et al.* (1992) injected an antibody against γ-tubulin into several mammalian cell lines. Injection into interphase cells did not markedly alter the microtubule array. If microtubules were first disassembled with cold or nocodazole, however, the antibodies did inhibit reassembly of microtubules upon warming or removal of nocodazole. If cells in the G_2 phase of the cell cycle were injected with the antibodies, the cells entered mitosis but normal mitotic spindles failed to form. There was a reduction of microtubules in some instances and an almost complete absence in others. Injection of antibodies into metaphase cells caused a reduction in microtubule numbers and blockage of anaphase.

Julian *et al.* (1993) injected anti-γ-tubulin antibodies into anaphase PtK_2 cells. Some injected cells were blocked in telophase while other cells progressed toward cytokinesis. Significantly, given the localization of γ-tubulin to midbodies, the injected antibodies inhibited midbody formation in all cells. These results suggest that γ-tubulin may be essential for midbody formation, but since γ-tubulin localization at the midbody appears to vary among cell lines and even among cells within a given cell line (Khodjakov and Rieder, 1999), it does not seem likely that γ-tubulin is a universal requirement for midbody formation.

Finally, Ahmad *et al.* (1994) examined the effects of anti-γ-tubulin antibodies on microtubule growth in neurons. They depolymerized microtubules in freshly plated neurons in culture with nocodazole. They then injected anti-γ-tubulin antibodies and removed the nocodazole. Microtubule assembly in general, and from the centrosome in particular, was inhibited in neurons injected with the antibody. Some microtubule assembly occurred, however (possibly due to antibody inactivation or assembly on to remnant small microtubule fragments). Axonal growth from the neurons was inhibited in antibody-injected cells and was abolished completely in the neurons in which microtubule nucleation was most severely inhibited. These data support the idea that γ-tubulin at the centrosome of neurons nucleates microtubule assembly, and the assembled microtubules detach from the centrosome and move into the developing axon.

C. *In Vitro* Studies

In vitro studies with several experimental systems and from several labs all indicate that γ-tubulin is very important, and probably essential, for microtubule assembly from centrosomes. The earliest studies were carried out with *X. laevis* egg extracts (Félix *et al.,* 1994; Stearns and Kirschner, 1994). In *X. laevis* embryos, centrosomes are derived from the centriole carried into the egg by the fertilizing sperm. The sperm centriole does not have the ability to nucleate microtubule assembly itself, but, after fertilization, proteins from the egg are recruited to form functional centrosomes capable of nucleating microtubule assembly. γ-Tubulin is one of the proteins that is recruited to the forming centrosome (Félix *et al.,* 1994; Stearns and Kirschner, 1994), and immunodepletion of γ-tubulin from egg extracts prevents the forming centrosomes from becoming competent to nucleate microtubule assembly (Félix *et al.,* 1994) as do neutralizing antibodies against γ-tubulin (Stearns and Kirschner, 1994).

The role of γ-tubulin in microtubule nucleation was studied in a similar fashion in *S. pombe.* The SPBs of *S. pombe* are able to nucleate microtubule assembly *in vitro,* and this ability is inhibited by an anti-γ-tubulin antibody (Masuda and Shibata, 1996).

These data, in combination with the genetic data and the *in vivo* antibody inhibition data discussed earlier, suggested very strongly that γ-tubulin played a key role in the nucleation of microtubule assembly from MTOCs. Final confirmation that this was the case came from two directions. First, Zheng *et al.* (1995) were able to purify large γ-tubulin complexes from *X. laevis.* These complexes (γ-TuRCs) appeared in the electron microscope to be circles with a single break or discontinuity. Most significantly, they had the ability to nucleate microtubule assembly *in vitro* and to bind to, and apparently cap, the minus ends of microtubules.

Second, experiments in which γ-tubulin is stripped from centrosomes and added back suggest very strongly that γ-tubulin and, in particular, γ-TuRCs are necessary for microtubule nucleation. Such experiments have been carried out in two systems, *D. melanogaster* (Moritz *et al.,* 1998) and *Spisula* (Schnackenberg *et al.,* 1998). *D. melanogaster* centrosomes were treated with 2.0 M KI, causing the removal of all or nearly all γ-tubulin as judged by Western blotting. Along with the loss of γ-tubulin, the ability of the centrosome to nucleate microtubule assembly was also lost. If KI-stripped *D. melanogaster* centrosomes were incubated in a *D. melanogaster* embryo extract, γ-tubulin from the extract became associated with the centrosome and the ability of the centrosome to nucleate microtubule assembly was restored. If γ-tubulin was depleted from the extract before centrosomes were treated with the extract, the ability of the centrosome to nucleate microtubule assembly was not restored. Gel filtration fractions containing γ-TuRCs were able to restore microtubule nucleation but, interestingly, pure γ-TuRCs did not restore microtubule nucleation. A factor with an estimated molecular mass of 220 kDa was required in addition to γ-TuRCs.

Similar results were obtained with *Spisula* (Schnackenberg *et al.,* 1998). One added feature of the *Spisula* experiments was that electron microscopy was used to determine that γ-tubulin rings (presumably γ-TuRCs) were present in the PCM before treatment, were removed by the KI treatment, and were restored by treatment with oocyte extracts. Again, the centrosomes were only capable of nucleating microtubule assembly when γ-TuRCs were present.

Interestingly, although γ-TuRCs are present in the cytoplasm of many cells, they do not nucleate microtubule assembly in this context. When they are purified, however, or when they bind to centrosomes, they gain the ability to nucleate microtubule assembly. This implies that γ-TuRCs change in some as yet undefined way when they are purified or bind to centrosomes. Many explanations are possible, but one interesing possiblity is that their ability to nucleate microtubule assembly is inhibited by a protein that binds to γ-TuRCs, and this protein is lost during purification and binding to the centrosome.

Additional important insights into γ-tubulin function come from the limited number of studies of γ-tubulin binding *in vitro* to microtubules. These studies have been carried out in two labs. Melki *et al.* (1993) first demonstrated that *in vitro* translated human γ-tubulin, or γ-tubulin produced in bacteria and folded *in vitro* by the TCP-1 complex, will bind to and cosediment with microtubules. The amount of γ-tubulin that cosedimented with microtubules depended on the number of microtubule ends present, and this suggested that the γ-tubulin might be binding to microtubule ends.

Li and Joshi (1995) demonstrated that *in vitro* translated human γ-tubulin binds specifically to the minus ends of axonemes (probably, but not certainly, to the minus ends of microtubules in the axonemes). They also demonstrated that binding of γ-tubulin to microtubules is saturable, that the amount of binding is proportional to the number of free microtubule ends, and that the stoichiometry of binding is 12.6 ± 4.9 γ-tubulin molecules per microtubule. They also determined the dissociation constant of γ-tubulin from microtubules to be $1.5\text{–}1.7 \times 10^{-10}$ M.

It must be remembered that *in vitro* translation mixtures contain many proteins, and it is quite conceivable that other proteins in the mixture may influence γ-tubulin binding. It seems unlikely, however, that proteins in the mixture would confer specific binding to the minus ends of microtubules. It is also worth noting that some of the binding studies carried out by Melki *et al.* (1993) were done with relatively pure, bacterially produced γ-tubulin. It thus seems clear that γ-tubulin binds specifically to the minus ends of microtubules although it is possible that the stoichiometry of binding and K_D determined by Li and Joshi may have been influenced by other factors in their *in vitro* translation mixtures.

V. γ-Tubulin and Microtubule Assembly: Current Knowledge and Unanswered Questions

In previous sections, I have discussed what is known about γ-tubulin synthesis and distribution and information as to the function of γ-tubulin. I now attempt to bring this information together to form a coherent picture of γ-tubulin synthesis and function in the cell, and then highlight some unanswered questions. I do not reiterate references for information that has been discussed earlier.

A. What We Know

γ-Tubulin is encoded by one or two genes in all organisms studied to date. After synthesis it is folded by the TCP-1 complex. Improperly folded γ-tubulin complexes with prefoldin and is recycled back to the TCP-1 complex for proper folding. Properly folded γ-tubulin associates with homologues of Spc97p and Spc98p to form γ-TuSCs. In *S. cerevisiae* γ-TuSCs enter the nucleus and bind to Spc110P on the inner face of the SPB. They also bind to Spc72p on the cytoplasmic face of the SPB. In *D. melanogaster* and probably in many other organisms, the γ-TuSC binds to other proteins to form the γ-TuRC. γ-TuRCS are present in the cytoplasm, but they do not nucleate microtubule assembly. At the onset of mitosis the amount of γ-tubulin at the centrosome increases. This apparently reflects increases in the binding capacity of PCM for γ-TuRCs. The components of the PCM to which γ-TuRCs bind are not known, but pericentrin is a very good candidate. At centrosomes γ-TuRCs nucleate microtubule assembly. A great volume of genetic and biochemical data points to the fact that γ-tubulin is essential for the nucleation of microtubule assembly from many or all MTOCs, including the centrosome.

B. Unanswered Questions

1. What Is the Mechanism of Microtubule Nucleation?

The exact mechanism by which γ-tubulin nucleates microtubule assembly at centrosomes remains undetermined. Two models have been proposed, but it is not clear which, if either, of the two models is correct.

The first model was suggested initially by Oakley *et al.* (1990) and greatly refined by Zheng *et al.* (1995). In this model (Fig. 3), the γ-TuRC serves as a template that nucleates microtubule assembly. This model postulates that tubulin dimers have a greater binding affinity for the γ-tubulin in

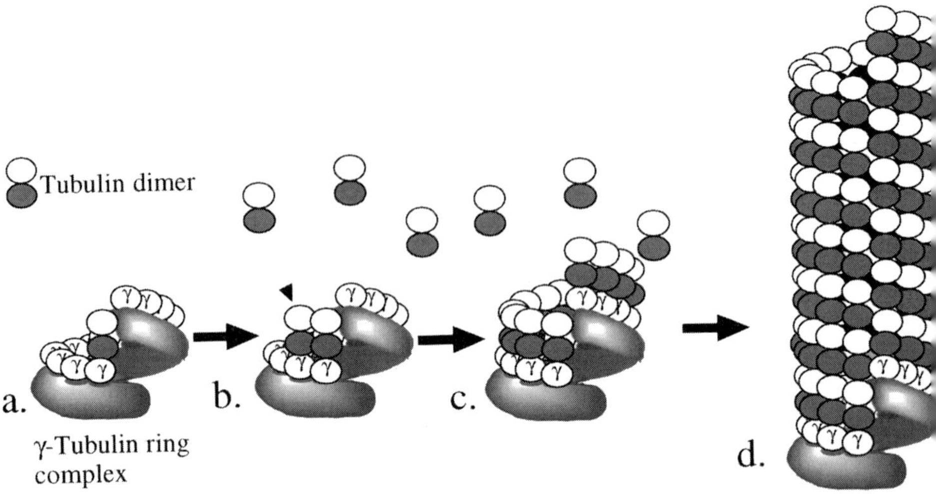

Figure 3 A model for microtubule nucleation by the γ-TuRC. (a) A γ-TuRC. In this model the γ-TuRC contains 13 γ-tubulin molecules and one tubulin dimer. Subsequent work suggests that the γ-TuRC may have an even number of γ-tubulin molecules and the presence of the tubulin dimer does not appear to be universal. (b) A tubulin dimer (arrowhead) binds to the γ-TuRC and the binding is stabilized by interaction with γ-tubulin and the adjacent tubulin dimer. (c) Additional dimers bind until in (d), a complete B-lattice microtubule is formed. Modified from Oakley (1995). Reproduced by permission.

γ-TuRCs than they have for each other. Tubulin dimers, thus, bind to γ-TuRCs, creating a layer of tubulin dimers. Additional dimers bind and a microtubule is formed. A template model such as the one shown would also explain how microtubule polarity is established by the centrosome. If α-tubulin binds to γ-tubulin, α-tubulin would be the ultimate subunit at the minus end and β-tubulin would be the ultimate subunit at the plus end. Finally, the number of γ-tubulin subunits in the γ-TuRC would establish the number of protofilaments in the microtubule.

It was originally suggested that γ-tubulin molecules might interact with β-tubulin molecules on the basis of genetic interactions between γ-tubulin alleles and β-tubulin alleles (Weil *et al.,* 1986; Oakley *et al.,* 1990). It is now clear that α-tubulin is the ultimate subunit at the minus end of microtubules (Nogales *et al.,* 1999), so if γ-tubulin nucleates microtubule assembly by a template mechanism such as the one shown in Fig. 3, it likely does so by interacting physically with α-tubulin. It is worth noting that this model predicts that γ-tubulin would not only interact longitudinally (along the protofilament) with α-tubulin but also laterally at the seam of the microtubule. If, as seems likely, γ-TuRCs are composed of γ-TuSCs (Oegema *et al.,* 1999), and each γ-TuSC contains two γ-tubulin molecules, then

γ-TuRCs should have an even number of γ-tubulin molecules. This would not permit the model shown in Fig. 3, but if, for example, there were seven γ-TuSCs in each γ-TuRC and there was an overlap of one γ-tubulin molecule, then 13 γ-tubulin molecules would be involved in microtubule nucleation and the complex could nucleate a 13-protofilament microtubule.

The model in Fig. 3 postulates no physical interaction between γ-tubulin and β-tubulin, so it is difficult, at first glance, to see how this model could be compatible with the observations of genetic interactions between β-tubulin and γ-tubulin that originally led to the discovery of γ-tubulin. We now know, however, that the genetic interactions observed by Weil *et al.* (1986) can not be explained by a simple physical interaction between γ-tubulin and β-tubulin. Each of the three original γ-tubulin (*mipA*) alleles displayed synthetic interactions with several β-tubulin alleles. These β-tubulin alleles have now been sequenced (Jung *et al.,* 1998), and when the mutated amino acids are placed onto the three-dimensional structure of β-tubulin (Nogales *et al.,* 1998), they are in three distinct regions. Since a single mutant amino acid in γ-tubulin cannot interact with three different, and spatially separate, regions of β-tubulin, the genetic interactions must reflect something other than a simple "lock and key" interaction.

The other model for γ-tubulin nucleation of microtubule assembly was proposed by Erickson and Stoffler (1996). It postulates that lateral interactions between γ-tubulin and both α-tubulin and β-tubulin serve to initiate the assembly of a protofilament. Other tubulin dimers would interact laterally with the protofilament to initiate additional protofilaments and eventually microtubules. This model predicts that γ-TuRCs would nucleate microtubule assembly but not establish protofilament number in the nucleated microtubules. It is not clear if this model predicts that γ-TuRCs would establish microtubule polarity. In fact, it appears that if this model is correct, γ-TuRCs might bind laterally to any protofilament extending from either end of a microtubule.

2. Is γ-Tubulin Required for Microtubule Assembly?

The simple answer to this question is no. Purified tubulin can assemble with little or no γ-tubulin present. Whether microtubules can assemble *in vivo* without γ-tubulin is a more open question, but the preponderance of data suggests that they can, at least to some extent. The genetic data are probably the most meaningful in this area. Repeated experiments have shown that γ-tubulin deletions or mutations cause a dramatic reduction in the number of microtubules present and failure of microtubule-rich structures such as mitotic spindles to assemble normally, but, invariably, some microtubule assembly occurs.

Taken at face value, these data suggest that γ-tubulin plays an important role in the nucleation of microtubule assembly but is not absolutely required. This is a very reasonable possibility. γ-Tubulin might be required for microtubule assembly at centrosomes and other MTOCs, but if γ-tubulin were absent, microtubules would self-nucleate and assemble to some extent. The microtubules that assembled, however, would be improperly positioned, and there would likely be fewer than the number required for proper assembly of structures such as the mitotic spindle.

One caveat with this interpretation is that with virtually all of the genetic data, the possibility cannot be ruled out that a small amount of functional γ-tubulin is present or that mutant γ-tubulin alleles have lost only a portion of their ability to nucleate microtubule assembly (discussed by Martin *et al.,* 1997). There are thus many data suggesting that microtubules can assemble *in vivo* without functional γ-tubulin, but none of the data are definitive. Finally, the possibility must be considered that in some systems, alternate normal pathways for microtubule assembly exist that do not require γ-tubulin in any way.

3. Does γ-Tubulin Have Functions Other Than Microtubule Nucleation?

The answer to this question is not known, but there are tantalizing hints that γ-tubulin may have additional functions. For example, the original γ-tubulin mutations were isolated as revertants of *benA33,* a heat-sensitive β-tubulin mutation of *A. nidulans. benA33* does not inhibit microtubule assembly. Rather, spindles form but do not function normally. The heat-sensitive blockage of microtubule function conferred by *benA33* is partially suppressed by several antimicrotubule agents and by α-tubulin mutations that apparently destabilize microtubules (Oakley and Morris, 1981; Gambino *et al.,* 1984; Oakley *et al.,* 1987). These and other data suggest that *benA33* hyperstabilizes microtubules. The fact that γ-tubulin mutations can suppress the heat sensitivity conferred by *benA33* suggests that γ-tubulin may have some role in the removal of tubulin dimers from microtubules. In this regard, it is worth noting that the three original γ-tubulin alleles reduced the benomyl resistance conferred by several β-tubulin mutants, not just *benA33.*

In *D. melanogaster* mutant γ-tubulin alleles show synthetic interactions with mutant alleles of the kinesin-like protein Ncd (Endow and Komma, 1998). These interactions may involve the transport of γ-tubulin, but it is possible that they reflect additional functions of γ-tubulin that involve interactions of γ-tubulin and Ncd.

It is also worth noting that we do not know what, if anything, γ-tubulin is doing in mitotic spindles. It is unlikely to be nucleating microtubule assembly. It is possible that microtubules in the spindles have detached from centrosomes bringing along a γ-TuRC cap at the minus end. It is possible that it is doing nothing and is simply an inactive cytoplasmic form that has become associated with the rather sticky mitotic spindle. It is also possible that it is doing something important that we do not yet understand.

VI. Concluding Comments

It was my very great fortune to discover γ-tubulin along with Elizabeth Oakley, and it is personally gratifying to see how much progress there has been in the field of γ-tubulin research in the past 10 years. There are many questions yet to be answered, however. I have already mentioned some of them, but there are others. It would be wonderful to know more about the structure of γ-TuRCs and the mechanism by which they nucleate microtubule assembly, and to know how the ability of γ-TuRCs to nucleate microtubule assembly is regulated and, in particular, why γ-TuRCs in the cytoplasm do not nucleate microtubule assembly. It will also be important to know how movement of γ-TuRCs to and from centrosomes is controlled, and what γ-tubulin does in the many types of cells that do not have morphologically differentiated MTOCs. Finally, there is tantalizing evidence that γ-tubulin may have functions in addition to microtubule nucleation, and it will be very exciting to determine if this is the case and, if so, what these functions are. These questions, and others we cannot even imagine at this point, will surely keep the field of γ-tubulin research active and exciting for years to come.

Acknowledgments

I thank C. Elizabeth Oakley for proofreading the manuscript and Dr. Tetsuya Horio (School of Medicine, University of Tokushima) for providing the computer drawings that served as a basis for Figs. 1 and 3. Supported by NIH grant GM31837 and NSF grant MCB-9808480.

References

Ahmad, F. J., Joshi, H. C., Centonze, V. E., and Baas, P. W. (1994). Inhibition of microtubule nucleation at the neuronal centrosome compromises axon growth. *Neuron* **12,** 271–280.

Akashi, T., Yoon, Y., and Oakley, B. R. (1997). Characterization of γ-tubulin complexes in *Aspergillus nidulans* and detection of putative γ-tubulin interacting proteins. *Cell Motil. Cytoskel.* **37,** 149–158.

Bergen, L. G., Kuriyama, R., and Borisy, G. G. (1980). Polarity of microtubules nucleated by centrosomes and chromosomes of chinese hamster ovary cells in vitro. *J. Cell Biol.* **84,** 151–159.

Binarová, P., Hause, B., Dolezel, J., and Draber, P. (1998). Association of γ-tubulin with kinetochore/centromeric region of plant chromosomes. *Plant J.* **14,** 751–757.

Burns, R. G. (1995). Identification of two new members of the tubulin family. *Cell Motil. Cytoskel.* **31,** 255–258.

Debec, A., Detraves, C., Montmory, C., Geraud, G., and Wright, M. (1995). Polar organization of γ-tubulin in acentriolar mitotic spindles of *Drosophila melanogaster* cells. *J. Cell Sci.* **108,** 2645–2653.

Dictenberg, J. B., Zimmerman, W., Sparks, C. A., Young, A., Vidair, C., Zheng, Y., Carrington, W., Fay, F. S., and Doxsey, S. J. (1998). Pericentrin and γ-tubulin form a protein complex and are organized into a novel lattice at the centrosome. *J. Cell Biol.* **141,** 163–174.

Doxsey, S. J., Stein, P., Evans, L., Calarco, P. D., and Kirschner, M. (1994). Pericentrin, a highly conserved centrosome protein involved in microtubule organization. *Cell* **76,** 639–650.

Dutcher, S. K., and Trabuco, E. C. (1998). The *UNI3* gene is required for assembly of basal bodies of *Chlamydomonas* and encodes δ-tubulin, a new member of the tubulin superfamily. *Mol. Biol. Cell* **9,** 1293–1308.

Endow, S. A., and Komma, D. J. (1998). Assembly and dynamics of an anastral:astral spindle: the meiosis II spindle of *Drosophila* oocytes. *J. Cell Sci.* **111,** 2487–2495.

Erickson, H. P., and Stoffler, D. (1996). Protofilaments and rings, two conformations of the tubulin family conserved from bacterial FtsZ to α/β and γ tubulin. *J. Cell Biol.* **135,** 5–8.

Félix, Marie-A., Antony, C., Wright, M., and Maro, B. (1994). Centrosome assembly in vitro: Role of γ-tubulin recruitment in *Xenopus* sperm aster formation. *J. Cell Biol.* **124,** 19–31.

Fuller, S. D., Gowen, B. E., Reinsch, S., Sawyer, A., Buendia, B., Wepf, R., and Karsenti, E. (1995). The core of the mammalian centriole contains γ-tubulin. *Curr. Biol.* **5,** 1384–1393.

Gambino, J., Bergen, L. G., and Morris, N. R. (1984). Effects of mitotic and tubulin mutations on microtubule architecture in actively growing protoplasts of *Aspergillus nidulans. J. Cell Biol.* **99,** 830–838.

Geissler, S., Siegers, K., and Schiebel, E. (1998). A novel protein complex promoting formation of functional α- and γ-tubulin. *EMBO J.* **17,** 952–966.

Goddard, R. H., Wick, S. M., Silflow, C. D., and Snustad, D. P. (1994). Microtubule components of the plant cell cytoskeleton. *Plant Physiol.* **104,** 1–6.

Gueth-Hallonet, C., Antony, C., Aghion, J., Santa-Maria, A., Lajoie-Mazenc, I., Wright, M., and Maro, B. (1993). γ-Tubulin is present in acentriolar MTOCs during early mouse development. *J. Cell Sci.* **105,** 157–166.

Heald, R., Tournebize, R., Blank, T., Sandaltzopoulos, R., Becker, P., Hyman, A., and Karsenti, E. (1996). Self-organization of microtubules into bipolar spindles around artificial chromosomes in *Xenopus* egg extracts. *Nature* **382,** 420–425.

Hoffman, J. C., Vaughn, K. C., and Joshi, H. C. (1994). Structural and immunocytochemical characterization of microtubule organizing centres in pteridophyte spermatogenous cells. *Protoplasma* **179,** 46–60.

Horio, T., and Oakley, B. R. (1994). Human γ-tubulin functions in fission yeast. *J. Cell Biol.* **126,** 1465–1473.

Horio, T., Uzawa, S., Jung, M. K., Oakley, B. R., Tanaka, K., and Yanagida, M. (1991). The fission yeast γ-tubulin is essential for mitosis and is localized at two different microtubule organizing centers. *J. Cell Sci.* **99,** 693–700.

Joshi, H. C., and Palevitz, B. A. (1996). γ-Tubulin and microtubule organization in plants. *Trends Cell Biol.* **6,** 41–44.

Joshi, H. C., Palacios, M. J., McNamara, L., and Cleveland, D. W. (1992). γ-Tubulin is a centrosomal protein required for cell cycle-dependent microtubule nucleation. *Nature* **356,** 80–83.

Julian, M., Tollon, Y., Lajoie-Mazenc, I., Moisand, A., Mazarguil, H., Puget, A., and Wright, M. (1993). γ-Tubulin participates in the formation of the midbody during cytokinesis in mammalian cells. *J. Cell Sci.* **105,** 145–156.

Jung, M. K., May, G. S., and Oakley, B. R. (1998). Mitosis in wild-type and β-tubulin mutant strains of *Aspergillus nidulans. Fungal Genet. Biol.* **24,** 146–160.

Kalt, A., and Schliwa, M. (1993). Molecular components of the centrosome. *Trends Cell Biol.* **3,** 118–127.

Khodjakov, A., and Rieder, C. L. (1999). The sudden recruitment of γ-tubulin to the centrosome at the onset of mitosis and its dynamic exchange throughout the cell cycle, do not require microtubules. *J. Cell Biol.* **146,** 585–596.

Kimble, M., and Kuriyama, R. (1992). Functional components of microtubule-organizing centers. *Int. Rev. Cytol.* **136,** 1–50.

Knop, M., and Schiebel, E. (1997). Spc98p and Spc97p of the yeast γ-tubulin complex mediate binding to the spindle pole body via their interaction with Spc110p. *EMBO J.* **16,** 6985–6995.

Knop, M., and Schiebel, E. (1998). Receptors determine the cellular localization of a γ-tubulin complex and thereby the site of microtubule formation. *EMBO J.* **17,** 3952–3967.

Knop, M., Pereira, G., Geissler, S., Grein, K., and Schiebel, E. (1997). The spindle pole body component Spc97p interacts with the γ-tubulin of *Saccharomyces cerevisiae* and functions in microtubule organization and spindle pole body duplication. *EMBO J.* **16,** 1550–1564.

Krauhs, E., Little, M., Kempf, T., Hofer-Warbinek, R., Ade, W., and Ponstingl, H. (1981). Complete amino acid sequence of β-tubulin from porcine brain. *Proc. Natl. Acad. Sci. USA* **78,** 4156–4160.

Lajoie-Mazenc, I., Tollon, Y., Detraves, C., Julian, M., Moisand, A., Gueth-Hallonet, C., Debec, A., Salles-Passador, I., Puget, A., Mazarguil, H., Raynaud-Messina, B., and Wright, M. (1994). Recruitment of antigenic γ-tubulin during mitosis in animal cells: presence of γ-tubulin in the mitotic spindle. *J. Cell Sci.* **107,** 2825–2837.

Lambert, A. M. (1993). Microtubule-organizing centers in higher plants. *Curr. Opin. Cell Biol.* **5,** 116–122.

Li, Q., and Joshi, H. C. (1995). γ-Tubulin is a minus end-specific microtubule binding protein. *J. Cell Biol.* **131,** 207–214.

Little, M., and Seehaus, T. (1988). Comparative analysis of tubulin sequences. *Comp. Biochem. Physiol.* **90B,** 655–670.

Liu, B., Joshi, H. C., Wilson, T. J., Silflow, C. D., Palevitz, B. A., and Snustad, D. P. (1994). γ-Tubulin in *Arabidopsis:* Gene sequence, immunoblot, and immunofluorescence studies. *Plant Cell* **6,** 303–314.

Liu, B., Joshi, H. C., and Palevitz, B. A. (1995). Experimental manipulation of γ-tubulin distribution in *Arabidopsis* using anti-microtubule drugs. *Cell Motil. Cytoskel.* **31,** 113–129.

Llamazares, S., Tavosanis, G., and Gonzales, C. (1999). Cytological characterization of the mutant phenotypes produced during early embryogenesis by null and loss-of-function alleles of the *γtub37C* gene in *Drosophila. J. Cell Sci.* **112,** 659–667.

Marschall, L. G., Jeng, R. L., Mulholland, J., and Stearns, T. (1996). Analysis of Tub4p, a yeast γ-tubulin-like protein: Implications for microtubule-organizing center function. *J. Cell Biol.* **134,** 443–454.

Martin, M. A., Osmani, S. A., and Oakley, B. R. (1997). The role of γ-tubulin in mitotic spindle formation and cell cycle progression in *Aspergillus nidulans. J. Cell Sci.* **110,** 623–633.

Martin, O. C., Gunawardane, R. N., Iwamatsu, A., and Zheng, Y. (1998). Xgrip109: A gamma tubulin-associated protein with an essential role in γ-tubulin ring complex (γTuRC) assembly and centrosome function. *J. Cell Biol.* **141,** 675–687.

Masuda, H., and Shibata, T. (1996). Role of γ-tubulin in mitosis-specific microtubule nucleation from the *Schizosaccharomyces pombe* spindle pole body. *J. Cell Sci.* **109,** 165–177.

Matthies, H. J. G., McDonald, H. B., Goldstein, L. S. B., and Theurkauf, W. E. (1996). Anastral meiotic spindle morphogenesis: Role of the Non-Claret Disjunctional kinesin-like protein. *J. Cell Biol.* **134,** 455–464.

McIntosh, J. R. (1983). The centrosome as organizer of the cytoskeleton. *Modern Cell Biol.* **2,** 115–142.

McIntosh, J. R., and Euteneuer, U. (1984). Tubulin hooks as probes for microtubule polarity: An analysis of the method and an evaluation of data on microtubule polarity in the mitotic spindle. *J. Cell Biol.* **98,** 525–533.

Meads, T., and Schroer, T. A. (1995). Polarity and nucleation of microtubules in polarized epithelial cells. *Cell Motil. Cytoskel.* **32,** 273–288.

Melki, R., Vainberg, I., Chow, R. L., and Cowan, N. J. (1993). Chaperonin-mediated folding of vertebrate actin-related protein and gamma-tubulin. *J. Cell Biol.* **122,** 1301–1310.

Merdes, A., and Cleveland, D. W. (1997). Pathways of spindle pole formation: Different mechanisms; conserved components. *J. Cell Biol.* **138,** 953–956.

Mogensen, M. M., Mackie, J. B., Doxsey, S. J., Stearns, T., and Tucker, J. B. (1997). Centrosomal deployment of γ-tubulin and pericentrin: Evidence for a microtubule-nucleating domain and a minus-end docking domain in certain mouse epithelial cells. *Cell Motil. Cytoskel.* **36,** 276–290.

Moritz, M., Braunfeld, M. B., Sedat, J. W., Alberts, B., and Agard, D. A. (1995). Microtubule nucleation by γ-tubulin containing rings in the centrosome. *Nature* **378,** 638–640.

Moritz, M., Zheng, Y., Alberts, B. M., and Oegema, K. (1998). Recruitment of the γ-tubulin ring complex to *Drosophila* salt-stripped centrosome scaffolds. *J. Cell Biol.* **142,** 775–786.

Murphy, S. M., Urbani, L., and Stearns, T. (1998). The mammalian γ-tubulin complex contains homologues of the yeast spindle pole body components Spc97p and Spc98p. *J. Cell Biol.* **141,** 663–674.

Nogales, E., Wolf, S. G., and Downing, K. H. (1998). Structure of the $\alpha\beta$ tubulin dimer by electron crystallography. *Nature* **391,** 199–203.

Nogales, E., Whittaker, M., Milligan, R. A., and Downing, K. H. (1999). High-resolution model of the microtubule. *Cell* **96,** 79–88.

Oakley, B. R. (1995). A nice ring to the centrosome. *Nature* **378,** 555–556.

Oakley, B. R., and Morris, N. R. (1981). A β-tubulin mutation in *Aspergillus nidulans* that blocks microtubule function without blocking assembly. *Cell* **24,** 837–845.

Oakley, C. E., and Oakley, B. R. (1989). Identification of γ-tubulin, a new member of the tubulin superfamily encoded by *mip*A gene of *Aspergillus nidulans. Nature* **338,** 662–664.

Oakley, B. R., Oakley, C. E., and Rinehart, J. E. (1987). Conditionally lethal *tubA* α-tubulin mutations in *Aspergillus nidulans. Mol. Gen. Genet.* **208,** 135–144.

Oakley, B. R., Oakley, C. E., Yoon, Y., and Jung, M. K. (1990). γTubulin is a component of the spindle-pole-body that is essential for microtubule function in *Aspergillus nidulans. Cell* **61,** 1289–1301.

Oegema, K., Wiese, C., Martin, O. C., Milligan, R. A., Iwamatsu, A., Mitchison, T. J., and Zheng, Y. (1999). Characterization of two related *Drosophila* γ-tubulin complexes that differ in their ability to nucleate microtubules. *J. Cell Biol.* **144,** 721–733.

Rizzolo, L. J., and Joshi, H. C. (1993). Apical orientation of the microtubule organizing center and associated γ-tubulin during the polarization of the retinal pigment epithelium in vivo. *Dev. Biol.* **157,** 147–156.

Schnackenberg, B. J., Khodjakov, A., Rieder, C. L., and Palazzo, R. E. (1998). The disassembly and reassembly of functional centrosomes *in vitro. Proc. Natl. Acad. Sci. USA* **95,** 9295–9300.

Sobel, S. G., and Snyder, M. (1995). A highly divergent γ-tubulin gene is essential for cell growth and proper microtubule organization in *Saccharomyces cerevisiae. J. Cell Biol.* **131,** 1775–1788.

Spang, A., Geissler, S., Grein, K., and Schiebel, E. (1996). γ-Tubulin-like Tub4p of *Saccharomyces cerevisiae* is associated with the spindle pole body substructures that organize microtubules and is required for mitotic spindle formation. *J. Cell Biol.* **134,** 429–441.

Stearns, T., and Kirschner, M. (1994). In vitro reconstitution of centrosome assembly and function: The central role of γ-tubulin. *Cell* **76,** 623–637.

Stearns, T., Evans, L., and Kirschner, M. (1991). γTubulin is a highly conserved component of the centrosome. *Cell* **65,** 825–836.

Sundberg, H. A., and Davis, T. N. (1997). A mutational analysis identifies three functional regions of the spindle pole component Spc110p in *Saccharomyces cerevisiae. Mol. Biol. Cell* **8,** 2575–2590.

Sunkel, C. E. Gomes, R., Sampaio, P., Perdigao, J., and Gonzalez, C. (1995). γ-Tubulin is required for the structure and function of the microtubule organizing centre in *Drosophila* neuroblasts. *EMBO J.* **14,** 28–36.

Tassin, Anne-M., Celati, C., Moudjou, M., and Bornens, M. (1998). Characterization of the human homologue of the yeast Spc98p and its association with γ-tubulin. *J. Cell Biol.* **141,** 689–701.

Tavosanis, G., Llamazares, S., Goulielmos, G., and Gonzales, C. (1997). Essential role for γ-tubulin in the acentriolar female meiotic spindle of *Drosophila. EMBO J.* **16,** 1809–1819.

Vainberg, I. E., Lewis, S. A., Rommelaere, H., Ampe, C., Vandekerckhove, J., Klein, H. L., and Cowan, N. J. (1998). Prefoldin, a chaperone that delivers unfolded proteins to cytosolic chaperonin. *Cell* **93,** 863–873.

Vaughn, K. C., and Harper, J. D. I. (1998). Microtubule-organizing centers and nucleating sites in plants. *Int. Rev. Cytol.* **181.** 75–149.

Vogel, J. M., Stearns, T., Rieder, C. L., and Palazzo, R. E. (1997). Centrosomes isolated from *Spisula solidissima* oocytes contain rings and an unusual stoichiometeric ratio of α/β tubulin. *J. Cell Biol.* **137,** 193–202.

Weil, C. F., Oakley, C. E., and Oakley, B. R. (1986). Isolation of *mip* (microtubule-interacting protein) mutations of *Aspergillus nidulans. Mol. Cell. Biol.* **6,** 2963–2968.

Wick, S. M. (1991). The preprophase band. *In* "The Cytoskeletal Basis of Plant Growth and Form," (Lloyd, C. W., Ed.), pp. 231–244. Academic Press, London.

Wilson, P. G., and Borisy, G. G. (1998). Maternally expressed γTub37CD in *Drosophila* is differentially required for female meiosis and embryonic mitosis. *Dev. Biol.* **199,** 273–290.

Wilson, P. G., Zheng, Y., Oakley, C. E., Oakley, B. R., Borisy, G. G., and Fuller, M. T. (1997). Differential expression of two γ-tubulin isoforms during gametogenesis and development in *Drosophila. Dev. Biol.* **184,** 207–221.

Wise, D. O., and Oakley, B. R. (1997). Expression patterns of two γ-tubulin genes in *Homo sapiens. Mol. Biol. Cell* **8,** 45a.

Zhang, D., and Nicklas, R. B. (1995). The impact of chromosomes and centrosomes on spindle assembly as observed in living cells. *J. Cell Biol.* **129,** 1287–1300.

Zheng, Y., Jung, M. K., and Oakley, B. R. (1991). γ-Tubulin is present in *Drosophila melanogaster* and *Homo sapiens* and is associated with the centrosome. *Cell* **65,** 817–823.

Zheng, Y., Wong, M. L., Alberts, B., and Mitchison, T. (1995). Nucleation of microtubule assembly by a γ-tubulin-containing ring complex. *Nature* **378,** 578–583.

3

γ-Tubulin Complexes and Their Role in Microtubule Nucleation

Ruwanthi N. Gunawardane,[1,2] *Sofia B. Lizarraga,*[1,2] *Christiane Wiese,*[1] *Andrew Wilde,*[1] and *Yixian Zheng*[1]
[1] Department of Embryology
Carnegie Institution of Washington
Baltimore, Maryland 21210

[2] Department of Biology
Johns Hopkins University
Baltimore, Maryland 21218

I. Introduction

Microtubules are dynamic polymers found in all eukaryotic cells. They are involved in diverse cellular functions, including mitosis, vesicular transport, and intracellular organization. Microtubules are long, hollow, cylindrical structures (25 nm in diameter) assembled from α- and β-tubulin heterodimers. The α- and β-tubulins each bind to a molecule of GTP, but only the β-tubulin bound GTP is exchangeable after dimerization. The head-to-tail assembly of tubulin dimers results in microtubules with a faster growing plus end and a slower growing minus end (Desai and Mitchison, 1998).

Microtubules can assemble *in vitro* from purified tubulins in the presence of GTP and Mg^{2+}. The assembly begins with a slow nucleation step followed by a more rapid microtubule elongation phase. Microtubule nucleation is

Current Topics in Developmental Biology, Vol. 49

0070-2153/00 $35.00

the rate-limiting step in which a number of tubulin dimers assemble into "seeds" from which microtubules elongate (Desai and Mitchison, 1998). Consequently, spontaneous microtubule nucleation *in vitro* shows a strong dependence on tubulin concentration, with efficient nucleation possible only at higher than the physiological tubulin concentration.

In vivo, the kinetic barrier of microtubule nucleation is overcome by the presence of morphologically distinct organelles collectively referred to as microtubule organizing centers (MTOCs). MTOCs contain activities that allow microtubule assembly to proceed at tubulin concentrations too low to support efficient spontaneous nucleation (Kellogg *et al.,* 1994).

The two well-studied MTOCs are the metazoan centrosomes and the fungal spindle pole bodies (SPB). Although performing similar functions, centrosomes and SPBs are structurally distinct. The centrosome consists of a pair of centrioles, made of nine triplet microtubules, and an electron-dense pericentriolar material (PCM) that harbors the microtubule-nucleating activity (reviewed in Kalnins, 1992). The SPB, in contrast, is a three-layered plaque structure embedded in the nuclear envelope. (For a detailed description of the structure and function of the SPB, refer to other chapters in this book.) In interphase, a centrosome, juxtaposed to the nuclear envelope, nucleates a radial array of microtubules. This centrosome duplicates once and only once during S phase, and the daughter centrosomes found at the spindle poles participate in nucleating and organizing spindle microtubules in mitosis (Kellogg *et al.,* 1994). Electron tomography reconstruction of purified centrosomes reveals that the PCM contains hundreds of ring-shaped structures approximately 25 nm in diameter (Moritz *et al.,* 1995b; Vogel *et al.,* 1997). The presence of these rings at the PCM correlates with microtubule nucleating activity (Schnackenberg *et al.,* 1998).

A major breakthrough in understanding the molecular nature of microtubule nucleation by the MTOC came from the discovery of γ-tubulin (Oakley and Oakley, 1989). Since its initial isolation and characterization in *Aspergillus nidulans,* γ-tubulin has been found to be as ubiquitous and highly conserved as the α- and β-tubulins. Unlike α- and β-tubulins, γ-tubulin does not appear to become incorporated into the microtubule lattice; instead it is localized to all the MTOCs examined to date, where it is involved in microtubule nucleation (reviewed in Oakley, 1992; Joshi, 1993; Wiese and Zheng, 1999; Zimmerman *et al.,* 1999). Genetic studies in *Aspergillus nidulans* (Oakley *et al.,* 1990), *Saccharomyces cerevisiae* (Sobel and Snyder, 1995), *Schizosaccharomyces pombe* (Horio *et al.,* 1991), and *Drosophila melanogaster* (Sunkel *et al.,* 1995) provide strong *in vivo* evidence that γ-tubulin plays an important role in microtubule nucleation from the MTOCs. This evidence is further supported by *in vitro* studies showing that an anti-γ-tubulin antibody can block microtubule nucleation from centrosomes (Joshi *et al.,* 1992).

Insight into the mechanism of microtubule nucleation by γ-tubulin comes from both biochemical and structural studies. γ-Tubulin exists as a complex with other proteins inside the cell, and in certain cell types, multiple γ-tubulin complexes of different sizes co-exist (Raff *et al.,* 1993; Stearns and Kirschner, 1994; Moritz *et al.,* 1998; Oegema *et al.,* 1999). To date, two types of γ-tubulin complexes, the γ-tubulin ring complex (γTuRC) and the γ-tubulin small complex (γTuSC), have been characterized in metazoa (Zheng *et al.,* 1995; Oegema *et al.,* 1999). In this chapter, we will first discuss the purification and characterization of γTuRC and γTuSC, and then turn our attention to studies addressing the function of γTuRC at the centrosome. We will end with a discussion of the proposed mechanism of γTuRC-mediated microtubule nucleation.

II. Biochemical and Structural Studies of γTuRC and γTuSC

A. Purification and Characterization of γTuRC and γTuSC

The biochemical study of γ-tubulin, like most rare cytoskeletal proteins, is hampered by the difficulty to overexpress γ-tubulin exogeneously. Both bacterial and baculovirus-expressed γ-tubulins are mostly insoluble. γ-Tubulin produced by *in vitro* translation behaves as a monomer (Melki *et al.,* 1993) and can bind to the minus ends of preformed microtubules *in vitro* (Li and Joshi, 1995). However, the lack of pure monomeric γ-tubulin prevents further biochemical analysis of its function.

The first indication that γ-tubulin is present in large complexes comes from sucrose-gradient sedimentation and gel-filtration analyses (Fig. 1) of the endogenous *Drosophila* and *Xenopus* γ-tubulin (Raff *et al.,* 1993; Stearns and Kirschner, 1994). Hydrodynamic studies of γ-tubulin in organisms ranging from fungi to humans further support the idea that γ-tubulin functions in a protein complex *in vivo* (Meads and Schroer, 1995; Akashi *et al.,* 1997). These initial biochemical studies of γ-tubulin emphasize the need to purify and study the endogenous γ-tubulin complex in order to understand its mechanism of action.

Endogenous γ-tubulin constitutes only ~0.001–0.01% of the total cellular protein (Stearns and Kirschner, 1994; Zheng *et al.,* 1995), making purification of the complex challenging. In addition, since the large γ-tubulin complex (γTuRC; see later discussion) in the extracts made from *Xenopus* eggs or *Drosophila* embryos dissociates in the presence of 500 mM NaCl, conventional chromatography methods are limited. To facilitate the purification of *Xenopus* γTuRC, we developed a purification procedure that combines conventional and antibody affinity chromatography steps (Zheng *et al.,* 1995, 1998). A similar protocol was later used to purify the γTuRC

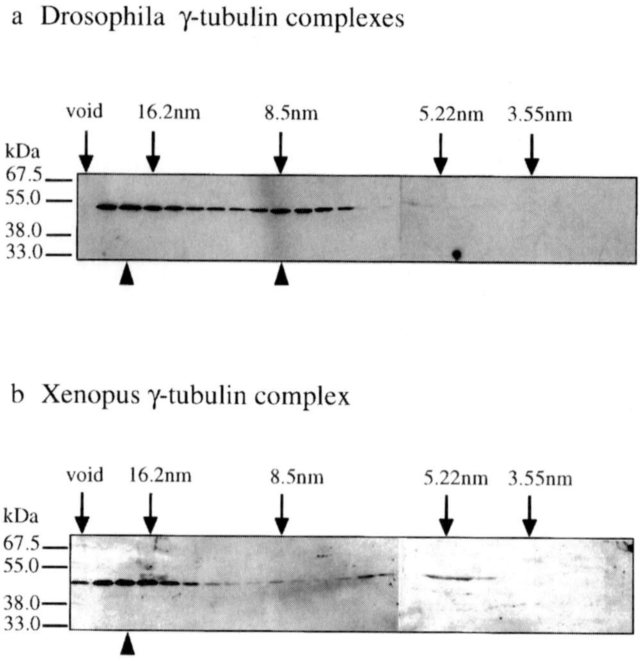

Figure 1 Gel filtration analysis of *Drosophila* and *Xenopus* γ-tubulin. γ-Tubulin in *Drosophila* embryo extracts and *Xenopus* egg extracts fractionates as large complexes on Superose 6 gel filtration columns. γ-Tubulin is visualized by Western blotting with antibodies against γ-tubulin. The numbers denote the Stokes radii of standards run on similar gel filtration columns. The arrowheads point to the peak fractions of γ-tubulin.

and the γTuSC from *Drosophila* embryo extracts (Fig. 2) (Oegema *et al.*, 1999). Antibody affinity purification is possible because the C-terminal 16–18 amino acids of γ-tubulin are highly immunogenic and are exposed on the surface of γ-tubulin complexes. The purification procedures used for *Xenopus* and *Drosophila* γTuRCs should be adaptable to the purification of γ-tubulin complexes of other organisms.

The purified *Xenopus* and *Drosophila* γTuRCs contain similar polypeptides, which we refer to as gamma ring proteins (grips) (Figs. 3a and 3b). γ-Tubulin complexes immunoprecipitated from extracts of mammalian tissue culture cells (Murphy *et al.*, 1998; Tassin *et al.*, 1998) or sheep brain (Detraves *et al.*, 1997) contain a similar group of proteins, suggesting that γTuRC is a universal complex in metazoa (Table I). Interestingly, whereas a small amount of α- and β-tubulins copurify with the *Xenopus* γTuRC and the putative γTuRC from sheep brain, γTuRC purified from *Drosophila* or immunoprecipitated from mammalian tissue culture cells do not contain

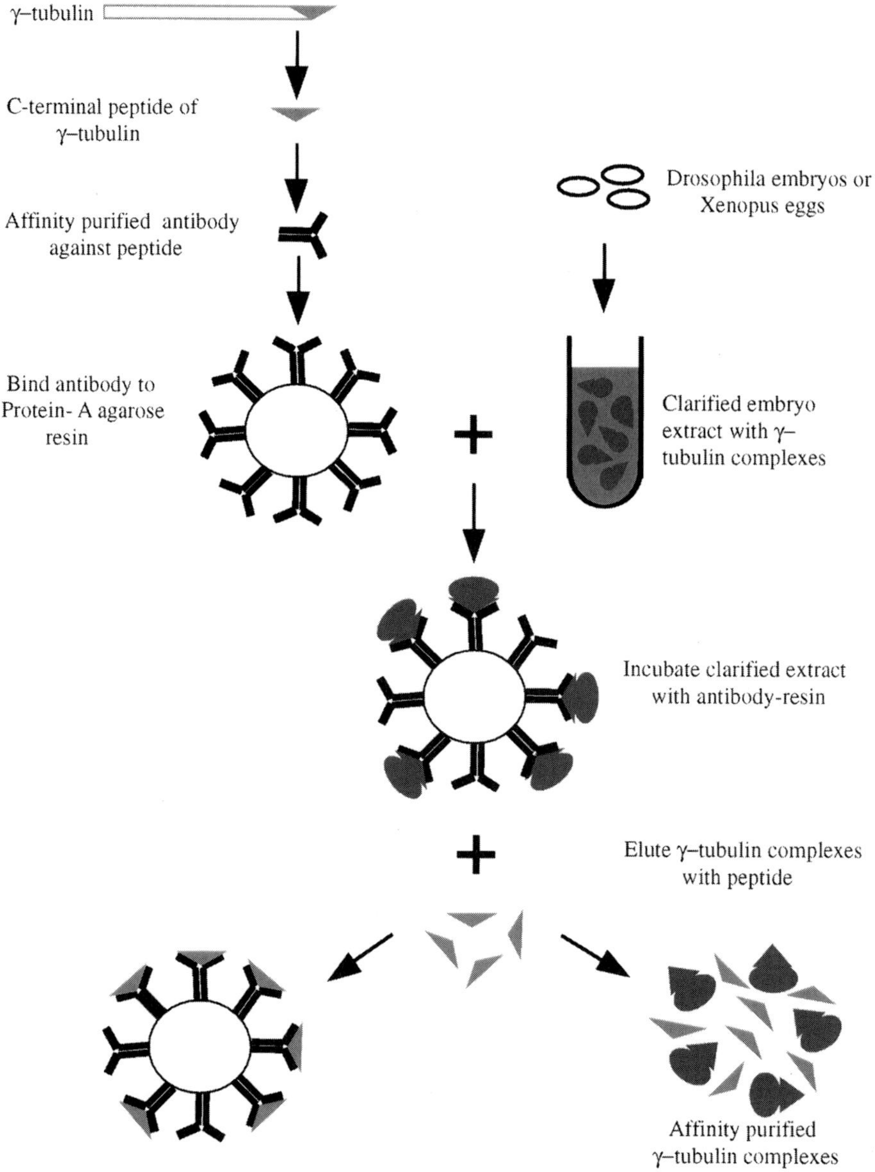

Figure 2 Schematic of antibody affinity purification of γ-tubulin complexes from *Drosophila* embryos and *Xenopus* egg extracts (figure is not drawn to scale).

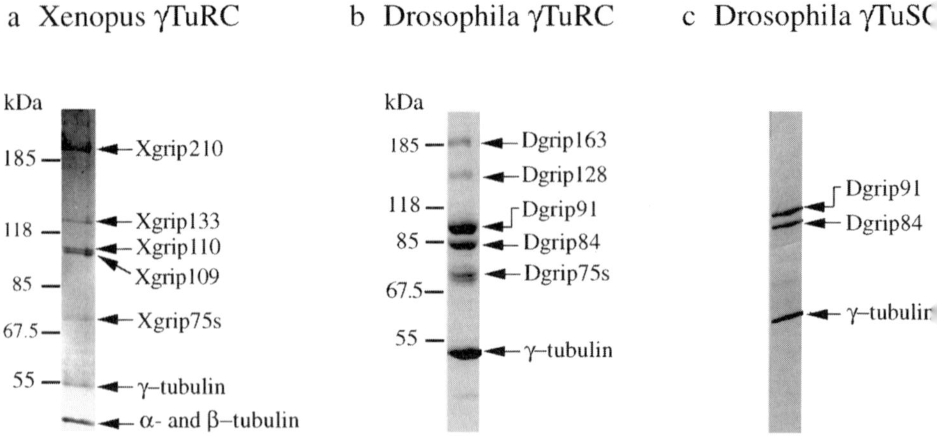

Figure 3 Protein components of the *Xenopus* and *Drosophila* γ-tubulin complexes. Purified γTuRCs from *Xenopus* (a) and *Drosophila* (b) consist of ~7 polypeptides referred to as grips for gamma ring proteins. In addition to γTuRC, a γ-tubulin small complex (γTuSC) containing the three most abundant proteins of γTuRC, Dgrip91, Dgrip84, and γ-tubulin, can also be purified from *Drosophila* embryo extracts (c). Dgrips refer to the *Drosophila* grips and Xgrips to the *Xenopus* grips. The numbers denote the apparent molecular mass of each protein.

α- and β-tubulin. It is not yet clear whether the presence of α- and β-tubulin in certain γTuRC preparations is significant.

To date, the following γTuRC subunits have been cloned: *Drosophila* Dgrip84 and Dgrip91 (Oegema *et al.*, 1999), human hGCP2 and hGCP3/HsSpc98 (Murphy *et al.*, 1998; Tassin *et al.*, 1998), and *Xenopus* Xgrip109

Table I Protein Components of Different Metazoan γ-Tubulin Complexes[a]

Drosophila	Xenopus	Sheep	Human
	α and β-tubulin	α and β-tubulin	
γ-tubulin	γ-tubulin	γ-tubulin	γ-tubulin
Dgrip75s	Xgrip75s	75	71
Dgrip84	Xgrip110	105	76
Dgrip91	Xgrip109	130	hGCP2
Dgrip128	Xgrip133	195	hGCP3/HsSpc98
Dgrip163	Xgrip210	250	

[a] γ-Tubulin exists in protein complexes with multiple polypeptides in metazoans. α-Tubulin and β-tubulin copurify with *Xenopus* and sheep γ-tubulin, but not with the *Drosophila* or human γ-tubulin (Zheng *et al.*, 1995; Detraves *et al.*, 1997; Tassin *et al.*, 1998; Murphy *et al.*, 1998; Oegema *et al.*, 1999).

(Martin *et al.,* 1998). Based on the degree of sequence homology, Dgrip91, hGCP3, and Xgrip109 belong to one family of conserved proteins, whereas Dgrip84 and hGCP2 are homologous to each other and define another family of proteins. Interestingly, the two families are also related to each other. For example, Dgrip84 and Dgrip91 are ~30% identical at the amino acid level. All of these γTuRC subunits colocalize with γ-tubulin at the centrosome (Martin *et al.,* 1998; Murphy *et al.,* 1998; Tassin *et al.,* 1998; Oegema *et al.,* 1999). In *Xenopus* egg extract, Xgrip109 interacts directly with γ-tubulin in the γTuRC and is required for assembly of the γTuRC *in vitro* (Martin *et al.,* 1998). Therefore, the two families of grips are likely to be essential for γTuRC function in metazoa.

To understand γTuRC assembly, Oegema *et al.* (1999) characterized the two major γ-tubulin complexes in *Drosophila* embryo extracts (Figs. 3b and 3c). When *Drosophila* embryo extract is treated with 0.5–1 M NaCl, the γTuRC dissociates, but γ-tubulin remains associated with Dgrip84 and Dgrip91. We refer to this stable subcomplex as the γ-tubulin small complex, γTuSC. Hydrodynamic and stoichiometric studies suggest that the γTuSC is a tetramer most likely containing two γ-tubulins, one Dgrip84, and one Dgrip91 (Oegema *et al.,* 1999). Since each γTuRC contains 10–15 γ-tubulins, one γTuRC can accommodate approximately six γTuSCs. Interestingly, the Tub4p complex, the only soluble γ-tubulin complex in *Saccharomyces cerevisiae,* is analogous to *Drosophila* γTuSC; the non γ-tubulin subunits of the Tub4p complex, Spc97p and Spc98p (Geissler *et al.,* 1996; Knop *et al.,* 1997), share sequence similarity with the Dgrip84 and Dgrip91 families of proteins, respectively (Table 2). It is not clear whether the Tub4p complex, like the γTuSC, also assembles into a higher order structure at the spindle pole body in *Saccharomyces cerevisiae.* For a more detailed discussion on the Tub4p complex, please refer to the other chapter in this

Table II Protein Components of the Tub4p Complex and the Metazoan γ-Tubulin Small Complexes[a]

Budding Yeast	Drosophila	Xenopus	Human
Tub4p	γ-tubulin	γ-tubulin	γ-tubulin
SPC97	Dgrip84	Xgrip110	hGCP2
SPC98	Dgrip91	Xgrip109	hGCP3/ HsSpc98

[a] The components of the metazoan γ-tubulin small complexes are homologous to one another as well as to the components of the only γ-tubulin complex found in yeast (Geissler *et al.,* 1996; Knop *et al.,* 1997; Martin *et al.,* 1998; Murphy *et al.,* 1998; Tassin *et al.,* 1998; Oegema *et al.,* 1999). The proteins with the highest sequence homologies are shown in the same row.

book and recent reviews (Pereira and Schiebel, 1997; Saunders, 1999; Wiese and Zheng, 1999).

B. The Structure of the γTuRC

The *Xenopus* and *Drosophila* γTuRCs share not only conserved protein subunits, but also appear as similar open ring or "lockwasher" structures with a diameter of approximately 25 nm (Figs. 4a and 4b) (Zheng *et al.*, 1995; Oegema *et al.*, 1999). γTuRCs from both organisms show a certain degree of heterogeneity when viewed either by cryo–electron microscopy (cryo-EM) or by negative-staining EM. It is not yet clear whether this

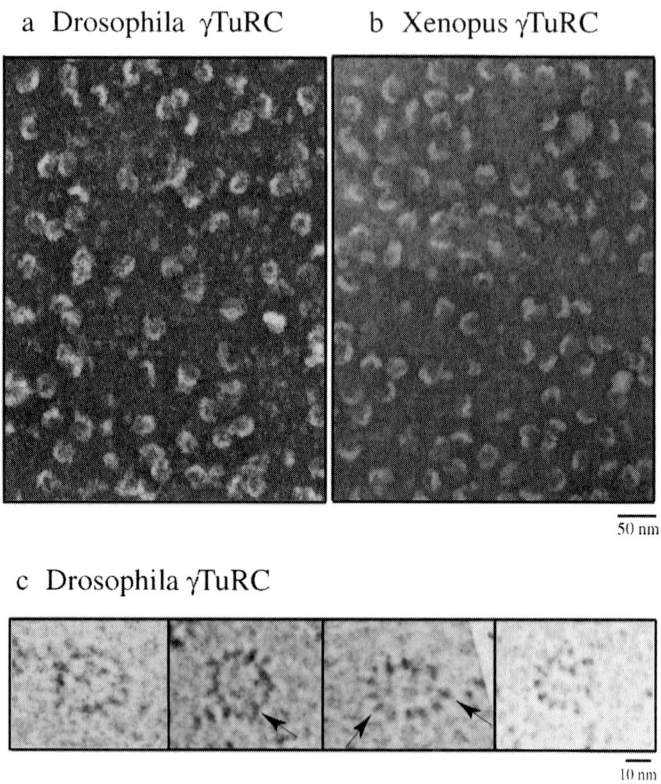

Figure 4 Electron micrographs of purified *Drosophila* and *Xenopus* γTuRC. Electron micrographs of negatively stained *Drosophila* (a) and *Xenopus* (b) γTuRCs. Cryo–electron micrographs of *Drosophila* γTuRC (c) reveal the modular structure of the γTuRC. The arrows point to the subunits that make up the ring wall.

reflects structural differences among the γTuRCs, or a result of damage during purification. Alternatively, the observed heterogeneity could reflect that the γTuRC is a flexible structure as suggested by the different degrees of openness of some rings (see Fig. 2d in Wiese and Zheng, 1999). Further structural analysis using three-dimensional reconstructions should help to resolve this issue.

Cryo-EM studies of *Drosophila* γTuRC have provided the highest resolution images of the ring structure thus far (Fig. 4c) (Oegema *et al.,* 1999). One obvious feature of the γTuRC is that it is made of 10–14 similarly shaped subunits surrounding a poorly defined central structure. Interestingly, the number of subunits corresponds to the estimated number of γ-tubulins per γTuRC. Since there are two γ-tubulins per γTuSC, two subunits seen on the ring could correspond to one γTuSC. A more detailed study of the γTuRC using a combination of cryo-EM and three-dimensional reconstruction is under way and promises to provide important insights into the assembly and function of this complex (D. Agard, personal communication).

C. The Guanine Nucleotide Binding Properties of *Drosophila* γTuSC and γTuRC

γ-Tubulin, like α- and β-tubulins, contains a GTP binding motif in its primary sequence (Burns, 1995). GTP is required for microtubule assembly, whereas hydrolysis of the β-tubulin bound GTP leads to microtubule depolymerization (Desai and Mitchison, 1998). Therefore, understanding the role of GTP in γ-tubulin assembly or function may shed light on the assembly and function of γTuRC. Oegema *et al.* (1999) found that γ-tubulin in the γTuSC or in the γTuRC can be UV crosslinked to GTP. In addition, they showed that γ-tubulin in the purified γTuSC has an exchangeable nucleotide-binding site that has a higher affinity for GDP than GTP. This finding is somewhat puzzling because the exchangeable site on β-tubulin has a higher affinity for GTP than GDP. The low yield of purified γTuRC did not allow these researchers to determine whether γ-tubulin in γTuRC binds to GTP or GDP preferentially. It will be important to determine whether GTP is required for γTuSC to assemble into γTuRC. This type of study, which requires assays for γTuRC assembly and large quantities of purified proteins, will depend on the development of a more efficient γTuRC purification procedure.

D. The Microtubule Nucleating Activity of γTuRC and γTuSC

Both *Xenopus* and *Drosophila* γTuRC can stimulate microtubule nucleation *in vitro* compared to buffer controls (Zheng *et al.,* 1995; Oegema *et al.*, 1999). Two pieces of evidence suggest that each γTuRC is able to nucleate one microtubule at a time. First, Zheng *et al.* (1995) showed, by negative staining EM, that a fraction of microtubules nucleated by *Xenopus* γTuRC has one γTuRC-like structure associated with one end. This end appears to be the minus end when the polarity of the nucleated microtubules is marked using brightly labeled rhodamine tubulin and viewed by fluorescence microscopy. Furthermore, Oegema *et al.* (1999) found that at a given tubulin concentration, the number of microtubules nucleated by γTuRC depends linearly on the concentration of γTuRC. γTuSC can also nucleate microtubules *in vitro,* but it is approximately 150-fold less active than γTuRC per mole of complex (Oegema *et al.,* 1999).

III. Evidence That γTuRC Is the Major Microtubule Nucleator at the Centrosome

Both genetic and cell biological studies show that γ-tubulin is required for microtubule nucleation from the centrosome. Does γ-tubulin function in the form of γTuRC at the centrosome? EM-tomography studies of purified *Drosophila* centrosomes provide the first indication that γTuRC-like ring structures are present in the PCM (Moritz *et al.,* 1995a). Using immunogold labeling in combination with EM-tomography, Moritz *et al.* (1995b) showed that these rings contain γ-tubulin and are localized at the base of microtubules nucleated from the centrosome. Additional support was provided by a later finding that the PCM of isolated *Spisula* centrosomes also contains a large number of γTuRC-like ring structures (Vogel *et al.,* 1997). These structural studies not only bring our view of the previously "electron dense, amorphous" PCM into much sharper focus, but also support the idea that γTuRC is somehow recruited to the centrosome to nucleate microtubules.

More direct evidence for γTuRC recruitment to the centrosome stems from several biochemical studies using two different assays for centrosome assembly. One assay, based on the original observation by Lohka and Masui (1983), uses the pair of centrioles provided by demembranated *Xenopus* sperm in combination with *Xenopus* egg extract (in the presence of an ATP regeneration system) to assemble a PCM, and therefore a centrosome, at the tip of the sperm nuclei. Using this assay, Felix *et al.* (1994) and Martin *et al.* (1998) showed that immunodepletion of γ-tubulin (and therefore γTuRC) from *Xenopus* egg extract blocks the formation of an active centrosome. Furthermore, the studies by Martin *et al.* (1998) suggest that intact γTuRC is required for centrosome function, since immunodepletion of Xgrip109, a subunit of γTuRC, not only blocks γTuRC assembly but also centrosome formation.

The second assay, based on the studies of Buendia *et al.* (1992), uses isolated centrosomes. Treating centrosomes with 1–2 M potassium iodide (KI) generates a centrosome core that does not nucleate microtubules. Moritz *et al.* (1998) and Schnackenberg *et al.* (1998) showed that the inactive centrosome core can regain microtubule-nucleating activity after incubating with extracts made from *Drosophila* embryos or *Spisula* oocytes, respectively. EM-tomography analysis revealed that treatment with KI leads to the loss of γTuRC-like rings from the centrosome (Schnackenberg *et al.*, 1998); the recovery of microtubule nucleating activity correlates with the reappearance of the rings at the PCM (Schnackenberg *et al.*, 1998). Furthermore, immunodepletion of γTuRC from the *Drosophila* embryo extract blocks its ability to complement KI-treated centrosomes; adding the purified γTuRC back to the depleted extract is sufficient to restore complementation activity (Moritz *et al.*, 1998).

Taken together, these studies strongly suggest that γTuRC is recruited to centrosomes to nucleate microtubules. The centrosome assembly assays just described should be useful to further study how γTuRC is recruited to and docked at the centrosome.

IV. The Mechanism of Microtubule Nucleation

A. Spontaneous Nucleation of Microtubules: *In Vitro* Studies

Microtubules, unlike other linear polymers such as F-actins, are made of more than 10 protofilaments. Consequently, microtubule nucleation proceeds with more complicated kinetics than that of other simpler polymers. Since the purification of tubulin and the discovery of *in vitro* assembly conditions, much effort has been directed toward understanding the dynamic behavior of microtubules. However, only a few published studies attempt to understand the kinetics of microtubule nucleation *in vitro*. Some of these studies base their analyses on the classic model for protein polymer nucleation by Oosawa and Asakura (1975), which suggests that a certain number of monomers must come together to make a seed that then elongates to form a polymer. Whereas the rate of seed formation is proportional to the monomer concentration to the $(n + 1)$ power (n is the number of monomers in the seed), the elongation rate of the polymer from the seed is directly proportional to the monomer concentration. Therefore, elongation of a microtubule is favored over nucleation, and nucleation is efficient only at high tubulin concentrations.

In a study by Voter and Erickson (1984), microtubule polymerization was recorded by measuring turbidity over time. Fitting the kinetic curves of microtubule assembly to computer-generated models, these researchers conclude that a microtubule seed is made of six to seven tubulin dimers. They propose that the seed is made of two parallel tubulin protofilaments that elongate both longitudinally and laterally to give rise to microtubules. Consistent with this idea, electron micrographs of negatively stained samples from early time points of microtubule polymerization show bundles of protofilament sheets (Voter and Erickson, 1984).

Fygenson *et al.* (1995) revisited the issue of microtubule nucleation under conditions of "marginal nucleation" using video-enhanced, differential-interference imaging. Based on a few simple assumptions and mathematical deductions, they propose that 12 ± 2 tubulin dimers are required to form a microtubule seed. The same number was reached when Fygenson *et al.* (1995) used their method to analyze the kinetic data of Erickson and Voter.

Although the exact size of a microtubule seed is controversial, these studies agree that a large number of tubulin heterodimers (6–14) are involved in seed formation. This clearly indicates the complexity and difficulty in studying microtubule nucleation. A better understanding will probably require the development of new methodologies in the future.

B. Microtubule Nucleation by γTuRC: The Two Nucleation Models

The spontaneous formation of microtubule polymer is dependent upon the concentration of tubulin dimer *in vitro. In vivo,* spontaneous assembly of microtubules is suppressed by a low endogenous tubulin concentration, and nucleation generally occurs from specific saturable nucleation sites on the centrosome (Kellogg *et al.,* 1994). Since increasing evidence shows that these nucleating sites are made of γTuRC, it is important to understand γTuRC-mediated microtubule nucleation.

Based on certain features of the microtubule structure, two models have been proposed to explain microtubule nucleation by γTuRC (Zheng *et al.,* 1995; Erickson and Stoffler, 1996). Microtubule protofilaments are arranged as a three-start helix with a seam running along the polymer length. The term "three-start" means that a complete turn, starting from the seam, ends three monomers above the starting point. In this type of lattice, α- and β-tubulins interact laterally with a tubulin subunit of their own kind, except at the seam, where an α-tubulin contacts a β-tubulin or vice versa (Fig. 5a, color insert). Cryo-EM studies show that the growing end of a microtubule appears as a protofilament sheet that eventually closes, presumably along the seam, to form a cylindrical microtubule (Chretien *et al.,* 1995; Desai and Mitchison, 1998). Therefore, microtubule polymerization is thought to proceed mainly by extension of the protofilament sheets. When shortening microtubules are viewed by EM, protofilaments appear to curl back from the microtubule wall—sometimes into rings (Mandelkow *et al.*, 1991; Erickson and Stoffler, 1996; Desai and Mitchison, 1998). It is believed that the curling of protofilaments is the main driving force for microtubule depolymerization (Chretien and Wade, 1991; Hyman *et al.*, 1995; Melki *et al.*, 1989; Desai and Mitchison, 1998).

The first γTuRC-mediated microtubule nucleation model, the seeded nucleation model (Fig. 5b, color insert) (Zheng *et al.*, 1995), is based on the observation that γTuRC resembles a lockwasher-like ring with a diameter of 25 nm. In addition, each γTuRC contains 10–15 γ-tubulin molecules, which roughly corresponds to the number of tubulin protofilaments found in a microtubule. Zheng *et al.* (1995) proposed that γTuRC could act as a seed to nucleate microtubules with each γ-tubulin in the γTuRC positioned at the minus end of one tubulin protofilament in the nucleated microtubule. The predictions of this seeded nucleation model are (1) γ-tubulins in the γTuRC would be arranged in the same way as one helical path of the tubulin monomers in a microtubule (Figs. 5a and 5b); (2) γTuRC could block the growth of the minus end of its nucleated microtubules; (3) γTuRC may nucleate microtubules with a fixed protofilament number as determined by the number of γ-tubulins involved in microtubule nucleation.

A very different model was proposed by Erickson and Stoffler (1996). In this model, which we will refer to as the protofilament-nucleation model, γ-tubulins in the γTuRC interact with each other in a head-to-tail fashion to form a curved tubulin protofilament (Fig. 5c). This idea builds upon the observations that both tubulin and FtsZ, a bacterial homolog of tubulin, can polymerize into straight protofilaments and ring-like structures (Erickson and Stoffler, 1996). This model predicts that the nucleating activity of γTuRC does not depend on its ringlike structure, but instead, requires γTuRC to uncurl to act as a precursor protofilament, promoting both lateral and longitudinal interactions with tubulin dimers. One important implication of this model is that γTuRC could not block the minus end elongation of microtubules. Therefore, following nucleation, microtubules would grow from both ends, with the γTuRC either falling off or incorporating into the microtubule lattice.

A few attempts at differentiating the two nucleation models have so far been inconclusive. Consistent with the seeded nucleation model, Zheng *et al.* (1995) found that microtubules nucleated by *Xenopus* γTuRC are partially capped at their minus ends. However, the purified *Drosophila* γTuRC does not seem to cap the minus end of its nucleated microtubules (K. Oegema and Y. Zheng, unpublished observation). It is not clear whether the Drosophila γTuRC is incorporated into the microtubules or has fallen off.

Consistent with the protofilament-nucleation model, microtubules nucleated by the *Drosophila* γTuRC have a range of tubulin protofilament number similar to that of spontaneously assembled microtubules (V. Guenebaut and D. Agard, personal communication). However, the minus ends of microtubules nucleated by *Drosophila* γTuRC, when reconstructed using EM-tomography, appeared to be capped by a cuplike structure (V. Guenebaut and D. Agard, personal communication). This observation is more consistent with the seeded-nucleation model. Clearly, we are just beginning to understand γTuRC-mediated microtubule nucleation. Although structural studies have not yet yielded conclusive information, efforts toward building the high-resolution structure of γTuRC will certainly help to solve this difficult problem.

C. Microtubule Nucleation by γTuRC: The Efficiency of Nucleation

Purified γTuRC clearly stimulates microtubule nucleation at tubulin concentrations that support little spontaneous microtubule assembly *in vitro.* One important question is how efficient γTuRC is as a nucleator. For instance, γTuRC could simply resemble a microtubule seed (like those found in a spontaneous nucleation reaction) to support microtubule elongation. In this scenario, γTuRC-mediated microtubule assembly should follow elongation kinetics, proceeding without a significant lag time. Alternatively, if γTuRC does not act as a seed, it could act to stabilize the labile spontaneously nucleated tubulin aggregates that are in equilibrium with tubulin dimers, thereby facilitating the formation of microtubule seeds. If this is the case, γTuRC-mediated microtubule nucleation would follow kinetics similar to that of spontaneously nucleated microtubules. We should note that the two γTuRC-mediated microtubule nucleation models discussed earlier can accommodate both of these possibilities.

Our preliminary kinetic analysis indicates that γTuRC-mediated microtubule nucleation shows a very strong tubulin concentration dependence, similar to spontaneous nucleation (C. Wiese and Y. Zheng, unpublished data). This observation suggests that the γTuRC may not be an efficient nucleator. A more careful kinetics analysis of microtubule nucleation by γTuRC is needed to distinguish the two possibilities. It is also important to determine how γTuRC interacts with tubulin dimers during nucleation.

D. Microtubule Nucleation from the Centrosome: The Control of Nucleating Activity

It may not be surprising that the γTuRC is not as efficient a nucleator as one might expect for a stable microtubule seed. Studies in various types of tissue culture cells show that up to 80% of γTuRC is soluble in the cytosol (Moudjou *et al.,* 1996), yet most microtubules in the cell are nucleated from the centrosome. Therefore, the cell may regulate γTuRC by activation only at the centrosome. Consistent with this idea, our preliminary studies show that *Drosophila* centrosomes are much more active in microtubule nucleation than the *Drosophila* γTuRC purified from the same stage embryos (C. Wiese and Y. Zheng, unpublished data).

How can γTuRC become activated after recruiting to the centrosome? We propose two possibilities. EM micrographs of the purified γTuRC show that the complex has a certain degree of flexibility (Oegema *et al.,* 1999). Therefore, one possibility is that binding to the PCM could confine γTuRC in a fixed conformation, rendering it more active. Alternatively, γTuRC could be activated by a centrosomal protein. For example, XMAP215 (and other proteins in this family) is known to stimulate microtubule elongation *in vitro* and localize to centrosomes *in vivo* (Gard and Kirschner, 1987; Vasquez *et al.,* 1994; Nabeshima *et al.,* 1995; Chen *et al.,* 1998). Genetic studies also suggest that the yeast homolog of XMAP215 interacts with the γ-tubulin complex at the spindle pole body (Knop and Schiebel, 1998). Therefore, XMAP215 could interact with the γTuRC at the centrosome to stimulate microtubule nucleation. RanBPM, a Ran binding protein localized to the centrosome, is another good candidate for activating γTuRC at the centrosome. Overexpression of RanBPM in tissue culture cells leads to ectopic microtubule aster formation in the cytosol. Both γ-tubulin and RanBPM are found at the center of the ectopic asters (Nakamura *et al.,* 1998). Several studies have shown that Ran, the small GTPase involved in regulating nuclear trafficking in interphase, stimulates microtubule aster and spindle formation in mitotic *Xenopus* egg extracts (Kalab *et al.,* 1999; Ohba *et al.,* 1999; Wilde and Zheng, 1999; Carazo-Salas *et al.,* 1999; Zhang *et al.,* 1999). Therefore, Ran could be involved in activating γTuRC through RanBPM at the centrosome. It is interesting to note that the Arp2/3 complex, which is involved in nucleating actin, is also a poor nucleator on its own but becomes more efficient in the presence of two regulators, the small GTPase cdc42, and WASP (Wiskott–Aldrich Syndrome Protein) (Ma *et al.,* 1998; Yarar *et al.,* 1999).

V. Concluding Remarks

The identification of γ-tubulin and its essential role in microtubule nucleation was a major breakthrough in the study of microtubules and centrosomes. The subsequent purification of γTuRC provided the much-needed tool to investigate the role of γ-tubulin in microtubule nucleation and centrosome assembly. Studies in several laboratories have established that γTuRC is essential for both microtubule nucleation and formation of a functional centrosome. Further characterization of the γTuRC components will allow us to study the assembly of γTuRC and its recruitment to the centrosome. These studies will be important in understanding centrosome assembly, an essential process for proper cell division.

Acknowledgment

We thank Ona C. Martin for her valuable comments in preparing this manuscript.

References

Akashi, T., Yoon, Y., and Oakley, B. R. (1997). Characterization of γ-tubulin complexes in *Aspergillus nidulans* and detection of putative γ-tubulin interacting proteins. *Cell Motil. Cytoskeleton* **37,** 149–158.

Buendia, B., Draetta, G., and Karsenti, E. (1992). Regulation of the microtubule nucleating activity of centrosomes in *Xenopus* egg extracts: role of cyclin A–associated protein kinase. *J. Cell Biol.* **116,** 1431–1442.

Burns, R. G. (1995). Analysis of the γ-tubulin sequences: implications for the functional properties of γ-tubulin. *J. Cell Sci.* **108,** 2123–2130.

Carazo-Salas, R. E., Guarguaglini, G., Gruss, O. J., Segref, A., Karsenti, E., and Mattaj, I. W. (1999). Generation of GTP-bound Ran by RCC1 is required for chromatin-induced mitotic spindle formation. *Nature* **400,** 178–181.

Chen, X. P., Yin, H., and Juffaker, T. C. (1998). The yeast spindle pole body component Spc72p interacts with Stu2p and is required for proper microtubule assembly. *J. Cell Biol.* **141,** 1169–1179.

Chretien, D., and Wade, R. H. (1991). New data on the microtubule surface lattice. *Biol. Cell* **71,** 161–174.

Chretien, D., Fuller, S. D., and Karsenti, E. (1995). Structure of growing microtubule ends: Two dimensional sheets close into tubes at variable rates. *J. Cell Biol.* **129,** 1311–1328.

Desai, A., and Mitchison, T. J. (1998). Microtubule polymerization dynamics. *Annu. Rev. Cell Dev. Biol.* **13,** 83–117.

Detraves, C., Mazarguil, H., Lajoie-Mazenc, I., Julian, M., Raynaud-Messina, B., and Wright, M. (1997). Protein complexes containing γ-tubulin are present in mammalian brain microtubule protein preparations. *Cell Motil. Cytoskeleton* **36,** 179–189.

Erickson, H., and Stoffler, D. (1996). Protofilaments and rings, two conformations of the tubulin family conserved from bacterial FtsZ to α/β and γ-tubulin. *J. Cell Biol.* **135,** 5–8.

Felix, M. A., Antony, C., Wright, M., and Maro, B. (1994). Centrosome assembly *in vitro:* role of γ-tubulin recruitment in *Xenopus* sperm aster formation. *J. Cell Biol.* **124,** 19–31.

Fygenson, D. K., Flyvberg, H., Sneppen, K., Libchaber, A., and Leibler, S. (1995). Spontaneous nucleation of microtubules. *Physical. Rev E.* **51,** 5058–5063.

Gard, D. L., and Kirschner, M. W. (1987). A microtubule-associated protein from *Xenopus* eggs that specifically promotes assembly at the plus-end. *J. Cell Biol.* **105,** 2203–2215.

Geissler, S., Pereira, G., Spang, A., Knop, M., Soues, S., Kilmartin, J., and Schiebel, E. (1996). The spindle pole body component Spc98p interacts with the γ-tubulin-like Tub4p of *Saccharomyces cerevisiae* at the sites of microtubule attachment. *EMBO J.* **15,** 3899–3911.

Horio, T., Uzawa, S., Jung, M. K., Oakley, B. R., Tanaka, K., and Yanagida, M. (1991). The fission yeast γ-tubulin is essential for mitosis and is localized at microtubule organizing centers. *J. Cell Sci.* **99,** 693–700.

Hyman, A. A., Chretien, D., Arnal, I., and Wade, R. H. (1995). Structural changes accompanying GTP hydrolysis in microtubules: information from a slowly hydrolyzable analogue guanylyl-(α, β)-methylene-diphosphonate. *J. Cell Biol.* **128,** 117–125.

Joshi, H. C. (1993). γ-Tubulin: the hub of cellular microtubule assemblies. *Bioessays* **15,** 637–643.

Joshi, H. C., Palacios, M. J., McNamara, L., and Cleveland, D. W. (1992). γ-Tubulin is a centrosomal protein required for cell cycle-dependent microtubule nucleation. *Nature* **356,** 80–83.

Kalab, P., Pu, R. T., and Dasso, M. (1999). The ran GTPase regulates mitotic spindle assembly. *Curr. Biol.* **9,** 481–484.

Kalnins, V. I. (1992). "The Centrosome." Academic Press, San Diego.

Kellogg, D. R., Moritz, M., and Alberts, B. M. (1994). The centrosome and cellular organization. *Annu. Rev. Biochem.* **63,** 639–674.

Knop, M., and Schiebel, E. (1998). Receptors determine the cellular localization of a γ-tubulin complex and thereby the site of microtubule formation. *EMBO J.* **17,** 3952–3967.

Knop, M., Pereira, G., Giessler, S., Grien, K., and Schiebel, E. (1997). The spindle pole body component Spc97p interacts with the γ-tubulin of *Saccharomyces cerevisiae* and functions in microtubule organization and spindle pole body duplication. *EMBO J.* **16,** 1550–1564.

Li, Q., and Joshi, H. (1995). γ-Tubulin is a minus end-specific microtubule binding protein. *J. Cell Biol.* **131,** 207–214.

Lohka, M. J., and Masui, Y. (1983). Formation *in vitro* of sperm pronuclei and mitotic chromosomes induced by amphibian ooplasmic components. *Science* **220,** 719–721.

Ma, L., Rohatgi, R., and Kirschner, M. W. (1998). The Arp2/3 complex mediates actin polymerization induced by the small GTP-binding protein Cdc42. *Proc. Natl. Acad. Sci. USA* **95,** 15362–15367.

Mandelkow, E. M., Mandelkow, E., and Milligan, R. A. (1991). Microtubule dynamics and microtubule caps: a time-resolved cryo-electron microscopy study. *J. Cell Biol.* **114,** 977–991.

Martin, O., Gunawardane, R., Iwamatsu, A., and Zheng, Y. (1998). Xgrip109: a γ-tubulin-associated protein with an essential role in γ-tubulin ring complex (γTuRC) assembly and centrosome function. *J. Cell Biol.* **141,** 675–687.

Meads, T., and Schroer, T. A. (1995). Polarity and nucleation of microtubules in polarized epithelial cells. *Cell Motil. Cytoskeleton* **32,** 273–288.

Melki, R., Carlier, M. F., Pantaloni, D., and Timasheff, S. N. (1989). Cold depolymerization of microtubules to double rings: geometric stabilization of assemblies. *Biochemistry* **28,** 9143–9152.

Melki, R., Vainberg, I., Chow, R., and Cowan, N. (1993). Chaperonin-mediated folding of vertebrate actin-related protein and γ-tubulin. *J. Cell Biol.* **122,** 1301–1310.

Moritz, M., Braunfeld, M. B., Fung, J. C., Sedat, J. W., Alberts, B. M., and Agard, D. (1995a). Three-dimensional structural characterization of centrosomes from early *Drosophila* embryos. *J. Cell Biol.* **130,** 1149–1159.

Moritz, M., Braunfeld, M., Sedat, J., Alberts, B., and Agard, D. (1995b). γ-Tubulin-containing rings in the centrosome. *Nature* **378,** 638–640.

Moritz, M., Zheng, Y., Alberts, B. M., and Oegema, K. (1998). Recruitment of the γ-tubulin ring complex to *Drosophila* salt-stripped centrosome scaffolds. *J. Cell Biol.* **142,** 775–786.

Moudjou, M., Bordes, N., Paintrand, M., and Bornens, M. (1996). γ-Tubulin in mammalian cells: the centrosomal and cytosolic forms. *J. Cell Sci.* **109,** 875–887.

Murphy, S. M., Urbani, L., and Stearns, T. (1998). The mammalian γ-tubulin complex contains homologues of the yeast spindle pole body components spc97p and spc98p. *J. Cell Biol.* **141,** 663–674.

Nabeshima, K., Kurooka, H., Takeuchi, M., Kinoshita, K., Nakaseko, Y., and Yanagida, M. (1995). p93^{dis1}, which is required for sister chromatid separation, is a novel microtubule and spindle pole body associating protein phosphorylated at the Cdc2 target sites. *Genes Dev.* **9,** 1572–1585.

Nakamura, M., Masuda, H., Horii, J., Kuma, K., Yokoyama, N., Ohba, T., Nishitani, H., Miyata, T., Tanaka, M., and Nishimoto, T. (1998). When overexpressed, a novel centrosomal protein, RanBPM, causes ectopic microtubule nucleation similar to γ-tubulin. *J. Cell Biol.* **143,** 1041–1052.

Oakley, B. R. (1992). γ-Tubulin: the microtubule organizer? *Trends in Cell Biol.* **2,** 1–5.

Oakley, C. E., and Oakley, B. R. (1989). Identification of γ-tubulin, a new member of the tubulin superfamily encoded by mipA gene of *Aspergillus nidulans. Nature* **338,** 662–664.

Oakley, B. R., Oakley, C. E., Yoon, Y., and Jung, M. K. (1990). γ-Tubulin is a component of the spindle pole body that is essential for microtubule function in *Aspergillus nidulans. Cell* **61,** 1289–1301.

Oegema, K., Wiese, C., Martin, O. C., Milligan, R. A., Iwamatsu, A., Mitchison, T. J., and Zheng, Y. (1999). Characterization of two related *Drosophila* γ-tubulin complexes that differ in their ability to nucleate microtubules. *J. Cell Biol.* **144,** 721–733.

Ohba, T., Nakamura, M., Nishitani, H., and Nishimoto, T. (1999). Self-organization of microtubule asters induced in *Xenopus* egg extracts by GTP-bound Ran. *Science* **284,** 1356–1358.

Oosawa, F., and Asakura, S. (1975). "Thermodynamics of the Polymerization of Protein." Academic press, New York.

Pereira, G., and Schiebel, E. (1997). Centrosome-microtubule nucleation. *J. Cell Sci.* **110,** 295–300.

Raff, J. W., Kellogg, D. R., and Alberts, B. M. (1993). *Drosophila* γ-tubulin is part of a complex containing two previously identified centrosomal MAPs. *J. Cell Biol.* **121,** 823–835.

Saunders, W. S. (1999). Action at the ends of microtubules. *Curr. Opin. Cell Biol.* **11,** 129–133.

Schnackenberg, B. J., Khodjakov, M. W., Rieder, C. L., and Palazzo, R. E. (1998). The disassembly and reassembly of functional centrosomes *in vitro. Proc. Natl. Acad. Sci. USA* **95,** 9295–9300.

Sobel, S., and Snyder, M. (1995). A highly divergent γ-tubulin gene is essential for cell growth and proper microtubule organization in *Saccharomyces cerevisiae. J. Cell Biol.* **131,** 1775–1788.

Stearns, T., and Kirschner, M. (1994). *In vitro* reconstitution of centrosome assembly and function: The role of γ-tubulin. *Cell* **76,** 623–637.

Sunkel, C., Gomes, R., Sampaio, P., Perdigao, J., and Gonzalez, C. (1995). γ-Tubulin is required for the structure and function of the microtubule organizing centre in *Drosophila* neuroblasts. *EMBO J.* **14,** 28–36.

Tassin, A. M., Celati, C., Moudjou, M., and Bornens, M. (1998). Characterization of the human homologue of the yeast SPC98p and its association with γ-tubulin. *J. Cell Biol.* **141,** 689–701.

Vasquez, R. J., Gard, D., and Cassimeris, L. (1994). XMAP from *Xenopus* eggs promotes rapid plus end assembly of microtubules and rapid microtuble polymer turnover. *J. Cell Biol.* **127,** 985–993.

Vogel, J., Stearns, T., Rieder, C., and Palazzo, R. (1997). Centrosomes isolated from *Spisula solidissima* oocytes contain rings and an unusual stoichiometric ratio of α/β-tubulin. *J. Cell Biol.* **137,** 193–202.

Voter, W., and Erickson, H. (1984). The kinetics of microtubule assembly. Evidence for a two-stage nucleation mechanism. *J. Biol. Chem.* **259,** 10430–10438.

Wiese, C., and Zheng, Y. (1999). γ-Tubulin complexes and their interaction with microtubule organizing centers. *Curr. Opin. Struct. Biol.* **9,** 250–259.

Wilde, A., and Zheng, Y. (1999). Stimulation of microtubule aster formation and spindle assembly by the small GTPase Ran. *Science* **284,** 1359–1362.

Yarar, D., To, W., Abo, A., and Welch, M. D. (1999). The Wiskott–Aldrich syndrome protein directs actin-based motility by stimulating actin nucleation with the Arp2/3 complex. *Curr. Biol.* **9,** 555–558.

Zhang, C., Hughes, M., and Clarke, P. R. (1999). Ran-GTP stabilises microtubule asters and inhibits nuclear assembly in *Xenopus* egg extracts. *J. Cell Sci.* **112,** 2453–2461.

Zheng, Y., Wong, M., Alberts, B., and Mitchison, T. (1995). Nucleation of microtubule assembly by a γ-tubulin-containing ring complex. *Nature* **378,** 578–583.

Zheng, Y., Wong, M. I., Alberts, B., and Mitchison, T. (1998). Purification and Assay of γ-tubulin ring complexes. *Methods Enzymol.* **298,** 218–228.

Zimmerman, W., Sparks, C. A., and Doxsey, S. J. (1999). Amorphous no longer: the centrosome comes into focus. *Curr. Opin. Cell Biol.* **11,** 122–128.

4

γ-Tubulin of Budding Yeast

Jackie Vogel and *Michael Snyder*
Department of Molecular, Cellular and Developmental Biology
Yale University
New Haven, Connecticut 06520-8103

I. Introduction

A. Microtubules and MTOCs

Microtubules are 25-nm-diameter biopolymers composed of α- and β-tubulins, GTP-binding proteins that are a major component of the cytoskeleton in all eukaryotic cells. Microtubules participate in many cellular processes, such as establishing cell shape and polarity, vectoral organelle transport, and chromosome segregation. To accomplish such diverse roles, microtubules are organized into a variety of different arrays during the cell cycle. For example, polarized arrays of cytoplasmic microtubules function in polarized secretion and nuclear migration, and antiparallel microtubule arrays characteristic of the spindle apparatus function during chromosome segregation. The structure of these arrays and many of their respective functions are conserved in cells as diverse as those of unicellular fungi and the highly specialized cells of multicellular organisms.

Microtubules are intrinsically polarized; microtubules have a fast-growing or plus end, and a slow-growing or minus end, and exhibit dynamic instabil-

Current Topics in Developmental Biology, Vol. 49

0070-2153/00 $35.00

ity (Mitchison and Kirschner, 1984a). In cells, the polarity of the arrangement of the majority of microtubules is dictated by microtubule organizing centers (MTOC), organelles that are required for microtubule nucleation *in vivo*. The plus ends of newly assembled microtubules are distal to the MTOC, where they may interact with kinetochores, microtubule attachment sites at the cell cortex, or other microtubules (Huitorel and Kirschner, 1988; Kaverina *et al.,* 1998; Miller and Rose, 1998; Miller *et al.,* 1999). Conversely, the minus ends of microtubules are proximal to the MTOC, where they are likely stabilized by the interaction with this organelle (Mitchison and Kirschner, 1984b; Moritz *et al.,* 1995a). Thus, the MTOC has a pivotal role in determining the structure of microtubule arrays.

B. Microtubules and Spindle Pole Bodies in Yeast

1. Overview of Microtubule Organization and Function in Yeast

Microtubules perform a variety of essential functions during the yeast cell cycle and life cycle, and consequently microtubule organization varies during these periods (Kilmartin and Adams, 1984; Adams and Pringle, 1984; Jacobs *et al.,* 1988). The yeast MTOC, the spindle pole body (SPB), is located in the nuclear membrane, which remains intact throughout the entire cell cycle (Byers, 1981). The organization of microtubules and their roles during the yeast cell cycle is shown in Fig. 1. During vegetative growth, the SPB begins to duplicate at the end of anaphase and duplication is completed by the beginning of S-phase of the cell cycle (Adams and Kilmartin, 1999). Cytoplasmic microtubules extend from the SPB and the half bridge that connects the SPB to its duplication satellite (Byers and Goetsch, 1975), and interact with the incipient bud site in unbudded cells and the bud cortex in budded cells (Brachat *et al.,* 1998; Miller and Rose, 1998; Miller *et al.,* 1999). Cortical microtubule capture sites in the bud are thought to interact with the plus ends of microtubules and help mediate nuclear migration and spindle positioning prior to mitosis (Snyder *et al.,* 1991; Carminati and Stearns, 1997; Shaw *et al.,* 1997; Lee, *et al.,* 2000; Kurinek, *et al.,* 2000). Nuclear microtubules, which are required for chromosome segregation, are present throughout the yeast cell cycle (O'Toole *et al.,* 1999). During S phase, after SPB duplication has been completed, a short (~1 γm) spindle is initially formed. During late mitosis the spindle elongates dramatically (~8 γm) while the pole-to-kinetichore microtubules shorten, driving nuclear division (Kilmartin *et al.,* 1984; O'Toole *et al.,* 1999). Yeast is extremely parsimonious; each SPB assembles 16 pole-to-kinetichore microtubules, e.g., one microtubule per kinetochore, and ~4 pole-to-pole interdigitizing microtubules (Peterson and Ris, 1976; Winey *et al.,* 1995; O'Toole *et al.,* 1999).

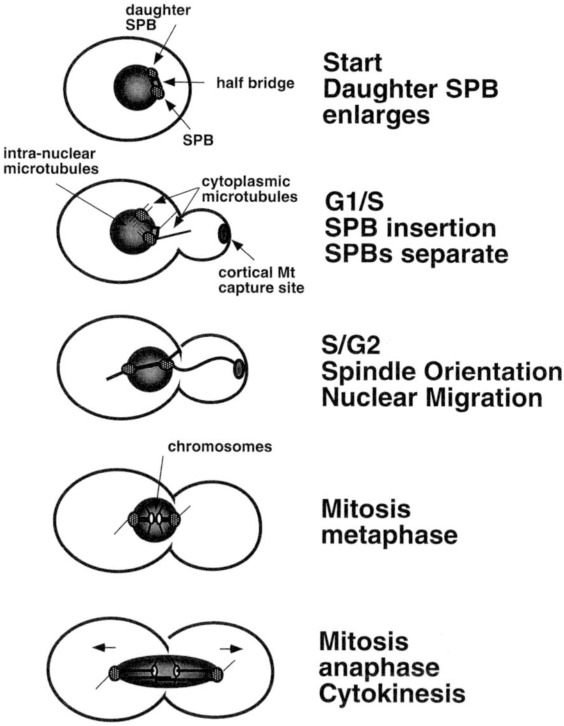

Figure 1 Microtubule functions during the yeast cell cycle. Unbudded yeast cells contain a single SPB and a SPB satellite that begins to recruit proteins at Start in G1 of the cell cycle. After the SPB duplicates, it inserts into the nuclear membrane and by S-phase a short spindle has formed. The plus ends of the cytoplasmic microtubules interact with proteins located at the bud cortex and help position the nucleus at the neck of the bud (nuclear migration), and by G2 the spindle is oriented such that it is parallel with the long axis of the mother bud. In early mitosis (metaphase), the nucleus penetrates the neck of the bud; later the spindle elongates (anaphase), followed by cytokinesis.

Microtubules are also important for nuclear fusion during mating (Byers and Goetsch, 1975; Maddox *et al.,* 1999; Pereira *et al.,* 1999). A schematic of the roles of microtubules during mating is shown in Fig. 2. When yeast cells mate, the two cells contact and fuse, resulting in the formation of a heterokaryon in which the two cells share a common cytoplasm but have not yet fused their nuclei. Microtubules emanating from the SPBs of the two opposing nuclei interdigitate, and the nuclei are drawn together using these cytoplasmic arrays. Ultimately the nuclei fuse at the site of the SPBs.

2. Overview of the Yeast Spindle Pole Body

Microtubule arrays in yeast cells are nucleated and organized by the SPBs (reviewed in Page and Snyder, 1993). The SPB is a trilaminar structure: A

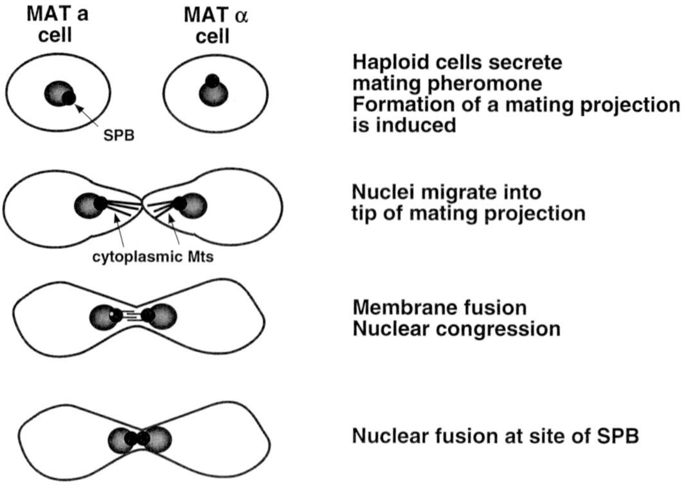

Figure 2 Microtubule functions during mating and karyogamy. Haploid cells of opposite mating types (MAT a or MAT α) secrete pheromone, stimulating the formation of mating projections. Cytoplasmic microtubules extend from the single SPB to the cortical region at the tip of the projection and facilitate the migration of the nucleus into the tip of the mating projection. The mating cells fuse, forming a heterokaryon, and the cytoplasmic microtubules from each cell interdigitate and pull the nuclei together. Ultimately the nuclei fuse at the site of their SPBs.

central electron-dense plaque is embedded in the nuclear membrane, which is flanked by inner and outer plaques (Byers, 1981). Nuclear and cytoplasmic microtubules emanate from the inner and outer plaques, respectively (Byers *et al.,* 1978; Hyams and Borisy, 1978). Duplication and insertion of the SPB into the nuclear membrane and the early stages of spindle assembly have been carefully documented in studies by Adams and Kilmartin (1999) and O'Toole *et al.* (1999). A daughter SPB satellite begins to form at the end of mitosis. At Start in G1, the satellite begins to enlarge and SPB duplication is completed during G1/S of the cell cycle; subsequently the daughter SPB inserts into the nuclear membrane (Byers and Goetsch, 1975). Immediately after its insertion, the daughter SPB assembles microtubules and together with the parent SPB begins to assemble a short spindle (O'Toole *et al.,* 1999). Cytoplasmic microtubules function during migration of the nucleus to the bud neck and position the spindle by the end of G2 (Sullivan and Huffaker, 1992; Palmer *et al.,* 1992; Carminati and Stearns, 1997; Shaw *et al.,* 1997). However, unlike most other eukaryotic cells, the SPB does not dramatically increase microtubule nucleation as cells enter mitosis, though the length of the spindle microtubules increases dramatically during anaphase (Kilmartin and Adams, 1984; Winey *et al.,* 1995; O'Toole *et al.,* 1999).

3. Molecular Composition of the Yeast Spindle Pole Body

The identification and characterization of protein components of SPBs has increased dramatically in the past several years, due in large part to the completion of the yeast genome sequencing project. The number of known SPB constituents now exceeds that for the analogous centrosome of animal cells. A comprehensive discussion of the structure of the SPB appears in a separate chapter of this volume (Francis and Davis). However, a short discussion of the composition of the SPB of budding yeast is presented here. The resident proteins of the core SPB structure of yeast (inner plaque, and the nuclear and cytoplasmic outer plaques) have been identified by a variety of approaches. These methods include (1) SPB purification in conjunction with monoclonal antibody production (Rout and Kilmartin, 1990), or MALD-MS (Wigge *et al.,* 1998), or purification of soluble complexes (Elliott *et al.,* 1999); (2) homology with MTOC components in other organisms (Sobel and Snyder, 1995); and (3) genetic studies (Geissler *et al.,* 1996; Knop *et al.,* 1997). A schematic diagram of the structure of the budding yeast SPB and its resident proteins is shown in Fig. 3. The SPB is composed of several distinct layers. The central plaque of the SPB is composed of a lattice of Spc42p (Donaldson and Kilmartin, 1996; Bullitt *et al.,* 1997). The nuclear side of the central plaque contains the coil–coil protein Nuf1p/Spc110p and calmodulin (Cmd1p) (Mirzayan *et al.,* 1992; Geiser *et al.,* 1993), and Spc29p, which appears to function as a bridge between the inner nuclear plaque complex and Spc42p (Elliott *et al.,* 1999).

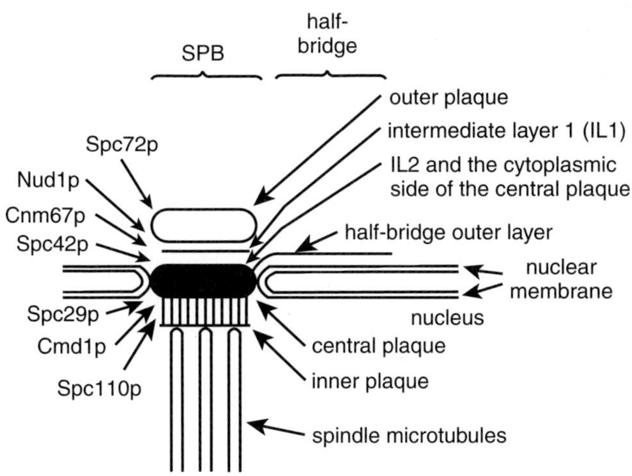

Figure 3 Molecular composition of the SPB. Protein constituents of the SPB of yeast. Reproduced from Adams and Kilmartin, *Journal of Cell Biology* (1999) **45,** 809–823, by copyright permission of the Rockefeller University Press.

The cytoplasmic side of the SPB contains Spc72p, Cnm67p, Nud1p, and other associated proteins (Knop and Schiebel, 1998; Adams and Kilmartin, 1999). Stu2p is a peripheral protein that binds Spc72 and the lateral walls of microtubules *in vitro* and is therefore likely to bridge microtubules and proteins resident in the outer cytoplasmic plaque (Wang and Huffaker, 1997). Finally, the half bridge, a structure that connects the SPB to the duplication satellite, contains Kar1p (Rose and Fink, 1987; Spang *et al.*, 1995) and the centrin homolog Cdc31p (Baum *et al.*, 1986; Spang *et al.*, 1993).

One complex of proteins that is particularly important for microtubule organization is the γ-tubulin complex. In budding yeast this complex is comprised of Tub4p, the γ-tubulin homolog of budding yeast (Sobel and Snyder, 1995; Spang *et al.*, 1996; Marschall *et al.*, 1996), and two other proteins, Spc98p (Rout and Kilmartin, 1990; Geissler *et al.*, 1996) and Spc97p (Knop *et al.*, 1997). This complex resides in both the inner and outer plaques and is thought to be the key element in nucleating microtubule assembly.

In summary, the SPB is a complex structure that undegoes dynamic changes during the cell cycle. This review summarizes our understanding of the γ-tubulin complex of yeast, its role in microtubule nucleation, and components that may regulate its function.

II. Tubulins of Yeast

There are a large number of well-defined mutations in yeast tubulins (Thomas *et al.*, 1985; Schatz *et al.*, 1988; Huffaker *et al.*, 1988; Spang *et al.*, 1996; Marschall *et al.*, 1996) that perturb specific aspects of spindle and/or cytoplasmic microtubule function. An overview of yeast tubulins (α, β, and γ), representative mutations, and genetic interactions is shown in Table I. These mutations, or phenotypes arising from treating cells with microtubule destabilizing drugs such as benomyl, have been used in conjunction with genetic screens to identify proteins involved in various aspects of microtubule assembly and/or function (Stearns and Botstein, 1988; Stearns *et al.*, 1990; Hoyt *et al.*, 1991; Knop *et al.*, 1997). In addition, large-scale disruption techniques are now being used to study proteins that are involved in microtubule organization. For example, 7680 mTn-insertion alleles affecting more than 1718 different genes were transformed into a haploid strain for phenotypic analysis. Transformed haploid strains were subsequently scored for 21 phenotypes after growth under different test conditions, including the microtubule destabilizing drug benomyl. Many noncharacterized ORFs whose mTn-insertion alleles present sensitivity or resistance to this drug were identified (Ross-MacDonald *et al.*, submitted; Coelho and Snyder, unpublished observations). Many of the mTn-mutagenized genes that ex-

Table I Overview of Yeast Tubulins

Locus	Essential	Mutant Phenotypes (Representative)	Suppressors	Colethal
TUB1 α Tub1p	Yes	*tub1-1* No microtubules	*RBL2*	*cin1,2,4Δ* *pac2Δ* *PAC2 (GAL)*
		tub1-724 α/β heterodimer unstable		*rbl2Δ* *PAC2 (GAL)*
TUB3 α Tub3p	No	*tub3Δ* Benomyl sensitive	*kar3Δ*	*tub1-1*
TUB2 β Tub2p	Yes	*GAL::TUB2* No microtubules	*RBL2*	
		tub2-401 No astral microtubules		
		tub2-406 No spindle microtubules	*STU1*	
		tub2-150, tub2-404 Hyperstable microtubules Spindle elongation defect	*JSN1*	*stu2-1, num1Δ*
TUB4 γ Tub4p	Yes	*GAL::TUB4* Depletion: long cytoplasmic mt, monopolar spindles, spindle orientation defect Overexpression: no phenotype		
		tub4Δ/TUB4 Benomyl sensitive		
		tub4-32, tub4-34 Localization defective *De novo* microtubule nucleation is defective Long cytoplasmic mt Monopolar spindles		
		tub4-1 Spindle elongation defect	*SPC98*	*gim2/ pac10Δ, gim3Δ, gim4Δ, gim5Δ, spc98-1, bub2Δ*

hibit benomyl-dependent phenotypes are expected to be involved in some aspect of microtubule-dependent processes in yeast.

A. α- and β-Tubulins

The tubulin heterodimer consists of a GTP-bound α-tubulin subunit, and a β-tubulin subunit with an exchangeable GTP binding site (reviewed in

Bayley *et al.*, 1994). Yeast have three genes encoding conventional (α or β) tubulins. The essential *TUB1* and non-essential *TUB3* genes encode α-tubulins (Tub1p and Tub3p; Schatz *et al.*, 1986), while the essential *TUB2* gene encodes β-tubulin (Tub2p; Neff *et al.*, 1983). Yeast α- and β-tubulins share ~70% sequence identity with their counterpart tubulins of other eukaryotes (Neff *et al.*, 1983; Schatz *et al.*, 1986). In particular, the GTP-binding site and catalytic residues are highly conserved in all tubulins (Sage *et al.*, 1995). This conservation is correlated with function, since yeast tubulin subunits polymerize into microtubules under conditions similar to those for tubulins isolated from mammalian cells; moreover, yeast tubulin coassembles with tubulin purified from mammalian cells (Kilmartin, 1981; Barnes *et al.*, 1992).

Whereas all tubulins share high overall sequence identity, yeast tubulins contain N-terminal and/or C-terminal regions of variability relative to tubulins of other eucaryotes (Neff *et al.*, 1983; Schatz *et al.*, 1986). The N-terminal variable region of α-tubulins (which is identical in Tub1p and Tub3p) can tolerate a variety of substitutions that do not affect function, suggesting that this region does not have a specific function in yeast (Schatz *et al.*, 1987). The C-terminal regions of Tub1p, Tub3p, and Tub2p are highly divergent among the different tubulins and between different species. For example, Tub2p has a short (11 residues) C-terminal extension not found in other yeast tubulins or in β-tubulins of other organisims (Neff *et al.*, 1983). Deletion of the Tub2p C-terminal extension results in increased sensitivity to microtubule destabilizing drugs (Katz and Solomon, 1988; Matsuzaki *et al.*, 1988), suggesting that it may be important for stabilizing microtubules in yeast. Since the C termini of β-tubulins of other organisms are known to interact with microtubule-associated proteins (MAPs) *in vitro* (Serrano *et al.*, 1984; de Pereda and Andreu, 1996), it is possible that the C-terminal extension of Tub2p may constitute part of a binding site for such a protein interaction in yeast.

Tubulins represent a major cytosolic protein in animal cells, composing as much as 3% of the total cellular protein in oocytes of the surf clam *Spisula* (Burnside *et al.*, 1973). In budding yeast, tubulin is much less abundant and accounts for only ~0.05% of the soluble protein (Kilmartin, 1981). A number of studies indicate that the steady-state level of tubulins is tightly regulated in yeast. A reduction of *TUB2* gene dosage in diploid cells by disrupting one of the two copies does not have a pronounced effect on microtubule organization as determined by indirect immunofluorescence using anti-tubulin antibodies, but does result in increased sensitivity to microtubule destabilizing drugs such as benomyl (Katz *et al.*, 1990). Interestingly, reducing *TUB1* gene dosage by one copy results in a corresponding ~50% reduction in Tub1p levels, but not in *TUB1* mRNA levels (Katz *et al.*, 1990). These results suggest that yeast cells do not maintain a substantial pool of free tubulin subunits for microtubule assembly. Increasing *TUB1* gene dosage has no phenotypic effect and results in a minor increase in *TUB1* mRNA levels and almost no alteration in α-tubulin protein levels (Burke *et al.,* 1989; Katz *et al.,* 1990). Conversely, a modest increase in β-tubulin gene dosage relative to α-tubulin in yeast has a number of phenotypic consequences. Increasing Tub2p levels is extremely toxic to cells; *TUB2* overexpression results in aneuploidy, and the disassembly of microtubule structures and cells arrest in G2 with a large-bud (presumably in mitosis) (Burke *et al.,* 1989; Weinstein and Solomon, 1990). Together, these studies indicate that the paucity of free tubulin in yeast is the result of tight controls at posttranscriptional and/or posttranslational levels. Overexpression of Tub2p can be suppressed by concomitant overexpression of either *TUB1* (Weinstein and Solomon, 1990) or *RBL2,* a gene whose product binds β-tubulin and is involved in the assembly of microtubules (Archer *et al.,* 1995, 1998). Thus, in many cases, genes that are either colethal with or supress mutations in yeast tubulins are involved in tubulin synthesis and/or folding. Examples of colethality and suppression interactions are summarized in Table I. Based on phenotypic analysis and Rb12p-mediated suppression, it is likely that Tub2p overexpression disrupts the stoichiometric ratio of limited components required for microtubule assembly.

B. γ-Tubulin

γ-Tubulin is a third member of the tubulin superfamily, sharing ~30% identity with α- and β-tubulins (Oakley and Oakley, 1989). Despite intensive effort, γ-tubulin was not readily identified in yeast, largely because of its low abundance and its unusually high sequence divergence relative to γ-tubulins of other organisms. It was identified relatively recently through the yeast genome sequencing project. A previously uncharacterized ORF with a predicted amino acid sequence with 29–38% identity to γ-tubulins of other organisms was identified and named *TUB4* (Sobel and Snyder, 1995). The divergent γ-tubulin encoded by *TUB4* (Tub4p) localizes to the SPB (Sobel and Snyder, 1995; Spang *et al.,* 1996; Marschall *et al.,* 1996). The region directly upstream of the TUB4 gene contains an MCB box, a DNA element associated with G1 expression (Sobel and Snyder, 1995), and analysis of mRNA levels throughout the cell cycle indicates that it is predominantly expressed in G1 (Marschall *et al.,* 1996; Cho *et al.,* 1998; Spellman *et al.,* 1998). As with γ-tubulins of other organisms (Oakley *et al.,* 1990; Stearns *et al.,* 1991; Horio and Oakley, 1994; Liu *et al.,* 1994; Sunkel *et al.,* 1995), the product of the *TUB4* gene is essential for viability (Sobel and Snyder, 1995). Interestingly, γ-tubulins of *Xenopus laevis* (Xgam) or *Schizosaccharomyces pombe* (Tug1+) do not restore viability of *tub4*Δ cells (Spang *et al.,* 1996; Marschall *et al.,* 1996). In contrast, human γ-tubulin can function in S. pombe cells (Horio and Oakley, 1994). The failure of Xgam or Tug1+ to complement *tub4*Δ cells suggests that some of the divergent residues of Tub4p are specifically required for its function in yeast. A highly divergent C-terminal extension is present in Tub4p and is absent in both Xgam and Tug1+ γ-tubulins; thus, this region is a candidate for specifying a distinct function of Tub4p. Depletion or mutation of Tub4p results in defects in microtubule organization (Sobel and Snyder, 1995; Spang *et al.,* 1996; Marschall *et al.,* 1996). Thus, based on both sequence homology and functional similarity, the *TUB4* gene product is the γ-tubulin of budding yeast. Since no other tubulin homologs are apparent from the completed yeast genome sequence, the presence of a γ-tubulin more similar to those of other organisms is not likely.

To summarize what is known regarding γ-tubulin function in yeast, we compare the consequences of three types of perturbations: (1) depletion or dosage reduction, (2) mutations, and (3) overexpression. As detailed earlier, use of these approaches has provided considerable insight into the function and/or regulation of α- and β-tubulins in yeast.

1. Depletion or Dosage Reduction

Depletion of γ-tubulin has two important consequences in mammalian cells (Joshi and Cleveland, 1992). First, cells treated with microtubule-depolymerizing drugs fail to reform microtubules when γ-tubulins have been sterically blocked by anti-γ-tubulin antibodies, demonstrating that *in vivo* microtubule assembly requires γ-tubulin as a nucleation factor. Second, when cytoplasmic stores of γ-tubulin are depleted in G2 or early M-phase, cells fail to assemble a spindle, linking the recruitment of soluble γ-tubulin to the burst of microtubule nucleation observed during centrosome maturation and G2/M-phase transition.

Depletion of Tub4p has dramatic effects on microtubule organization in yeast. By controlling *TUB4* expression with the inducible *GAL1/10* promoter region (Johnston and Davis, 1984), Sobel and Snyder demonstrated that depletion of Tub4p results in a number of microtubule defects. Depletion of Tub4p would be expected to cause a loss of microtubules; however, long cytoplasmic microtubules were observed. The formation of long cytoplasmic microtubules may be due to loss of nucleation sites and the polymerization of free heterodimers into the remaining microtubules. In addition to defects in cytoplasmic microtubules, spindles were often monopolar as a result of the SPBs failing to separate. The few bipolar spindles observed did not elongate. Thus, depletion of Tub4p has an effect on the assembly of cytoplasmic and spindle microtubules similar to that described for γ-tubulin depletion in mammalian cells.

The amount of soluble γ-tubulin in the filamentous fungus *Aspergillus* has been determined to be ~30% of the total γ-tubulin content of the cell (Akashi *et al.,* 1997). This result suggests that in *Aspergillus,* the majority of γ-tubulin is associated with the SPB. Because yeast cells do not maintain substantial amounts of free tubulin dimer, reducing *TUB1* or *TUB2* gene dosage in diploid yeast cells results in increased sensitivity to the microtubule destabilizing drug benomyl (Ben^s). Similarly, we have found that reducing *TUB4* gene dosage (*TUB4/tub4Δ*) in diploid cells confers a moderate Ben^s phenotype, suggesting a reduction in the number and/or stability of microubules. Since a moderate reduction of *TUB4* gene dosage is sufficient to confer a Ben^s phenotype, it is unlikely that yeast cells maintain a significant cytoplasmic pool of Tub4p. Interestingly, we have identified an extragenic mutation in the *AME1* locus that confers benomyl resistance and presumably stabilizes microtubules. This mutation can suppress the Ben^s phenotype of *TUB4* heterozygotes (Vogel, Liszt, and Snyder, unpublished results). One possibility is that *amel* mutation promotes microtubule assembly and/or stability.

2. Mutation

To determine the function of an uncharacterized yeast protein, it is common to create a lesion in the gene of interest and determine the resulting phenotype of a strain carrying this mutation. Phenotypes arising from a null mutation can be investigated if thc gene is not essential for viability. However, if the gene is essential, conditional mutations that permit growth at the permissive condition, but result in inviability at the restrictive condition (low or high temperature, or sensitivity to a drug such as benomyl) are employed.

In the case of *TUB4,* there are three conditional mutations that have been characterized to date: *tub4-1* (Spang *et al.,* 1996), *tub4-32,* and *tub4-34* (Marschall *et al.,* 1996). A comparison of the phenotypes of the previously described perturbations of Tub4p protein levels with those of *tub4* conditional alleles is shown in Table I. The predominant phenotype of *tub4-32* and *tub4-34* cells is the formation of monopolar spindles and long cytoplasmic microtubules. These phenotypes are similar to those observed when Tub4p is depleted (Sobel and Snyder, 1995). In addition, *tub4-32* and *tub4-34* mutants cannot reassemble microtubules, indicating that the nucleation capacity of the SPB is lost or greatly reduced as a result of these mutations (Marschall *et al.,* 1996). In contrast, the *tub4-1* mutation does not appear to be defective in initiating microtubule nucleation, since a short spindle is assembled at the restrictive temperature. Furthermore, defects in spindle structure are not observed in *tub4-1* cells arrested in G2 by treatment with hydroxyurea and examined by electron microscopy. These results indicate that the *tub4-1* defect occurs after DNA replication and, based on spindle length, prior to anaphase. In addition, *tub4-1* is colethal with disruption of a gene required for a cell cycle checkpoint that monitors spindle function, *BUB2* (Hoyt *et al.,* 1991). *BUB2* is required for the large budded arrest of *tub4-1* cells, and *bub2Δ tub4-1* double mutants have reduced viability (Spang *et al.,* 1996). Together, these results suggest the *tub4-1* mutation results in a defect in the interaction between Tub4p and proteins of the inner and/or outer plaque proteins of the SPB (Spang *et al.,* 1996). Since the *tub4-1* defect is restricted to mitosis, this would suggest that the interaction between microtubules and the SPB is stable unless force is exerted, such as during chromosome segregation.

Spindle structure and function are monitored by cell cycle checkpoint proteins encoded by the MAD and BUB genes (Li and Murray, 1991; Hoyt *et al.,* 1991). The spindle checkpoint has a dual role in monitoring microtubule interactions with kinetochores and at SPBs (Tavormina and Burke, 1998; Li, 1999; Alexandru *et al.,* 1999). The colethality observed in *tub4-1bub2Δ* double mutants is particularly interesting in the light of recent studies indicating that Bub2p localizes to the SPB and is thought to function independently of the *MAD*-dependent checkpoint (Li, 1999; Fraschini *et al.,* 1999). It is significant that, in contrast to *tub4-1* or Tub4p depletion, the *tub4-32* or *tub4-34* mutations do not appear to trigger the spindle checkpoint at the restrictive temperature. *tub4-32* and *tub4-34* cells are defective in Tub4p and Spc98p localization to the SPB and *de novo* microtubule nucleation and therefore lack microtubules. In contrast, SPBs in *tub4-1* or Tub4p depleted cells have defective or reduced Tub4p, respectively, and display defects in microtubule structure. Thus, we speculate the *BUB2*-dependent checkpoint cannot monitor defects in SPB function in the absence of Tub4p/Spc98p and/or microtubules.

The effect of *tub4* mutations on other residents of the SPB inner and outer plaques provides possible insight into their respective defects in function. The localization of a peripheral component of the inner and outer plaques, Spc98p (Rout and Kilmartin, 1990), is unaffected by the *tub4-1* mutation (Spang *et al.,* 1996). In contrast, the SPB localization of Spc98p in *tub4-32* and *tub4-34* mutants is defective (Marschall *et al.,* 1996). Electron micrographs of spindle microtubules in *tub4-1* cells confirms that the minus ends are released (Spang *et al.,* 1996). Spc98p and Tub4p interact physically (Knop *et al.,* 1997), suggesting that in the *tub4-1* mutant the interaction between Tub4-1p and Spc98p can be forced apart by the tension exerted by the spindle. In contrast, in *tub4-32* and *tub4-34* mutants the complex containing Tub4p and Spc98p is likely to be inherently unstable and therefore fails in initiating *de novo* microtubule nucleation. Thus, the *tub4-32* and *tub4-34* mutations result in microtubule defects during SPB separation and other microtubule-dependent processes early in the cell cycle.

The phenotypes resulting from perturbations and mutations of Tub4p indicate that it has a crucial role in microtubule assembly and function in yeast. The failure of *tub4-32* and *tub4-34* cells to assemble microtubules (Marschall *et al.,* 1996) demonstrates that Tub4p is required for *de novo* microtubule nucleation. The spindle defect of the *tub4-1* mutant (Spang *et al.,* 1996) indicates that Tub4p has an additional role in anchoring the minus ends of spindle microtubules during chromosome segregation. In support of this function, three-dimensional reconstructions of the yeast spindle demonstrate that the minus ends of nuclear microtubules are capped, while the plus ends are flared (O'Toole *et al.,* 1999), suggesting the residents of the inner and perhaps also the outer plaques contribute to stability of microtubule minus ends. In a later section of this chapter, we examine the role of proteins that coassemble with Tub4p in the yeast γ-tubulin complex.

3. Overexpression

In yeast, the expression of α- and β-tubulin genes is tightly controlled (see previous section). In contrast, as much as 100-fold overexpression of *TUB4* under the control of the exogenous *GAL1* promoter appears to have no deleterious effect in budding yeast (Sobel and Snyder, 1995; Marschall *et al.,* 1996). This observation is particularly interesting since *TUB4* expression is cell cycle regulated (Marschall *et al.,* 1997). When overexpressed, the levels of Tub4p remain high, indicating that, unlike yeast α-tubulin, excess *TUB4* mRNA or Tub4 protein is not degraded. There are several nonexclusive possibilities for why overexpression of *TUB4* is not toxic to yeast cells. First, other proteins that are required for Tub4p function may be limiting, such that the excess protein is inactive. It has been shown that Spc98p, a protein which binds to Tub4p and is part of the γ-tubulin complex of yeast, is less than or similar to Tub4p in abundance and therefore is limiting (Geissler *et al.,* 1996). In addition, overexpression of Spc98p is toxic in the absence of *TUB4* overexpression, suggesting that Spc98p might compete for other proteins involved Tub4p activation or function. Second, excess Tub4p may be sequestered from interacting with other proteins required for its function. In support of this hypothesis, when *TUB4* is overexpressed its protein product accumulates in a soluble multiprotein complex containing Gim1p/Yke2p, Gim2p/Pac10p, Gim3p, and Gim5p (Geissler *et al.,* 1998). Overexpression of *TUB4* in *gim1Δ/yke2Δ* cells results in accumulation of large-budded cells, indicating the activation of the spindle assembly checkpoint (Geissler *et al.,* 1998). The arrest phenotype of *gim1Δ/yke2Δ* cells overexpressing *TUB4* suggests that Gim1p/Yke2p normally sequesters excess Tub4p in the cytoplasm and prevents the formation of an active complex. The *gim1Δ/yke2Δ* mutation is colethal with many mutations in cytoskeletal proteins (e.g., α-,β-tubulins, actin) and proteins involved in the biogenesis of cytoskeletal structures (e.g., Rb12p, Bin2p, Bin3p) (Geissler *et al.,* 1998). When overexpressed *GIM1/YKE2* can suppress the *tub4-1* mutation and some temperature-sensitive *spc97* and *spc98* alleles (Geissler *et al.,* 1998). Together, these results suggest that Gim1p/Yke2p may have a ubiquitous role in assembling the cytoskeletion. A third possibility is that Tub4p must be posttranslationally modified to be functional. Overexpression of *TUB4* would presumably result in the accumulation of unmodified and therefore inactive protein. To date, no *in vivo* posttranslational modifications of Tub4p have been described, though multiple isoforms of the same gene product that are specific for centrosomal vs cytoplasmic compartments have been found (Moudjou *et al.,* 1996; Vogel and Palazzo, unpublished results).

In a series of experiments to determine if excess soluble Tub4p is posttranslationally modified, we found that overexpression results in several highly reproducible modifications of Tub4p (Vogel and Snyder, unpublished results). *TUB4* gene dosage was increased ~30-fold under the control of its own promoter on high-copy 2γ plasmids. Three distinct Tub4p isoforms ranging from most acidic to most basic ($\gamma1, \gamma2, \gamma3$) are detected in two-dimensional gels when *TUB4* is overexpressed in this manner. In contrast, a single Tub4p isoform ($\gamma2$) is detected when *TUB4* is expressed at endogenous levels. Interestingly, the mutation of a conserved tyrosine residue to phenylalanine (Tyr445Phe) in the otherwise hypervariable carboxy terminus of Tub4p either eliminates or dramatically reduces two of the three Tub4p isoforms, whereas a mutation at the adjacent serine residue (Ser444-Ala) does not. Substitution of Tyr445 with a acidic residue (Tyr445Asp) results in defects in microtubule function during mitosis, and 2μ overexpression of this allele decreases viability (Vogel and Snyder, unpublished results). It is not known yet if or how these modifications may activate or inhibit Tub4p function; however, these results suggest two possibilities. First, the uniform conservation of this carboxy terminal tyrosine residue in an otherwise highly variable region suggests that this residue is important in all γ-tubulins. Second, the carboxy terminus of Tub4p may be involved in some aspect of γ-tubulin function distinct from microtubule nucleation in yeast.

III. Interactions between Tubulins

A. γ/γ Interactions

Although all members of the tubulin superfamily (α-, β-, and γ- tubulins) share sequence homology (Burns, 1991) and the capacity to bind guanidine phosphonucleotides (Oegema *et al.,* 1999), γ-tubulin is functionally distinct from α- and β-tubulins in many respects. In contrast with α- and β-tubulins, γ-tubulin is much less abundant, comprising less than 0.01% of soluble protein (Stearns *et al.,* 1991). Furthermore, γ-tubulin is concentrated at MTOCs (Stearns *et al.,* 1991; Zheng *et al.,* 1991; Liu *et al.,* 1994; Moritz *et al.,* 1995b; Vogel *et al.,* 1997; Silflow *et al.,* 1997) or other sites of microtubule assembly (Gard, 1994). γ-Tubulin does not coassemble with α- and β-tubulins, but is essential for the efficient nucleation of microtubules and has been shown to bind the minus ends of microtubules *in vitro* (Li and Joshi, 1995; Zheng *et al.,* 1995; Moritz *et al.,* 1995b). Whether γ-tubulin forms polymers in a manner similar to α- and β-tubulins is still uncertain. Tubulins and the related bacterial protein FtsZ form rings during the polymerization phase of assembly (reviewed in Erickson and Stoffler, 1996). Tubulins and FtsZ have similar tertiary structures (Nogales *et al.,* 1998; Lowe and Amos, 1998). The closely related γ-tubulin is likely to share this tertiary structure that supports polymerization and ring formation. γ-Tubulin forms rings *in vivo,* but these ring structures contain other proteins that may contribute in forming the ring structure (Zheng *et al.,* 1995). However, overexpression of γ-tubulin in mammalian cells results in the formation of unusually large tubular structures that are labeled by anti-γ-tubulin antibodies (Shu and Joshi, 1995). Whether γ-tubulin forms ring structures under these conditions is not clear; however, ectopic nucleation of microtubules does occur, suggesting that excess γ-tubulin molecules may coalesce into complexes that promote microtubule assembly.

The yeast two-hybrid protein interaction assay has been used to test for intermolecular interactions between γ-tubulin molecules. Tub4p has been reported to self associate in one study (Nguyen *et al.,* 1998); however, Knop *et al.* (1997) report no interaction between Tub4p molecules in two hybrid assays. Thus, based on the results of these assays, it is not clear whether Tub4p self-associates in yeast. Coprecipitation experiments of the γ-tubulin complex of yeast and mammalian cells using differential epitope tags to identify independent molecules suggest that coprecipition of γ-tubulin molecules in the complex is dependent on the integrity of the entire complex (Knop *et al.,* 1997; Murphy *et al.,* 1998). This would suggest that γ-tubulins do not form high-affinity dimers or higher order structures at physiological concentrations. The formation of γ-tubulin rings may therefore be dependent on other components of the γ-tubulin complex.

B. α/β and γ-Tubulin Interactions

α/β and γ-tubulins are present together in TCP–chaperonin complexes of mammalian cells (Melki *et al.,* 1993) and the large γ-tubulin ring complex of *Xenopus* (Zheng *et al.,* 1995). Tub1p and Tub4p are both resident in the GIM complex of yeast, which appears to be important for α-tubulin biosynthesis and has a separate role in Tub4p function. Since *gim* mutations are colethal with mutations in *tub4, spc97,* and *spc98,* the Gim proteins may play a role in assembly of the Tub4p complex (Geissler *et al.,* 1998). Thus, in many organisms α-, β-, and γ-tubulins form multiprotein complexes that may represent the first steps in assembling nucleation complexes.

A central question that remains unresolved concerns how α-, β-, and γ-tubulins interact during microtubule nucleation. Two models for how γ-tubulin initiates microtubule assembly have been proposed. In the first model, a "template nucleation" mechanism seeds protofilament assembly and is constrained by the molecules of γ-tubulin in the ring complex (Zheng *et al.,* 1995). In the second model, an open ring composed of γ-tubulin stabilizes tubulin protofilaments via lateral interactions (Erickson and Stoffler, 1996). Two lines of evidence suggest that mincrotubule minus ends terminate with an exposed α-tubulin subunit (Mitchison, 1993; Nogales *et al.,* 1998). Labeled GTP analogs that bind to the exchangeable GTP catalytic site of β-tubulin are detected at the plus ends of microtubules assembled *in vitro,* suggesting that free microtubule plus ends terminate with β-tubulin (Mitchison, 1993). In addition, the crystal structure of the tubulin heterodimer suggests that microtubule plus ends are likely to terminate with β-tubulin (Nogales *et al.,* 1998). Since γ-tubulin interacts with microtubule minus ends (Li and Joshi, 1995; Zheng *et al.,* 1995; Moritz *et al.,* 1995b), in a "template nucleation" scenario the minus end subunit in closest proximity to γ-tubulin is expected to be α-tubulin. However, a mutation in γ-tubulin was isolated as an extragenic suppressor of a β-tubulin mutation in *Aspergillus,* suggesting that β- and γ-tubulins physically interact together (Oakley and Oakley, 1989). This apparent discrepancy can be resolved if γ/β complexes are present at nucleation sites of MTOCs. In support of this hypothesis, quantification of α-, β-, and γ-tubulins copurifying with centrosomes isolated from activated (M-phase) *Spisula* oocytes revealed the stochiometric ration of $\alpha:\beta$ tubulins is $\sim$1:5 rather than 1:1, indicating excess β-tubulin (Vogel *et al.,* 1997). A 1:1 ratio of α- and β-tubulin is expected for microtubules of centrioles (Kochanski and Borisy, 1990) or short microtubules that may copurify with centrosomes. High-resolution electron microscopy and three-dimensional reconstructions of isolated *Spisula* centrosomes show no microtubules, but reveal a single centriole and numerous 25-nm ring structures (Vogel *et al.,* 1997) similar to γ-tubulin-containing rings associated with centrosomes isolated from *Drosophila* embryos (Moritz *et al.,* 1995a,b). These results suggest that in centrosomes, β-tubulin is present either as a monomer or in a complex with a protein other than α-tubulin. A strong candidate for such an interaction is γ-tubulin.

How α-, β- and γ-tubulins might interact is unknown. However, examination of the variable C-termini of yeast tubulins reveals that this region of

Tub4p most closely resembles the C-termini of yeast α-tubulins Tub1p and Tub3p (γ/α: 32% identity; γ/β: 23% identity). An intriguing possibility is that this region of Tub4p mimics Tub1p and may interact with Tub2p during nucleation. Deletion of the acidic C termini of mammalian tubulins stabilizes microtubules and promotes assembly (Maccioni *et al.,* 1986; Vera and Maccioni, 1989); this region is also involved in vinblastine-mediated tubulin assembly (Rai and Wolff, 1998). We have found that removing the last 30 residues from the Tub4p C terminus confers a conditional Ben^R phenotype and defects in mitosis, suggesting this mutation can similarly stabilize microtubules. Furthermore, a short internal deletion that removes five Tub4p C-terminal residues (DSYLD) that are conserved between Tub1p and Tub4p (see Fig. 4) is sufficient to confer this conditional Ben^R phenotype, results in enhanced cytoplasmic microtubule arrays, and perturbs microtubule function during mitosis (Vogel and Snyder, unpublished results). One possibility is that the C-terminal region of Tub4p has a role similar to that of the acidic C termini of α- andβ-tubulin in controlling microtubule assembly.

IV. The γ-Tubulin Complex in Yeast

MTOCs are complex organelles composed of many proteins. For example, purified *Spisula* centrosomes contain a minimum of ~30 proteins that can be resolved as discrete bands (20–210 kDa) in one-dimensional gels (Vogel *et al.,* 1997). However, ~400 protein isoforms (20–210 kDa) can be resolved in two-dimensional gels, the vast majority of which remain uncharacterized (Vogel and Palazzo, unpublished results). Of these proteins, a minor subset are likely to be directly involved in microtubule nucleation. In order to identify proteins that interact with γ-tubulin and function in microtubule assembly, biochemical methods have been employed to isolate γ-tubulin-containing complexes from *Drosophila* embryos, *Xenopus* oocytes, and mammalian cells (Raff *et al.,* 1993; Zheng *et al.,* 1995; Murphy *et al.,* 1998). These preparations have been instrumental in identifying proteins that function in microtubule nucleation.

The SPB is similarly complex when compared to centrosomes; partially purified SPBs contain ~50 proteins that can be resolved in one-dimensional gels (Wigge *et al.,* 1998). A subset of these proteins (seven) are known to be resident in the SPB proper (see Fig. 3; Adams and Kilmartin, 1999). In contrast to centrosomes, the identities of most of the core residents of the SPB have been determined, and many can be separated into functional categories: structural (e.g., Spc42p), involved in duplication (e.g., Kar1p, Cdc31p), regulatory (e.g., Bub2p), or directly involved in microtubule organization (e.g., Spc72p, Stu2p, Tub4p, Spc98p, Spc97p). Thus, not only have the proteins that interact with Tub4p been identified, their interactions with other components of the SPB are rapidly being defined.

A. The 6S Tub4p Complex of Yeast

The γ-tubulin complex of yeast is a 6S complex composed of Spc97p, Spc98p, and at least two molecules of Tub4p (Geissler *et al.,* 1996) (Fig. 4). Homologs of proteins resident in the γ-tubulin complex of yeast have been identified in other organisms, indicating the conservation of nucleation complexes among eukaryotic cells (Tassin *et al.,* 1997, 1998; Murphy *et al.,* 1998; Martin *et al.,* 1998). Spc98p was identified by two methods: first as a 90-kDa resident of the nuclear and cytoplasmic SPB plaques using monoclonal antibodies generated against purified SPBs (Rout and Kilmartin, 1990), and second as a high-copy suppressor of the *tub4-1* mutation. This latter result raised the possibility that the two proteins interact physically (Geissler *et al.,* 1996). Similarly, *SPC97* was identified as a high-copy suppressor of the *spc98-2* mutation (Knop *et al.,* 1997). Both *SPC97* and *SPC98* are essential genes, and temperature-sensitive mutations exhibit synthetic lethal phenotypes with *spc110* and *spc72* mutations (Nguyen *et al.,* 1998; Knop and Schiebel, 1998) and affect SPB separation (*spc97-14*), SPB duplication, microtubule organization, and spindle elongation (Geissler *et al.,* 1996; Knop *et al.,* 1997).

A series of yeast two-hybrid assays coupled with coprecipitation experiments demonstrated that Tub4p, Spc97p, and SPC98p interact physically (Geissler *et al.,* 1996; Knop *et al.,* 1997). The phenotypes of *spc98-1,*

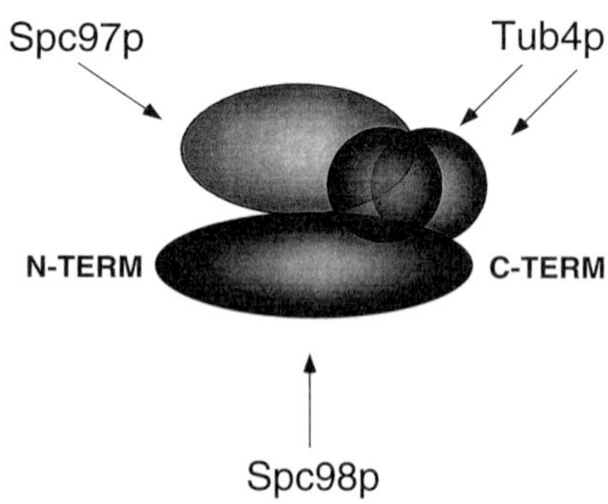

Figure 4 The 6S Tub4p complex. The γ-tubulin complex of yeast is composed of at least two molecules of Tub4p, and Spc98p and Spc97p. Tub4p interacts with the carboxy terminal region of Spc98p. A central domain of Spc98p interacts with Spc97p. Spc98p contains a nuclear localization sequence that directs import of the Tub4p complex into the nucleus.

spc97-14, and *spc97-20* mutants are correlated with defects in their direct (*spc98-1, spc97-14*) or indirect (*spc97-20*) physical interaction with Tub4p.

Interaction with Spc98p requires both N-terminal and C-terminal residues of Tub4p. However, the N-terminal 390 residues of Spc98p are not required for its interaction with Tub4p. Thus, interactions between Tub4p and Spc98p are likely to be mediated through the C terminus of Spc98p. A central domain of Spc98p (residues 147–551) is sufficient for its interaction with Spc97p; however, an overlapping truncation of Spc98p (residues 1–324) does not react in the two-hybrid assay. A cartoon of the predicted interactions between Tub4p, Spc97p, and Spc98p is shown in Fig. 4.

B. How the Tub4p Complex Binds to the Spindle Pole Body

In contrast to *TUB4,* overexpression of either *SPC97* or *SPC98* is toxic. *SPC97* or *SPC98* overexpression-induced lethality can be rescued by concomitant overexpression of *TUB4;* further, *SPC97* overexpression is not lethal when *SPC98* is cooverexpressed (Geissler *et al.,* 1996; Knop *et al.,* 1997). These results suggest the following: first, that Spc97p and Spc98p limit the number of functional nucleation sites, and second, that excess protein in the absence of Tub4p may block the SPB-binding sites and/or form partial, nonfunctional complexes with other proteins needed for initiating microtubule assembly. Together with the observation that Spc98p fails to localize to the SPB in *tub4-1* cells, these phenotypes suggest that Spc97 and Spc98 are likely to both be involved in mediating the attachment of Tub4p to the inner and outer plaques of the SPB.

The Tub4p complex is most likely assembled in the cytoplasm (Pereira *et al.,* 1998; Geissler *et al.,* 1998) and thus must be imported into the nucleus before it can attach to the inner plaque of the SPB. Spc98p contains a bifurcated nuclear localization sequence (NLS; Pereira *et al.,* 1998). Deletion or mutation of the NLS of Spc98p inhibits import of both Tub4p and Spc97p into the nucleus, indicating that the NLS of Spc98p is sufficient to promote the import of the Tub4p complex into the nucleus (Pereira *et al.,* 1998). In addition, the Mps1p kinase (Winey *et al.,* 1991; Lauze *et al.,* 1995) is likely to phosphorylate Spc98p; Spc98p is hypophosphorylated in temperature-sensitive *mps1-1* cells incubated at the restrictive temperature, and Spc98p is phosphorylated by Mps1p *in vitro* (Pereira *et al.,* 1998). The phosphorylation of Spc98p may be required to maintain the complex at the nuclear plaque, though how this modification contributes to the compartmentalization of the complex is unclear.

The Tub4p complex binds to the nuclear face of the SPB by its association with Nuf1p/Spc110p (Knop and Schiebel, 1997; Nguyen *et al.,* 1998). A complete discussion of the functions of Nuf1p/Spc110p is presented in a separate chapter in this volume, and so these functions are not discussed further here. However, Nuf1p/Spc110p is found only at the inner nuclear plaque of the SPB (Rout and Kilmartin, 1990; Kilmartin *et al.,* 1993), indicating that a different protein must mediate Tub4p complex binding at the cytoplasmic plaque. Using a two-hybrid screen for proteins that interact with the Tub4p complex, Knop and Schiebel (1998) identified a SPB protein, Spc72p, that is specifically associated with the periphery of the cytoplasmic plaque. The *SPC72* gene is essential (Chen *et al.,* 1998; Knop and Schiebel, 1998). *spc72* mutants have few cytoplasmic microtubules at the restrictive temperature, nuclear migration, and spindle positioning defects consistent with the loss of cytoplasmic microtubule function (Knop and Schiebel, 1998; Soues and Adams, 1998). In addition, overexpression of *SPC72* increases the number of cytoplasmic microtubules (Knop and Schiebel, 1998). These results indicate that Spc72p is required for the formation of cytoplasmic, but not spindle, microtubules. Finally, the Tub4p complex binds to GST:Spc72p *in vitro* and interacts with Spc72p in two-hybrid assays. Thus, Spc72p appears to be essential for localizing the Tub4p complex at the cytoplasmic plaque and is important for organization of cytoplasmic microtubules. It has been reported that *spc72* mutants do not elongate the spindle but complete cytokinesis (Chen *et al.,* 1998). This latter observation suggests that cytoplasmic microtubules may be important for some aspect of microtubule function during mitosis separate from anaphase B, which is not dependent on astral microtubules in yeast (Sullivan and Huffaker, 1992).

The *STU2* gene product has been shown to interact with Spc72p by coprecipitation and two-hybrid assays (Chen *et al.,* 1998). Interestingly, Stu2p is an essential 100-kDa coiled-coil protein that also binds the lateral walls of microtubules at tandem repeats within its sequence that act cooperatively to bind with an affinity of $\sim 5 \times 10^{-7}$M (Wang and Huffaker, 1997). The *stu2-1* allele is colethal with *tub2-404,* a mutation that increases the stability of microtubules (Wang and Huffaker, 1997). Thus, the colethality observed between *stu2-1* and *tub2-404* suggests that Stu2p can influence microtubule stability at the SPB. *STU2* is predominantly expressed in S-phase (Cho *et al.,* 1998; Spellman *et al.,* 1998); therefore, Stu2p is likely to stabilize existing microtubules but not function in their nucleation. Stu2p has homologs in *Caenorhabditis elegans* and mammals. ZYG-9, a C. elegans Stu2p homolog, promotes the formation of long microtubules, and mutants have phenotypes consistent with defects in nuclear positioning and migration (Matthews *et al.,* 1998). Finally, MPP-1, a mammalian Stu2p homolog, is phosphorylated in M-phase (Westendorf *et al.,* 1994). Although the microtubule-binding domain of Stu2p does not contain the consensus phosphopeptides of MPP-1, it is possible that the activity of Stu2p could similarly be cell cycle regulated. An attractive hypothesis is that Stu2p plays a crucial role in anchoring the minus ends of the cytoplasmic microtubules to Spc72p, perhaps during nuclear migration and spindle positioning. Alteration of Stu2p function could lead to release of the cytoplasmic microtubules and promote their rearrangement with the onset of spindle elongation. A schematic of the predicted cytoplasmic microtubule binding site, including the anchoring protein Stu2p, is shown in Fig. 5. Note that this complex is peripheral to core SPB proteins such as Cnm67p.

V. Future Directions

The intermolecular interactions within the γ-tubulin complex and between the complex and the SPB are extremely well characterized in yeast. This system will be extremely useful in elucidating the mechanisms that regulate Tub4p function during microtubule nucleation. To study the function of the Tub4p-complex under defined conditions, it will be necessary to develop an *in vitro* nucleation assay for the Tub4p complex. Such assays have been used with great success in higher eukaryotes (Zheng *et al.,* 1995, 1998;

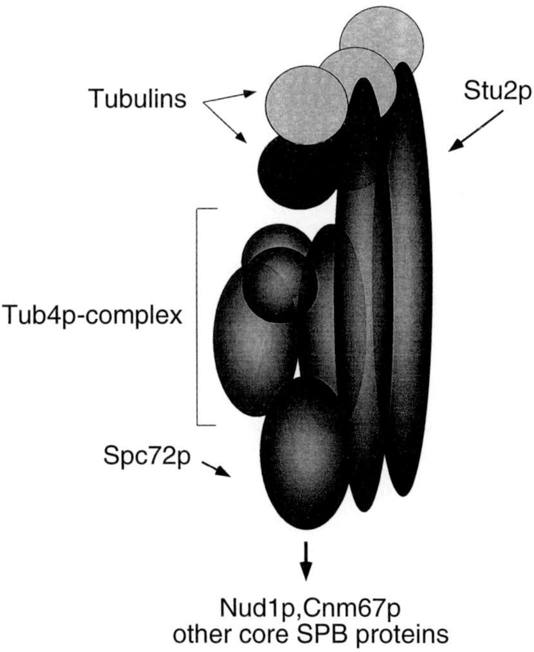

Figure 5 Composition of the putative cytoplasmic nucleation complex. The Tub4p complex binds to the cytoplasmic face of the SPB via its interaction with Spc72p. Stu2p anchors the interaction between the nucleation complex and the minus ends of microtubules by binding both Spc72p and the lateral walls of microtubules.

Oegema *et al.,* 1999). One of the barriers to developing such an assay in yeast is the divergence between Tub4p and γ-tubulins from organisms generally used to isolate purified tubulin, such as mammalin brain. Mammalian tubulins cannot assemble from isolated SPBs in the absence of preexisting yeast microtubules (Rout and Kilmartin, 1990), and heterologous γ-tubulin cannot complement *tub4Δ* (Spang *et al.,* 1996; Marschall *et al.,* 1996). Although methods for purifying yeast tubulins have been developed (Kilmartin, 1981; Barnes *et al.,* 1992), they are less practical than similar methods used for tubulin purification from brain (reviewed in Williams and Lee, 1982) or marine oocytes (Suprenant and Rebhun, 1984). Thus, it will be necessary to isolate Tub4p-complexes with attached tubulin oligomers, which can serve to seed the assembly of puified tubulins derived from other cell types.

In addition to initiating microtubule assembly, the yeast γ-tubulin complex may also be involved in regulating minus end dynamics in response to cell cycle progression. Phenotypes of tubulin mutants such as *tub2-150, tub2-404,* which increase the stability of microtubules, indicate that dynamic microtubules are required for spindle elongation and the completion of mitosis (Huffaker *et al.,* 1988; Machin *et al.,* 1995). Furthermore, deletion of the minus end kinesinlike motor protein Kar3p (Meluh and Rose, 1990) or the cytoplasmic dynein motor protein Dyn1p (Eshel *et al.,* 1993) results in defects in nuclear migration and spindle elongation. Kar3p has been shown to depolymerize microtubules *in vitro;* thus, Kar3p may also function to depolymerize microtubules at their minus ends (Endow *et al.,* 1994). In addition, *kar3Δ* and *dhc1/dyn1Δ* cells have excess cytosplamic microtubules (Saunders *et al.,* 1997) and defects in spindle function or nuclear migration (Carminati and Stearns, 1997). Together, these studies suggest that minus end dynamics are important for microtubule-dependent processes in yeast. Minus end release has been shown to occur in neurons and in cultured epithelial cells (Baas and Yu, 1996; Keating *et al.,* 1997; Rodinov and Borisy, 1997). Given its importance in microtubule assembly and its position at the minus ends of microtubules, γ-tubulin and its associated proteins may play an important role in controlling these dynamics. Our isolation of *tub4* mutants that stabilize microtubules is consistent with such a role. Given the conservation of γ-tubulin complexes among different cell types, it is likely that any postnucleation functions of the Tub4p complex in yeast may also be conserved in other eukaryotic cells. Thus, study of budding yeast presents a unique opportunity to uncover novel aspects of γ-tubulin function.

Acknowledgments

The authors thank Paulo Coelho and Brendan Manning (Yale) and Trish Davis (University of Washington) for their comments on the manuscript, P. Coelho for unpublished results, and members of the Snyder laboratory for many helpful discussions. This work has been supported by a NIH Fellowship GM17941-03 (J. Vogel) and NIH grant GM52197-04 (M. Snyder).

References

Adams, I. R., and Kilmartin, J. V. (1999). Localization of core spindle pole body (SPB) components during SPB duplication in *Saccharomyces cerevisiae. J. Cell Biol.* **145,** 809–823.

Adams, A. E., and Pringle, J. R. (1984). Relationship of actin and tubulin distribution to bud growth in wild-type and morphogenetic-mutant *Saccharomyces cerevisiae. J. Cell Biol.* **98,** 934–945.

Akashi, T., Yoon, Y., and Oakley, B. R. (1997). Characterization of γ-tubulin complexes in *Aspergillus nidulans* and detection of putative γ-tubulin interacting proteins. *Cell Motil. Cytoskeleton* **37,** 149–158.

Alexandru, G., Zachariae, W., Schleiffer, A., and Nasmyth, K. (1999). Sister chromatid separation and chromosome re-duplication are regulated by different mechanisms in response to spindle damage. *EMBO J.* **18,** 2707–2721.

Archer, J. E., Vega, L. R., and Solomon, F. (1995). Rbl2p, a yeast protein that binds to β-tubulin and participates in microtubule function in vivo. *Cell* **82,** 425–434.

Archer, J. E., Magendantz, M., Vega, I. R., and Solomon, F. (1998). Formation and function of the Rbl2p–β-tubulin complex. *Mol. Cell Biol.* **18,** 1757–1762.

Bass, P. W., and Yu, W. (1996). A composite model for establishing the microtubule arrays of the neuron. *Mol. Neurobiol.* **12,** 145–161.

Barnes, G., Louie, K. A., and Botstein, D. (1992). Yeast proteins associated with microtubules in vitro and in vivo. *Mol. Biol. Cell* **3,** 29–47.

Baum, P., Furlong, C., and Byers, B. (1986). Yeast gene required for spindle pole body duplication: homology of its product with Ca^{2+}-binding proteins. *Proc. Natl. Acad. Sci. USA* **83,** 5512–5516.

Bayley, P., Sharma, K., and Martin, S. (1994). Microtubule dynamics in vitro. In "Microtubules" (Hyams, J. S., and Lloyd, C. W., Eds.), pp. 111–137. Wiley-Liss, New York.

Brachat, A., Kilmartin, J. V., Wach, A., and Philippsen, P. (1998). *Saccharomyces cerevisiae* cells with defective spindle pole body outer plaques accomplish nuclear migration via half-bridge-organized microtubules. *Mol. Biol. Cell* **9,** 977–991.

Bullitt, E., Rout, M. P., Kilmartin, J. V., and Akey, C. W. (1997). The yeast spindle pole body is assembled around a central crystal of Spc42p. *Cell* **89,** 1077–1086.

Burke, D., Gasdaska, P., and Hartwell, L. (1989). Dominant effects of tubulin overexpression in *Saccharomyces cerevisiae. Mol. Cell Biol.* **9,** 1049–1059.

Burns, R. G. (1991). α-, β-, and γ-tubulins: sequence comparisons and structural constraints. *Cell Motil. Cytoskel.* **20,** 181–189.

Burnside, B., Kozak, C., and Kafatos, F. C. (1973). Tubulin determination by an isotope dilution–vinblastine precipitation method. The tubulin content of *Spisula* eggs and embryos. *J. Cell Biol.* **59,** 755–762.

Byers, B. (1981). Cytology of the yeast cell cycle. In "The Molecular Biology of the Yeast *Saccharomyces:* Life Cycle and Inheritance" (J. N. Strathern, E. W. Jones, and J. R. Broach, Eds.), pp. 59–56. Cold Spring Harbor Press, Cold Spring Harbor, NY.

Byers, B., and Goetsch, L. (1975). Behavior of spindles and spindle plaques in the cell cycle and conjugation of *Saccharomyces cerevisiae. J. Bacteriol.* **124,** 511–523.

Byers, B., Shriver, K., and Goetsch, L. (1978). The role of spindle pole bodies and modified microtubule ends in the initiation of microtubule assembly in *Saccharomyces cerevisiae*. *J. Cell Sci.* **30,** 331–352.

Carminati, J., and Stearns, T. (1997). Microtubules orient the mitotic spindle in yeast through dynein-dependent interactions with the cell cortex. *J. Cell Biol.* **138,** 629–641.

Chen, X. Y. P., Yin, H. W., and Huffaker, T. C. (1998). The yeast spindle pole body component Spc72p interacts with Stu2p and is required for proper microtubule assembly. *J. Cell Biol.* **141,** 1169–1179.

Cho, R. J., Campbell, M. J., Winzeler, E. A., Steinmetz, L., Conway, A., Wodicka, L., Wolfsberg, T. G., Gabrielian, A. E., Landsman, D., Lockhart, D. J., and Davis, R. W. (1998). A genome-wide transcriptional analysis of the mitotic cell cycle. *Mol. Cell* **2,** 65–73.

de Pereda, J., and Andreu, J. (1996). Mapping surface sequences of the tubulin dimer and taxol-induced microtubules with limited proteolysis. *Biochemistry* **35,** 14184–14202.

Donaldson, A. D., and Kilmartin, J. V. (1996). Spc42p: a phosphorylated component of the *S. cerevisiae* spindle pole body (SPD) with an essential function during SPB duplication. *J. Cell Biol.* **132,** 887–901.

Elliott, S., Knop, M., Schlenstedt, G., and Schiebel, E. (1999). Spc29p is a component of the Spc110p subcomplex and is essential for spindle pole body duplication. *Proc. Natl. Acad. Sci. USA* **96,** 6205–6210.

Endow, S. A., Kang, S. J., Satterwhite, L. L., Rose, M. D., Skeen, V. P., and Salmon, E. D. (1994). Yeast Kar3 is a minus-end microtubule motor protein that destabilizes microtubules preferentially at the minus ends. *EMBO J.* **13,** 2708–2713.

Erickson, H. P., and Stoffler, D. (1996). Protofilaments and rings, two conformations of the tubulin family conserved from bacterial FtsZ to alpha/beta and gamma tubulin. *J. Cell Biol.* **135,** 5–8.

Eshel, D., Urrestarazu, L. A., Vissers, S., Jauniaux, J. C., van Vliet-Reedijk, J. C., Planta, R. J., and Gibbons, I. R. (1993). Cytoplasmic dynein is required for normal nuclear segregation in yeast. *Proc. Natl. Acad. Sci. USA* **90,** 11172–11176.

Fraschini, R., Formenti, E., Lucchini, G., and Piatti, S. (1999). Budding yeast Bub2 is localized at spindle pole bodies and activates the mitotic checkpoint via a different pathway from Mad2. *J. Cell Biol.* **145,** 979–991.

Gard, D. L. (1994). γ-Tubulin is asymmetrically distributed in the cortex of *Xenopus* oocytes. *Devl. Biol.* **161,** 131–140.

Geiser, J. R., Sundberg, H. A., Chang, B. H., Muller, E. G., and Davis, T. N. (1993). The essential mitotic target of calmodulin is the 110-kilodalton component of the spindle pole body in *Saccharomyces cerevisiae*. *Mol. Cell Biol.* **13,** 7913–7924.

Geissler, S., Pereira, G., Spang, A., Knop, M., Soues, S., Kilmartin, J., and Schiebel, E. (1996). The spindle pole body component Spc98p interacts with the γ-tubulin-like Tub4p of *Saccharomyces cerevisiae* at the sites of microtubule attachment. *EMBO J.* **15,** 3899–3911.

Geissler, S., Siegers, K., and Schiebel, E. (1998). A novel protein complex promoting formation of functional α- and γ-tubulin. *EMBO J.* **17,** 952–966.

Horio, T., and Oakley, B. R. (1994). Human γ-tubulin functions in fission yeast. *J. Cell Biol.* **126,** 1465–1473.

Horio, T., Uzawa, S., Jung, M. K., Oakley, B. R., Tanaka, K., and Yanagida, M. (1991). The fission yeast γ-tubulin is essential for mitosis and is localized at microtubule organizing centers. *J. Cell Sci.* **99,** 693–700.

Hoyt, M. A., Toti, L., and Roberts, BT. (1991). *S. cerevisiae* genes required for cell cycle arrest in response to loss of microtubule function. *Cell* **66,** 507–517.

Huffaker, T. C., Thomas, J. H., and Botstein, D. (1988). Diverse effects of β-tubulin mutations on microtubule formation and function. *J. Cell Biol.* **106,** 1997–2010.

Huitorel, P., and Kirschner, M. W. (1988). The polarity and stability of microtubule capture by the kinetochore. *J. Cell Biol.* **106,** 151–159.

Hyams, J. S., Borisy, G. G. (1978). Nucleation of microtubules in vitro by isolated spindle pole bodies of the yeast *Saccharomyces cerevisiae. J. Cell Biol.* **78,** 401–414.

Jacobs, C. W., Adams, A. E., Szaniszlo, P. J., and Pringle, J. R. (1988). Functions of microtubules in the *Saccharomyces cerevisiae* cell cycle. *J. Cell Biol.* **107,** 1409–1426.

Johnston, M., and Davis, R. W. (1984). Sequences that regulate the divergent GAL1-GAL10 promoter in *Saccharomyces cerevisiae. Mol. Cell Biol.* **4,** 1440–1448.

Joshi, H. C., Palacios, M. J., McNamara, L., and Cleveland, D. W. (1992). γ-Tubulin is a centrosomal protein required for cell cycle-dependent microtubule nucleation. *Nature* **356,** 80–83.

Katz, W. S., and Solomon, F. (1988). Diversity among β-tubulins: a carboxy-terminal domain of yeast β-tubulin is not essential in vivo. *Mol. Cell Biol.* **8,** 2730–2736.

Katz, W., Weinstein, B., and Solomon, F. (1990). Regulation of tubulin levels and microtubule assembly in *Saccharomyces cerevisiae:* consequences of altered tubulin gene copy number. *Mol. Cell Biol.* **10,** 5286–5294.

Kaverina, I., Rottner, K., and Small, J. V. (1998). Targeting, capture, and stabilization of microtubules at early focal adhesions. *J. Cell Biol.* **142,** 181–190.

Keating, T. J., Peloquin, J. G., Rodionov, V. I., Momcilovic, D., and Borisy, G. G. (1997). Microtubule release from the centrosome. *Proc. Natl. Acad. Sci. USA* **94,** 5078–5083.

Kilmartin, J. V. (1981). Purification of yeast tubulin by self-assembly *in vitro. Biochemistry* **20,** 3629–3633.

Kilmartin, J. V., and Adams, A. E. (1984). Structural rearrangements of tubulin and actin during the cell cycle of the yeast *Saccharomyces. J. Cell Biol.* **98,** 922–933.

Kilmartin *et al.* (1981) p. 121, 122

Kilmartin, J. V., Dyos, S. L., Kershaw, D., and Finch, J. T. (1993). A spacer protein in the *Saccharomyces cerevisiae* spindle poly body whose transcript is cell cycle-regulated. *J. Cell Biol.* **123,** 1175–1184.

Kochanski, R. S., and Borisy, G. G. (1990). Mode of centriole duplication and distribution. *J. Cell Biol.* **110,** 1599–1605.

Korineck, W. S., Copeland, M. S., Chaudhuri, A., and Chant, J. (2000). Molecular linkage underlying microtubule orientation toward cortical sites in yeast. *Science,* **287,** 2257–2259.

Knop, M., and Schiebel, E. (1997). Spc98p and Spc97p of the yeast γ-tubulin complex mediate binding to the spindle pole body via their interaction with Spc110p. *EMBO J.* **16,** 6985–6995.

Knop, M., Pereira, G., Geissler, S., Grein, K., and Schiebel, E. (1997). The spindle pole body component Spc97p interacts with the γ-tubulin of *Saccharomyces cerevisiae* and functions in microtubule organization and spindle pole body duplication. *EMBO J.* **16,** 1550–1564.

Knop, M., and Schiebel, E. (1998). Receptors determine the cellular localization of a γ-tubulin complex and thereby the site of microtubule formation. *EMBO J.* **17,** 3952–3967.

Lauze, E., Stoelcker, B., Luca, F. C., Weiss, E., Schutz, A. R., and Winey, M. (1995). Yeast spindle pole body duplication gene MPS1 encodes an essential dual specificity protein kinase. *EMBO J.* **14,** 1655–1663.

Lee, L., Tirnauer, J. S., Li, J., Schuyler, S. C., Liu, J. Y., and Pellman, D. (2000). Positioning of the mitotic spindle by cortical–microtubule capture mechanism. *Science,* **287,** 2260–2262.

Li, R. (1999). Bifurcation of the mitotic checkpoint pathway in budding yeast. *Proc. Natl. Acad. Sci. USA* **96,** 4989–4994.

Li, Q., and Joshi, H. (1995). γ-Tubulin is a minus end-specific microtubule binding protein. *J. cell Biol.* **131,** 207–214.

Li, R., and Murray, A. W. (1991). Feedback control of mitosis in budding yeast. *Cell* **66,** 519–531.

Liu, B., Joshi, H. C., Wilson, T. J., Silflow, C. D., Palevitz, B. A., and Snustad, D. P. (1994). γ-Tubulin in *Arabidopsis:* Gene sequence, immunoblot, and immunofluorescence studies. *Plant Cell* **6,** 303–314.

Lowe, J., and Amos, L. A. (1998). Crystal structure of the bacterial cell-division protein FtsZ. *Nature* **391,** 203–206.

Maccioni, R., Serrano, L., Avila, J., and Cann, J. (1986). Characterization and structural aspects of the enhanced assembly of tubulin after removal of its carboxyl-terminal domain. *Eur. J. Biochem.* **156,** 375–381.

Machin, N. A., Lee, J. M., and Barnes, G. (1995). Microtubule stability in budding yeast: characterization and dosage suppression of a benomyl-dependent tubulin mutant. *Mol. Biol. Cell* **6,** 1241–1259.

Maddox, P., Chin, E., Mallavarapu, A., Yeh, E., Salmon, E. D., and Bloom, K. (1999). Microtubule dynamics from mating through the first zygotic division in the budding yeast *Saccharomyces cerevisiae. J. Cell Biol.* **144,** 977–987.

Marschall, L., Jeng, R., Mulholland, J., and Stearns, T. (1996). Analysis of Tub4p, a yeast γ-tubulin-like protein: implications for microtubule-organizing center function. *J. Cell Biol.* **134,** 443–454.

Martin, O. C., Gunawardane, R. N., Iwamatsu, A., and Zheng, Y. (1998). Xgrip109: a γ tubulin-associated protein with an essential role in γ tubulin ring complex (γTuRC) assembly and centrosome function. *J. Cell Biol.* **141,** 675–687.

Matsuzaki, F., Matsumoto, S., and Yahara, I. (1988). Truncation of the carboxy-terminal domain of yeast β-tubulin causes temperature-sensitive growth and hypersensitivity to antimitotic drugs. *J. Cell Biol.* **107,** 1427–1435.

Matthews, L. R., Carter, P., Thierry-Mieg, D., and Kemphues, K. (1998). ZYG-9, a *Caenorhabditis elegans* protein required for microtubule organization and function, is a component of meiotic and mitotic spindle poles. *J. Cell Biol.* **141,** 1159–1168.

Melki, R., Vainberg, I. E., Chow, R. L., and Cowan, N. J. (1993). Chaperonin-mediated folding of vertebrate actin-related protein and γ-tubulin. *J. Cell Biol.* **122,** 1301–1310.

Meluh, P. B., and Rose, M. D. (1990). *KAR3,* a kinesin-related gene required for yeast nuclear fusion. *Cell* **60,** 1029–1041.

Miller, R., and Rose, M. (1998). Kar9p is a novel cortical protein required for cytoplasmic microtubule orientation in yeast. *J. Cell Biol.* **140,** 377–390.

Miller, R. K., Matheos, D., and Rose, M. D. (1999). The cortical localization of the microtubule orientation protein, kar9p, is dependent upon actin and proteins required for polarization. *J. Cell Biol.* **144,** 963–975.

Mirzayan, C., Copeland, C. S., and Snyder, M. (1992). The NUF1 gene encodes an essential coiled-coil related protein that is a potential component of the yeast nucleoskeleton. *J. Cell Biol.* **116,** 1319–1332.

Mitchison, T. (1993). Localization of an exchangeable GTP binding site at the plus end of microtubules. *Science* **261,** 1044–1047.

Mitchison, T., and Kirschner, M. (1984a). Dynamic instability of microtubule growth. *Nature* **312,** 237–242.

Mitchison, T. J., and Kirschner, M. (1984b). Microtubule assembly nucleated by isolated centrosomes. *Nature* **213,** 232–237.

Moritz, M., Braunfeld, M., Fung, J., Sedat, J., Alberts, B., and Agard, D. (1995a). Three-dimensional structural characterization of centrosomes from early *Drosophila* embryos. *J. Cell Biol.* **130,** 1149–1159.

Moritz, M., Braunfeld, M. B., Sedat, J. W., Alberts, B., Agard, D. A., Moritz, M., Zheng, Y., and Alberts, B. M. (1995b). Microtubule nucleation by γ-tubulin-containing rings in the centrosome. *Nature* **378,** 638–640.

Moritz, M., Zheng, Y., Alberts, B. M., and Oegema, K. (1998). Recruitment of the γ-tubulin ring complex to *Drosophila* salt-stripped centrosome scaffolds. *J. Cell Biol.* **142,** 775–786.

Moudjou, M., Bordes, N., Paintrand, M., and Bornens, M. (1996). γ-Tubulin in mammalian cells: the centrosomal and the cytosolic forms. *J. Cell Sci.* **109,** 875–887.

Murphy, S. M., Urbani, L., and Stearns, T. (1998). The mammalian γ-tubulin complex contains homologues of the yeast spindle pole body components spc97p and spc98p. *J. Cell Biol.* **141,** 663–674.

Neff, N. F., Thomas, J. H., Grisafi, P., and Botstein, D. (1983). Isolation of the β-tubulin gene from yeast and demonstration of its essential function in vivo. *Cell* **33,** 211–219.

Nguyen, T. V. D., Crawford, D. K., and Davis, T. N. (1998). A genetic analysis of interactions with Spc110p reveals distinct functions of Spc97p and Spc98p, components of the yeast γ-tubulin complex. *Mol. Biol. Cell* **9,** 2201–2216.

Nogales, E., Wolf, S., and Downing, K. (1998). Structure of the α β tubulin dimer by electron crystallography. *Nature* **391,** 199–203.

Oakley, C. E., and Oakley, B. R. (1989). Identification of γ-tubulin, a new member of the tubulin superfamily encoded by mipA gene of *Aspergillus nidulans. Nature* **338,** 662–664.

Oakley, B., Oakley, E., Yoon, Y., and Jung, M. K. (1990). γ-Tubulin is a component of the spindle pole body that is essential for microtubule function in *Aspergillus nidulans. Cell* **61,** 1289–1301.

Oegema, K., Wiese, C., Martin, O. C., Milligan, R. A., Iwamatsu, A., Mitchison, T. J., and Zheng, Y. (1999). Characterization of two related *Drosophila* γ-tubulin complexes that differ in their ability to nucleate microtubules. *J. Cell Biol.* **144,** 721–733.

O'Toole, E. T., Winey, M., and McIntosh, J. R. (1999). High-voltage electron tomography of spindle pole bodies and early mitotic spindles in the yeast *Saccharomyces cerevisiae. Mol. Biol. Cell.* **10,** 2017–2031.

Page, B. D., and Snyder, M. (1993). Chromosome segregation in yeast. *Annu. Rev. Microbiol.* **4,** 231–261.

Palmer, R. E., Sullivan, D. S., Huffaker, T., and Koshland, D. (1992). Role of astral microtubules and actin in spindle orientation and migration in the budding yeast, *Saccharomyces cerevisiae. J. Cell Biol.* **119,** 583–593.

Pereira, G., Knop, M., and Schiebel, E. (1998). Spc98p directs the yeast γ-tubulin complex into the nucleus and is subject to cell cycle-dependent phosphorylation on the nuclear side of the spindle pole body. *Mol. Biol. Cell* **9,** 775–793.

Pereira, G., Grueneberg, U., Knop, M., and Schiebel, E. (1999). Interaction of the yeast γ-tubulin complex-binding protein Spc72p with Kar1p is essential for microtubule function during karyogamy. *EMBO J.* **18,** 4180–4195.

Peterson, J. B., and Ris, H. (1976). Electron-microscopic study of the spindle and chromosome movement in the yeast *Saccharomyces cerevisiae. J. Cell Sci.* **22,** 219–242.

Raff, J. W., Kellogg, D. R., and Alberts, B. M. (1993). *Drosophila* γ-tubulin is part of a complex containing two previously identified centrosomal MAPs. *J. Cell Biol.* **121,** 823–835.

Rai, S., and Wolff, J. (1998). The C terminus of β-tubulin regulates vinblastine-induced tubulin polymerization. *Proc. Natl. Acad. Sci. USA* **95,** 4253–4257.

Rodinov, V. I., and Borisy, G. G. (1997). Microtubule treadmilling in vivo. *Science* **275,** 215–218.

Rose, M. D., and Fink, G. R. (1987) KAR1, a gene required for function of both intranuclear and extranuclear microtubules in yeast. *Cell* **48,** 1047–1060.

Rout, M. P., and Kilmartin, J. V. (1990). Components of the yeast spindle and spindle pole body. *J. Cell Biol.* **111,** 1913–1927.

Sage, C. R., Dougherty, C. A., Davis, A. S., Burns, R. G., Wilson, L., and Farrell, K. W. (1995). Site-directed mutagenesis of putative GTP-binding sites of yeast β-tubulin: evidence that α-, β-, and γ-tubulins are atypical GTPases. *Biochemistry* **34,** 7409–7419.

Saunders, W., Hornack, D., Lengyel, V., and Deng, C. (1997). The *Saccharomyces cerevisiae* kinesin-related motor Kar3p acts at preanaphase spindle poles to limit the number and length of cytoplasmic microtubules. *J. Cell Biol.* **137,** 417–431.

Schatz, P. J., Pillus, L., Grisafi, P., Solomon, F., and Botstein, D. (1986a). Two functional α-tubulin genes of the yeast *Saccharomyces cerevisiae* encode divergent proteins. *Mol. Cell Biol.* **6,** 3711–3721.

Schatz, P. J., Solomon, F., and Botstein, D. (1986b). Genetically essential and nonessential α-tubulin genes specify functionally interchangeable proteins. *Mol. Cell Biol.* **6,** 3722–3733.

Schatz, P. J., Georges, G. E., Solomon, F., and Botstein, D. (1987). Insertions of up to 17 amino acids into a region of α-tubulin do not disrupt function in vivo. *Mol. Cell Biol.* **7,** 3799–3805.

Schatz, P. J., Solomon, F., and Botstein, D. (1988). Isolation and characterization of conditional-lethal mutations in the TUB1 a-tubulin gene of the yeast *Saccharomyces cerevisiae.* Genetics **120,** 681–695.

Serrano, L., Avila, J., and Maccioni, R. (1984). Controlled proteolysis of tubulin by subtilisin: localization of the site for MAP2 interaction. *Biochemistry* **23,** 4675–4681.

Shaw, S., Yeh, E., Maddox, P., Salmon, E., and Bloom, K. (1997). Astral microtubule dynamics in yeast: a microtubule-based searching mechanism for spindle orientation and nuclear migration into the bud. *J. Cell Biol.* **139,** 985–994.

Shu, H. B., and Joshi, H. C. (1995). γ-Tubulin can both nucleate microtubule assembly and self-assemble into novel tubular structures in mammalian cells. *J. Cell Biol.* **130,** 1137–1147.

Silflow, C. D., Liu, B., LaVoie, M., Richardson, E. A., and Palevitz, B. A. (1999). γ-Tubulin in *Chlamydomonas:* characterization of the gene and localization of the gene product in cells. *Cell Motil. Cytoskel.* **42,** 285–297.

Sobel, S. G., and Snyder, M. (1995). A highly divergent γ-tubulin gene is essential for cell growth and proper microtubule organization in *Saccharomyces cerevisiae. J. Cell Biol.* **131,** 1775–1788.

Snyder, M., Gehrung, S., and Page, B. D. (1991). Studies concerning the temporal and genetic control of cell polarity in *Saccharomyces cerevisiae. J. Cell Biol.* **114,** 515–532.

Soues, S., and Adams, I. R. (1998). SPC72: a spindle pole component required for spindle orientation in the yeast *Saccharomyces cerevisiae. J. Cell Sci.* **111,** 2809–2818.

Spang, A., Courtney, I., Fackler, U., Matzner, M., and Schiebel, E. (1993). The calcium-binding protein cell division cycle 31 of *Saccharomyces cerevisiae* is a component of the half bridge of the spindle pole body. *J. Cell Biol.* **123,** 405–416.

Spang, A., Courtney, I., Grein, K., Matzner, M., and Schiebel, E. (1995). The Cdc31p-binding protein Kar1p is a component of the half bridge of the yeast spindle pole body. *J. Cell Biol.* **128,** 863–877.

Spang, A., Grein, K., Schiebel, E. (1996). γ-Tubulin-like Tub4p of *Saccharomyces cerevisiae* is associated with the spindle pole body substructures that organize microtubules and is required for mitotic spindle formation. *J. Cell Biol.* **134,** 429–441.

Spellman, P. T., Sherlock, G., Zhang, M. Q., Iyer, V. R., Anders, K., Eisen, M. B., Brown, P. O., Botstein, D., and Futcher, B. (1998). Comprehensive identification of cell cycle-regulated genes of the yeast *Saccharomyces cerevisiae* by microarray hybridization. *Mol. Biol. Cell* **9,** 3273–3297.

Stearns, T., and Botstein, D. (1988). Unlinked noncomplementation: isolation of new conditional-lethal mutations in each of the tubulin genes of *Saccharomyces cerevisiae. Genetics* **119,** 249–260.

Stearns, T., Hoyt, M. A., and Botstein, D. (1990). Yeast mutants sensitive to antimicrotubule drugs define three genes that affect microtubule function. *Genetics* **124,** 251–262.

Stearns, T., Evans, L., and Kirschner, M. (1991). γ-Tubulin is a highly conserved component of the centrosome. *Cell* **65,** 825–836.

Sullivan, D. S., and Huffaker, T. C. (1992). Astral microtubules are not required for anaphase B in *Saccharomyces cerevisiae. J. Cell Biol.* **119,** 379–388.

Sunkel, C. E., Gomes, R., Sampaio, P., Perdigao, J., and Gonzalez C. (1995). γ-Tubulin is required for the structure and function of the microtubule organizing centre in *Drosophila* neuroblasts. *EMBO J.* **14,** 28–36.

Suprenant, K. A., Rebhun, L. J. (1984). Purification and characterization of oocyte cytoplasmic tubulin and meiotic spindle tubulin of the surf clam *Spisula solidissima. J. Cell Biol.* **98,** 253–266.

Tassin, A. M., Celati, C., Paintrand, M., Bornens, M. (1997). Identification of an Spc110p-related protein in vertebrates. *J. Cell Sci.* **110,** 2533–2545.

Tassin, A. M., Celati, C., Moudjou, M., and Bornens, M. (1998). Characterization of the human homologue of the yeast spc98p and its association with γ-tubulin. *J. Cell Biol.* **141,** 689–701.

Tavormina, P. A., and Burke, D. J. (1998). Cell cycle arrest in cdc20 mutants of *Saccharomyces cerevisiae* is independent of Ndc10p and kinetochore function but requires a subset of spindle checkpoint genes. *Genetics* **148,** 1701–1713.

Thomas, J. H., Neff, N. F., and Botstein, D. (1985). Isolation and characterization of mutations in the β-tubulin gene of *Saccharomyces cerevisiae. Genetics* **111,** 715–734.

Vera, J. C., and Maccioni, R. B. (1989). Biochemical dissection of the role of the one-kilodalton carboxyl-terminal moiety of tubulin in its assembly into microtubules. *Biochemistry* **28,** 333–339.

Vogel, J., Stearns, T., Rieder, C., and Palazzo, R. (1997). Centrosomes isolated from *Spisula solidissima* oocytes contain rings and an unusual stoichiometric ratio of α/β tubulin. *J. Cell Biol.* **137,** 193–202.

Wang, P. J., and Huffaker, T. C. (1997). Stu2p: A microtubule-binding protein that is an essential component of the yeast spindle pole body. *J. Cell Biol.* **139,** 1271–1280.

Weil, C. F., Oakley, C. E., and Oakley, B. R. (1986). Isolation of mip (microtubule-interacting protein) mutations of *Aspergillus nidulans. Mol. Cell. Biol.* **6,** 2963–2968.

Weinstein, B., and Solomon, F. (1990). Phenotypic consequences of tubulin overproduction in *Saccharomyces cerevisiae:* differences between α-tubulin and β-tubulin. *Mol. Cell Biol.* **10,** 5295–5304.

Westendorf, J. M., Rao, P. N., and Gerace, L. (1994). Cloning of cDNAs for M-phase phosphoproteins recognized by the MPM2 monoclonal antibody and determination of the phosphorylated epitope. *Proc. Natl. Acad. Sci. USA* **91,** 714–718.

Wigge, P. A., Jensen, O. N., Holmes, S., Soues, S., Mann, M., and Kilmartin, J. V. (1998). Analysis of the *Saccharomyces* spindle pole by matrix-assisted laser desorption/ionization (MALDI) mass spectrometry. *J. Cell Biol.* **141,** 967–977.

Williams, R. C., Jr., and Lee, J. C. (1982). Preparation of tubulin from brain. *Methods Enzymol.* **85b,** 376–385.

Winey, M., Goetsch, L., Baum, P., and Byers, B. (1991). MPS1 and MPS2: novel yeast genes defining distinct steps of spindle pole body duplication. *J. Cell Biol.* **114,** 745–754.

Winey, M., Mamay, C. L., O'Toole, E. T., Mastronarde, D. N., Giddings, T. H., Jr., McDonald, K. L., and McIntosh, J. R. (1995). Three-dimensional ultrastructural analysis of the *Saccharomyces cerevisiae* mitotic spindle. *J. Cell Biol.* **129,** 1601–1615.

Zheng, Y., Jung, M. K., and Oakley, B. (1991). γ-Tubulin is present in *Drosophila melanogaster* and *Homo sapiens* and is associated with the centrosome. *Cell* **65,** 817–823.

Zheng, Y., Wong, M. L., Alberts, B., and Mitchison, T. (1995). Nucleation of microtubule assembly by a γ-tubulin-containing ring complex. *Nature* **378,** 578–583.

Zheng, Y., Wong, M. L., Alberts, B., Mitchison, T. (1998). Purification and assay of gamma tubulin ring complex. *Methods Enzymol.* **298,** 18–28.

5

The Spindle Pole Body of *Saccharomyces cerevisiae:* Architecture and Assembly of the Core Components

Susan E. Francis and *Trisha N. Davis*
Department of Biochemistry
University of Washington
Seattle, Washington 98195

I. Introduction

The spindle pole body (SPB) functions as the microtubule organizing center (MTOC) in the budding yeast *Saccharomyces cerevisiae,* in a manner analogous to the centrosome of animal cells. It is the only MTOC in the yeast cell, serving to initiate nuclear microtubules that form the mitotic spindle as well as cytoplasmic microtubules that position the nucleus and align the spindle during cell division. Given the functional similarities between centrosomes and SPBs, a remarkable level of structural diversity exists between these organelles. However, the degree of protein conservation documented between yeast and other eukaryotes suggests that centrosomal homologs will be found for many SPB proteins. Moreover, despite gross structural differences, similar patterns will likely be apparent at the level of assembly, organization, and regulation of these components. To some extent this prediction has already been realized. γ-Tubulin, first found in the fungus *Aspergillus nidulans* (Oakley and Oakley, 1989), has also been

Current Topics in Developmental Biology, Vol. 49

0070-2153/00 $35.00

found in *Saccharomyces cerevisiae, Schizosaccharomyces pombe, Drosophila, Xenopus, Spisula,* and humans, where it is crucial part of the γ-tubulin complex required for microtubule nucleation (Horio *et al.,* 1991; Li and Joshi, 1995; Marschall *et al.,* 1996; Sobel and Snyder, 1995; Spang *et al.,* 1996a; Stearns and Kirschner, 1994; Vogel *et al.,* 1997; Zheng *et al.,* 1991, 1995). The yeast γ-tubulin complex is composed of three proteins, Tub4p (γ-tubulin), Spc98p, and Spc97p (Knop *et al.,* 1997; Knop and Schiebel, 1997; Nguyen *et al.,* 1998). Homologs of Spc97p and Spc98p have recently been identified in humans and *Xenopus* (Martin *et al.,* 1998; Murphy *et al.,* 1998; Tassin *et al.,* 1998). Although yeast cells have no structure resembling a centriole, they have a form of centrin, a known component of centrioles. Thus, ultrastructural differences do not necessitate distinctions at the molecular level.

II. SPB Duplication and Assembly of the Mitotic Spindle

The budding yeast SPB is a multilayered cylindrical organelle that is embedded in the nuclear envelope throughout the cell cycle (Byers, 1981b). Like the centrosome, the SPB is duplicated once and only once. Duplication of the SPB occurs before the initiation of DNA synthesis and bud emergence. The first visible stage of SPB duplication occurs during G1, when a spherical satellite forms in the cytoplasm attached to the half-bridge, about 150 nm from the extant SPB (Byers, 1981a). Recent data indicate that the satellite is constructed, at least in part, from proteins also found on the mature SPB (Adams and Kilmartin, 1999). The satellite gives rise to the duplication plaque, an elongated structure that is closely associated with the cytoplasmic side of the half-bridge and is the immediate precursor of the SPB. The duplication plaque is inserted into the nuclear membrane, adjacent to the half-bridge, via a nuclear pore–like structure, thereby allowing completion of SPB duplication (Adams and Kilmartin, 1999). The SPBs remain in a side-by-side configuration, connected by the bridge, until the bud is approximately one-third the size of the mother cell. At about this time, DNA synthesis is completed, and the SPBs move apart, severing the bridge so that a half-bridge is associated with each SPB.

Once the SPB has duplicated and DNA synthesis is completed, mitosis begins. Time-lapse video microscopy of mitosis in *S. cerevisiae* reveals distinct stages with discrete kinetic and morphological properties (Straight *et al.,* 1997; Yeh *et al.,* 1995). At the end of S phase, the SPBs separate in the nuclear membrane and a short bipolar spindle (<1 μm) forms between them. At the ulterastructural level, kinetochore microtubules (one per kinetochore) and interpolar microtubules are observed, although no ordered microtubule array is apparent (O'Toole *et al.,* 1999; Winey *et al.,*

1995). In contrast to animal and plant chromosomes, yeast chromosomes do not congress to form a metaphase plate; instead, attached sister chromatids move freely along the spindle (Straight *et al.*, 1997; Winey *et al.*, 1995). The spindle lengthens to 1.5–2.0 μm and remains that length for 10% of the cell cycle until anaphase. Anaphase takes about 25% of the cell cycle and occurs in four stages. Anaphase A is largely completed in less than a minute (Straight *et al.*, 1997). The first phase of anaphase B is a rapid elongation of the spindle to about 6 μm with one SPB in each cell. The elongation stops and the nucleus oscillates over 1–2 γm within the bud neck. In the final stage of anaphase B, the spindle elongates slowly, reaching a maximum size of about 10–12 μm, thus positioning the spindle poles at the ends of each cell. After anaphase, the spindle poles migrate to the center of their respective cells, and cytokinesis occurs, producing two unbudded cells (Yeh *et al.*, 1995).

III. Structure of the SPB

In addition to forming the spindle poles during cell division, the SPB plays an important role in mating and spore wall formation. Analysis of how the SPB is organized and modified for these different roles is just beginning, but a detailed structural description of the mitotic SPB has been completed. Early transmission EM studies revealed a trilaminate structure consisting of electron-dense outer, central, and inner plaques (Byers, 1981a; Byers and Goetsch, 1974, 1975; Moens and Rapport, 1971). These observations have been substantially augmented by two analyses: one of isolated SPBs using cryoelectron microscopy and image processing and another of SPBs in fast-frozen and freeze-substituted cells using electron tomography. Six distinct layers can now be discerned in the SPB (Bullitt *et al.*, 1997; O'Toole *et al.*, 1999). Proceeding from the cytoplasm into the nucleus are the outer plaque followed by intermediate layers 1 and 2, the central plaque, the inner plaque, and the nuclear microtubule caps (Fig. 1). The central plaque is in the plane of the nuclear membrane and is anchored to the membrane by short hooklike structures (O'Toole *et al.*, 1999). The membrane adjacent to the SPB is modified forming the bridge or half-bridge, depending on the stage of the cell cycle. Microtubules are nucleated from both sides of the nuclear membrane. The minus ends, proximal to the SPB, have distinctive tapered capped ends, whereas the plus ends, distal to the SPB, are flared (O'Toole *et al.*, 1999). Cytoplasmic microtubules radiate from the outer plaque to the cell cortex and are required to position the nucleus in the cell (Sullivan and Huffaker, 1992). The cytoplasmic microtubules are similar to the astral microtubules in other eukaryotic centrosomes. Nuclear microtubules emanate from the inner plaque to form the mitotic spindle (Byers

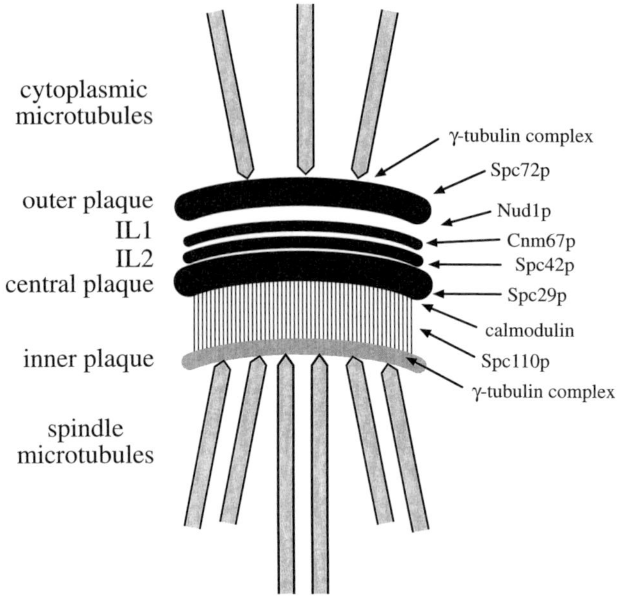

Figure 1 The spindle pole body of *S. cerevisiae* (adapted from Adams and Kilmartin, 1999; Bullitt *et al.,* 1997; Elliott *et al.,* 1999; O'Toole *et al.,* 1999; Wigge *et al.,* 1998).

and Goetsch, 1974, 1975; Moens and Rapport, 1971; O'Toole *et al.,* 1999). Interestingly, kinetochore microtubules appear to be associated with the SPB throughout the cell cycle (O'Toole *et al.,* 1999).

Diploid SPBs have a mass of approximately 1.0–1.5 gigadaltons (Bullitt *et al.,* 1997). The diameter of the SPB is correlated with chromosome number, such that the greater the chromosome number, the larger the structure. How the size of the SPB is altered to accommodate a change in chromosome number is not known. SPBs also vary in diameter in a way that does not depend on chromosome number. Isolated SPBs from a diploid strain fall into two size classes with diameters of 118 and 151 nm. Interestingly, only lateral growth of the SPB is detected; the same vertical arrangement of layers with fixed dimensions is seen regardless of SPB diameter (Bullitt *et al.,* 1997). One possibility is that the SPBs grow in diameter after the initial assembly process and insertion into the envelope is completed. The purpose of the lateral growth is not clear, but is unlikely to be for the attachment of kinetochore microtubules, which remain attached throughout the cell cycle (O'Toole *et al.,* 1999).

IV. SPB Core Components

A major breakthrough in the analysis of SPBs occurred when Rout and Kilmartin (1990) developed a method to obtain highly enriched preparations of SPBs. Monoclonal antibodies prepared against this enriched SPB fraction led to the identification of three abundant and essential SPB constituents: Spc110p, Spc42p, and Spc98p (Rout and Kilmartin, 1990, 1991). Once the yeast genome was sequenced, the highly enriched preparations of SPBs were characterized by mass spectrometry to identify 23 spindle pole components, 11 not previously localized to the pole (Wigge *et al.,* 1998). Surprisingly few of these (6) make up the core structure of the SPB, which remains once the microtubules along with the inner and outer plaques are removed by treatment with heparin. This core structure contains approximately 1000 copies each of Nud1p, Cnm67p, Spc42p, Spc29p, calmodulin, and Spc110p (Adams and Kilmartin, 1999). Harsher treatment of the intact SPBs with nonionic detergent and DEAE dextran removes Nud1p, leaving a core with the 5 other components (D. Schieltz, J. Yates, and T. Davis, unpublished observation). This review focuses on the assembly, organization, and regulation of the core of the SPB. Another critical part of the SPB is the γ-tubulin complex consisting of Spc97p, Spc98p, and Tub4p and found in both the inner and outer plaques (Geissler *et al.,* 1996; Knop *et al.,* 1997). The yeast γ-tubulin complex is reviewed separately in this volume (Vogel and Snyder).

A composite picture of the structure of the SPB core can be drawn from immunolocalization, biochemical, and genetic data (Fig. 1). A brief description of the organization of the core complex is given before each component is discussed in detail. Cnm67p, Spc42p, and Spc110p are predicted to have significant coiled-coil domains, a feature shared by many known centrosomal proteins and thought to be an important determinant of SPB assembly that, in part, produces the structural regularity characteristic of the organelle. Spc42p is located in IL2 at the cytoplasmic side of the central plaque where it forms a nearly crystalline core layer (Bullitt *et al.,* 1997). Extraction of intact SPBs yields two Spc42p-containing fractions, one of which has the cytoplasmic components Cnm67p and Nud1p/Spc94p, and the other has central plaque components Spc110p, calmodulin, and Spc29p (Elliott *et al.,* 1999; Knop and Schiebel, 1998). By immunoelectron microscopy, Cnm67p and Nud1p localize to the outer plaque, although recent evidence places Cnm67p closer to IL1 (Adams and Kilmartin, 1999; Wigge *et al.,* 1998). Also at the outer plaque is Spc72p, the tether for the γ-tubulin complex on the cytoplasmic side of the SPB (Chen *et al.,* 1998; Knop and Schiebel, 1998; Soues and Adams, 1998; Wigge *et al.,* 1998).

Although Spc72p is not strictly a part of the core of the SPB, we include it in our review for completeness. Calmodulin and Spc29p bind to Spc110p, which links the central plaque to the nuclear γ-tubulin complex in the inner plaque (Adams and Kilmartin, 1999; Elliott *et al.,* 1999; Kilmartin and Goh, 1996; Knop and Schiebel, 1997; Nguyen *et al.,* 1998; Stirling *et al.,* 1996; Sundberg *et al.,* 1996).

A. Spc42p: The Crystalline Core

SPC42 is predicted to encode a 363 residue polypeptide with a short region of dimer-forming coiled coil between residues 64 and 137 (Berger *et al.,* 1995; Wolf *et al.,* 1997). The details of how Spc42p molecules interact to form the IL2 crystalline layer are not yet complete. Early crosslinking experiments on Spc42p crystals revealed Spc42p monomers, dimers, tetramers, and hexamers (Bullitt *et al.,* 1997). Cryoelectron microscopy of isolated SPBs suggests a hexagonal packing arrangement, consistent with a subunit structure that could be produced by either three dimers or two timers of Spc42p. High-voltage electron microscopy of fast-frozen freeze-substituted cells suggests a square packing arrangement of IL2 subunits (O'Toole *et al.,* 1999). In either case, Spc42p creates the internal scaffolding of the SPB. Vertical striations can be seen connecting IL2 to the central plaque (O'Toole *et al.,* 1999). These striations could represent the coiled coil of Spc42p, but their identity has not yet been established. Two-hybrid analyses place Spc42p between Cnm67p and Spc110p. The C-terminal region of Spc42p binds the C-terminal region of Cnm67p. The N-terminal region of Spc42p binds Spc29p, and the C-terminal region of Spc110p in the central plaque (discussed in detail below) (Fig. 2) (Adams and Kilmartin, 1999; Elliott *et al.,* 1999). Thus, Spc42p apparently forms a single parallel layer of filaments linking IL2 to the central plaque.

Spc42p is an essential SPB component also present on the satellite and the duplication plaque, precursors of the SPB (Adams and Kilmartin, 1999). Temperature-sensitive mutations in *SPC42* block SPB duplication and give rise to a variety of dead-end structures, which cannot be converted to functional SPBs by return to the permissive temperature (Donaldson and Kilmartin, 1996). The mutations substitute proline at positions 86, 110, or 121 in the coiled coil. Thus, altering the coiled coil interrupts the packing arrangement of the Spc42p polymer and disrupts SPB assembly.

The morphology of IL2 is readily manipulated by altering Spc42p. Overexpression of Spc42p produces a two-dimensional dome-shaped structure that forms when a sheet of Spc42p polymer stretches across the cytoplasmic surface of the nuclear envelope (Donaldson and Kilmartin, 1996). This overproduced Spc42p forms a double layer in contrast to the single layer

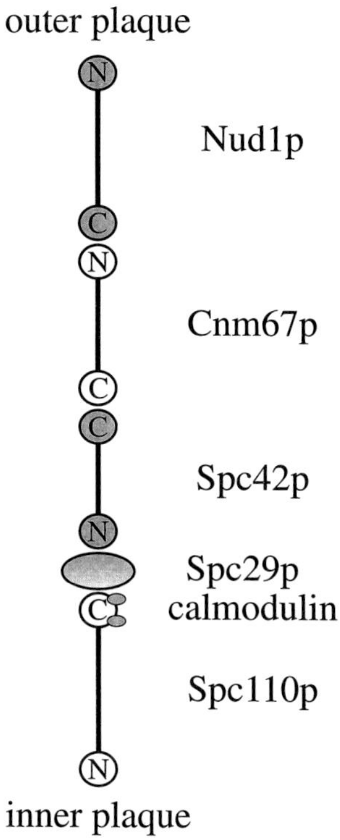

Figure 2 Arrangement of the core proteins of the SPB. Spc110p, Spc42p, and Cnm67p contain substantial regions of coiled coil. Nud1p is a leucine-rich repeat protein. Calmodulin associates with the C-terminal region of Spc110p. (Adapted from Adams and Kilmartin, 1999; Elliott *et al.*, 1999.)

normally formed in wild-type SPBs, presumably because the contacts required to keep Spc42p in a single layer are limiting (Bullitt *et al.,* 1997). A mutant Spc42p in which all 34 serines are changed to alanine also displays striking structural polymorphisms. For example, prolonged exposure to α-factor causes the central plaque to curl into a ball in cells expressing S:A Spc42p but not in cells expressing the wild-type protein (Adams and Kilmartin, 1999). Overexpressed S:A Spc42p forms a large electron-dense ball that contrasts markedly with the two-dimensional structure formed by overexpression of the wild-type protein.

B. The Nuclear Side of the SPB: Spc110p, Spc29p, and Calmodulin

SPC110 is predicted to encode a 944-residue protein with globular domains separated by a central region, residues 155–798, predicted to form a

71-kDa coiled coil (Fig. 3) (Berger *et al.,* 1995). Consistent with this prediction, expression of fragment containing residues 227–756 produces a rod of 790 Å, the expected length of a coiled coil at 1.5 Å/residue (Kilmartin *et al.,* 1993). Immunoelectron microscopy places Spc110p between the central and inner plaques where vertical strands bridging the central plaque to the capped ends of the microtubules can be seen (Kilmartin *et al.,* 1993; O'Toole *et al.,* 1999; Rout and Kilmartin, 1990). Expression of truncated forms of Spc110p that are deleted in the coiled-coil domain, decreases the distance between the central and inner plaques proportionately, indicating that Spc110p assumes an extended configuration and that it links the central and inner plaques (Kilmartin *et al.,* 1993). Kilmartin and others have speculated Spc110p may act as a flexible spacer, which both ensures an optimal separation between the two plaques and enables the SPB to withstand stresses produced during mitosis (Bullitt *et al.,* 1997; Kilmartin *et al.,* 1993).

1. Regulation of Spc110p by Calmodulin

Spc110p is the essential mitotic target of calmodulin, the ubiquitous Ca^{2+}-binding protein that converts changes in the intracellular concentrations of Ca^{2+} into cellular responses (Geiser *et al.,* 1993; Stirling *et al.,* 1994). As shown by a two-hybrid analysis, residues 900–927 are sufficient for binding wild-type calmodulin (Geiser *et al.,* 1993). A 13-residue segment within this region (902–913) is predicted to form an amphipathic α-helix typical of Ca^{2+}-dependent calmodulin-binding sites (Stirling *et al.,* 1994). Since mutations in this segment abolish calmodulin binding, this segment represents the core of the calmodulin-binding site (Stirling *et al.,* 1994). Moreover, a modified peptide derived from residues 897–917 binds calmodulin with an affinity of 70 nM (Geier *et al.,* 1996).

Although the core site resembles a Ca^{2+}-dependent calmodulin-binding site, substantial evidence suggests calmodulin binding to Spc110p is not Ca^{2+} dependent. First, the affinity of the peptide for calmodulin is the same at 0.1 μM Ca^{2+} as at 1.35 μM Ca^{2+}. Second, a mutant calmodulin, Cmd1-6p (3D-A), in which all the Ca^{2+}-binding sites have been inactivated, local-

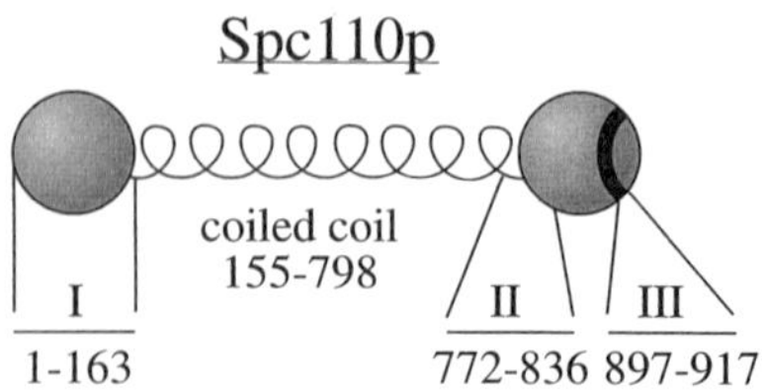

Figure 3 Three regions of Spc110p defined by a mutational analysis (Sundberg and Davis, 1997).

izes to the SPB as measured by immunofluorescence and using a GFP fusion (Moser *et al.,* 1997). Moreover, two-hybrid analysis indicates that Cmd1-6p binds well to the C-terminal region of Spc110p (Geiser *et al.,* 1993). Interestingly, binding of Cmd1-6p requires more than just the core calmodulin-binding site. Binding at levels equivalent to wild-type calmodulin requires approximately 150 residues in the C-terminal region of Spc110p (H. Sundberg and T. Davis, unpublished results). Finally, wild-type calmodulin can bind to the C-terminus of Spc110p produced in bacteria in the presence of Ca^{2+} or EGTA (T. Davis, unpublished results). Thus, the calmodulin-binding domain of Spc110p appears to be a core peptide, which may have Ca^{2+}-dependent characteristics, embedded in a Ca^{2+}-independent site.

Calmodulin binding to Spc110p is Ca^{2+}-independent, but Ca^{2+} binding could still affect the function of the calmodulin-Spc110p complex. Calmodulin may always be bound to Spc110p but adopt a different conformation in the presence of Ca^{2+}. Precedence for this comes from phosphorylase kinase. Calmodulin is an integral subunit of phosphorylase kinase even in the presence of EGTA. Ca^{2+} binding to calmodulin changes the enzyme to an active conformation (Cohen, 1988). The structural effects of Ca^{2+}-binding to calmodulin-Spc110p have not yet been examined. However, the functional effects must be subtle because the mutant calmodulin with inactivated Ca^{2+}-binding sites (Cmd1-6p) supports the growth of yeast. No phenotype has yet been detected during growth or sporulation, both processes that require the SPB.

Calmodulin associates with Spc110p at the central plaque of the SPB throughout the cell cycle (Geiser *et al.,* 1993; Kilmartin and Goh, 1996; Spang *et al.,* 1996b; Sundberg *et al.,* 1996). The observation that calmodulin localizes only to the central plaque defines the orientation of Spc110p in the SPB as C-terminus at the central plaque and N-terminus at the inner plaque. Thus, Spc110p, like Spc42p, is aligned as parallel filaments. This orientation was confirmed using antibodies against the N-terminal region of Spc110p, which exclusively label the inner plaque (Spang *et al.,* 1996b).

2. Regulation of Spc110p by Phosphorylation

Spc110p is phosphorylated in a cell-cycle-dependent manner and is phosphorylated on multiple residues. Two isoforms of Spc110p are distinguished by SDS–polyacrylamide gel electrophoresis. The slower-migrating 120-kDa isoform occurs by additional phosphorylation of the 112-kDa isoform on serine and/or threonines. In synchronous cultures, the 120-kDa isoform appears when the SPBs separate and disappears as cells enter anaphase. Cell-cycle-dependent phosphorylation depends on Clb5p and Clb6p, two cyclins induced at the end of G1. Three canonical sites for the cyclin-dependent kinase are present in the N-terminus of Spc110p (Mirzayan *et al.*, 1992). However, mutating all of these sites does not abolish the mobility shift resulting from phosphorylation (Friedman *et al.*, 1996). Cell-cycle dependent phosphorylation of Spc110p depends on Mps1p, the dual-specificity kinase required for SPB duplication (Weiss and Winey, 1996). Moreover, Mps1p phosphorylates Spc110p *in vitro* (D. Friedman, M. Winey, and T. Davis, unpublished observations). Thus, Spc110p is likely to be a direct target of Mps1p.

Efforts to map the phosphorylated serine/threonine(s) by deletion analysis of Spc110p showed that the region between residues 828 and 856 is required for cell-cycle-dependent phosphorylation (Friedman *et al.*, 1996). When all the serines and threonines between 828 and 856 are replaced with alanine, cell-cycle-dependent phosphorylation still occurs, indicating that the region between residues 828 and 856 contains information required for phosphorylation, but does not contain the phosphorylated residues. As discussed later, this region very likely interacts with the central plaque. We speculate that, the observed cell-cycle-dependent phosphorylation may be dependent on proper assembly of Spc110p into the SPB.

3. Overexpression of Spc110p

Overexpression of Spc110p produces a novel nuclear polymer consisting of a large spheroidal array of Spc110p and calmodulin (Kilmartin and Goh, 1996). These particles are usually positioned near the SPB and have microtubules associated with them. The regular structures result from antiparallel interactions between Spc110p molecules, whereas only parallel interactions normally occur at the SPB. This may suggest that interactions of Spc110p at either the amino or, more likely, the carboxy terminal regions dictate the orientation of the filaments. In the overexpression strains, Spc110p is in excess of the number of binding sites in the SPB and is free to assume another configuration.

4. A Mutational Analysis Defines Three Functional Regions of Spc110p: Region I, Region II, and Region III

Mutational analysis of Spc110p has yielded yeast strains with highly informative phenotypes that help to reconstruct the process of SPB assembly (Kilmartin and Goh, 1996; Stirling *et al.*, 1994, 1996; Sundberg and Davis, 1997). Three functional regions have been identified by complementation studies of *SPC110* mutants (Fig. 3)(Sundberg and Davis, 1997). Region I is located at the N-terminus between residues 1 and 163 and is involved in interaction with the γ-tubulin complex (Knop and Schiebel, 1997; Nguyen *et al.*, 1998; Sundberg and Davis, 1997). Region II includes residues 772–836 and is important in maintaining the integrity of the SPB. Region III defines the calmodulin-binding site, between residues 897 and 917, a region critical for proper assembly and function of the SPB (Kilmartin and Goh, 1996; Stirling *et al.,* 1996; Sundberg and Davis, 1997). Alleles with mutations in one region of *SPC110* can be complemented by alleles in either of the two other regions (Sundberg and Davis, 1997). Thus, a functional SPB can be made from hybrid multimers with complementing mutations.

a. Region III of Spc110p: The Calmodulin-Binding Site. Temperature-sensitive mutations in either calmodulin or the calmodulin-binding site of Spc110p interfere with the attachment of Spc110p to the SPB during SPB duplication (Kilmartin and Goh, 1996; Sundberg *et al.,* 1996). Such mutations also cause a variety of abnormal spindle structures, including broken spindles, presumably because the SPBs do not have a full complement of microtubules (Davis, 1992; Kilmartin and Goh, 1996; Stirling *et al.,* 1996; Sundberg *et al.,* 1996). In a strain relying on half-calmodulin, SPBs disintegrate, yielding monopolar spindles after long incubations at the restrictive temperature (Sun *et al.,* 1991).

The cellular consequences of a conditional mutation in the calmodulin-binding site were studied at the ultrastructural level for two mutants, *spc110-1* (K594T, I759S, C911R) and *spc110-220* (S853G, C911R) (Kilmartin and Goh, 1996; Sundberg *et al.,* 1996). Both alleles have the same mutation in the core calmodulin-binding site (C911R), which interferes with calmodulin-binding (Sundberg *et al.,* 1996). The alleles have different mutations outside this site, perhaps explaining the slightly different phenotypes observed. For both mutants, cells synchronized by elutriation at G1 and shifted to the nonpermissive temperature complete SPB duplication but are plagued with a progressive deterioration of the mitotic apparatus. The cells arrest with large buds and a G2 content of DNA and display a dramatic loss of viability when spindle formation is attempted. The *spc110-1* mutant cells generally arrest with unbalanced spindles where one SPB has fewer associated microtubules than the other (Kilmartin and Goh, 1996). Thus, the mutations in *spc110-1* interfere with attachment of microtubules to the SPB, presumably by blocking attachment of Spc110p to the central plaque.

Further evidence for this model comes from the characterization of the other mutant (Sundberg *et al.,* 1996). At the nonpermissive temperature, the *spc110-220* mutant cells have less asymmetry in microtubule distribution between the two SPBs; nevertheless, broken spindles are common. Most striking is the presence of a transient electron-dense intranuclear structure that is present during SPB assembly and disappears as the duplicated SPBs separate. This structure has been called the IMO, for intranuclear microtubule organizer, because during its short life span, microtubules extend from it as well as from the SPB. The IMOs contain Spc110-220p and the components of the γ-tubulin complex, but no calmodulin. The presence of IMOs in calmodulin mutants provides additional evidence that they are caused by decreased interactions between calmodulin and Spc110p. How the IMO forms is somewhat mysterious. A reasonable possibility is that Spc110-220p cannot be converted to its assembly-competent form and instead associates in a mislocalized protein aggregate. Curiously, only one IMO is ever observed per cell (Sundberg *et al.,* 1996).

Interestingly, truncations of Spc110p that completely remove the calmodulin-binding site are viable. In addition, truncations of Spc110p up to residue 856 act as dominant suppressors of *cmd1-1,* a temperature-sensitive calmodulin mutant (Geiser *et al.,* 1993). It thus appears that deletion of the calmodulin-binding site relieves a requirement for calmodulin. This apparent paradox is most easily explained by modeling calmodulin regulation of Spc110p on calmodulin activation of enzymes. In the case of myosin light chain kinase, calmodulin binding induces a conformational change that repositions an inhibitory region, thereby allowing enzyme activation. By analogy, calmodulin may bind to the carboxy terminus of Spc110p, alter its conformation, and reveal sites that are required for Spc110p function and thus viability of the organism. Since calmodulin-binding mutants of Spc110p fail to properly assemble at the SPB, a reasonable hypothesis is that calmodulin relieves inhibition that prevents binding of Spc110p to the central plaque.

b. Region II of Spc110p and Spc29p. To determine how Spc110p connects to the central plaque, a combination of genetic, biochemical, and two-hybrid approaches have been pursued. Deletions of large portions of the coiled-coil domain of Spc110p reveal that much of it is expendable. Specifically, deletions from residues 216 to 710, which constitute about 70% of the coiled coil, shorten the distance between the central and inner plaques, but produce no obvious deleterious consequences. However, a smaller deletion that extends into the globular C-terminal domain, from residues 489–809, results in a severe growth phenotype. Deletion from 489–824 is lethal (Kilmartin *et al.,* 1993). Truncation of the C terminus of Spc110p at residue 856 results in a viable strain, whereas truncation at residue 828 is lethal (Geiser *et al.,* 1993). Taken together, these studies reveal an essential region between residues 711 and 856 encompassing the essential region II (residues 772–836) defined by mutational analysis (Sundberg and Davis, 1997).

The function of region II can be deduced from analysis of the temperature-sensitive region II mutant *spc110-226* (L772M, R795L, D797V, N823S, L836P) (Sundberg and Davis, 1997). When cultures of the *spc110-226* mutant are synchronized and shifted to the restrictive temperature, the mutant cells appear to duplicate their SPBs properly and proceed to form normal short bipolar spindles. However, aberrant structures, which are produced by SPB disintegration, accumulate as mitosis progresses. When examined by electron microscopy, many cells have monopolar spindles, as well as what look to be fragmented SPBs that have pulled away from the nuclear membrane. These nuclear structures contain Spc110p, Spc98p, and calmodulin and are associated with microtubules (Sundberg and Davis, 1997; O'Toole and Davis, unpublished observations). Aberrant cytoplasmic structures, possibly arising from detached pieces of the outer plaque, are also seen. The breakdown in SPB integrity seen in region II mutants is consistent with the idea that this region is required for stable attachment to other central plaque proteins.

Proteins that interact with the C-terminal region of Spc110p were identified by cell fractionation. SPBs were collected in a high-speed pellet and subjected to a series of extractions that disrupt the intact SPBs while retaining protein associations within some of the subcomplexes. Three proteins are part of the Spc110p subcomplex: Spc42p, Spc29p, and calmodulin (Elliott *et al.,* 1999; Knop and Schiebel, 1997). Two-hybrid analyses indicate that the C-terminal region of Spc110p interacts with itself, the N-terminal region of Spc42p, full-length Spc29p and calmodulin (Adams and Kilmartin, 1999; Elliott *et al.,* 1999; Geiser *et al.,* 1993; Knop and Schiebel, 1997). Spc29p and Spc42p also interact (Elliott *et al.,* 1999). As noted earlier; a variety of methods demonstrate a direct interaction between calmodulin and Spc110p. A recent result that calmodulin must bind to Spc110p to reveal the Spc29p binding site (Elliott *et al.,* 1999) provides solid evidence for the earlier model that calmodulin relieves inhibition that prevents Spc110p from binding to the central plaque (Geiser *et al.,* 1993; Sundberg *et al.,* 1996). The two-hybrid analyses place the Spc29p binding site adjacent to the calmodulin-binding site overlapping a portion of region II (811–898).

Whether both Spc42p and Spc29p directly bind Spc110p has not yet been established. The two-hybrid analyses alone cannot resolve this issue because endogenous nuclear proteins can act as linkers between the fragments being tested and confound the assay. Nevertheless, a model that has Spc42p linked to a calmodulin–Spc110p complex by Spc29p seems easiest to visualize. First, Spc42p has been clearly placed in IL2 on the cytoplasmic face of the central plaque, while Spc29p is present in the central plaque (Adams and Kilmartin, 1999). Second, overexpressed Spc29p accumulates in the nucleus, whereas overexpressed Spc42p accumulates in the cytoplasm (Elliott *et al.,* 1999). These patterns are consistent with an order Spc42p–Spc29p–Spc110p. Alternatively, the Spc110p–calmodulin complex may interact directly with both Spc42p and Spc29p.

Consistent with the physical interactions already described, genetic interactions between *SPC110, SPC42, SPC29,* and *CMD1* are well documented (Adams and Kilmartin, 1999; Elliott *et al.,* 1999; Sundberg and Davis, 1997).

Synthetic lethal interactions have been noted between *SPC110* alleles and alleles of *SPC42, SPC29,* and *CMD1*. Synthetic lethality also occurs between *SPC29* and *SPC42* alleles. *CMD1* and *SPC42* are dosage-dependent suppressors of conditional mutants of *SPC110.* Interestingly, moderate overexpression of *SPC29* worsens the phenotype of some *SPC110* alleles, suggesting the importance of stoichiometry of components in building the SPB.

Overexpression of single components from the Spc110p–Spc42p–Spc29p complex does not alter the diameter of the SPB. However, cooverexpression of all three components produces an enlarged SPB. Cooverexpression of calmodulin is not necessary because it is normally present in 100-fold excess over the SPB components. These results demonstrate that the amounts of the three components are critical in determining the size of the SPB and suggest a possible mechanism of SPB size regulation in response to ploidy (Elliott *et al.,* 1999).

Mutations in SPC29 lead to failures in SPB duplication (Adams and Kilmartin, 1999; Elliott *et al.,* 1999). *spc29-20* (E166D, I220T) cells synchronized with α-factor and released under nonpermissive conditions duplicate their SPBs (Adams and Kilmartin, 1999). The daughter SPB is inserted in the nuclear envelope and has a normal half-bridge but is significantly smaller than the maternal SPB. Adams and Kilmartin suggest that the preassembled Spc29-20p is stable on the satellite present in the synchronized cells. At the restrictive temperature, additional Spc29-20p cannot assemble on the satellite and a small SPB is formed. At late times, this small SPB is unstable and disintegrates. Electron microscopy of cells with the *spc29-2* (R161S) or *spc29-3* (L158Q) alleles showed only one SPB structure per cell, consistent with a defect in SPB duplication. The single SPB structure displays a variety of defective duplication intermediates, including partially duplicated and fused SPBs, as well as SPBs with an elongated half-bridge (Elliott *et al.,* 1999). All three alleles are characterized by a transient or incomplete cell cycle arrest that leads to an increase in ploidy (Adams and Kilmartin, 1999; Elliott *et al.,* 1999).

c. Region I of Spc110p: Interaction with the γ-Tubulin Complex. Immunoelectron microscopy localizes the amino terminus of Spc110p to the inner plaque of the SPB in close proximity to the ends of spindle microtubules (Spang *et al.,* 1996b). A synthetic lethal screen using the region I allele, *spc110-221* (M15T, T30M, D62E, S67P, K99N, K110E, L119S, N163Y), revealed genetic interactions between *SPC110, SPC98,* and *SPC97* (Nguyen *et al.,* 1998). *SPC98* was identified as a dosage-dependent suppressor of *tub4-1* (Geissler *et al.,* 1996), and *SPC97* was identified as a dosage-dependent suppressor of *spc98-2* (Knop *et al.,* 1997). Tub4p, Spc98p, and Spc97p form the soluble 6S γ-tubulin complex (Geissler *et al.,* 1996; Knop and Schiebel, 1997) and are discussed in detail in a different review in this volume (Vogel and Snyder).

The interaction between the γ-tubulin complex and Spc110p is direct because purified γ-tubulin complex binds to a GST fusion of the N-terminal region of Spc110p (1–204) purified from *Escherichia coli* (Knop and Schiebel, 1997). All evidence indicates that Spc98p forms a crucial part of this direct linkage. Allele-specific interactions between *SPC98* and N-terminal *spc110* mutants provide good genetic evidence for a direct protein–protein interaction (Nguyen *et al.,* 1998). Moreover, Spc98p interacts strongly with the N-terminal region of Spc110p as measured in three different two-hybrid systems (Knop and Schiebel, 1997; Nguyen *et al.*, 1998). In contrast, similar types of experiments demonstrate that the binding of Spc97p to Spc110p requires Spc98p (Nguyen *et al.,* 1998). Spc98p could act as an intermediary between Spc97p and Spc110p or could change the conformation of Spc97p to a form able to bind Spc110p.

Substantial evidence, albeit circumstantial, suggests that nucleation of spindle microtubules occurs only by association of the γ-tubulin complex with Spc110p. Most convincing is the observation that the substantial soluble pool of the γ-tubulin complex is not associated with microtubules, whereas a requirement for Spc110p has been clearly shown for microtubule nucleation at the SPB (Kilmartin and Goh, 1996). Overexpression of Spc110p produces spheroidal structures that are associated with microtubules (Kilmartin and Goh, 1996). In contrast, overexpression of wild-type and mutant Spc42p produces dramatic structures, but they do not associate with microtubules (Adams and Kilmartin, 1999; Donaldson and Kilmartin, 1996). Furthermore, aberrant structures that are associated with microtubules contain Spc110p. For example, mutations in the calmodulin-binding site of Spc110p result in the formation of IMOs, which contain Spc110p, the γ-tubulin complex, and microtubules (Sundberg *et al.,* 1996). Mutations in region II of Spc110p results in fragments of the SPB that contain Spc110p, the γ-tubulin complex, and microtubules (Sundberg *et al.,* 1996; O'Toole and Davis, unpublished observation). Thus, there is a strong correlation between the association of Spc110p with the γ-tubulin complex and the presence of microtubules. Future experiments will identify the specific molecular events that lead to nucleation.

C. The Cytoplasmic Side of the SPB: Spc72p, Cnm67p, and Nud1p

1. Spc72p: The Cytoplasmic Anchor for the γ-Tubulin Complex

Since Spc110p is found exclusively in the nucleus, a cytoplasmic counterpart to Spc110p must exist to anchor the γ-tubulin complex to the cytoplasmic side of the SPB. A candidate protein, Spc72p, has been identified and characterized by three groups (Chen *et al.,* 1998; Knop and Schiebel, 1998; Soues and Adams, 1998). Knop and Schiebel found Spc72p in a two-hybrid screen for proteins that specifically bind the γ-tubulin complex. Spc72p coimmunoprecipitates with the γ-tubulin complex, although the majority of immunoprecipitated Spc72p is not associated with the complex, suggesting that there may be a substantial soluble fraction (Knop and Schiebel, 1998). In two-hybrid analyses, the N-terminus of Spc72p interacts with both Spc97p and Spc98p as well as with Tub4p provided that Spc97p and Spc98p are cooverexpressed.

Spc72p was also identified by its interaction with Stu2p (Chen *et al.,* 1998). Stu2p is an essential SPB protein that binds laterally to microtubules *in vitro.* Stu2p localizes to the SPB even in the presence of nocodazole, indicating that this protein is an integral component of the SPB (Wang and Huffaker, 1997). These properties suggest that Stu2p may play a role in attachment of microtubules to the SPB, perhaps providing lateral links between the microtubules and Spc72p in addition to those provided to the ends via the γ-tubulin complex.

Mutations in *SPC72* cause reduced numbers of cytoplasmic microtubules and defects in nuclear migration. SPBs in mutant strains duplicate normally and short bipolar spindles form; however, the spindles fail to elongate and nuclear division does not occur. The mutant cells do not arrest but instead proceed through the cell cycle, resulting in aploid and polyploid cells (Chen *et al.,* 1998; Knop and Schiebel, 1998). At restrictive temperature, the SPBs in the *spc72* mutant cells contain altered outer plaques with discontinuities or small gaps. Moreover, the distance between the central and outer plaque is about twice as large as in wild-type SPBs (Chen *et al.,* 1998). The loss of cytoplasmic microtubules and defects in nuclear migration caused by mutations in *SPC72* are very similar to the phenotype of a tubulin mutant that cannot make cytoplasmic microtubules (Sullivan and Huffaker, 1992). These aspects of the mutant phenotype, along with the defects in the outer plaque in the mutant cells, support the model that Spc72p is the cytoplasmic counterpart to Spc110p and binds the γ-tubulin complex to the cytoplasmic side of the SPB. However, *spc72* mutants also fail in spindle elongation, a process that involves nuclear microtubules but not cytoplasmic microtubules (Sullivan and Huffaker, 1992). This suggests that Spc72p plays a role, perhaps indirect, in spindle elongation. Moreover, deletion of *SPC72* is lethal in some strain backgrounds (Chen *et al.,* 1998; Knop and Schiebel, 1998) but causes slow growth in others (Soues and Adams, 1998). In the strains that can survive deletion of *SPC72,* a few cytoplasmic microtubules remain, so some other protein must support the nucleation of cytoplasmic microtubules at a low efficiency (Soues and Adams, 1998).

Because the defect in spindle elongation in *spc72* mutants was unexpected, Knop and Schiebel (1998) performed two additional experiments.

The results of both provide further evidence for the role of Spc72p in cytoplasmic microtubule organization. Overexpression of Spc72p results in increased numbers of cytoplasmic microtubules. Overexpression of Spc72p deleted for the γ-tubulin binding region decreased the number of cytoplasmic microtubules and produced other phenotypes associated with loss of cytoplasmic microtubules. In conclusion, Spc72p appears to be the cytoplasmic receptor for the γ-tubulin complex, although certain complexities of the mutant phenotypes remain to be explained.

To begin to address how microtubule nucleation differs in the nucleus and cytoplasm, Knop and Schiebel exchanged the γ-tubulin-binding N-terminal domains between Spc110p and Spc72p (Knop and Schiebel, 1998). Deletion of the N-terminal γ-tubulin binding domains from either Spc72p or Spc110p is lethal. (An *spc72* null is lethal in the strain used by these investigators.) An *spc72* null can be rescued by the chimeric Spc72p/Spc110p protein from the N-terminus of Spc110p and the C-terminus of Spc72p, suggesting that the fusion protein can substitute for Spc72p. In contrast, replacing the N terminus of Spc110p with the N terminus of Spc72p does not rescue an *spc110* null, and overexpression of this fusion protein is lethal. Presumably, the fusion protein acts in a dominant negative manner (Knop and Schiebel, 1998). These results suggest that control of spindle microtubule nucleation is particularly exacting.

2. Cnm67p and Nud1p: Links between Cytoplasmic Microtubules and Spc42p

Cytoplasmic microtubules have some flexibility in how they interact with the SPB. Deletion of *CNM67* diminishes or completely depletes the outer plaque. Under these circumstances, cytoplasmic microtubules extend from the half-bridge of the SPB, reminiscent of an earlier developmental stage when astral microtubules extend from the bridge prior to SPB separation (Adams and Kilmartin, 1999; Brachat *et al.,* 1998; Byers and Goetsch, 1975). In a *cnm67* deletion strain, Spc72p also localizes to the half-bridge. Nud1p, which is usually found in association with Cnm67p, is partly (about 10%) relocalized to the half-bridge while the remainder is not associated with the SPB (Adams and Kilmartin, 1999). When these strains are rescued with *CNM67,* proper localization of both Spc72p and Nud1p occurs.

Cells with mutations in *NUD1* also produce SPBs with depleted outer plaques; the cytoplasmic microtubules extend from the half-bridge. Thus, like Cnm67p, Nud1p is required to form an outer plaque and attach cytoplasmic microtubules to the SPB. Unlike Cnm67p, Nud1p is essential and is required to exit mitosis (Adams and Kilmartin, 1999). Two-hybrid analyses indicate that the N-terminal region of Cnm67p binds to the C-terminal region of Nud1p. Since localization of Nud1p and Spc72p depends on

Cnm67p and given the two-hybrid interactions between Cnm67p and Spc42p, a likely order starting from the central plaque is Spc42p–Cnm67p–Nud1p–Spc72p (Fig. 2). It is worth noting that the *NUD1* nomenclature is confusing. Two *S. cerevisiae* genes are named *NUD1.* The essential SPB gene is YOR373W on chromosome XV or *NUD1* (also called *SPC94*). The *NUD1* name has also been used for the gene *RNC1/NUC2,* which is YKR056W on chromosome XI. Rnc1p is an exonuclease involved in recombination and has no known relation to the SPB (Asefa *et al.,* 1998; Semionov *et al.,* 1999).

V. Assembly of the Core SPB Components

Recent advances in visualizing SPB duplication intermediates allow for a description of SPB assembly at the molecular level (Adams and Kilmartin, 1999). The central plaque components Spc42p, Spc29p, and the cytoplasmic components Cnm67p and Nud1p assemble to form a satellite on the distal end of the cytoplasmic side of the half-bridge. The satellite already has the plaquelike morphology found in IL2 and the central plaque of the SPB, the layers associated with Spc42p and Spc29p (Adams and Kilmartin, 1999; O'Toole *et al.,* 1999). The satellite maintains this plaquelike structure as it enlarges to a duplication plaque by the further addition of these four core components. The outer plaque including Tub4p begins to be assembled on the duplication plaque. Prior to that, the outer plaque components Spc72p and Tub4p as well as cytoplasmic microtubules are associated with the half-bridge. Nuclear Spc110p is added as the duplication plaque is inserted into the nuclear envelope (Adams and Kilmartin, 1999). Characterization of Spc110p mutants suggests four stages in the assembly of Spc110p onto the SPB: (a) Spc110p associates with calmodulin and oligomerizes; (b) Spc110p binds to the central plaque; (c) the attachment to the central plaque is stabilized; and (d) the γ-tubulin complex attaches to the N-terminal region of Spc110p (Sundberg and Davis, 1997). The order of these stages has not yet been established. Once the two SPBs are in the side-by-side configuration in the envelope, lateral growth of both SPBs may occur (Bullitt *et al.,* 1997).

The role of the half-bridge in duplication is emphasized by the requirement of two half-bridge components, Cdc31p and Kar1p, for the duplication process (Byers, 1981b; Rose and Fink, 1987; Spang *et al.,* 1993, 1995). Cdc31p is a centrin homolog and Kar1p binds Cdc31p to the half-bridge (Baum *et al.,* 1986; Biggins and Rose, 1994). The duplication plaque lies along a portion of the half-bridge, which has a distinctive bent conformation. The side of the duplication plaque attached to the bridge ultimately associates with the Spc110p–calmodulin complex after insertion into the envelope. In the duplication plaque, Spc42p and Spc29p presumably interact with components of the half-bridge instead of Spc110p. In agreement with this arrangement, overexpression of a cytoplasmic version of Spc110p disrupts the interaction between the duplication plaque and the half-bridge (Adams and Kilmartin, 1999). Adams and Kilmartin (1999) propose that during the normal insertion process, the half-bridge retracts across the nuclear face of the duplication plaque, exposing Spc42p and Spc29p for interaction with the Spc110p–calmodulin complex in the nucleus.

Additional proteins are required for the duplication process, including two ubiquitinlike proteins, Dsk2p and Rad23p, and a proteosome cap subunit, Pcs1p (Biggins *et al.,* 1996; McDonald and Byers, 1997). Mps1p is a dual specificity protein kinase required to convert the satellite into a duplication plaque (Lauzé *et al.,* 1995; Winey *et al.,* 1991). In *mps1* mutants, the half-bridge enlarges but the satellite is not stable and a duplication plaque does not form (Winey *et al.,* 1991). Mps1p has additional functions, including activation of the Mad1p-dependent mitotic checkpoint and phosphorylation of Spc110p (Schutz and Winey, 1998; Weiss and Winey, 1996; Friedman, Winey, and Davis, unpublished observation). Mps2p and Ndc1p are required for insertion of the duplication plaque into the nuclear membrane (Thomas and Botstein, 1986; Winey *et al.,* 1991, 1993). Ndc1p has several potential transmembrane domains and localizes close to the nuclear envelope of SPBs and nuclear pores, suggesting that these large organelles may insert into the nuclear envelope by a common mechanism (Chial *et al.,* 1998).

The timing of the duplication events during the cell cycle has been established. Cdc31p acts very early, probably during the previous cell cycle, to allow satellite formation (Winey *et al.,* 1991). The requirement for proteins in the previous cell cycle for the subsequent duplication process is reminiscent of the formation of the prereplicative complex required for DNA replication. Like DNA replication, SPB duplication occurs once and only once per cell cycle, and so a mechanism must exist to prevent reduplication. An attractive hypothesis is that the proteosome component and the ubiquitin-like proteins are required to destroy the block to reduplication and allow Cdc31p to prime the half-bridge. Dsk2p acts before Cdc31p and aids its assembly onto the half-bridge (Biggins and Rose, 1994). Once the half-bridge is primed, the satellite is formed during G1, before Start (Byers and Goetsch, 1975). After Start, the satellite grows into a duplication plaque and inserts into the nuclear envelope such that the entire process is completed by the beginning of S phase. The SPBs separate and the spindle forms at the end of S phase (Byers and Goetsch, 1975; Sundberg *et al.,* 1996).

VI. Transcriptional Regulation of Core SPB Components

Most genes encoding spindle pole components are induced during G1 of the cell cycle. The recently published microarray data allows the comparison of the regulation of all the spindle pole genes in a single experiment (Spellman *et al.*, 1998). Interestingly, the relative timing of induction of the genes correlates with the relative timing of incorporation of the proteins into the SPB. The transcription of *SPC42, SPC29,* and *NUD1* is induced threefold in mid-G1, shortly after *CLN1* induction. Transcription of *CNM67* is not as highly induced but seems to be regulated similarly to *SPC42. SPC110* is induced threefold late in G1 just before induction of the histone genes. Expression of the calmodulin gene does not change during the cell cycle. *CNM67, SPC42, SPC29,* and *SPC110* have MluI cell cycle boxes in their promoters within 250 base pairs upstream of the start codon (using the degenerate MCB NCGCGT). The timing of the peaks of *SPC42, SPC29,* and *CNM67* expression is the same as that seen for other genes regulated by MCB. However, the peak of expression of *SPC110* is later, so additional factors may regulate *SPC110* transcription. The fork head transcription factor Hcm1p is required for full induction of *SPC110* transcription but does not seem to be involved in regulating the timing of induction (Zhu and Davis, 1998).

VII. Homologs of Spindle Pole Components in Other Organisms

In budding yeast, the nuclear membrane separates spindle and astral microtubules. Although true in other fungi, this separation does not exist during cell division in most other eukaryotic cells. How other eukaryotic cells nucleate different classes of microtubules is not known. As in yeast, different proteins may be required to nucleate different classes of microtubules, or one protein may be modified in such a way that it can fulfill these functions. In either event, we predict that an Spc110p-related protein will be found in other eukaryotic cells. Using yeast as a starting point for analyzing how the centrosome might be assembled, we speculate that Spc110p will be found at the core of the centrosome, close to and somehow interacting with the microtubule nucleation apparatus in the mammalian cell. Evidence already suggests that calmodulin plays an essential role at the centrosome of all eukaryotic cells. First, calmodulin localizes to the SPB of budding and fission yeast, the MTOC of *Dictyostelium* and *Paramecium,* and the pericentriolar region in cultured mammalian cells (Geiser *et al.*, 1993; Moser *et al.*, 1997; Willingham *et al.*, 1983; Zhu and Clarke, 1992). Calmodulin localizes to the pericentriolar region in PTK1 cells after nocodazole treatment indicating that calmodulin is part of the core centrosome (Sweet and Welsh, 1988). Second, calmodulin is required for cell proliferation in cultured fibroblasts (Rasmussen and Means, 1989). Expression of antisense calmodulin RNA in mammalian tissue culture cells leads to a delay during mitosis (Rasmussen *et al.*, 1990). It seems entirely possible an interaction between calmodulin and an Spc110p-like molecule is responsible for these effects.

Possible Spc110p-related proteins have been identified using monoclonal antibodies that recognize two different epitopes of yeast Spc110p (Tassin *et al.*, 1997). These antibodies bind to a 100-kDa protein in extracts enriched for centrosomes from human cells and to a 116-kDa protein from *Xenopus* egg extracts. Several observations suggest that these immunologically related proteins are the higher eukaryotic versions of Spc110p. First, Spc110p antibodies recognize centrosomes in both *Xenopus* and human cells. Second, as is seen for Spc110p, the 100-kDa protein from human cells is phosphorylated during mitosis. Third, a portion of the Spc110p-related protein copurifies with a 25S γ-TuRC-like complex from *Xenopus* extracts. Furthermore, antibodies against Spc110p inhibit nucleation by centrosomes isolated from *Xenopus*. Taken together, these results suggest that an Spc110p-related protein is involved in microtubule nucleation in other eukaryotes. The finding that an Spc110p-related protein associates with the γ-TuRC is intriguing, since it implies that there is conservation of the essential microtubule-nucleating components.

Functionally, Spc110p resembles pericentrin. Pericentrin and γ-tubulin are closely associated in the centrosome, and antibodies against pericentrin disrupt microtubule nucleation (Doxsey *et al.*, 1994). However, when examined at the molecular level, the relationship between Spc110p and pericentrin seems strained. Like Spc110p, pericentrin has a large coiled-coil region, but this is characteristic of many centrosomal proteins (Stearns and Winey, 1997). At 220 kDa, pericentrin is much larger than Spc110p or Spc110p-related proteins and thus does not appear to be the protein identified by Tassin and co-workers. Also, careful scrutiny of the pericentrin sequence does not reveal a calmodulin-binding site. As mentioned, because calmodulin regulation is so central for proper assembly of Spc110p into the SPB, it seems likely that a mammalian Spc110p homolog will retain aspects of this relationship, especially since calmodulin is found at the centrosome. An alluring possibility is that the mammalian homolog of Spc110p interacts with pericentrin by means of its coiled region, but we can only speculate until the mammalian Spc110p homolog is cloned and its behavior studied.

VIII. Future Directions

Significant progress has been made in describing both the structure and assembly of the yeast SPB. Perhaps the greatest surprise has been that such a large and complex organelle is composed, at least at its core, of only six highly repeated components. How closely the centrosome will parallel the

SPB in that regard remains to be seen. Of the core SPB proteins, only calmodulin and an Spc110p-related protein have been found in centrosomes. About 100 proteins are estimated to be in the centrosome (Kellogg *et al.,* 1994). How many of these form a core structure compared to the number associated with the microtubules remains to be determined. The obvious lack of structural similarity to the SPB suggests the centrosome is a markedly divergent organelle. On the other hand, at least one important theme from SPB organization is emerging in the centrosome literature. Many centrosomal proteins are predicted to form coiled coils (Bouckson-Castaing *et al.,* 1996; Doxsey *et al.,* 1994; Fry *et al.,* 1998; Heuer *et al.,* 1995; Witczak *et al.,* 1999). Thus, like the proteins of the SPB, coiled-coil interactions may dictate how the PCM is organized. Recent advances in imaging suggest that while lacking the obvious structural regularity of the SPB, the pericentriolar material of the centrosome is not amorphous as had previously been described. The PCM from *Spisula* centrosomes extracted with KI produces a filamentous network of 10–15 nm filaments that lack centrioles when visualized by tomography (Schnackenberg *et al.,* 1998). The proteins that comprise this network have not been identified. Examination of pericentrin stained centrosomes by optical sectioning microscopy also reveals an underlying latticelike network that could be formed by repeating subunits interacting by means of their coiled coil (Dictenberg *et al.,* 1998). It seems reasonable to suggest that just as the SPB is a highly ordered structure organized as layers of filamentous proteins, the centrosome is a highly ordered structure organized as reticula of filamentous proteins.

Now that the major core SPB proteins have been identified, the first phase of SPB analysis is nearing completion, and difficult and interesting questions can be addressed. The SPB has essential functions in cell division, mating, and spore wall formation, but almost nothing is known about how the SPB is modified to fulfill these different roles. In fact, even basic questions are currently unanswered. It is not known whether new SPB components are dispersed to both mother and daughter SPBs during duplication or just to the daughter SPB. We have clues about the order of SPB assembly, but do not know whether the SPB is assembled by joining together subcomplexes or by adding proteins one by one onto the SPB core. Several regulatory proteins that direct duplication and assembly have been identified, but understanding their contributions still lies ahead. Although the arrangement of the major components is known, there are likely to be minor components yet to be added to the picture. Related to the question of assembly is the issue of SPB repair. It is not known whether the SPB can be remodeled in response to miscues in assembly or to damage. The near crystalline packing arrangements of some of the core components might argue against remodeling of the core structure, but the association of some of the components such as Tub4p and Spc72p with the SPB is quite dynamic (Chen *et al.,* 1998; Marschall *et al.,* 1996).

The matter of how the cell regulates the size of the SPB is particularly fascinating. When cells mate, their SPBs fuse to make a structure about 50% larger than in the haploid, so the increase in size is easy to rationalize, but how the cell regulates the size of the daughter SPB is unclear. Similarly, during the equational division of meiosis, both SPBs are reduced in size even though at the outset of division one of them is the characteristic size of the diploid organelle. Equally perplexing is the observation that the SPBs in mitotic cells exist in two different size classes. It would seem that the pathway for enlarging the SPB must be different from that for SPB duplication/assembly, since in the latter case assembly occurs mostly in the cytoplasm, whereas in the former case the SPB is embedded in the nuclear membrane. It seems likely that the nuclear membrane imposes special constraints and requires special modifications to the SPB assembly pathway used during duplication.

The study of how Spc110p and Spc72p modulate microtubule nucleation is just beginning. Since there is no ectopic nucleation of microtubules from the γ-tubulin complex and the complex is localized to the inner and outer plaques, it follows that interaction with nuclear Spc110p and cytoplasmic Spc72p is a critical step in microtubule formation. When binding of the γ-tubulin complex occurs is not known for either protein. Neither are the modifications to Spc110p or Spc72p that lead to nucleation. It is not known whether the cytoplasmic and nuclear γ-tubulin complexes undergo compartment-specific modifications that potentiate nucleation, although it seems likely that nucleation of microtubules would be differentially regulated in the cytoplasm and nucleus.

Finally, the maternal and daughter SPBs are fundamentally different (Vallen *et al.,* 1992). Kinetochore microtubules are associated with the maternal SPB throughout the cell cycle. What role these microtubules play in spindle formation is not known. They could be severed and new microtubules nucleated from the maternal SPB, although no evidence exists for this. Alternatively, they could be regulated to elongate at the appropriate time during mitosis, in which case nucleation would be specific to the daughter SPB. The number of microtubules that form the mitotic spindle appears to be constant, but it is unknown whether the number of microtubules is regulated by SPB size, or whether the size of the SPB is controlled by the number of microtubules. Recent advances in defining the composition and assembly of the SPB give us the molecular tools to answer many of these interesting questions about SPB function and behavior.

References

Adams, I. R., and Kilmartin, J. V. (1999). Localization of core spindle pole body (SPB) components during SPB duplication in *Saccharomyces cerevisiae. J. Cell Biol.* **145,** 809–823.

Asefa, B., Kauler, P., Cournoyer, D., Lehnert, S., and Chow, T. Y. (1998). Genetic analysis of the yeast NUD1 endo-exonuclease: a role in the repair of DNA double-strand breaks. *Curr. Genet.* **34,** 360–7.

Baum, P., Furlong, C., and Byers, B. (1986). Yeast gene required for spindle pole body duplication: homology of its product with Ca^{2+} -binding proteins. *Proc. Natl. Acad. Sci. USA* **83,** 5512–5516.

Berger, B., Wilson, D. B., Wolf, E., Tonchev, T., Milla, M., and Kim, P. S. (1995). Predicting coiled coils by use of pairwise residue correlations. *Proc. Natl. Acad. Sci. USA* **92,** 8259–8263.

Biggins, S., and Rose, M. D. (1994). Direct interaction between yeast spindle pole body components: Kar1p is required for Cdc31p localization to the spindle pole body. *J. Cell Biol.* **125,** 843–852.

Biggins, S., Ivanovska, I., and Rose, M. D. (1996). Yeast ubiquitin-like genes are involved in duplication of the microtubule organizing center. *J. Cell Biol.* **133,** 1331–1346.

Bouckson-Castaing, V., Moudjou, M., Ferguson, D. J., Mucklow, S., Belkaid, Y., Milon, G., and Crocker, P. R. (1996). Molecular characterisation of ninein, a new coiled-coil protein of the centrosome. *J. Cell Sci.* **109,** 179–190.

Brachat, A., Kilmartin, J. V., Wach, A., and Philippsen, P. (1998). *Saccharomyces cerevisiae* cells with defective spindle pole body outer plaques accomplish nuclear migration via half-bridge-organized microtubules. *Mol. Biol. Cell* **9,** 977–991.

Bullitt, E., Rout, M. P., Kilmartin, J. V., and Akey, C. W. (1997). The yeast spindle pole body is assembled around a central crystal of Spc42p. *Cell* **89,** 1077–1086.

Byers, B. (1981a). Cytology of the yeast life cycle. *In* "The Molecular Biology of the Yeast *Saccharomyces:* Life Cycle and Inheritance" (J. N. Strathern, E. W. Jones, and J. R. Broach, Eds.), pp. 59–96. Cold Spring Harbor Laboratory, Cold Spring Harbor, NY.

Byers, B. (1981b). Multiple roles of the spindle pole bodies in the life cycle of *Saccharomyces cerevisiae. In* "Molecular Genetics in Yeast, Alfred Benzon Symposium 16" (D. von Wettstein, J. Friis, M. Kielland-Brandt, and A. Stenderup, Eds.), pp. 119–134. Munksgaard, Copenhagen.

Byers, B., and Goetsch, L. (1974). Duplication of spindle plaques and integration of the yeast cell cycle. *Cold Spring Harb. Symp. Quant. Biol.* **38,** 123–31.

Byers, B., and Goetsch, L. (1975). Behavior of spindles and spindle plaques in the cell cycle and conjugation of *Saccharomyces cerevisiae. J. Bacteriol.* **124,** 511–523.

Chen, X. P., Yin, H., and Huffaker, T. C. (1998). The yeast spindle pole body component Spc72p interacts with Stu2p and is required for proper microtubule assembly. *J. Cell Biol.* **141,** 1169–79.

Chial, H. J., Rout, M. P., Giddings, T. H., and Winey, M. (1998). *Saccharomyces cerevisiae* Ndc1p is a shared component of nuclear pore complexes and spindle pole bodies. *J. Cell Biol.* **143,** 1789–1800.

Cohen, P. (1988). The regulation of phosphorylase kinase activity by calmodulin and troponin. *In* "Calmodulin" (P. Cohen and C. B. Klee, Eds.), Vol. 5, pp. 123–144. Elsevier, Amsterdam.

Davis, T. N. (1992). A temperature-sensitive calmodulin mutant loses viability during mitosis. *J. Cell Biol.* **118,** 607–617.

Dictenberg, J. B., Zimmerman, W., Sparks, C. A., Young, A., Vidair, C., Zheng, Y., Carrington, W., Fay, F. S., and Doxsey, S. J. (1998). Pericentrin and gamma-tubulin form a protein complex and are organized into a novel lattice at the centrosome. *J. Cell Biol.* **141,** 163–174.

Donaldson, A. D., and Kilmartin, J. V. (1996). Spc42p: A phosphorylated component of the *S. cerevisiae* spindle pole body (SPB) with an essential function during SPB duplication. *J. Cell Biol.* **132,** 887–901.

Doxsey, S. J., Stein, P., Evans, L., Calarco, P. D., and Kirschner, M. (1994). Pericentrin, a highly conserved centrosome protein involved in microtubule organization. *Cell* **76,** 639–650.

Elliott, S., Knop, M., Schlenstedt, G., and Schiebel, E. (1999). Spc29p is a component of the Spc110p subcomplex and is essential for spindle pole body duplication. *Proc. Natl. Acad. Sci. USA* **96,** 6205–6210.

Friedman, D. B., Sundberg, H. A., Huang, E., and Davis, T. N. (1996). The 110-kD spindle pole body component of *Saccharomyces cerevisiae* is a phosphoprotein that is modified in a cell cycle-dependent manner. *J. Cell Biol.* **132,** 903–914.

Fry, A. M., Mayor, T., Meraldi, P., Stierhof, Y. D., Tanaka, K., and Nigg, E. A. (1998). C-Nap1, a novel centrosomal coiled-coil protein and candidate substrate of the cell cycle-regulated protein kinase Nek2. *J. Cell Biol.* **141,** 1563–74.

Geier, B. M., Wiech, H., and Schiebel, E. (1996). Binding of centrins and yeast calmodulin to synthetic peptides corresponding to binding sites in the spindle pole body components Kar1p and Spc110p. *J. Biol. Chem.* **271,** 28366–28374.

Geiser, J. R., Sundberg, H. A., Chang, B. H., Muller, E. G. D., and Davis, T. N. (1993). The essential mitotic target of calmodulin is the 110-kilodalton component of the spindle pole body in *Saccharomyces cerevisiae. Mol. Cell. Biol.* **13,** 7913–7924.

Geissler, S., Pereira, G., Spang, A., Knop, M., Soues, S., Kilmartin, J., and Schiebel, E. (1996). The spindle pole body component Spc98p interacts with the gamma-tubulin-like Tub4p of *Saccharomyces cerevisiae* at the sites of microtubule attachment. *Embo J.* **15,** 3899–3911.

Heuer, J. G., Li, K., and Kaufman, T. C. (1995). The *Drosophila* homeotic target gene centrosomin (cnn) encodes a novel centrosomal protein with leucine zippers and maps to a genomic region required for midgut morphogenesis. *Development* **121,** 3861–3876.

Horio, T., Uzawa, S., Jung, M. K., Oakley, B. R., Tanaka, K., and Yanagida, M. (1991). The fission yeast gamma-tubulin is essential for mitosis and is localized at microtubule organizing centers. *J. Cell Sci.* **99,** 693–700.

Kellogg, D. R., Moritz, M., and Alberts, B. M. (1994). The centrosome and cellular organization. *Annu. Rev. Biochem.* **63,** 639–674.

Kilmartin, J. V., and Goh, P.-Y. (1996). Spc110p: Assembly properties and role in the connection of nuclear microtubules to the yeast spindle pole body. *EMBO J.* **15,** 4592–4602.

Kilmartin, J. V., Dyos, S. L., Kershaw, D., and Finch, J. T. (1993). A spacer protein in the *Saccharomyces cerevisiae* spindle pole body whose transcript is cell cycle-regulated. *J. Cell Biol.* **123,** 1175–1184.

Knop, M., and Schiebel, E. (1997). Spc98p and Spc97p of the yeast γ-tubulin complex mediate binding to the spindle pole body via their interaction with Spc110p. *EMBO J.* **16,** 6985–6995.

Knop, M., and Schiebel, E. (1998). Receptors determine the cellular localization of a gamma-tubulin complex and thereby the site of microtubule formation. *Embo J.* **17,** 3952–3967.

Knop, M., Pereira, G., Geissler, S., Grein, K., and Schiebel, E. (1997). The spindle pole body component Spc97p interacts with the γ-tubulin of *Saccharomyces cerevisiae* and functions in microtubule organization and spindle pole body duplication. *EMBO J.* **16,** 1550–1564.

Lauzé, E., Stoelcker, B., Luca, F. C., Weiss, E., Schutz, A. R., and Winey, M. (1995). Yeast spindle pole body duplication gene *MPS1* encodes an essential dual specificity protein kinase. *EMBO J.* **14,** 1655–1663.

Li, Q., and Joshi, H. C. (1995). γ-Tubulin is a minus end-specific microtubule binding protein. *J. Cell Biol.* **131,** 207–214.

Marschall, L. G., Jeng, R. L., Mulholland, J., and Stearns, T. (1996). Analysis of Tub4p, a yeast γ-tubulin-like protein: Implications for microtubule-organizing center function. *J. Cell Biol.* **134,** 443–454.

Martin, M. C., Gunawardane, N., Iwamatsu, A., and Zheng, Y. (1998). Xgrip 109: A γ-Tubulin-associated protein with an essential role in γ-tubulin ring complex (γ-TuRC) assembly and centrosome function. *J. Cell Biol.* **141,** 675–687.

McDonald, H. B., and Byers, B. (1997). A proteasome cap subunit required for spindle pole body duplication in yeast. *J. Cell Biol.* **137,** 539–53.

Mirzayan, C., Copeland, C. S., and Snyder, M. (1992). The *NUF1* gene encodes an essential coiled-coil related protein that is a potential component of the yeast nucleoskeleton. *J. Cell Biol.* **116,** 1319–1332.

Moens, P. B., and Rapport, E. (1971). Spindles, spindle plaques, and meiosis in the yeast *Saccharomyces cerevisiae* (Hansen). *J. Cell Biol.* **50,** 344–61.

Moritz, M., Braunfeld, M. B., Fung, J. C., Sedat, J. W., Alberts, B. M., and Agard, D. A. (1995). Three-dimensional structural characterization of centrosomes from early *Drosophila* embryos. *J. Cell Biol.* **130,** 1149–1159.

Moritz. M., Zheng, Y., Alberts, B. M., and Oegema, K. (1998). Recruitment of the gamma-tubulin ring complex to *Drosophila* salt-stripped centrosome scaffolds. *J. Cell Biol.* **142,** 775–786.

Moser, M. J., Flory, M. R., and Davis, T. N. (1997). Calmodulin localizes to the spindle pole body of *Schizosaccharomyces pombe* and performs an essential function in chromosome segregation. *J. Cell Science* **110,** 1805–1812.

Murphy, S. M., Urbani, L., and Stearns, T. (1998). The mammalian gamma-tubulin complex contains homologues of the yeast spindle pole body components spc97p and spc98p. *J. Cell Biol.* **141,** 663–674.

Nguyen, T., Vinh, D. B. N., Crawford, D. K., and Davis, T. N. (1998). A genetic analysis of interactions with Spc110p reveals distinct functions of Spc97p and Spc98p, components of the yeast gamma-tubulin complex. *Mol. Biol. Cell* **9,** 2201–2216.

Oakley, C. E., and Oakley, B. R. (1989). Identification of γ-tubulin, a new member of the tubulin superfamily encoded by *mip A* gene of *Aspergillus nidulans. Nature* **338,** 662–664.

O'Toole, E. T., Winey, M., and McIntosh, J. R. (1999). High-voltage electron tomography of spindle pole bodies and early mitotic spindles in the yeast *Saccharomyces cerevisiae Mol. Biol. Cell* **10,** 2017–2031.

Rasmussen, C. D., and Means, A. R. (1989). Calmodulin is required for cell-cycle progression during G1 and mitosis. *EMBO J.* **8,** 73–82.

Rasmussen, C. D., Means, R. L., Lu, K. P., May, G. S., and Means, A. R. (1990). Characterization and expression of the unique calmodulin gene of *Aspergillus nidulans. J. Biol. Chem.* **265,** 13767–13775.

Rose, M. D., and Fink, G. R. (1987). *KAR1,* a gene required for function of both intranuclear and extranuclear microtubules in yeast. *Cell* **48,** 1047–1060.

Rout, M. P., and Kilmartin, J. V. (1990). Components of the yeast spindle and spindle pole body. *J. Cell Biol.* **111,** 1913–1927.

Rout, M. P., and Kilmartin, J. V. (1991). Yeast spindle pole body components. *Cold Spring Harbor Symp. Quant. Biol.* **56,** 687–691.

Schnackenberg, B. J., Khodjakov, A., Rieder, C. L., and Palazzo, R. E. (1998). The disassembly and reassembly of functional centrosomes in vitro. *Proc. Natl. Acad. Sci. USA* **95,** 9295–9300.

Schutz, A. R., and Winey, M. (1998). New alleles of the yeast MPS1 gene reveal multiple requirements in spindle pole body duplication. *Mol. Biol. Cell* **9,** 759–774.

Semionov, A., Cournoyer, D., and Chow, T. Y. (1999). The effect of the *Saccharomyces cerevisiae* endo-exonuclease NUD1 gene expression on the resistance of HeLa cells to DNA-damaging agents. *Mutat. Res.* **433,** 169–181.

Sobel, S. G., and Snyder, M. (1995). A highly divergent γ-tubulin gene is essential for cell growth and proper microtubule organization in *Saccharomyces cerevisiae. J. Cell Biol.* **131,** 1775–1788.

Soues, S., and Adams, I. R. (1998). SPC72: a spindle pole component required for spindle orientation in the yeast *Saccharomyces cerevisiae. J. Cell Sci.* **111,** 2809–2818.

Spang, A., Courtney, I., Fackler, U., Matzner, M., and Schiebel, E. (1993). The calcium-binding protein cell division cycle 31 of *Saccharomyces cerevisiae* is a component of the half bridge of the spindle pole body. *J. Cell Biol.* **123,** 405–416.

Spang, A., Courtney, I., Grein, K., Matzner, M., and Schiebel, E. (1995). The Cdc31p-binding protein Kar1p is a component of the half bridge of the yeast spindle pole body. *J. Cell Biol.* **128,** 863–877.

Spang, A., Geissler, S., Grein, K., and Schiebel, E. (1996a). γ-Tubulin-like Tub4p of *Saccharomyces cerevisiae* is associated with the spindle pole body substructures that organize microtubules and is required for mitotic spindle formation. *J. Cell Biol.* **134,** 429–441.

Spang, A., Grein, K., and Schiebel, E. (1996b). The spacer protein Spc110p targets calmodulin to the central plaque of the yeast spindle pole body. *J. Cell Sci.* **109,** 2229–2237.

Spellman, P. T., Sherlock, G., Zhang, M. Q., Iyer, V. R., Anders, K., Eisen, M. B., Brown, P. O., Botstein, D., and Futcher, B. (1998). Comprehensive identification of cell cycle-regulated genes of the yeast *Saccharomyces cerevisiae* by microarray hybridization. *Mol. Biol. Cell* **9,** 3273–3297.

Stearns, T., and Kirschner, M. (1994). In vitro reconstitution of centrosome assembly and function: the central role of γ-tubulin. *Cell* **76,** 623–637.

Stearns, T., and Winey, M. (1997). The cell center at 100. *Cell* **91,** 303–309.

Stirling, D. A., Welch, K. A., and Stark, M. J. R. (1994). Interaction with calmodulin is required for the function of Spc110p, an essential component of the yeast spindle pole body. *EMBO J.* **13,** 4329–4342.

Stirling, D. A., Rayner, T. F., Prescott, A. R., and Stark, M. J. R. (1996). Mutations which block the binding of calmodulin to Spc110p cause multiple mitotic defects. *J. Cell Sci.* **109,** 1297–1310.

Straight, A. F., Marshall, W. F., Sedat, J. W., and Murray, A. W. (1997). Mitosis in living budding yeast: anaphase A but no metaphase plate. *Science* **277,** 574–578.

Sullivan, D. S., and Huffaker, T. C. (1992). Astral microtubules are not required for anaphase B in *Saccharomyces cerevisiae. J. Cell Biol.* **119,** 379–388.

Sun, G. H., Ohya, Y., and Anraku, Y. (1991). Half-calmodulin is sufficient for cell proliferation. Expressions of N- and C-terminal halves of calmodulin in the yeast *Saccharomyces cerevisiae. J. Biol. Chem.* **266,** 7008–7015.

Sundberg, H. A., and Davis, T. N. (1997). A mutational analysis identifies three functional regions of the spindle pole component Spc110p in *Saccharomyces cerevisiae. Mol. Biol. Cell* **8,** 2575–2590.

Sundberg, H. A., Goetsch, L., Byers, B., and Davis, T. N. (1996). Role of calmodulin and Spc110p interaction in the proper assembly of spindle pole body components. *J. Cell Biol.* **133,** 111–124.

Sweet, S. C., and Welsh, M. J. (1988). Calmodulin colocalization with cold-stable and nocodazole-stable microtubules in living PtK1 cells. *Eur. J. Cell Biol.* **47,** 88–93.

Tassin, A. M., Celati, C., Paintrand, M., and Bornens, M. (1997). Identification of an Spc110p-related protein in vertebrates. *J. Cell Sci.* **110,** 2533–2545.

Tassin, A. M., Celati, C., Moudjou, M., and Bornens, M. (1998). Characterization of the human homologue of the yeast Spc98p and its association with gamma-tubulin. *J. Cell Biol.* **141,** 689–701.

Thomas, J. H., and Botstein, D. (1986). A gene required for the separation of chromosomes on the spindle apparatus in yeast. *Cell* **44,** 65–76.

Vallen, E. A., Scherson, T. Y., Roberts, T., van Zee, K., and Rose, M. D. (1992). Asymmetric mitotic segregation of the yeast spindle pole body. *Cell* **69,** 505–515.

Vogel, J. M., Stearns, T., Rieder, C. L., and Palazzo, R. E. (1997). Centrosomes isolated from *Spisula solidissima* oocytes contain rings and an unusual stoichiometric ratio of alpha/beta tubulin. *J. Cell Biol.* **137,** 193–202.

Wang, P. J., and Huffaker, T. C. (1997). Stu2p: A microtubule-binding protein that is an essential component of the yeast spindle pole body. *J. Cell Biol.* **139,** 1271–1280.

Weiss, E., and Winey, M. (1996). The *Saccharomyces cerevisiae* spindle pole body duplication gene *MPS1* is part of a mitotic checkpoint. *J. Cell Biol.* **132,** 111–123.

Wigge, P. A., Jensen, O. N., Holmes, S., Soues, S., Mann, M., and Kilmartin, J. V. (1998). Analysis of the *Saccharomyces* spindle pole by matrix-assisted laser desorption/ ionization (MALDI) mass spectrometry. *J. Cell Biol.* **141,** 967–977.

Willingham, M. C., Wehland, J., Klee, C. B., Richert, N. D., Rutherford, A. V., and Pastan, I. H. (1983). Ultrastructural immunocytochemical localization of calmodulin in cultured cells. *J. Histochem. Cytochem.* **31,** 445–461.

Winey, M., Goetsch, L., Baum, P., and Byers, B. (1991). *MPS1* and *MPS2:* Novel yeast genes defining distinct steps of spindle pole body duplication. *J. Cell Biol.* **114,** 745–754.

Winey, M., Hoyt, M. A., Chan, C., Goetsch, L., Botstein, D., and Byers, B. (1993). *NDC1:* A nuclear periphery component required for yeast spindle pole body duplication. *J. Cell Biol.* **122,** 743–751.

Winey, M., Mamay, C. L., O'Toole, E. T., Mastronarde, D. N., Giddings Jr., T. H., McDonald, K. L., and McIntosh, J. R. (1995). Three-dimensional ultrastructural analysis of the *Saccharomyces cerevisiae* mitotic spindle. *J. Cell Biol.* **129,** 1601–1615.

Witczak, O., Skalhegg, B. S., Keryer, G., Bornens, M., Tasken, K., Jahnsen, T., and Orstavik, S. (1999). Cloning and characterization of a cDNA encoding an A-kinase anchoring protein located in the centrosome, AKAP450. *Embo J.* **18,** 1858–1868.

Wolf, E., Kim, P. S., and Berger, B. (1997). MultiCoil: a program for predicting two- and three-stranded coiled coils. *Protein Sci.* **6,** 1179–1189.

Yeh, E., Skibbens, R. V., Cheng, J. W., Salmon, E. D., and Bloom, K. (1995). Spindle dynamics and cell cycle regulation of dynein in the budding yeast, *Saccharomyces cerevisiae. J. Cell Biol.* **130,** 687–700.

Zheng, Y., Jung, M. K., and Oakley, B. R. (1991). γ-Tubulin is present in *Drosophila melanogaster* and *Homo sapiens* and is associated with the centrosome. *Cell* **65,** 817–823.

Zheng, Y., Wong, M.-L., Alberts, B., and Mitchison, T. (1995). Nucleation of microtubule assembly by a γ-tubulin-containing ring complex. *Nature* **378,** 578–583.

Zhu, Q., and Clarke, M. (1992). Association of calmodulin and an unconventional myosin with the contractile vacuole complex of *Dictyostelium discoideum. J. Cell Biol.* **118,** 347–358.

Zhu, G., and Davis, T. N. (1998). The fork head transcription factor Hcm1p participates in the regulation of SPC110, which encodes the calmodulin-binding protein in the yeast spindle pole body. *Biochim. Biophys. Acta* **1448,** 236–244.

6

The Microtubule Organizing Centers of *Schizosaccharomyces pombe*

Iain M. Hagan and *Janni Petersen*
School of Biological Sciences
University of Manchester
Manchester M13 9PT, United Kingdom

I. Introduction

The fission yeast *Schizosaccharomyces pombe* is assuming an increasingly higher profile in the study of diverse aspects of microtubule function during mitosis and in cellular morphogenesis. The radical differences in the cell cycle organization, morphology, and great evolutionary distance that distinguishes *S. pombe* from the most intensively studied model eukaryote, *Saccharomyces cerevisiae,* means that valuable lessons will be learned from comparative studies of these two unicellular systems. There are three distinct microtubule organizing centers in *S. pombe.* One of these, the spindle pole body (SPB) is present throughout the entire life cycle, whereas the other two are transient structures. The EMTOC forms at the end of mitosis at the site of cell division; the TAM forms at the cell tip during mating. Thus, the system not only offers many insights into the molecular basis of microtubule nucleation in general, but lends itself to the study of *de novo* MTOC formation.

Current Topics in Developmental Biology, Vol. 49

0070-2153/00 $35.00

A. Microtubule Organization in *S. pombe*

As with many other eukaryotes, DNA replication and chromosome segregation occur in distinct cell cycle phases (Nurse, 1985). During interphase, cytoplasmic microtubules extend along the long axis of the cell between the two cell tips to generate an apparently random array (Fig. 1) (Hagan and Hyams, 1988). Most of the microtubules in this array do not extend the full length of the cell. These microtubules play a key role in maintaining the direction of polarized growth (Toda *et al.,* 1983; Verde *et al.,* 1995; Beinhauer *et al.,* 1997; Mata and Nurse, 1997; Sawin and Nurse, 1997; Hagan, 1998). Once a critical size threshold has been attained, growth by cell tip extension ceases and mitosis is initiated (Mitchison and Nurse, 1985). Mitotic commitment is accompanied by the rapid loss of cytoplasmic microtubules and the nucleation of microtubules from the intranuclear face of the SPBs (McCully and Robinow, 1971; Ding *et al.,* 1997). A short spindle forms between the two SPBs and elongates to span the nucleus. Bundles of microtubules also begin to emerge from the cytoplasmic side of the SPB during this spindle elongation (Hagan and Hyams, 1988). These astral microtubules arise from completely different nucleating sites to those that generate the intranuclear spindle (Tanaka and Kanbe, 1986; Ding *et al.,* 1997). Astral microtubules are particularly prominent in anaphase B when they extend tangentially from the SPB until they contact the cell wall (Fig. 1; Section III, A; Fig. 4d–f). Although there is generally only one astral bundle per SPB, multiple bundles can be seen extending in different directions from some anaphase SPBs, indicating that astral microtubules may be dynamic. Occasionally astral microtubules associate with the plasma membrane bending the spindle, leading to the speculation that they may be interacting with plasma membrane associated motor proteins to assist in pole separation (Hagan and Hyams, 1996). The number of microtubules in the central spindle decreases while their average length increases as anaphase B progresses (Ding *et al.,* 1993). In wild-type cells anaphase B continues until the nuclei reach the cell tips, although the duration of anaphase B is subject to timer and cell volume dependent regulation (Hagan *et al.,* 1990). The spindle then breaks down and microtubules are nucleated from three points in the cell: the cytoplasmic face of each SPB and a novel microtubule organizing center that forms at the cell equator (Hagan and Hyams, 1988; Hagan and Yanagida, 1997; Horio *et al.,* 1991). The SPBs then move around each nucleus to lead them back along cytoplasmic microtubules toward the cell center until they reach a point that will constitute the middle of the daughter cells after division. The migrating nuclei often adopt a pointed morphology with the SPB at the apex as if the point at which they experience maximum force during migration toward the cell center is at the SPB (Hagan and Yanagida, 1997). Upon completion of

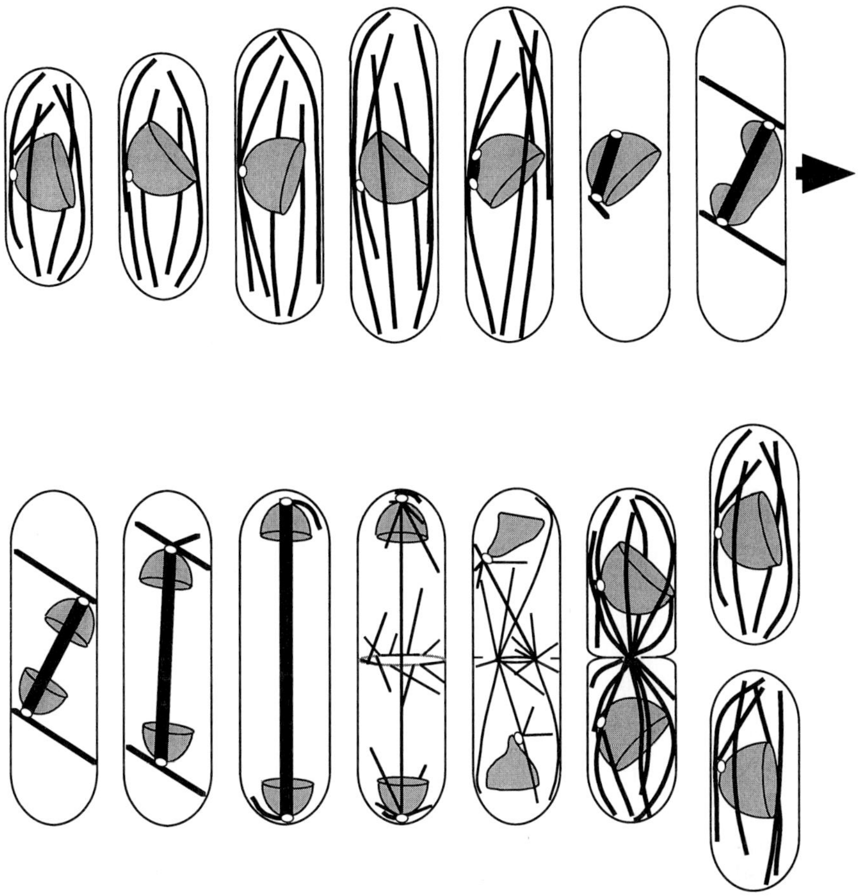

Figure 1 Changes in the organization of the fission yeast microtubule cytoskeleton that accompany progression through the mitotic cell cycle. Microtubules are shown as black lines. The large gray objects represent the chromatin and the open circles the SPB. Cell cycle progression is from left to right and from the top row to the bottom. During interphase the microtubules form a basket that runs along the cell periphery apparently in association with or just below the plasma membrane. During interphase the nucleus gently rocks back and forth, presumably because the SPB is oscillating around the middle of the cell. Upon commitment to mitosis the random array of interphase microtubules is replaced by a short spindle that elongates to span the nucleus. Astral microtubules extend tangentially from the cytoplasmic face of the SPB during this spindle elongation. Once the chromosomes are correctly aligned, the spindle elongates again to extend the length of the cell. Multiple astral microtubules are sometimes seen during anaphase B. Upon completion of anaphase B the spindle breaks down and the cytoplasmic face of the SPB and the EMTOC become very active as microtubule organizing centers. The faint ring around the center of the cell at this time indicates that the exact nature of the EMTOC is still unknown. The microtubules of the EMTOC and the cytoplasmic face of the SPB often appear to interact as the SPB leads the nucleus toward the middle of the two daughter cells as the cytoplasmic microtubule network becomes more extensive. Microtubule nucleation is apparently random after cytokinesis is complete. Adapted with permission from Hagan (1998).

cytokinesis, microtubule nucleation is again random. A full discussion of this microtubule cytoskeleton can be found in Hagan (1998); details of the kinetics of discrete stages of spindle formation and elongation can be found in Nabeshima *et al.* (1998).

II. Cytoplasmic Microtubule Organizing Centers

A. The Equatorial MTOC

A particularly interesting feature of fission yeast mitosis is the formation, *de novo,* of an MTOC at the cell equator at the end of mitosis. This structure is referred to as the equatorial MTOC (EMTOC). The EMTOC is reminiscent of the preprophase band seen in plant systems and in some respects may be likened to the midbody. However, whereas the midbody may arise from antiparallel sliding of microtubules and microtubule nucleation, the EMTOC forms in the absence of any other microtubules (Heitz and Hagan, unpublished). Although the EMTOC forms near the site of MT overlap in the late anaphase spindle, the nuclear envelope remains intact throughout mitosis in fission yeast and so separates the overlap zone from the site of EMTOC formation on the plasma membrane. The independence of EMTOC formation and the position of the overlap zone are underscored by mutants in which the spindle wanders into one half of the cell but the EMTOC remains centrally placed (Hagan and Hyams, 1988).

A number of studies suggest that the EMTOC is a ring structure that resembles the cytokinetic ring (Horio *et al.*, 1991; Nabeshima *et al.*, 1995; Pichova *et al.*, 1995; Beinhauer *et al.*, 1997; Heitz and Hagan, unpublished). Chimeric GFP-dis1 fusion proteins stain both the SPB and a constricting ring in the center of the cell during cytokinesis (Nabeshima *et al.*, 1995). Similarly, GFP/Mal3 fusion proteins stain microtubules and give a similar equatorial ring staining during cytokinesis (Beinhauer *et al.*, 1997). Finally, confocal microscopy shows ring-shaped microtubule configurations in the cell equator at the end of mitosis (Pichova *et al.*, 1995). It is therefore a strong possibility that a component of the cytokinetic ring will emerge as a template for the recruitment of γ-tubulin-containing complexes to form the EMTOC.

The limited electron microscopy of the sites of microtubule nucleation in late anaphase has provided little detail of EMTOC structure. The best images to date show five microtubules extending randomly from a very small amorphous structure within the main body of the cell (Kanbe *et al.*, 1989; Horio *et al.*, 1991). This pattern is consistent with immunofluorescence images that originally led to the conclusion that the EMTOC was composed of discrete MTOC nucleation points rather than a ring (Hagan and Hyams, 1988). It is therefore possible that the EMTOC is a noncontinuous structure or that microtubules are released by the EMTOC after they have been nucleated and that motor activity generates the aggregates seen in these EM studies.

In the absence of mutants that specifically disrupt EMTOC formation without affecting any other aspects of the cell cycle, it is currently difficult to definitively determine EMTOC function. Colocalization of the SPB, microtubules, and chromatin in a range of different cell cycle mutant backgrounds supports the view that the position of the interphase nucleus is determined by SPB microtubule interactions (Hagan and Yanagida, 1997). It is therefore possible that nucleation of microtubules from the site of cytokinesis is a means of giving the nucleus every opportunity to move away from a danger zone where it could be randomly cleaved by the constricting septation machinery. The potential for such catastrophic events in the logarithmically growing cells used for most studies in the literature is slight; however, a response to nutrient limitation is a reduction in the size threshold for commitment to mitosis (Nasmyth, 1979). This reduction in cell size greatly increases the likelihood that one or other of the postmitotic nuclei may be randomly cut in two by the septum. In the natural environment where nutrient starvation is thought to be a frequent occurrence, the ability to ensure that nuclei rapidly gain the correct position after mitosis would confer a distinct advantage.

B. The Tip Associated MTOC That Forms during Conjugation

Fission yeast cells switch between one of two mating types (minus, h^-, or plus, h^+). Starvation of a population of mixed mating type results in growth of cells of opposite mating type toward one another (Yamamoto *et al.,* 1997). After the two cell tips touch, the cell walls separating the two cytoplasms are degraded and the nuclei migrate toward one another before fusing at their SPBs (Robinow, 1977; Hirata and Tanaka, 1982; Chikashige *et al.,* 1994, 1997; Svoboda *et al.,* 1995; Ding *et al.,* 1998). Localization of microtubules during these rearrangements revealed two foci for microtubule nucleation: one at the SPB and the other at the cell tip (Figs. 2a–c,f; see color insert) (Petersen *et al.,* 1998a). The tip-associated MTOC (TAM) is only seen in 20% of touching prezygotes. This frequency is in remarkable accord with the frequency with which the formin Fus1 is seen at the tips of these cells (Petersen *et al.,* 1998b), suggesting that similar controls act to recruit TAM components and Fus1 once cells have touched. Moreover, the departure of centromeres from the SPBs also requires cell-to-cell contact of the prezygotes (Section III,D, Fig. 6C) (Chikashige *et al.,* 1997). Together these data suggest that cell contact produces a signal that generates changes in the organization of the actin and microtubule cytoskeletons and the nucleus.

The mechanism determining where and when the TAM will form still remains to be established; however, it may share a common mechanism with EMTOC formation. The F-actin cytoskeleton rearranges during mating to form a highly polarized structure at the tip of conjugating cells that involves the function of multiple molecules, including profilin and a formin (Petersen *et al.,* 1998b, c). Similar molecules execute essential functions in the actin ring that forms at the cell equator to execute cytokinesis (Gould and Simanis, 1997), raising the distinct possibility that the TAM forms by association with similar template molecules to those that may be responsible for the formation of the EMTOC. Whether or not TAM requires an F-actin cytoskeleton for assembly, F-actin is not required to maintain association of this structure with the tip (Figs. 2d,e). In contrast, cold shock, which, among other things, disrupts microtubules, can result in TAM disassociation from the tip (Fig. 2f). These data are summarized in cartoon form in Fig. 3.

III. The Spindle Pole Body

A. SPB Structure

In common with some other fungi, the interphase fission yeast SPB is composed of two component parts separated by the nuclear membrane (McCully and Robinow, 1971; Health, 1981; Ding *et al.,* 1997). A large,

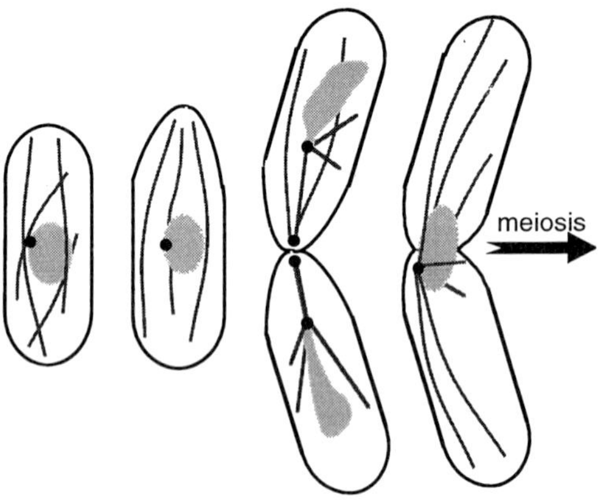

Figure 3 *S. pombe* microtubule cytoskeleton during mating. Microtubules (lines), chromatin (gray), and regions containing γ-tubulin (dots) are shown. Reproduced with permission from *Current Biology* (Petersen *et al.,* 1998a).

highly organized structure is found on the cytoplasmic side; a much less well defined amorphous mass is associated with the nuclear side of the membrane [Figs. 4a (arrow) and 5a] (Ding *et al.,* 1997). The cytoplasmic component resembles the budding yeast SPB in that it is a multilayered structure that is associated with a smaller bridge structure that extends over the nuclear envelope. Although these structures appear to be distinct, thin filaments appear to run through the membrane and connect these two parts of the SPB [Figs. 4a (filaments) and 5]. The connection between these two elements is likely to be responsible for maintaining a connection between the two parts of the SPB to enable spindle formation and genome positioning in interphase (Sections II,A, III,D).

SPB duplication in fission yeast contrasts with the conservative generation of a new SPB in budding yeast (O'Toole and McIntosh, 1999). In fission yeast, a single, large, interphase cytoplasmic SPB structure is replaced by two smaller cytoplasmic bodies that are connected by a bridge structure (Fig. 5b) (Ding *et al.,*1997). The smaller structures then expand until they have regained the size of the original SPB. The bridge continues to connect the two SPBs until mitosis (Figs. 4b, 5c), when SPB organization undergoes radical alterations; the nuclear envelope falls away from the cytoplasmic domain and fragments; the region beneath the cytoplasmic component fills with large, granular, amorphous, osmophilic structures (Figs. 4c, 5d). The nuclear envelope beneath the two SPBs then completely disperses, allowing the cytoplasmic and nuclear SPB components to unite to form a single structure (Figs. 4d, 4e, 5e). Microtubule nucleation from the inner face of each SPB follows, and microtubule interdigitation forces, presumably driven by microtubule motor proteins such as the kinesin-related protein Cut7, push the two SPBs apart (Fig. 5f) (Hagan and Yanagida, 1990, 1992). As anaphase B proceeds, the fenestra in the nuclear envelope decreases in diameter (Figs. 4f, 5g) resulting in the separation of the two components once again by an intact nuclear envelope at the end of mitosis.

B. Regulation of the SPB Cycle

Surprisingly little is known about the regulation of the SPB duplication cycle in fission yeast. The timing of SPB duplication has been judged by electron microscopic analysis of cells synchronized with respect to cell cycle progression to be in G2 (Ding *et al.,* 1997). The only candidates for regulators of the SPB duplication cycle that have been identified to date are *mph1*$^+$ and *dph1*$^+$ (He *et al.,* 1998). Both are implicated in SPB regulation because of the role played by the homologous proteins in budding yeast. Mph1 is related to the budding yeast gene *MPS1*$^+$, which is required for both the spindle assembly checkpoint and SPB duplication (Lauze *et al.,*

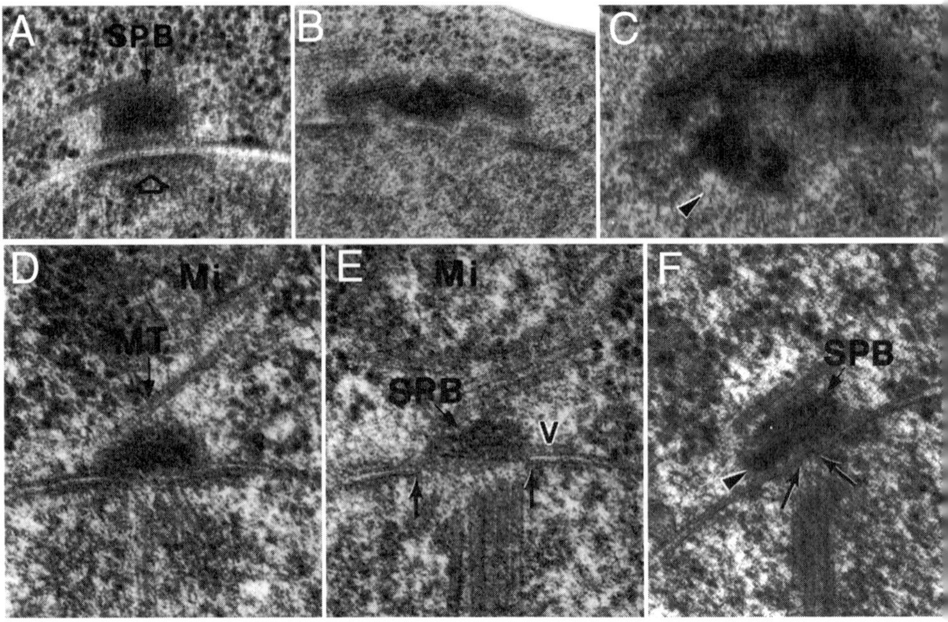

Figure 4 The spindle pole body integrates into the nuclear envelope during mitosis in fission yeast. The figure shows a number of images of the SPB reproduced with permission from the manuscript by Ding *et al.* (1997) in which they use freeze substitution electron microscopy to prepare cells for the definitive electron microscopic analysis of SPB structure throughout the *S. pombe* cell cycle. The interphase SPB is composed of a large cytoplasmic structure (labeled SPB) which sits on the opposite side of the nuclear envelope to an amorphous osmophillic region that contains γ-tubulin (open arrow in a). This nucleoplasmic region is presumably also where the centromeres associate with the SPBs in interphase cells, although this awaits confirmation. As the SPBs duplicate and remain connected by a bridge structure, the nucleoplasmic material enlarges to form a ring structure around the SPBs (b). Around the time of mitosis the nuclear envelope in the SPBs fragments and the region between the SPB and the nuclear envelope fills with large amorphous structures (arrowhead in c). (d, e) Serial sections through an early anaphase spindle show that the SPB sits within a fenestra in the nuclear envelope (arrows in e). A bridge structure extends along the cytoplasmic surface of the nuclear envelope (v in e). Multiple astral microtubules associate tangentially with the SPB. A mitochondrion is often seen in the region around the SPB (McCully and Robinow, 1971; Kanbe *et al.*, 1989; Ding *et al.*, 1997). As anaphase B progresses the gap in the nuclear envelope gets smaller (small arrows in f) and the larger SPB component becomes cytoplasmic once more. The arrowhead in f indicates the presence of a dark half bridge structure.

1995; Hardwick *et al.*, 1996). Dph1 is related to the budding yeast gene DSK2^{+}, which was isolated in a screen for suppressors of a *kar1*^{ts} mutant (Biggins *et al.*, 1996). Dsk2p has extensive similarity to ubiquitin at its amino terminus. *KAR1*$^{+}$ encodes a structural component of the budding yeast half-bridge, and *kar1* mutants have an extended SPB structure that

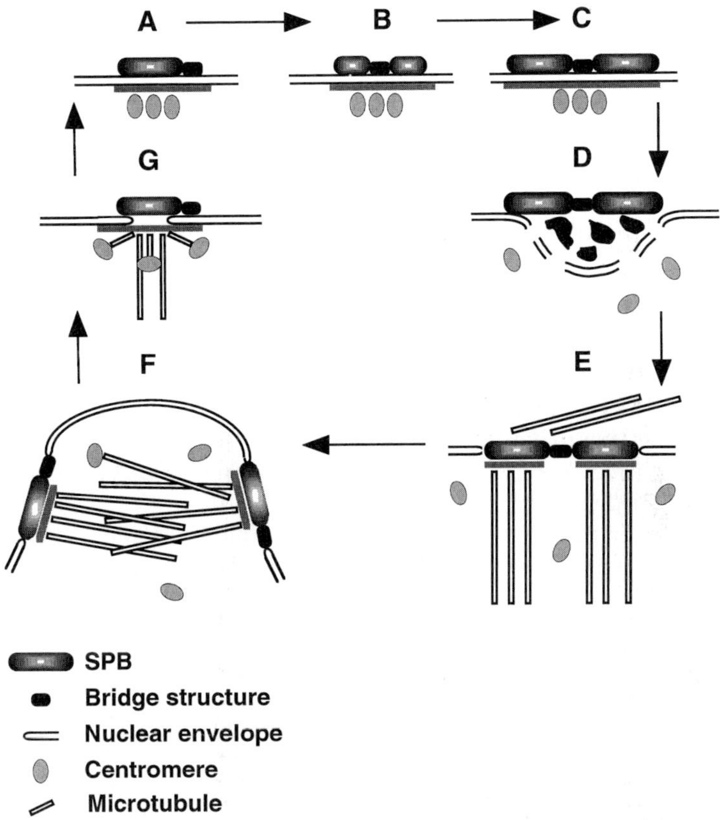

Figure 5 A cartoon summarizing the changes in the SPB structure and function that accompany progression of the mitotic cell cycle in *S. pombe.* See Section III,A for details.

lacks a half-bridge (Rose and Fink, 1987). Thus, Dph1 could be expected to play some role in SPB structure or regulation in fission yeast. Although this is backed up by a minor defect in spindle formation upon overexpression, the gene is not essential. Similarly, *mph1*$^+$ also differs from its budding yeast relative is not being essential. Moreover, manipulations of the gene carried out to date have failed to influence the SPB cycle. Thus, neither gene product appears to completely mimic the function of its budding yeast counterpart.

A biochemical approach toward understanding SPB activation has been used by Masuda and colleagues. They have a permeabilized cell system as an assay to fractionate *Xenopus* egg extracts to identify fractions that activate microtubule nucleation on interphase SPBs (Masuda *et al.,* 1992). An activity has been identified in mitotic but not interphase extracts. Interestingly this activity appears to associate with the amino terminus of γ-tubulin (Masuda and Shibata, 1996) and has been identified as ribonucleotide reductase (H. Masuda, personal communication).

C. The SPB and Cell Cycle Regulation

1. The SPB and Regulation of the G2/M Transition

The first cell cycle regulators to be found at the SPB were the two components of MPF, the protein kinase complex that regulates commitment to mitosis $p34^{cdc2}$ and Cdc13/cyclin B (Alfa *et al.,* 1990). A strong association of the complex with the SPB was seen in mitotic cells up until the destruction of cyclin B early in anaphase B. The characterization of the *cut12/stf1*$^{+}$ gene suggests that events at the SPB association may make a crucial contribution to the correct regulation of the activation of MPF at the G2/M transition (Bridge *et al.,* 1998).

During normal cell cycle progression MPF is inhibited by phosphorylation of a tyrosine residue at position 15 by Wee1/Mik1 kinase (Nurse, 1990; Dunphy, 1994). This inhibitory phosphate is removed by a phosphatase that is encoded by the *cdc25*$^{+}$ gene. The timing of the commitment to mitosis is therefore largely determined by the balance of activities of Wee1 and Cdc25. Mutations in the *stf1*$^{+}$ gene were identified because they conferred the ability to grow in the absence of Cdc25 activity (Hudson *et al.,* 1990, 1991). The *cut12.1* mutation, on the other hand, was studied because it resulted in a temperature-sensitive inability to form a spindle (Bridge *et al.,* 1998). Light and electron microscopy showed that Cut12 encodes and essential component of the SPB. Suprisingly, *cut12*$^{+}$ is allelic to *stf1*$^{+}$. The ability of mutations in a structural component of the SPB to alter the requirement for an otherwise essential MPF regulator suggest that events at the SPB play a crucial role in regulating cell cycle progression. It is tempting to speculate that a cell cycle regulatory role may underly the parthenogenetic activity of centrosomes (see Chapter 1).

2. The SPB and Regulation of Cytokinesis

Cytokinesis is effected by the constriction, at the end of anaphase B, of an F-actin ring that forms upon mitotic commitment. Constriction of the ring and the associated events of septation are controlled by a protein kinase network that is governed by the activity of an SPB-associated G protein called Spg1 (Schmidt *et al.,* 1997). SPB-associated Spg1 is found in its inactive, GDP-bound state during interphase, but is converted into the active GTP-bound form upon commitment to mitosis (Sohrmann *et al.,* 1998). Spg1 activation results in recruitment of the protein kinase Cdc7 to the separating SPBs. Cdc7 has a high affinity for the SPB until early anaphase B, when it is lost from one of the two SPBs. Consistently, antibodies that can differentiate between the GTP and GDP bound, forms of Spg1 show that the SPB that loses Cdc7 has GDP-bound inactive Spg1. In common with other G proteins of the ras super-family, the GTPase activity of Spg1 is stimulated by a GAP activity and the exchange of GDP for GTP is presumably stimulated by a GEF activity (although the GEF for Spg1 awaits identification). The GAP activity for Spg1 is provided by the combined action of the products of two genes: *byr4*$^+$ and *cdc16*$^+$ (Fankhauser *et al.*, 1993; Furge *et al.*, 1998). As expected, Cdc16 and Byr4 both associate with the interphase SPB; however, the precise details of their mitotic distribution reveals further levels of complexity to the regulation of the Spg1 cascade (Cerutti and Simanis, 1999).

Byr4 is found on both SPBs of the early mitotic spindle (Jwa and Song, 1998; Cerutti and Simanis, 1999; Jwa *et al.*, 1999). Byr4 is then localized on only one of the two SPBs during anaphase B. However, the asymmetry in Byr4 distribution is established before Cdc7 distribution becomes asymmetric. Epitope-tagged versions of Cdcl6 are found on the SPB throughout interphase, but the staining becomes weak upon commitment to mitosis (Cerutti and Simanis, 1999). Staining remains weak until the final stages of anaphase B after both Cdc7 and Byr4 have become asymmetric when a strong signal was seen on one SPB. As would be expected for Spg1 GAP proteins, Byr4 and Cdc16 both associated with the late anaphase B SPB that lacked Cdc7 (Cerutti and Simanis, 1999). Thus, anaphase B can now be divided into three distinct stages depending upon the ability to detect cytokinesis regulators to the SPB.

The localization of the Spg1 cascade to one SPB at the end of anaphase B raises a number of interesting questions, not least of which are, why is the cascade regulated from the SPB and from only one of the two SPBs? It is currently not possible to determine whether it is essential that the network be activated from only one of the two SPBs, as we have insufficient knowledge of the system to be able to activate the pathway on both spindle poles other than by deregulating the system completely by abolishing Cdc16 GAP activity (Sohrmann *et al.*, 1998). If, however, we assume that it is important, it may be that the coordination of cytokinesis with other aspects of cell cycle progression would be most easily achieved by direct cross-talk between pathways. Given the importance of the SPB in MPF regulation, the SPB would appear a prime candidate for such a coordinating center. In this respect the SPB can be likened to an air traffic control center that coordinates a large number of remote events without ever getting physically involved in them. Another possible explanation for the localization of this cascade on the SPB is suggested by the fact that the Spg1 pathway is part of the checkpoint control system that blocks mitotic exit if spindle integrity is compromised (Fankhauser *et al.,* 1993; Murone and Simanis, 1996). Thus, the localization of the cascade to the SPB may be required to couple the signal of spindle damage with the cell cycle machinery.

Perhaps one of the most intriguing observations to emerge from these studies from an SPB perspective was the demonstration that the SPBs inherited by the two daughter cells following division may be inherently different. Cerutti and Simanis raised this provocative suggestion after conducting a simple experiment to determine the consequences of activation of Cdc16 at discrete times during interphase (Cerutti and Simanis, 1999). They found that around 50% of cells that had been arrested in S phase by growth in hydroxyurea underwent septation upon inactivation of Cdc16 and presumably the subsequent activation of the Spg1 cascade. They suggest that the reason that the Spg1 pathway is active in only half and not in all of the cells is because only half of the cells inherit an SPB that had an active cascade in the previous division. In other words, the SPBs of the daughter cells are nonequivalent. In turn, this necessitates the regeneration of equivalence through SPB maturation before the next mitosis.

A final tantalizing question is posed by the distance between the SPB and the F-actin ring when cytokinesis is initiated. The distance could not be greater. How, then, can events at the SPB influence the F-actin ring function? Several studies suggest that active proteins/complexes either diffuse, or actively migrate, to the ring from either the SPB or another nuclear structure. The possibility of the pole as the source of the signal is supported by the localization of a downstream component of the Spg1 pathway, Sid2 (Sparks *et al.,* 1999). Sid2 is found on the SPB throughout the cell cycle. At the end of anaphase B it also forms a ring at the cell equator. A second possibility comes from the analysis of the mitotic exit network (MEN) in the budding yeast *Saccharomyces cerevisiae* (Bachant and Elledge, 1999). This network is highly related to the fission yeast Spg1 cascade and contains direct homologs of the fission yeast proteins, For example, budding yeast Tem1p is closely related to Spg1, and budding yeast Cdc 15 is so closely related to fission yeast Cdc7 that Cdc7 can partially supress a *cdc15* mutant (Shirayama *et al.,* 1994; Fankhauser and Simanis, 1994). These studies suggest that the activation of the MEN releases a protein phosphatase from tight association with the nucleolar proteins and that this phosphatase reverses the mitotic phosphorylation events to drive exit from mitosis (Shou *et al.,* 1999; Straight *et al.,* 1999; Visintin *et al.,* 1999).

3. The Polo-Related Kinase Plo1 and the SPB

The fission yeast polo kinase Plo1 is involved in both cytokinesis and spindle formation. Fission yeast cells lacking Plo1 are unable to form the bipolar spindle and execute cytokinesis (Ohkura *et al.,* 1995), whilst some conditional Plo1 mutants misplace the F-actin ring (Bahler *et al.*, 1998). Over production of Plo1 forces cells to execute cytokinesis inappropriately during interphase cells (Ohkura *et al.*, 1995; Mulvihill *et al.*, 1999). Like the components of the Spg1 pathway and MPF, Plo1 associates with the mitotic SPB (Bahler *et al.*, 1998; Mulvihill *et al.*, 1999). Plo1 SPB association persists from commitment to mitosis until early anaphase B when the intensity of Plo1 signals at the SPB diminishes. Overexpression of Plo1 shows that it lies upstream of the Spg1 cascade (Mulvihill *et al.*, 1999). The mechanism by which Plo1 associates with the SPB is currently unclear; however, it involves the Stf1/Cut12 protein, as Plo1 prematurely associates with the interphase SPB of 20% of cells containing the *stf1.1* mutation. The involvement of Stf1/Cut12 in SPB recruitment is likely to be indirect because Plo1 can still associate with the SPB of cells that are unable to execute mitosis because the *stf1/cut12* gene has been deleted. Although this observation is consistent with the growing evidence of a role for Plo1 in G2/M commitment, the true significance of this phenomenon awaits determination.

4. Does SPB Localization of a Cell Cycle Regulator Mean That It Is an SPB Resident?

There is one important and unresolved issue that has considerable impact upon interpretation of the localization of cell cycle regulators to the SPB: Do they merely associate transiently with the SPB, or are they permanent residents? Transient association would be consistent with the formation of active complexes at the SPB that then diffuse away to act at other sites. If they are permanently anchored to the SPB for the duration of the signal it would suggest that a limited number of molecules are making the major decisions in cell cycle regulation and that the MTOC is an organizer in the regulatory as well as the structural sense of the word. Without the application of real-time imaging and photobleaching GFP fusion proteins in living cells, it is difficult to differentiate between transient and permanent association, but with the current pace of progress the application of these technologies is probably just around the corner.

D. Centromere and Telomere Association with the SPB and Genome Positioning

Upon sexual differentiation, a change in nuclear positioning accompanies cell fusion. The SPBs lead the nuclei of each mating pair toward one another along an elaborate track of microtubules before they both undertake a series of paired migrations back and forth until the nuclei fuse at the SPB to generate a single diploid nucleus (Fig. 6a) (Hirata and Tanaka, 1982; Bahler *et al.,* 1993; Chikashige *et al.,* 1994). The molecular basis of the fusion event remains to be elucidated, but it requires the activity of a type 1 membrane protein called Tht1 (Tange *et al.,* 1998). If the immediate environment of the cell after nuclear fusion is nutrient rich, the diploid cell can reenter a mitotic cell cycle. If, however, nutrients remain sparse, meiosis will occur (Yamamoto *et al.,* 1997). Meiotic prophase is accompanied by a remarkable series of nuclear migrations as the SPB leads the nucleus from one end of the cell toward the other, which again requires an elaborate array of microtubules emanating from the SPB (Fig. 6b) (Chikashige *et al.,* 1994; Svoboda *et al.,* 1995; Hagan and Yanagida, 1995; Ding *et al.,* 1998). The morphology of the nucleus during this migration was termed the "horsetail" stage by Robinow (Robinow, 1977). Similarly, the mitotic nucleus appears to follow the SPB along microtubule tracks to maintain the central position of the nucleus throughout interphase (Hagan and Yanagida, 1997).

The mitotic and meiotic migrations all share the common feature that the SPB directs nuclear movement along a network of microtubules (Fig. 6b) (Svoboda *et al.,* 1995; Hagan and Yanagida, 1995; Ding *et al.,* 1998). The ability of the entire genome to follow the SPB in these migrations indicates a strong association of all of chromatin with the SPB. During the mitotic cell cycle this is achieved by a tight association of the centromere sequences with the interphase SPB, whereas in meiosis the telomeres replace the centromeres at the SPB (Funabiki *et al.,* 1993; Chikashige *et al.,* 1994, 1997; Saitoh *et al.,* 1997). The switch in SPB association that accompanies sexual differentiation occurs in two stages (Chikashige *et al.,* 1997). In response to nutrient limitation and exposure to mating phero-

---

Figure 6 The *S. pombe* SPB and sexual differentiation. (a) Following fusion of the two prezygotes to form the zygote (Figs. 1 and 2), the two nuclei undergo a protracted series of migrations during which the two SPBs appear to use microtubule-tracks to oscillate between the two ends of the cells before the nuclei eventually fuse (Hirata and Tanaka, 1982; Chikashige *et al.,* 1994; Ding *et al.,* 1998). The nucleus then embarks on another set of microtubule-mediated migrations to drag the tip of the nucleus from one end of the cell to the other (Robinow, 1977; Chikashige *et al.* 1994; Ding *et al.,* 1998). Throughout the horsetail the entire microtubule cytoskeleton extends from the SPB and undergoes a number of characteristic changes that are described in detail in Ding *et al.,* (1998). (b) The migrating nucleus is preceded by a fan-shaped array that extends from the SPB to contact the cell cortex (left-hand cell in b). A rigid bundle of microtubules extends behind the migrating nucleus and maintains a stable association with the cortex until the direction of migration switches. When this happens the stable attachment between these distal microtubules and the cortex appears to break down and the microtubules extend around the end of the cell as the fan tail behind the migrating nucleus becomes a thick bundle. Once the nucleus reaches the end of its migration, the cycle is repeated many times. (c) Heterochromatin association with the SPB. See Section III,D for details.

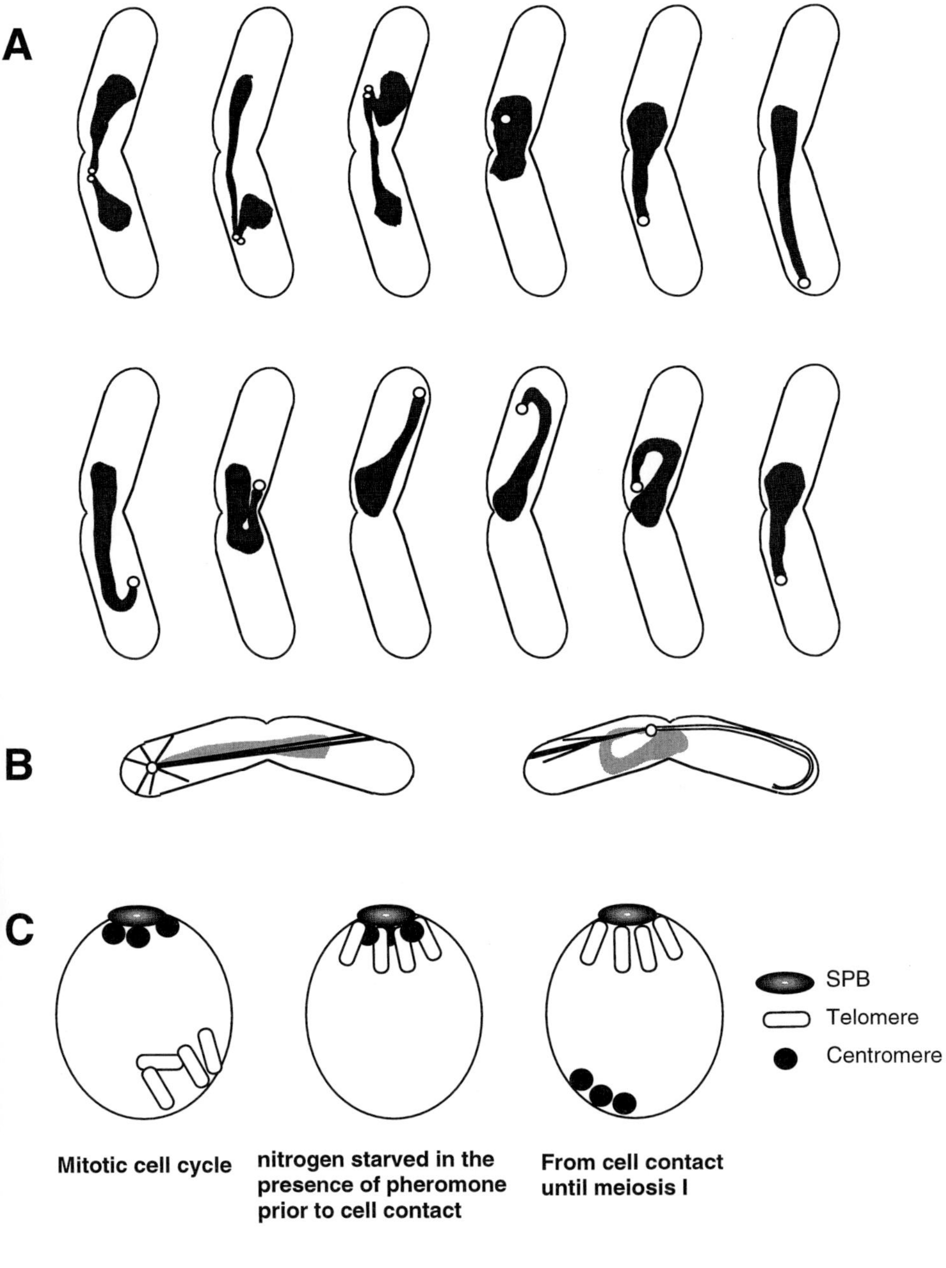
A
B
C
SPB
Telomere
Centromere
Mitotic cell cycle
nitrogen starved in the presence of pheromone prior to cell contact
From cell contact until meiosis I

mones, the telomeres join the centromeres on the SPB of each cell of the mating pair (Fig. 6c). The centromeres are then released following cell contact and the telomeres remain SPB associated until nuclear division. The molecular basis for this association of specific sets of heterochromatin at discrete stages of the life cycle with the SPB remains unclear, however, clues are beginning to emerge from the study of some mutant strains.

Telomere association with the SPB is disrupted by mutation of the *kms1*$^{+}$ *lot2*$^{+}$ and *taz1*$^{+}$ genes (Shimanuki *et al.*, 1997; Cooper *et al.*, 1998; Nimmo *et al.*, 1998). *lot2*$^{+}$ and *taz1*$^{+}$ are both required to maintain normal telomere length during vegetative growth (Cooper *et al.*, 1998; Nimmo *et al.*, 1998), whereas *kms1*$^{+}$ is likely to encode an SPB component as GFP Kms1 fusions got to the SPB (Tange *et al.*, 1998). The association of telomeres with the SPB is more severely disrupted in *lot2* than in *taz1* mutants. In both mutants SPB migration between the cell ends is unaffected by the mutation, but the ability of the chromatin to follow the SPB is severely compromised. This suggests that telomere/SPB association is essential for nuclear migration. These two mutants also provide a clue as to why so much energy is invested in such dramatic nuclear migrations. The frequency of meiotic recombination is radically reduced in both *lot2.s17* and *taz1.uv3*. The reduction in recombination is proportional to the degree of reduction in telomere association with the horsetail SPB.

The lack of recombination when the ability of the telomeres to associate with the SPB is compromised suggested one model whereby the migration of the nuclei helps align the chromosomes as it repeatedly shakes the chromosomes. This notion is further supported by the reduction in recombination frequency and homologous chromosome pairing in cells disrupted for the *dhc1*$^{+}$ gene. *dhc1*$^{+}$ encodes a fission yeast homolog of the heavy chain of the minus end directed microtubule motor, dynein (Yamamoto *et al.*, 1999). *dhc1*$^{+}$ is absolutely required for horsetail movement of the SPB, but not for interphase nuclear positioning or nuclear migration during karyogamy. Microtubule nucleation is unaffected by disruption of *dhc1*$^{+}$, indicating that horsetail movement is likely to be directly powered by the motor function of dynein, although how this is achieved is currently unknown. It is probably significant that thick bundles trailing behind migrating nuclei are particularly stable (Ding *et al.*, 1998). Dynein heavy chain is enriched at the SPB and on the microtubules that contact the cell cortex. At present it is not clear whether Dhc1 is tethered to the SPB or is simply there because it has migrated toward the minus end of the microtubules.

Much less is known about the molecular basis of the association of centromere sequences with the SPB during interphase of the mitotic cell cycle beyond the fact that it is unaffected by mutations in the *swi6*$^{+}$, *clr4*$^{+}$, and *rik1*$^{+}$ genes. Deletion of these genes results in radical remodeling of centromere heterochromatin and severe kinetochore defects during mitosis

(Ekwall *et al.,* 1996). Similarly, the mechanism that regulates the maintenance of the position of the interphase nucleus in the midpoint of the cell during tip growth remains a mystery. Weak clues have come from the study of the SPB component Sad1 and the protein phosphatase Ppbl, as overexpression of either gene results in misplaced SPBs and deformed nuclei (Yoshida *et al.,* 1994; Hagan and Yanagida, 1995). Interestingly, mutations in the *C. elegans* Unc84 protein, which exhibits homology to Sad1 and, like Sad1, associates with the nuclear envelope, is required for nuclear positioning in this system (Malone *et al.,* 1999).

E. The SPB and Sporulation

During metaphase II of meiosis the cytoplasmic face of the SPB differentiates. A number of extra layers appear as the SPB expands into the cytoplasm (Hirata and Tanaka, 1982; Tanaka and Hirata, 1982). The forespore membrane then forms from the outer limit of this structure and starts to grow around the nucleus (Fig. 7). By the time meiosis II is complete, the nucleus is surrounded by the membrane, which then differentiates further to become the spore wall. Staining for the Sad1 protein at this time shows that a transient change from a dot into a crescent when the differentiation is initiated (Hagan and Yanagida, 1995). A number of *spo*$^-$ mutants that are unable to form spores have defects in different aspects of SPB structure at this time, the most severe of which is a complete loss of the entire SPB (Hirata and Shimoda, 1994). The *spo15*$^+$ gene encodes an SPB component that is required for the transformation of the Sad1 dot to a crescent during meiosis II, but that is not required for mitotic growth (Ikemoto *et al.,* 1999). The molecular identity of other *spo*$^+$ gene products is likely to reveal much about SPB differentiation.

IV. Molecular Characterization of SPB Components

In contrast to our comprehensive understanding of the molecular composition and architecture of the budding yeast SPB, only a handful of fission yeast SPB components have been identified.

γ-Tubulin was the first fission yeast SPB component to be identified (Horio *et al.,* 1991). The molecule closely resembles its counterparts in higher systems. Indeed, a human γ-tubulin can substitute for the fission yeast protein (Horio and Oakley, 1994). Immunogold staining shows that γ-tubulin is found both on the cytoplasmic face of the SPB and in the thin nucleoplasmic layer that underlies the nuclear envelope during interphase (Ding *et al.,* 1997) (Fig. 4a).

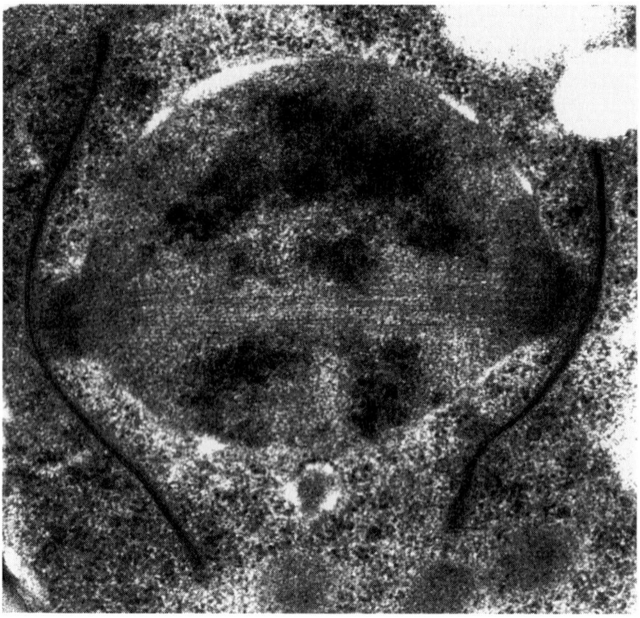

Figure 7 The outer face of the SPB differentiates during metaphase of meiosis II to produce promote the formation of the forespore membrane. The figure is reproduced with kind permission from Prof. J. S. Hyams and Dr. T. Chappell (University College London). It shows an image of a cell prepared by freeze substitution following induction of a *pat1.118* diploid synchronized meiosis (Bahler *et al.,* 1991). Both SPBs have clearly differentiated to generate an extended structure that has instigated the formation of the spore membrane (the thick dark structure), which will eventually become the spore wall.

The SPB component Sad1 was identified in a screen for mutants that were unable to form a bipolar spindle (Hagan and Yanagida, 1995). Unlike γ-tubulin, Sad1 only associates with the SPB and not the EMTOC or TAM (Fig. 3a) (Petersen *et al.,* 1998a). Its predicted amino acid sequence suggests a modular structure. A third of the way along the protein from the amino terminus is a short stretch of highly hydrophobic amino acids, which is long enough to span the nuclear membrane. Consistently, overexpression of the protein results in its association with the entire nuclear periphery. This may indicate that the wild-type protein spans the nuclear envelope. The carboxy-terminal portion of the molecule shows similarity to a number of reading frames in the EST databases from species as diverse as rat, rice, and *C. elegans* (Malone *et al.,* 1999; Sheldrick and Hagan, unpublished). Surprisingly, the same motif is not found in any reading frames predicted from the *S. cerevisiae* genome sequence.

The case for a role in nuclear envelope association of the SPB is much stronger for the Cut11 protein (West *et al.,* 1998). *cut11*$^{+}$ encodes a protein with multiple membrane spanning motifs. Fusions between Cut11 and GFP localize to the nuclear pore complex during interphase and show strong SPB association during the early stages of mitosis. Significantly, this staining diminishes during anaphase B, as the SPB is excluded from the nuclear envelope. Electron microscope images of temperature sensitive *cut11* mutants revealed mitotic SPBs that appeared to have fallen through a gap in the nuclear envelope to reside entirely within the nucleoplasm. Together these data strongly support a role for Cut11 in mediating the interaction between the mitotic SPB and the nuclear envelope.

At the level of fluorescence microscopy a temperature-sensitive mutation in the gene encoding the SPB component Cut12/Stf1 has a terminal phenotype that is indistinguishable from that of *cut11* mutants. Further support for a similar role is provided by the synthetic lethality between *cut12.1* and *cut11* mutants (West *et al.,* 1998). However, in contrast to *cut11*$^+$, the *stf1*$^+$/*cut12*$^+$ sequence does not predict any hydrophobic membrane-spanning domains and the protein is only found at the SPB. The only features of note in Cut12/Stf1 are two regions predicted to form a coiled-coil motif. Genetic data discussed in greater detail earlier suggest that Cut12/Stf1 plays a key role in the regulation of commitment to mitosis (Bridge *et al.,* 1998). It is therefore possible that the *cut12.1* mutation resembles *cut11*^{ts} defect because it cannot regulate the integration of the SPB into the nuclear envelope during the commitment to mitosis.

As is the case with budding yeast, the small calcium-binding protein calmodulin associates with the fission yeast SPB throughout the cell cycle (Moser *et al.,* 1997). Antibodies to human centrin 3, a small Ca^{2+}-binding protein related to Cdc31p from budding yeast found in MTOCs of many different systems, stain fission yeast SPBs (Baum *et al.,* 1986; Middendrop *et al.,* 1997; Anne Paoletti and Michel Bornens, personal communication). Encouragingly, preliminary analysis of an *S. pombe* centrin homolog (embl accession number: CAA20670) shows that it is an essential protein that is required for spindle formation (Anne Paoletti, Fred Chang, and Michel Bornens, personal communication).

A number of other molecules show SPB association, but have little additional data for a functional role in the generation or maintenance of SPB structure. Dis1 is a member of a conserved family of microtubule-associated proteins, including the Stu2 protein of budding yeast, XMAP215 from *Xenopus,* and the product of the *Drosophila mini spindles* gene (Nabeshima *et al.,* 1995; Wang and Huffaker, 1997; Charrasse *et al.,* 1998; Matthews *et al.,* 1998; Cullen *et al.,* 1999). GFP-Dis1 fusion protein associates with microtubules and both the SPB and the EMTOC. The carboxyl terminus alone is sufficient to direct GFP to the SPB (Nabeshima *et al.,* 1995). Association of members of this protein family with spindle poles may be a common theme as Stu2 was found in budding yeast SPB preparations (Wigge *et al.,* 1998).

The analysis of these budding yeast SPB preparations also identified, among other proteins, the budding yeast protein bim1 or yEB1 and a protein called Ndc80p (Schwartz *et al.,* 1997; Wigge *et al.,* 1998). The mammalian counterpart of bim1 is called EB1, and it associates with the protein that has been implicated in colorectal cancer, Adenamatous polyposis coli (APC) (Su *et al.,* 1995). As the fission yeast homolog of this protein, Mal3, associates with microtubules (Beinhauer *et al.,* 1997) and, like Dis1, forms an equatorial ring in anaphase cells, it seems quite likely that further analysis of the Mal3 protein will reveal an affinity for the SPB. Ndc80p associates with the SPB throughout the cell cycle. Interestingly, database comparisons reveal that the protein shares similarity with a human kinetochore protein called HEC and an open reading frame in *S. pombe* (Q10198) (Wigge *et al.,* 1998; Chen *et al.,* 1997). It will be interesting to see whether the protein encoded by Q10198 associates with the fission yeast SPB or the kinetochore.

The kinesin-related protein Cut7 has been localized to the spindle with an enrichment near the spindle termini (Hagan and Yanagida, 1992; Drummond and Hagan, 1998). This is a surprising location for a protein whose higher eukaryotic counterparts are plus end directed motor proteins. However, the closely related budding yeast proteins Cin8p and Kip1p have been found in budding yeast SPB preparations (Wigge *et al.,* 1998). Although it remains unclear whether there is an absolute SPB association of any of these molecules, these observations raise interesting conceptual frameworks for considering the mechanism of spindle formation.

A. Structural Components of the F-Actin Ring That Also Show SPB Association

The first stage of cytokinesis in fission yeast is the formation of an equatorial F-actin ring upon commitment to mitosis. This structure remains unchanged throughout mitosis until the end of anaphase B, when is constricts to cleave the cell in two (Marks *et al.,* 1986; Kitayama *et al.,* 1997; Demeter and Sazer, 1998).

Fission yeast cytokinesis mutants were classified by Nurse and colleagues as either early or late septation mutants depending upon whether any attempt is made to lay down a septum (Nurse *et al.,* 1976) (see earlier discussion). The early septation mutants are components of the Spg1 regulatory network. The late septation mutants have defects in structural components of the cytokinetic ring (for review, see Gould and Simanis, 1997) and include fission yeast homologs of profilin, tropomyosin, and a potential myosin light chain. Four of these structural proteins, Rng2, Cdc12, and two myosin II isoforms, have been found to associate with both the cytokinesis ring and a single spot on the nucleus (Chang *et al.,* 1997; Kitayama *et al.,*

1997; Bezanilla *et al.,* 1997; Eng *et al.,* 1998). The *cdc12⁺* gene encodes a formin that is first seen as a dot on the G2 nucleus before it forms a ring in anticipation of cytokinesis. This association of Cdc12.GFP with a spot on the nucleus and the misplacement of the septum in strains in which the interphase nucleus was misplaced led to a model for the selection of the site of cytokinesis. In this model it was proposed that an interaction between this spot and the cortex during the G2/M transition marked the site on the cortex from which the cytokinetic ring will form (Chang and Nurse, 1996; Chang *et al.,* 1996). Subsequently, a discrete domain of Cdc12 was shown to associate with the SPB (Petersen *et al.,* 1998b). This suggested that the Cdc12.GFP spot on the nucleus probably corresponded to the SPB. Similarly, the interphase Rng2 spot colocalizes with the SPB marker Sad1 (Eng *et al.,* 1998). Thus, the position of the SPB at the G2/M boundary may determine the position of the septum (Chang and Nurse, 1996). The analysis of strains in which septation has been deregulated by constitutive activation of the Spg1 cascade do, however, challenge one aspect of this model (Minet *et al.,* 1981; Fankhauser *et al.,* 1993; Fankhauser and Simanis, 1994; Gould and Simanis, 1997). After undergoing a normal mitosis, these strains undergo repeated septation events without any further nuclear divisions. The second septum is laid down adjacent to the initial centrally positioned septum and not directly over the SPB of one of the two nuclei as should happen according to the model. Thus, if the SPB can determine the site of septation, it follows that this activity must be restricted to interphase SPBs.

V. Perspectives

The completion of the *S. pombe* genome sequence has made it a particularly attractive organism for the analysis of a range of issues at the heart of cell biology. This is particularly apparent with the analysis of the microtubule cytoskeleton. The study of *S. pombe* is likely to make a major contribution to our understanding of how microtubules can control morphogenesis (Verde *et al.,* 1995; Mata and Nurse, 1997). The formation of discrete microtubule organizing centers at distinct stages of the life cycle also offers an attractive system for the study of *de novo* MTOC formation. Furthermore, the lack of any clear *S. cerevisiae* homologs for two of the handful of essential components of the *S. pombe* SPB that have been identified to date also suggests that *S. pombe* may reveal much about the degree of conservation of MTOC function between species.

The next few years are likely to see a rapid expansion of our understanding of microtubule organizing center function in fission yeast. The visualization of GFP microtubules should reveal where cytoplasmic microtubules come from and how dynamic they are. Do cytoplasmic microtubules polymerize at random, or does the EMTOC nucleate microtubules that remain stable throughout the cell cycle? Alternatively, are microtubules being constantly released from the SPB throughout interphase? The nature of the interaction between heterochromatin and the SPB is likely to be rapidly uncovered by the numerous genetic approaches to identify mutants with altered SPB, telomere, or centromere function. It should also become apparent whether there is an intimate connection between the actin cytoskeleton and cytoplasmic MTOC formation in *S. pombe.* Given the extensive history of cell cycle studies in *S. pombe,* the analysis of the unexpected role played by the SPB in coordinating key cell cycle transitions is likely to be very productive. In short, *S. pombe* is coming of age and, like so many other systems, the application of emerging technologies will revolutionize our perception of microtubule organization in this fascinating organism.

References

Alfa, C. E., Ducommun, B., Beach, D., and Hyams, J. S. (1990). Distinct nuclear and spindle pole body populations of cyclin-cdc2 in fission yeast. *Nature* **347,** 680–682.

Bachant, J. B., and Elledge, S. J. (1999). Mitotic treasures in the nucleolus. *Nature* **398,** 757–758.

Bähler, J., Schuchert, P., Grimm, C., and Kohli, J. (1991). Synchronized meiosis and recombination in fission yeast—observations with *pat1-114* diploid-cells. *Curr. Genet.* **19,** 445–451.

Bähler, J., Wyler, T., Loidl, J., and Kohli, J. (1993). Unusual nuclear structures in meiotic prophase of fission yeast: a cytological analysis. *J. Cell Biol.* **121,** 241–256.

Bähler, J., Steever, A. B., Wheatley, S., Wang, Y.-L., Pringle, J.R., Gould, K. L., and McCollum, D. (1998). Role of polo kinase and Mid1p in determining the site of cell division in fission yeast *J. Cell Biol.* **143,** 1603–1616.

Baum, P., Furlong, C., and Byers, B. (1986). Yeast gene required for spindle pole body duplication—homology of its product with Ca^{2+} binding proteins. *Proc. Natl. Acad. Sci. USA* **83,** 5512–5516.

Beinhauer, J. D., Hagan, I. M., Hegemann, J. H., and Fleig, U. (1997). Mal3, the fission yeast homolog of the human APC-interacting protein EB-1 is required for microtubule integrity and the maintenance of cell form. *J. Cell Biol.* **139,** 717–728.

Bezanilla, M., Forsburg, S. L., and Pollard, T. D. (1997). Identification of a second myosin-II in *Schizosaccharomyces pombe:* Myp2p is conditionally required for cytokinesis. *Mol. Biol. Cell* **8,** 2693–2705.

Biggins, S., Ivanovska, I., and Rose, M. D. (1996). Yeast ubiquitin-like genes are involved in duplication of the microtubule-organizing center. *J. Cell Biol.* **133,** 1331–1346.

Bridge, A. J., Morphew, M., Bartlett, R., and Hagan, I. M. (1998). The fission yeast SPB component Cut12 links bipolar spindle formation to mitotic control. *Genes & Dev.* **12,** 927–942.

Cerutti, L., and Simanis, V. (1999). Asymmetry of the spindle pole bodies and Spg1p GAP segregation during mitosis in fission yeast. *J. Cell Sci.* **112,** 2313–2321.

Chang, F., and Nurse, P. (1996). How fission yeast fission in the middle. *Cell* **84,** 191–194.

Chang, F., Woollard, A., and Nurse, P. (1996). Isolation and characterization of fission yeast mutants defective in the assembly and placement of the contractile actin ring. *J. Cell Sci.* **109,** 131–142.

Chang, F., Drubin, D., and Nurse, P. (1997). cdc12p, a protein required for cytokinesis in fission yeast, is a component of the cell division ring and interacts with profilin. *J. Cell Biol.* **137,** 169–182.

Charrasse, S., Schroeder, M., GauthierRouviere, C., Ango, F., Cassimeris, L., Gard, D. L., and Larroque, C. (1998). The TOGp protein is a new human microtubule-associated protein homologous to the *Xenopus* XMAP215. *J. Cell Sci.* **111,** 1371–1383.

Chen, Y., Riley, D. J., Chen, P.-L., and Lee, W.-H. (1997). HEC a novel protein with leucine heptad repeats specifically involved in mitosis. *Mol. Cell Biol.* **17,** 6049–6056.

Chikashige, Y., Ding, D. Q., Funabiki, H., Haraguchi, T., Mashiko, S., Yanagida, M., and Hiraoka, Y. (1994). Telomere-led premeiotic chromosome movement in fission yeast. *Science* **264,** 270–273.

Chikashige, Y., Ding, D. Q., Imai, Y., Yamamoto, M., Haraguchi, T., and Hiraoka, Y. (1997). Meiotic nuclear reorganization: Switching the position of centromeres and telomeres in the fission yeast *Schizosaccharomyces pombe. EMBO J.* **16,** 193–202.

Cooper, J. P., Watanabe, Y., and Nurse, P. (1998). Fission yeast Tazl protein is required for meiotic telomere clustering and recombination. *Nature* **392,** 828–831.

Cullen, C. F., Deak, P., Glover, D. M., and Ohkura, H. (1999). *mini spindles:* a gene encoding a conserved microtubule-associated protein required for the integrity of the mitotic spindle. *J. Cell Biol.* **146,** 1005–1018.

Demeter, J., and Sazer, S. (1998). imp2, a new component of the actin ring in the fission yeast *Schizosaccharomyces pombe. J. Cell Biol.* **143,** 415–427.

Ding, R., McDonald, K. L., and McIntosh, J. R. (1993). 3-Dimensional reconstruction and analysis of mitotic spindles from the yeast, *Schizosaccharomyces-pombe. J. Cell Biol.* **120,** 141–151.

Ding, R., West, R. R., Morphew, M., and McIntosh, J. R. (1997). The spindle pole body of *Schizosaccharomyces pombe* enters and leaves the nuclear envelope as the cell cycle proceeds. *Mol. Biol. Cell,* 1461–1479.

Ding, D.-Q., Chikashige, Y., Haraguchi, T., and Hiraoka, Y. (1998). Oscillatory nuclear movement in fission yeast meiotic prophase is driven by astral microtubules, as revealed by continuous observation of chromosomes and microtubules in living cells. *J. Cell Sci.* **111,** 701–712.

Drummond, D. R., and Hagan, I. M. (1998). Mutations in the bimC box of Cut7 indicate divergence of regulation within the bimC family of kinesin related proteins. *J. Cell Sci.* **111,** 853–865.

Dunphy, W. G. (1994). The decision to enter mitosis. *Trends Cell Biol.* **4,** 202–207.

Ekwall, K., Nimmo, E. R., Javerzat, J. P., Borgstrom, B., Egel, R., Cranston, G., and Allshire, R. (1996). Mutations in the fission yeast silencing factors *clr4*$^+$ and *rik1*$^+$ disrupt the localization of the chromo domain protein Swi6p and impair centromere function. *J. Cell Sci.* **109,** 2637–2648.

Eng, K., Naqvi, N. I., Wong, K. C. Y., and Balasubramanian, M. K. (1998). Rng2p, a protein required for cytokinesis in fission yeast, is a component of the actomyosin ring and the spindle pole body. *Curr. Biol.* **8,** 611–621.

Fankhauser, C., and Simanis, V. (1994). The Cdc7 protein-kinase is a dosage-dependent regulator of septum formation in fission yeast. *EMBO J.* **13,** 3011–3019.

Fankhauser, C., Marks, J., Reymond, A., and Simanis, V. (1993). The *S. pombe cdc 16* gene is required both for maintenance of $p34^{cdc2}$ kinase-activity and regulation of septum formation—a link between mitosis and cytokinesis. *EMBO J.* **12,** 2697–2704.

Funabiki, H., Hagan, I., Uzawa, S., and Yanagida, M. (1993). Cell cycle-dependent specific positioning and clustering of centromeres and telomeres in fission yeast. *J. Cell Biol.* **121,** 961–976.

Furge, K. A., Wong, K., Armstrong, J., Balasubramanian, M., and Albright, C. F. (1998). Byr4 and Cdc16 form a two component GTPase activating protein for the Spg1 GTPase that controls septation in fission yeast. *Curr. Biol.* **8,** 947–954.

Gould, K., and Simanis, V. (1997). The control of septum formation in fission yeast. *Genes & Dev.* **11,** 2939–51.

Hagan, I. M. (1998). Fission yeast microtubules. *J. Cell Sci.* **111,** 1603–1612.

Hagan, I. M., and Hyams, J. S. (1988). The use of cell-division cycle mutants to investigate the control of microtubule distribution in the fission yeast *Schizosaccharomyces pombe. J. Cell Sci.* **89,** 343–357.

Hagan, I. M., and Hyams, J. S. (1996). Forces acting on the fission yeast anaphase spindle. *Cell Motil. Cytoskel.* **34,** 69–75.

Hagan, I., and Yanagida, M. (1990). Novel potential mitotic motor protein encoded by the fission yeast *cut7*$^+$ gene. *Nature* **347,** 563–566.

Hagan, I., and Yanagida, M. (1992). Kinesin-related Cut7 protein associates with mitotic and meiotic spindles in fission yeast. *Nature* **356,** 74–76.

Hagan, I., and Yanagida, M. (1995). The product of the spindle formation gene *sad1*$^+$ associates with the fission yeast spindle pole body and is essential for viability. *J. Cell Biol.* **129,** 1033–1047.

Hagan, I., and Yanagida, M. (1997). Evidence for cell cycle-specific, spindle pole body-mediated, nuclear positioning in the fission yeast *Schizosaccharomyces pombe. J. Cell Sci.* **110,** 1851–1866.

Hagan, I. M., Riddle, P. N., and Hyams, J. S. (1990). Intramitotic controls in the fission yeast *Schizosaccharomyces pombe*—the effect of cell-size on spindle length and the timing of mitotic events. *J. Cell Biol.* **110,** 1617–1621.

Hardwick, K. G., Weiss, E., Luca, F. C., Winey, M., and Murray, A. W. (1996). Activation of the budding yeast spindle assembly checkpoint without mitotic spindle disruption. *Science* **273,** 953–956.

He, X. W., Jones, M. H., Winey, M., and Sazer, S. (1998). mph1, a member of the Mps1-like family of dual specificity protein kinases, is required for the spindle checkpoint in *S. pombe. J. Cell Sci.* **111,** 1635–1647.

Heath, I. B. (1981). Nucleus-associated organelles in fungi. *Int. Rev. Cyt.* **69,** 191–221.

Hirata, A., and Shimoda, C. (1994). Structural modification of spindle pole bodies during meiosis II is essential for the normal function of ascospores in *Schizosaccharomyces pombe:* Ultrastructural analysis of *spo* mutants. *Yeast.* **10,** 173–183.

Hirata, A., and Tanaka, K. (1982). Nuclear behaviour during conjugation and meiosis in the fission yeast *Schizosaccharomyces pombe. J. Gen Appl. Micro.* **28,** 263–274.

Horio, T., and Oakley, B. R. (1994). Human gamma-tubulin functions in fission yeast. *J. Cell Biol.* **126,** 1465–1473.

Horio *et al.* (1981) p. 210

Horio, T., Uzawa, S., Jung, M. K., Oakley, B. R., Tanaka, K., and Yanagida, M. (1991). The fission yeast y-tubulin is essential for mitosis and is localized at microtubule organizing centers. *J. Cell Sci.* **99,** 693–700.

Hudson, J. D., Feilotter, H., and Young, P. G. (1990). *stf1:* non wee mutations epistatic to *cdc25* in the fission yeast *Schizosaccharomyces pombe. Genetics* **126,** 309–315.

Hudson, J. D., Feilotter, H., Lingner, C., Rowley, R., and Young, P. (1991). *stfl:* a new supressor of the mitotic control gene *cdc25* in *Schizosaccharomyces pombe. Cold Symp. Quant. Biol.* **56,** 599–604.

Ikemoto, S., Nakamura, T., Kubo, M., and Shimoda, C. (2000). *S. pombe* sporulation—specific protein Spo 15p is localised to the spindle pole body and essential for its modification. J. Cell Sci. **113,** 545–554.

Jwa, M., and Song, K. (1998). Byr4, a dosage-dependent regulator of cytokinesis in *S. pombe* interacts with a possible small GTPase pathway including Spg1 and Cdc16. *Mol & Cells* **8,** 240–245.

Jwa, M., Shin, S., Albright, C. F., and Song, K. (1999). Byr4p, a possible regulator of mitosis and cytokinesis in fission yeast, localizes to the spindle pole body by its C-terminal domains. *J. Biochem. Mol. Biol.* **32,** 92–97.

Kanbe, T., Kobayashi, I., and Tanaka, K. (1989). Dynamics of cytoplasmic organelles in the cell cycle of the fission yeast *Schizosaccharomyces pombe:* three dimensional reconstruction from serial sections. *J. Cell Sci.* **94,** 647–656.

Kitayama, C., Sugimoto, A., and Yamamoto, M. (1997). Type II myosin heavy chain encoded by the *myo2* gene composes the contractile ring during cytokinesis in *Schizosaccharomyces pombe. J. Cell Biol.* **137,** 1309–1319.

Lauze, E., Stoelcker, B., Luca, F. C., Weiss, E., Schutz, A. R., and Winey, M. (1995). Yeast spindle pole body duplication gene Mps1 encodes and essential dual-specificity protein-kinase. *EMBO J.***14,** 1655–1663.

Malone, C. J., Fixsen, W. D., Horvitz, H. R., and Han, M. (1999). UNC-84 localizes to the nuclear envelope and is required for nuclear migration and anchoring during *C. elegans* development. *Development* **126,** 3171–3181.

Marks, J., Hagan, I. M., and Hyams, J. S. (1986). Growth polarity and cytokinesis in fission yeast—the role of the cytoskeleton. *J. Cell Sci.,* 229–241.

Masuda, H., and Shibata, T. (1996). Role of gamma-tubulin in mitosis-dpecific microtubule nucleation from the *Schizosaccharomyces pombe* spindle pole body. *J. Cell Sci.* **109,** 165–177.

Masuda, H., Sevik, M., and Cande, W.Z. (1992). *In vitro* microtubule nucleating activity of spindle pole bodies in fission yeast *Schizosaccharomyces pombe*—cell cycle-dependent activation in *Xenopus* cell-free-sxtracts. *J. Cell Biol.* **117,** 1055–1066.

Mata, J., and Nurse, P. (1997). Tea 1 and the microtubular cytoskeleton are important for generating global spatial order within the fission yeast cell. *Cell* **89,** 939–949.

Matthews, L. R., Carter, P., ThierryMieg, D., and Kemphues, K. (1998). ZYG-9, a *Caenorhabditis elegans* protein required for microtubule organization and function, is a component of meiotic and mitotic spindle poles. *J. Cell Biol.* **141,** 1159–1168.

McCully, K., and Robinow, C. F. (1971). Mitosis in the fission yeast *Schizosaccharomyces pombe:* a comparative study with light and electron microscopy. *J. Cell Sci.* **9,** 475–507.

Middendorp, S., Paoletti, A., Schiebel, E., and Bornens, M. (1997). Identification of a new mammalian centrin gene, more closely related to *Saccharomyces cerevisiae* CDC31 gene. *Proc. Natl. Acad. Sci. USA* **94,** 9141–9146.

Minet, M., Nurse, P., Thuriaux, P., and Mitchison, M. (1981). Uncontrolled septation in a cell division cycle mutant of the fission yeast *Schizosaccharomyces pombe. J. Bacteriol.* **137,** 440–446.

Mitchison, J. M., and Nurse, P. (1985). Growth in cell length in the fission yeast *Schizosaccharomyces pombe. J. Cell Sci.* **75,** 357–376.

Moser, M. J., Flory, M. R., and Davis, T. N. (1997). Calmodulin localizes to the spindle pole body of *Schizosaccharomyces pombe* and performs an essential function in chromosome segregation. *J. Cell Sci.* **110,** 1805—1812.

Mulvihill, D. M., Petersen, J., Ohkura, H., Glover, D. M., and Hagan, I. M. (1999). Plo1 kinase recruitment to the spindle pole body and its role in cell division in *Schizosaccharomyces pombe. Mol. Biol. Cell* **10,** 2771–2785.

Murone, M., and Simanis, V. (1996). The fission yeast *dma1* gene is a component of the spindle assembly checkpoint, required to prevent septum formation and premature exit from mitosis if spindle function is compromised. *EMBO J.* **15,** 6605–6616.

Nabeshima, K., Kurooka, H., Takeuchi, M., Kinoshita, K., Nakaseko, Y., and Yanagida, M. (1995). $p93^{Dis1}$, which is required for sister-chromatid separation, is a novel microtubule and spindle pole body-associating protein phosphorylated at the Cdc2 target sites. *Genes & Dev.* **9,** 1572–1585.

Nabeshima, K., Nakagawa, T., Straight, A. F., Murray, A., Chikashige, Y., Yamashita, Y. M., Hiraoka, Y., and Yanagida, M. (1998). Dynamics of centromeres during metaphase–anaphase transition in fission yeast: Dis1 is implicated in force balance in metaphase bipolar spindle. *Mol. Biol. Cell* **9,** 3211–3225.

Nasmyth, K. (1979). The control acting over the initiation of DNA replication in the yeast *Schizosaccharomyces pombe. J. Cell Sci.* **36,** 215–233.

Nimmo, E. R., Pidoux, A. L., Perry, P. E., and Allshire, R. C. (1998). Defective meiosis in telomere-silencing mutants of *Schizosaccharomyces pombe. Nature* **392,** 825–828.

Nurse, P. (1985). Cell cycle controls in yeast. *Trends. Genet.* **1,** 51–55.

Nurse, P. (1990). Universal control mechanism regulating onset of M-phase. *Nature* **344,** 503–508.

Nurse, P., Thuriaux, P., and Nasmyth, K. (1976). Genetic control of the cell division cycle in the fission yeast *Schizosaccharomyces pombe. Mol. Gen. Genet.* **146,** 167–178.

Ohkura, H., Hagan, I. M., and Glover, D. M. (1995). The conserved *Schizosaccharomyces pombe* kinase Plo1, required to form a bipolar spindle, the actin ring, and septum, can drive septum formation in G1 and G2 cells. *Genes & Dev.* **9,** 1059–1073.

O'Toole, E. T., and McIntosh, J. R. (1999). Life cycles of yeast spindle pole bodies: getting microtubules into a closed nucleus. *Biology of the Cell* **91,** 305–312.

Petersen, J., Heitz, M. J., and Hagan, I. M. (1998a). Conjugation in *S. pombe:* Identification of a microtubule organising centre, a requirement for microtubules and a role for Mad2. *Curr. Biol.* **8,** 963–966.

Petersen, J., Nielsen, O., Egel, R., and Hagan, I. M. (1998b). FH3, a domain found in formins targets the fission yeast formin Fus1 to the projection tip during conjugation. *J. Cell Biol.* **141,** 1217–1228.

Petersen, J., Nielsen, O., Egel, R., and Hagan, I. M. (1998c). F-actin distribution and function during sexual differentiation in *Schizosaccharomyces pombe. J. Cell Sci.* **111,** 867–876.

Pichova, A., Kohlwein, S. D., and Yamamoto, M. (1995). New arrays of cytoplasmic microtubules in the fission yeast *Schizosaccharomyces pombe. Protoplasma* **188,** 252–257.

Robinow, C. F. (1977). The number of chromosomes in *Schizosaccharomyces pombe:* light microscopy of stained preparations. *Genetics* **87,** 491–497.

Rose, M. D., and Fink, G. R. (1987). Karl, a gene required for function of both intranuclear and extranuclear microtubules in yeast. *Cell* **48,** 1047–1060.

Saitoh, S., Takahashi, K., and Yanagida, M. (1997). Mis6, a fission yeast inner centromere protein, acts during G1/S and forms specialized chromatin required for equal segregation. *Cell* **90,** 131–143.

Sawin, K. E., and Nurse, P. (1997). Role of microtubules and actin in reorienting cell polarity in fission yeast. *Mol. Biol. Cell* **8,** 1579–1579.

Schmidt, S., Sohrmann, M., Hofmann, K., Woolard, A., and Simanis, V. (1997). The Spg1 GTPase is an essential dosage-dependent inducer of septum formation in *Schizosaccharomyces pombe. Genes & Dev.* **11,** 1519–1534.

Schwartz, K., Richards, K., and Botstein, D. (1997). Bim1 encodes a microtubule-binding protein in yeast. *Mol. Biol. Cell* **8,** 2677–2691.

Shimanuki, M., Miki, F., Ding, D. Q., Chikashige, Y., Hiraoka, Y., Horio, T., and Niwa, O. (1997). A novel fission yeast gene, *kms1*$^{+}$, is required for the formation of meiotic prophase-specific nuclear architecture. *Mol. Gen. Genet.* **254,** 238–249.

Shirayama, M., Matsui, Y., and Toh, E. A. (1994). The yeast TEM1 gene, which encodes a GTP-binding protein is involved in termination of M phase. *Mol. Cell Biol.* **14,** 7476–7482.

Shou, W. Y., Seol, J.H., Shevchenko, A., Baskerville, C., Moazed, D., Chen, Z.W.S., Jang, J., Charbonneau, H., and Deshaies, R. J. (1999). Exit from mitosis is triggered by Tem1-

dependent release of the protein phosphatase Cdc14 from nucleolar RENT complex. *Cell* **97,** 233–244.

Sohrmann, M., Schmidt, S., Hagan, I., and Simanis, V. (1998). Asymmetric segregation on spindle poles of the *Schizosaccharomyces pombe* septum-inducing protein kinase Cdc7p. *Genes & Dev.* **12,** 84–94.

Sparks, C. A., Morphew, M., and McCollum, D. (1999). Sid2p, a spindle pole body kinase that regulates the onset of cytokinesis. *J. Cell Biol.* **146,** 777–790.

Straight, A. F., Shou, W. Y., Dowd, G. J., Turck, C. W., Deshaies, R. J., Johnson, A. D., and Moazed, D. (1999). Net1, a Sir2-associated nucleolar protein required for rDNA silencing and nucleolar integrity. *Cell* **97,** 245–256.

Su, L.,-K., Burrell, M., Hill, D. E., Gyuris, J., Brent, R., Wilshire, R., Trent, J., Vogelstein, B., and Kinzler, K. (1995). APC binds to the novel protein EBI. *Cancer Res.* **55,** 2972–2977.

Svoboda, A., Bahler, J., and Kohli, J. (1995). Microtubule-driven nuclear movements and linear elements as meiosis-specific characteristics of the fission yeasts *Schizosaccharomyces versitalis* and *Schizosaccharomyces pombe. Chromosoma* **104,** 203–214.

Tanaka, K., and Hirata, A. (1982). Ascospore development in the fission yeast *Schizosaccharomyces pombe* and *S. japonicus. J. Cell Sci.* **56,** 263–279.

Tanaka, K., and Kanbe, T. (1986). Mitosis in the fission yeast *Schizosaccharomyces pombe* as revealed by freeze-substitution electron-microscopy. *J. Cell Sci.* **80,** 253–268.

Tange, Y., Horio, T., Shimanuki, M., Ding, D.-Q., Hiraoka, Y., and Niwa, O. (1998). A novel fission yeast gene, *tht1+*, is required for the fusion of nuclear envelopes during karyogamy. *J. Cell Biol.,* 247–258.

Toda, T., Umesono, K., Hirata, A., and Yanagida, M. (1983). Cold-sensitive nuclear division arrest mutants of the fission yeast *Schizosaccharomyces pombe. J. Mol. Biol.* **168,** 251–270.

Verde, F., Mata, J., and Nurse, P. (1995). Fission yeast-cell morphogenesis—identification of new genes and analysis of their role during the cell-cycle. *J. Cell Biol.* **131,** 1529–1538.

Visintin, R., Hwang, E. S., and Amon, A. (1999). Ctf1 prevents premature exit from mitosis by anchoring Cdc14 phosphatase in the nucleolus. *Nature* **398,** 818–824.

Wang, P. J. J., and Huffaker, T. C. (1997). Stu2p: A microtubule-binding protein that is an essential component of the yeast spindle pole body. *J. Cell Biol.* **139,** 1271–1280.

West, R. R., Vaisberg, E. V., Ding, R., Nurse, P., and McIntosh, J. R. (1998). *cut11*$^+$: A gene required for cell cycle-dependent spindle pole body anchoring in the nuclear envelope and bipolar spindle formation in *Schizosaccharomyces pombe. Mol. Biol. Cell* **9,** 2839–2855.

Wigge, P. A., Jensen, O. N., Holmes, S., Soues, S., Mann, M., and Kilmartin, J.V. (1998). Analysis of the *Saccharomyces* spindle pole by matrix-assisted laser desorption/ ionization (MALDI) mass spectrometry. *J. Cell Biol.* **141,** 967–977.

Yamamoto, M., Imai, Y., and Watanabe, Y. (1997). Mating and sporulation in *Schizosaccharomyces pombe.* In *The Molecular and Cellular Biology of the Yeast Saccharomyces* (J. R. Pringle, J. R. Broach and E. W. Jones, eds.). New York: Cold Spring Harbor Laboratory Press, pp. 1036–1106.

Yamamoto, A., West, R. R., McIntosh, J. R., and Hiraoka, Y. (1999). A cytoplasmic dynein heavy chain is required for oscillatory nuclear movement of meiotic prophase and efficient meiotic recombination in fission yeast. *J. Cell Biol.* **145,** 1233–1251.

Yoshida, T., Toda, T., and Yanagida, M. (1994). A calcineurin-like gene *ppb1*$^+$ in fission yeast—mutant defects in cytokinesis, cell polarity, mating and spindle pole body positioning. *J. Cell Sci.* **107,** 1725–1735.

7

Comparative Structural, Molecular, and Functional Aspects of the *Dictyostelium discoideum* Centrosome

Ralph Gräf, Nicole Brusis, Christine Daunderer, Ursula Euteneuer, Andrea Hestermann, Manfred Schliwa, and *Masahiro Ueda*
Adolf-Butenandt-Institut/Zellbiologie
Ludwig-Maximilians-Universität München
D-80336 Munich, Germany

I. Introduction

For decades, the cellular slime mold *Dictyostelium discoideum* has been a well-established model organism for the study of basic aspects of development, cell motility, and phagocytosis (for review, see Maeda *et al.,* 1997). The reasons for this popularity are quite obvious: *Dictyostelium* cells show a simple form of development, i.e., in case of starvation they form fruiting bodies consisting of only two cell types that originate from prestalk cells and prespore cells (Takeuchi *et al.,* 1986). Furthermore, they exhibit fast ameboid motility (Fukui, 1993) and, since they feed on bacteria or yeast, they are well suited for the study of phagocytosis (Maniak *et al.,* 1995). Since the cytoskeleton plays a key role in cell motility, chemotaxis, and phagocytosis, *Dictyostelium* has become a useful model for cytoskeleton research. Especially the roles of actin and actin-binding proteins have been

* Correspondence should be addressed to: Ralph Gräf, Adolf-Butenandt-Institut/Zellbiologie, Ludwig-Maximilians-Universität München, Schillerstr. 42, D-80336 München, Germany.

Current Topics in Developmental Biology, Vol. 49

0070-2153/00 $35.00

thoroughly investigated (Noegel and Luna, 1995). However, the tubulin field could not keep pace with the progress in the actin/myosin field, although α and β-tubulin, which are encoded by single genes, have been cloned and sequenced (Trivinos-Lagos *et al.*, 1993) and microtubule dynamics has been examined at the microscopic level (Jensen *et al.*, 1991; Kitanishi-Yumura and Fukui, 1987; Neujahr *et al.*, 1998; Roos *et al.*, 1984, 1986; Rubino *et al.*, 1984). In addition, the lack of a large-scale purification protocol for native *Dictyostelium* tubulins (White *et al.*, 1983) has hampered the identification and characterization of microtubule-associated proteins.

The analysis of the centrosome, also called the nucleus-associated body (Roos, 1975), shared the same unfortunate fate as the biochemical and molecular analysis of *Dictyostelium* microtubules: There was no method to isolate centrosomes in high quantity and purity because they could not be dissociated from the nucleus without disintegration (Omura and Fukui, 1985). It took until 1998 to obtain the first molecular marker for the *Dictyostelium* centrosome, the cDNA coding for γ-tubulin (Euteneuer *et al.*, 1998). But recently, a mass isolation protocol for *Dictyostelium* centrosomes was established (Gräf *et al.*, 1998), allowing the effective production of a series of monoclonal antibodies (Gräf *et al.*, 1999). This, together with the successful cDNA sequencing project at the University of Tsukuba (Morio *et al.*, 1998) and the promising start of the *Dictyostelium* genome project (http://genome.imb-jena.de/Dictyostelium), makes *Dictyostelium* much more attractive as a model system for the study of centrosomal structure and function. Now we are able to identify potential centrosomal components in these data libraries by sequence similarity to known components from other species or by the use of mass spectrometric data obtained from individual, isolated centrosomal proteins. Thus, many of the key prerequisites that have made *Saccharomyces cerevisiae* a leading model system in centrosome research (Wigge *et al.*, 1998; Winsor and Schiebel, 1997) are now fulfilled for *Dictyostelium* as well. Besides the general advantages of *Dictyostelium* amebae as a model organism in cell biology, which include simple cultivation, good microscopic accessibility, and well-established genetic engineering techniques (Mann *et al.*, 1998), there also are several specific reasons for considering them as a model for centrosome research along with the well-established systems *Saccharomyces cerevisiae, Xenopus, Drosophila* eggs, and mammalian cells (for review, see Balczon, 1996). First, motility and differentiation of Dictyostelium cells allow studies of the centrosome's role in cell migration and development (Daunderer *et al.*, 1999; Schliwa *et al.*, 1998). Second, *Dictyostelium* centrosomes can be considered as representative for other lower eukaryotes, including fungi and protists, and thus may occupy an intermediary position between the highly specialized yeast spindle pole body (SPB) and centriole-containing centrosomes. It is hoped, therefore, that comparative analyses of these three, morphologically very different types of centrosomal structures may help elucidate common functional principles of this complex, highly dynamic organelle.

In the following two sections we will review in more detail the comparative morphology and molecular biology of the *Dictyostelium* centrosome, the *S. cerevisiae* SPB, and the animal centriole-containing centrosome. Furthermore, we will report on recent progress and future perspectives in *Dictyostelium* centrosome research that, we hope, will further this organism as a valuable model for centrosome research.

II. Comparative Morphology of the *Dictyostelium* Centrosome

A. Centrosomal Functions

Perhaps the best-known function of centrosomes is the organization of the interphase microtubule array and the mitotic spindle apparatus (Kalnins, 1992). A spindle consists of microtubules that are nucleated at the two spindle poles formed by duplicated centrosomes. Indeed, microtubule organization is the key function of this organelle and the basis for most, if not all, centrosomal functions known. The centrosome provides microtubule nucleation sites not only for spindle microtubules but also for the interphase microtubule cytoskeleton. Interphase microtubules, in turn, are required for the transport of other organelles along microtubules and the maintenance of cell architecture. For example, the Golgi apparatus is localized close to the centrosome because of the action of the microtubule minus-end directed motor dynein (Corthesy Theulaz *et al.,* 1992) and, at least in certain animal cells, by the linker protein GMAP-210, which connects the cis-Golgi compartment to microtubule minus ends (Infante *et al.,* 1999). Even a role of the centrosome in ameboid cell movement may be attributed to its microtubule-nucleating function. Employing a γ-tubulin-GFP mutant cell line with green fluorescent centrosomes (Fig. 1), Ueda *et al.* (1997) could show that the centrosome does not act as the "steering wheel" of the cell that determines the sites of pseudopod protrusion, as suggested earlier (for review, see Singer and Kupfer, 1986). Rather, pseudopod extension always preceded centrosome reorientation to the new leading edge of the cell, whereas new pseudopods were retracted if centrosome reorientation did not occur. Thus, centrosome reorientation is required for the maintenance of leading pseudopods, and the microtubule cytoskeleton mediates the communication between the cell periphery where microtubules interact with the cortical actin filament network (Neujahr *et al.,* 1998).

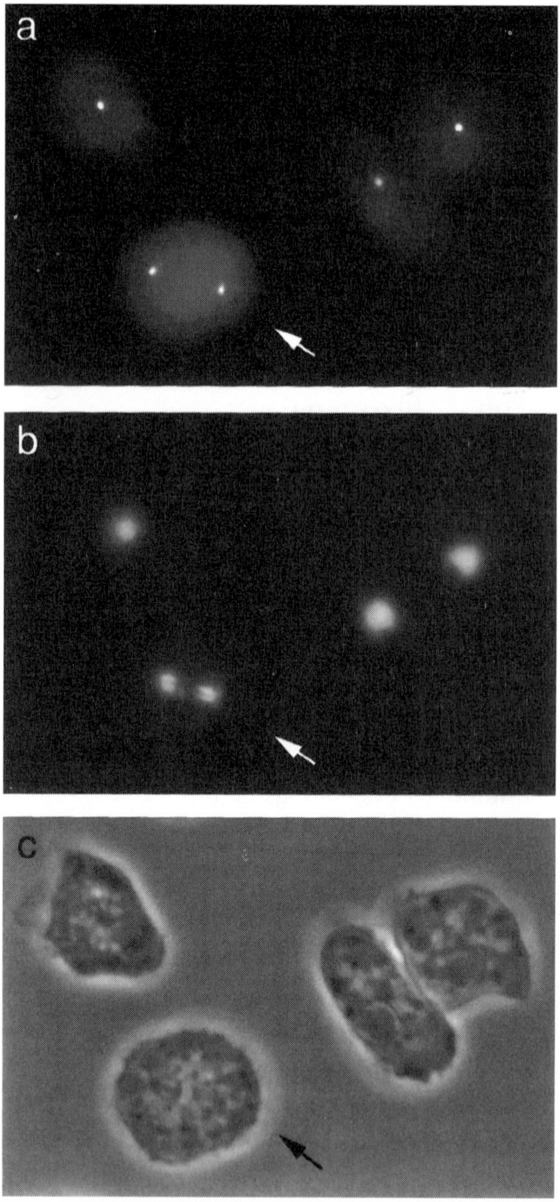

Figure 1 γ-Tubulin-GFP labeled cells in interphase and mitosis. Cells were fixed with methanol. The arrow points to a cell in late anaphase; the other cells are in interphase. Fluorescence images show (a) GFP localization at centrosomes and (b) staining of nuclei with 4,6-diamidino-2-phenylindole. (c) Phase contrast image.

B. Centrosomal Morphologies

Most fundamental centrosomal functions are shared between all eukaryotic cells, but the morphology of their centrosomes varies considerably. Common features include a central core structure (centrioles or layered plaques) embedded in more or less amorphous material, and a relatively tight association with the nucleus (Fig. 2). In most animal cells, the most conspicuous component of the centrosome is a pair of centrioles, highly geometric microtubule assemblies with the same morphologies as basal bodies (Mignot, 1996). They are embedded in a cagelike structure, a constituent of the pericentriolar matrix (PCM), which is formed mainly by pericentrin and γ-tubulin ring complexes (γ-TuRC) (Dictenberg *et al.,* 1998). γ-TuRCs are the sites of microtubule nucleation at the centrosome and, unexpectedly, were found in even larger number in the cytosol of animal species from clams to frogs (Moritz *et al.,* 1995; Oegema *et al.,* 1999; Vogel *et al.,* 1997; Zheng *et al.,* 1995). However, cytosolic γ-TuRCs obviously do not nucleate microtubules *in vivo.* Rather, the cytosolic pool of γ-TuRCs, also called gammasomes (Stearns and Kirschner, 1994), ensures rapid recruitment of γ-TuRCs to the centrosome if the cell has to increase the number of MT-nucleation sites (Raff *et al.,* 1993; Stearns and Kirschner, 1994). In addition to γ-tubulin, γ-TuRCs include at least six protein components, some of which are already known at the molecular level (see Section IV,A) (Oegema *et al.,* 1999). Since microtubule-nucleating γ-TuRCs appear to be the key component for centrosomal function, one might wonder about the function of the many, possibly up to 100 (Kalt and Schliwa, 1993; Kellogg *et al.,* 1994), other centrosomal proteins that also reside at the centrosome. Some of these proteins may build a scaffold for the γ-TuRCs; some, for example several kinases, are only temporary "guests" at the centrosome and may be involved in the regulation of centrosomal functions; some may use the centrosome solely as a vehicle to ensure their even distribution into daughter cells after mitosis; finally, some are components of the centrioles, whose role still remains enigmatic (Lange and Gull, 1996). There is evidence that centrioles are required for cell reproduction in somatic cells (Maniotis and Schliwa, 1991); however, they are absent in the first few meiotic divisions of mouse oocytes (Calarco Gillam *et al.,* 1983), in the female meiotic spindle of *Drosophila* (Sawin and Endow, 1993) and in certain *Drosophila* cell lines (Debec and Abbadie, 1989). Thus, despite the absence of centrioles, neither microtubule nucleation nor centrosome duplication are impaired in these cases.

In *Saccharomyces cerevisiae,* the centrosomal equivalent, called the spindle pole body (SPB), is embedded in the nuclear envelope (Winsor and Schiebel, 1997). It consists of three major plaques and a specialized region of the nuclear envelope, called the half-bridge, appearing as a projection

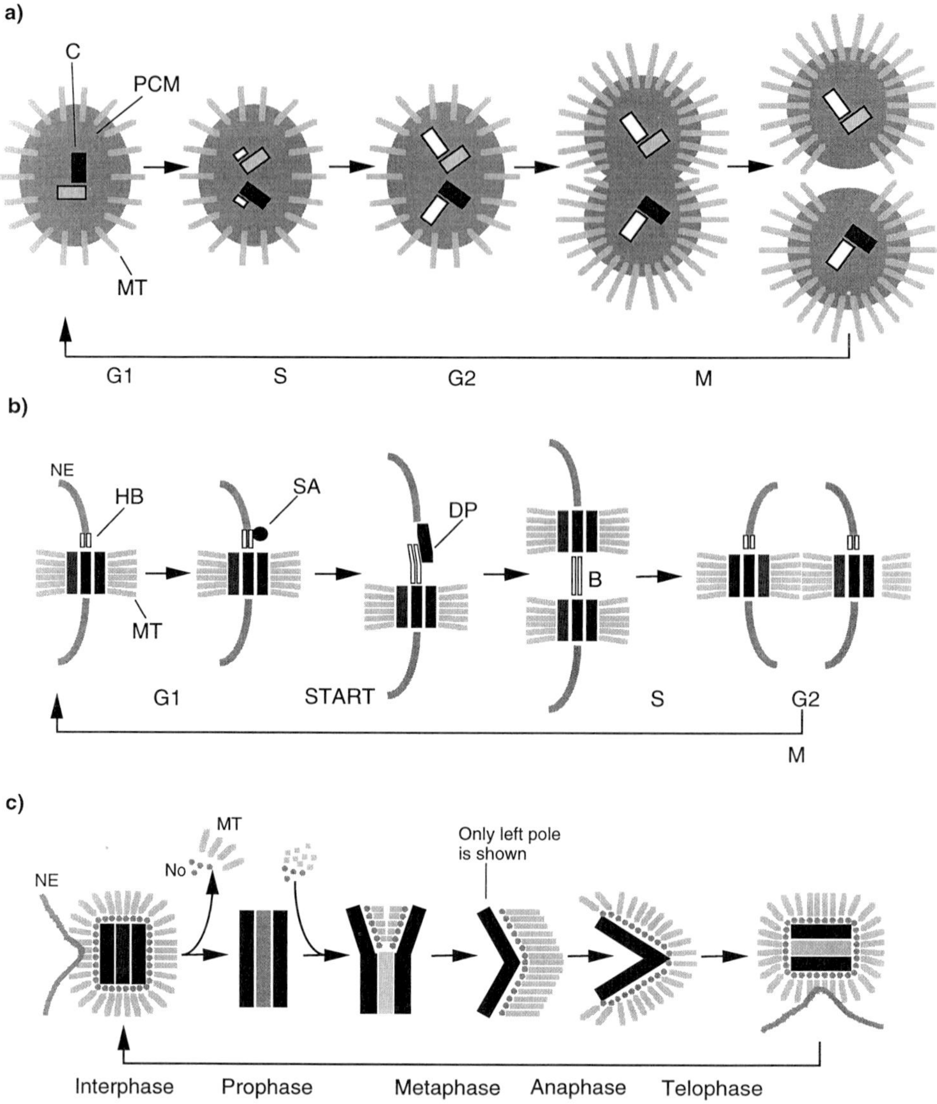

Figure 2 Schematic representation of centrosomal structure and their duplication: Duplication of (a) animal centrosomes, (b) *S. cerevisiae* spindle pole bodies, and (c) *Dictyostelium* centrosomes is depicted with respect to the cell cycle stage or mitotic phases of relevant events. Abbreviations: C, centrioles; PCM, pericentriolar matrix; MT, microtubules; NE, nuclear envelope; HB, half-bridge; SA, satellite; DP, duplication plaque; B, bridge; No, Nodules.

attached to the central plaque, while inner and outer plaques face the nuclear matrix and cytosol, respectively. Additionally, there is a thin intermediate layer between the outer and central plaque. Microtubule-nucleating complexes are located at the distal faces of the inner and outer plaque, but they seem to be much simpler than γ-TuRCs (Pereira and Schiebel, 1997) (see Section IV,A).

By contrast, in interphase, the *Dictyostelium* centrosome is entirely located in the cytosol and is connected to the nucleus by a fibrous linkage (Omura and Fukui, 1985). It consists of a box-shaped, layered core structure (dimensions approximately 130 × 250 × 280 nm), surrounded by a corona (thickness ~70–80 nm) with regularly spaced dense nodules embedded in an amorphous matrix (Moens, 1976; Roos, 1975). In vegetative interphase cells, 30–40 microtubules emanate from these nodules, which, as shown by electron microscopic immuno-gold labeling, contain γ-tubulin. Therefore, the corona likely corresponds to the pericentriolar matrix of animal centrosomes and the nodules are homologous to γ-TuRCs (Euteneuer *et al.*, 1998). The corona's resemblance to the pericentriolar matrix places the *Dictyostelium* centrosome closer to animal centrosomes than to yeast SPBs, which do not really contain a similar structure. However, with respect to the layered core structure, the *Dictyostelium* centrosome shows features found also in yeast SPBs, but the layers of these two organisms are most likely not homologous (see Section III). The entirely cytoplasmic localization of the *Dictyostelium* centrosome, as opposed to the yeast SPB's integration within the nuclear envelope and, even more importantly, the mode and time course of duplication, clearly distinguishes these two centrosomal structures.

III. Comparative Analysis of *Dictyostelium* Centrosome Duplication

The unusual mode of centrosome duplication in *Dictyostelium* is best illustrated when compared to the modes of centrosome duplication in animal and yeast cells. These two systems are therefore discussed first. After mitosis, in early G1, an animal cell possesses a centrosome with one pair of orthogonally oriented centrioles (Fig. 2a): a mature centriole generated at least two cell cycles before and an immature centriole acquired in the previous cell cycle. At the G1/S transition, the mother centrioles separate slightly and become disoriented, and procentrioles form at each centriole that grow to their full size usually in late G2 (Chretien *et al.*, 1997; Lange and Gull, 1996; Vorobjev and Nadezhdina, 1987). At the G2/M transition, the PCM expands and microtubule nucleating activity is promoted, most likely due to the acquisition of cytosolic γ-tubulin complexes (Felix *et al.*,

1994; Stearns and Kirschner, 1994; Zheng *et al.,* 1995). Maturation of the immature mother centriole occurs during mitosis and is characterized by the appearance of the centriolar appendages and satellites (Vorobjev and Chentsov-Yu, 1982) and at the molecular level by the acquisition of the 96-kDa protein cenexin in prophase (Lange and Gull, 1995). Upon centrosome separation, a process not yet fully understood, centrioles are distributed in a semiconservative manner, meaning that each daughter centrosome contains one of the two older centrioles (Fig. 2a) (Kochanski and Borisy, 1990).

Based on electron microscopic data (Byers and Goetsch, 1975), the first step of spindle pole body duplication in *S. cerevisiae* (Fig. 2b) occurs in late G1 at the cytoplasmic face of the half-bridge with appearance of the satellite that is the precursor of the new SPB (for review, see Hoyt and Geiser, 1996; Sobel, 1997; Winey and Byers, 1992). At the beginning of a new cell cycle, called "START," the satellite is very rapidly converted into a new SPB connected to the old one by the full bridge. Only recently, Adams and Kilmartin (1999) elucidated the role of the satellite in more detail and described intermediate stages of the duplication process: As a first new intermediate, a larger, so-called duplication plaque replaces the satellite at the distal end of the half-bridge. Both satellite and duplication plaque share several SPB components, suggesting strongly that both structures are precursors of the new SPB. In subsequent steps, the half-bridge grows underneath the duplication plaque and bends in such a way that the duplication plaque is inserted into the nuclear envelope. After assembly of the outer and inner plaque, scission of the bridge leads to two SPBs, each with one half-bridge (Fig. 2b). Because of interactions between the nuclear microtubules emanating from the two SPBs, they migrate to opposite positions and form a spindle between them that elongates until late G2. During the whole cell cycle, the old SPB remains buried within the nuclear envelope and the asymmetry of inner and outer plaques is preserved. Thus, like daughter centrioles, new SPBs seem to arise from a template-based process (Adams and Kilmartin, 1999) from mother centrioles or mother SPBs, respectively. Furthermore, the initial events of duplication in both animal cells and *S.cerevisiae* occur at G1/S or G1 with the formation of procentrioles or the emergence of the satellite, respectively. The whole duplication and separation procedure lasts through a considerable fraction of the entire cell cycle. However, the duplication event itself is very quick in *S.cerevisiae,* and unlike the semiconservative distribution of centriole pairs into the new daughter centrosomes, the new SPB does not seem to contain parts of the old SPB.

In *Dictyostelium,* the entire duplication procedure takes place during mitosis (Fig. 2c). This is the outcome of a recent study of *Dictyostelium* centrosomes, involving electron microscopic analysis of wild-type cells and video analysis of *Dictyostelium* mutants expressing a γ-tubulin-GFP fusion protein (Ueda *et al.,* 1997, 1999). The *Dictyostelium* centrosome cycle starts in early prophase with the expansion of the layered core along all three axes to appoximately twice the size of the interphase core. In late prophase, the corona with its microtubule nucleating nodules disappears, and concomitantly the microtubules emanating from the centrosome are lost (Kitanishi-Yumura and Fukui, 1987). Next, the central layer changes its structure and finally disappears. At the transition from prophase to prometaphase, the two outer layers peel apart and become inserted into the nuclear envelope. Instantly microtubules are nucleated from the nuclear surface of the two layers, which has now acquired γ-tubulin (Euteneuer *et al.,* 1998). These microtubules form a spindle whose elongation separates the spindle poles, which finally are opposed to each other in metaphase. Interestingly, microtubules do not seem to be required for duplication and initial separation, since both steps are unaffected by the presence of microtubule-depolymerizing drugs (Kitanishi *et al.,* 1984; Unger *et al.,* 1992; Welker and Williams, 1980). Beginning in metaphase, the edges of the spindle poles bend away from the nucleus, the first indication of a folding process, and microtubules now also extend into the cytosol from the edges of the poles. As mitosis proceeds and the spindle elongates, the curvature becomes more prominent. Finally, in telophase each centrosome folds back onto itself in such a way that the former cytoplasmic surface becomes buried inside a new, layered structure and the microtubule nucleating surface faces the outside, organizing the new interphase microtubule cytoskeleton. Maturation of the central layer to its interphase appearance starts in late telophase. Thus, in each centrosome cycle, the two outer layers of the core structure undergo an inside/outside reversal and the γ-tubulin complexes have to be redistributed from the outer cytoplasmic to the inner nucleoplasmic surface of each layer. These electron microscopic data are beautifully reflected by the dynamics of γ-tubulin-GFP, which labels the centrosome throughout the whole cycle. Even the enlargement of the centrosome in early prophase and the loss of the corona at the prophase/prometaphase transition become apparent by an increase or a decrease, respectively, in γ-tubulin-GFP fluorescence. Thus, an exact time scale completed our view of the morphogical events during the duplication cycle. In the strain used here, the whole cell cycle requires about 8–9 hr (Weijer *et al.,* 1984), but mitosis takes only approximately 15 min (Roos and Camenzind, 1981), and the centrosome splitting process and the initial separation events are completed within less than a minute.

With regard to the exact time course of the centrosome cycle and the splitting process, *Dictyostelium* seems unique, but there are some striking similarities to the SPB duplication cycle in the fission yeast *Schizosaccharomyces pombe* (Ding *et al.,* 1997; Hagan, 1998): *S. pombe* also displays a closed mitosis, and as in *Dictyostelium,* but unlike *S. cerevisiae,* the SPB is located in the cytoplasm in association with the nuclear envelope during most of interphase. The *S. pombe* SPB appears as a largely amorphous ellipsoid with a central lamina (McCully and Robinow, 1971; Tanaka and Kanbe, 1986). Its duplication occurs in late G2, and it comes to pass in a somewhat unclear manner. Once duplicated, the SPBs enter a fenestra formed after invagination of the nuclear envelope at the onset of mitosis (Ding *et al.,* 1997). In anaphase, after SPB separation, the fenestrae close and the SPBs are relocated to the cytoplasmic face of the nucleus. Thus, in both organisms the centrosome enters and leaves the nuclear envelope in the course of the centrosome cycle and duplication is initiated late in the cell cycle.

The comparison of centrosomal structures in animals, *S. cerevisiae,* and *Dictyostelium* reviewed so far allows us to draw the following conclusions: First, the *Dictyostelium* centrosome is unique with regard to the splitting process of the core structure and the timing of duplication events within the cell cycle. Second, the *Dictyostelium* centrosome and the *S. cerevisiae* SPB share some morphological features, but in *Dictyostelium,* the centrosomal core appears to be build symmetrically, and its outer layers most likely consist of the same proteins. By contrast, the mother SPB of *S. cerevisiae* seems to be unaltered in the course of SPB duplication and is asymmetrical in the structure and protein composition of its inner and outer plaques. For example, the spindle pole layer proteins Spc110p, Spc42p, and Spc72p are distributed asymmetrically over the three layers, and Kar1p and Cdc31p are found at the half-bridge only (Adams and Kilmartin, 1999; Elliott *et al.,* 1999). Of these, homologs of Spc110p and Cdc31p were immunolocalized to the corona instead of the layered core structure in the *Dictyostelium* centrosome (M. Ueda, R. Gräf, unpublished observation). Third, in *S. cerevisiae,* the mother SPB is embedded in the nuclear envelope during the entire cell cycle. Only the satellite and duplication plaque localize to the cytoplasmic face of the nucleus, whereas during interphase, animal and *Dictyostelium* centrosomes reside in the cytoplasm. Fourth, as in centrioles, the distribution of the outer layer components of the *Dictyostelium* core occurs in a semiconservative manner, since each new centrosome originates from one of the two old outer layers. In contrast, the new *S. cerevisiae* SPB seems to consist only of newly assembled components. Taken together, the *Dictyostelium* centrosome is unique in several aspects; however, it shows some similarities to animal centrosomes but relatively little resemblance to *S. cerevisiae* SPBs. The closer similarity to animal centrosomes becomes even more obvious when the molecular properties of known centrosomal components are compared.

IV. Comparative Molecular Biology of the *Dictyostelium* Centrosome

A. Approaches to the Molecular Analysis of Centrosomal Components

The key prerequisite for a comparative analysis of centrosomal structures at the molecular level and the understanding of centrosomal functions is the identification of as many centrosomal components as possible. But until recently, molecular analysis of centrosomal proteins has been a very difficult task. The first bona-fide centrosomal component, the CDC31 gene product (the yeast homolog of centrin), was cloned in 1986 (Baum *et al.,* 1986). The discovery and cloning of γ-tubulin in 1989 (Oakley and Oakley, 1989) and the elucidation of its role in microtubule nucleation (Oakley *et al.,* 1990) were a breakthrough for the molecular analysis of centrosomal functions. Nevertheless, cloning of new centrosomal components remained burdensome because of their low expression levels. Furthermore, cloning of centrosomal components identified at the protein level was often hampered by their weak representation in cDNA libraries. For example, in the case of pericentrin, only one clone out of 5×10^6 clones was isolated by immunoscreening with an autoantiserum from a scleroderma patient (Doxsey *et al.,* 1994). Fortunately, in recent years, the molecular analysis of the centrosome was accelerated rapidly by several new approaches and the extensive use of the yeast model system (Winsor and Schiebel, 1997): (i) The yeast two-hybrid system (Chien *et al.,* 1991) facilitated cloning of binding partners of known centrosomal components; (ii) completion of the yeast genome project circumvented time-consuming sequencing work and allowed fast cloning approaches; (iii) mass spectrometry in combination with genome project data allowed identification of yeast SPB components from minute amounts of starting material (Adams and Kilmartin, 1999; Wigge *et al.,* 1998); (iv) and finally, functional analyses of SPB components were facilitated by genetic studies using temperature-sensitive mutants (reviewed by Winsor and Schiebel, 1997). At the molecular level, the *S. cerevisiae* SPB is now the most thoroughly investigated MTOC. The other centrosomal model systems profited from the developments in yeast. After sufficient progress in the genome and cDNA sequencing projects in other organisms was made, several important centrosomal components, for example the human homologs of Spc97p and Spc98p, constituents of the γ-TuRC (Murphy *et al.,* 1998; Tassin *et al.,* 1998), were identified by their homology to the respective yeast proteins. However, it turned out that the similarity of yeast SPB components to their mammalian homologs is often very low, reflecting the large evolutionary distance and differences in centrosomal morphology. In many cases, mammalian homologs of well-known yeast

SPB components may in fact exist but are not detected in DNA data libraries because of the low level of conservation. One of these proteins is *S. cerevisiae* Spc110p, which seems to be a ubiquitous MTOC component since it was immunologically localized to mammalian and *Dictyostelium* centrosomes using yeast Spc110p-specific antibodies (Gräf *et al.,* 1998; Tassin *et al.,* 1997). So far it could not be found in any DNA database from *Dictyostelium* or higher organisms. Even within the phylum of fungi, amino acid identity of *S. cerevisiae* proteins to their homologs in other fungi such as *S. pombe, Aspergillus nidulans,* or *Neurospora crassa* is often rather low. For example, the primary structure of γ-tubulin is highly conserved in representative organisms of all phyla, yet the yeast homolog Tub4p is so aberrant that some considered it a member of a novel tubulin subfamily (discussed in Burns, 1996). By contrast, *Dictyostelium* γ-tubulin (Euteneuer *et al.,* 1998) is closely related to γ-tubulins of higher plants and animals (Fig. 3a), despite the relatively large evolutionary distance to all these model organisms (Baldauf and Doolittle, 1997).

Figure 3 Comparison of amino acid sequence identity of all currently known *Dictyostelium* centrosomal proteins with homologs of other representative species. Black bars refer to sequences of higher animals, dark shaded bars to protist sequences, gray bars to plant sequences, light shaded bars to fungal sequences. Amino acid identities were calculated using the BLAST, FASTA, GAP, or Bestfit programs of the GCG software package (Genetics Computer Group, University of Wisconsin). Numbers in the bars refer to the length of pairwise amino acid sequence comparisons; numbers above the columns indicate the percentage of amino acid identity. Abbreviations: Ap, *Anemia phyllitidis;* An, *Aspergillus nidulans;* Ce, *Caenorhabditis elegans;* Cr, *Chlamydomonas reinhardii;* Dd, *Dictyostelium discoideum;* Dm, *Drosophila melanogaster;* Hs, *Homo sapiens;* Mm, *Mus musculus;* Nc, *Neurospora crassa;* Ng, *Naegleria gruberi;* Pf, *Plasmodium falciparum;* Sc, *Saccharomyces cerevisiae;* Sp, *Schizosaccharomyces pombe;* Xl, *Xenopus laevis.* (**a** Identity of *Dictyostelium* γ-tubulin (AJ000492) to γ-tubulins of Ap (P34785), Dm (P42271), Hs (P23258), Sp (P25295), Nc (P53377), Pf (P34787), Ce (P34475), and Sc (P53378). (**b**) Identity of the Dd-Spc97 (JAX4a171e01 and JAX4b07a06) to Spc97 of Hs (AF042379), Dm (AF118379), Sp (AB026664), and Sc (U00027). (**c**) Identity of the Dd-Spc98 (JAX4a195e03) to Spc98 of Hs (AF042378), XI (AF052663), Dm (AF118380), and Sc (P53540). (**d**) Identity of *Dictyostelium* centrin (C84755) to centrins of Ng (U21725), Hs3 (Centrin 3; Y12473), Cr (X12634), Hs2 (Centrin 2; X72964), Hs1 (U03270), XI (U37538), Sc (P06704), (**e**) Identity of DdCP224 (AJ012088) to two sequence regions (numbers refer to the amino acid positions of the DdCP224 sequence) defined by expressed sequence tags (AI403129 and AI 405294) of *Drosophila.* (**f**) Identity of DdCP224 to region 1 (Matthews *et al.,* 1998) of the corresponding sequences of Hs (S68176), Sp (D55635), Ce (1a/1b refer to the first and second, duplicated region 1; AF035197) and Sc (P46675). (**g**) Identity of the Dd-Nek2 (SLD805) catalytic domain to Nek2 of Mm (AF013166), Hs (P51955), XI (AB019556), and NIMA-like kinases of An (NIMA; P11837), Sc (KIN3; P22209), Mm (Nek1; P51954), Nc (NIM1; P48479), Ce (YLK3; P41951). (**h**) Identity of the Dd-Nek2 full-length sequence to Nek2 of XI, and Hs and to NIMA-like kinases of Nc and An. Numbers in brackets are Swissprot/NBRF-PIR, EMBL/Genbank accession numbers, or clone numbers of the *Dictyostelium* genome (http://genome.imb-jena.de/Dictyostelium) or cDNA sequencing project (Morio *et al.,* 1998).

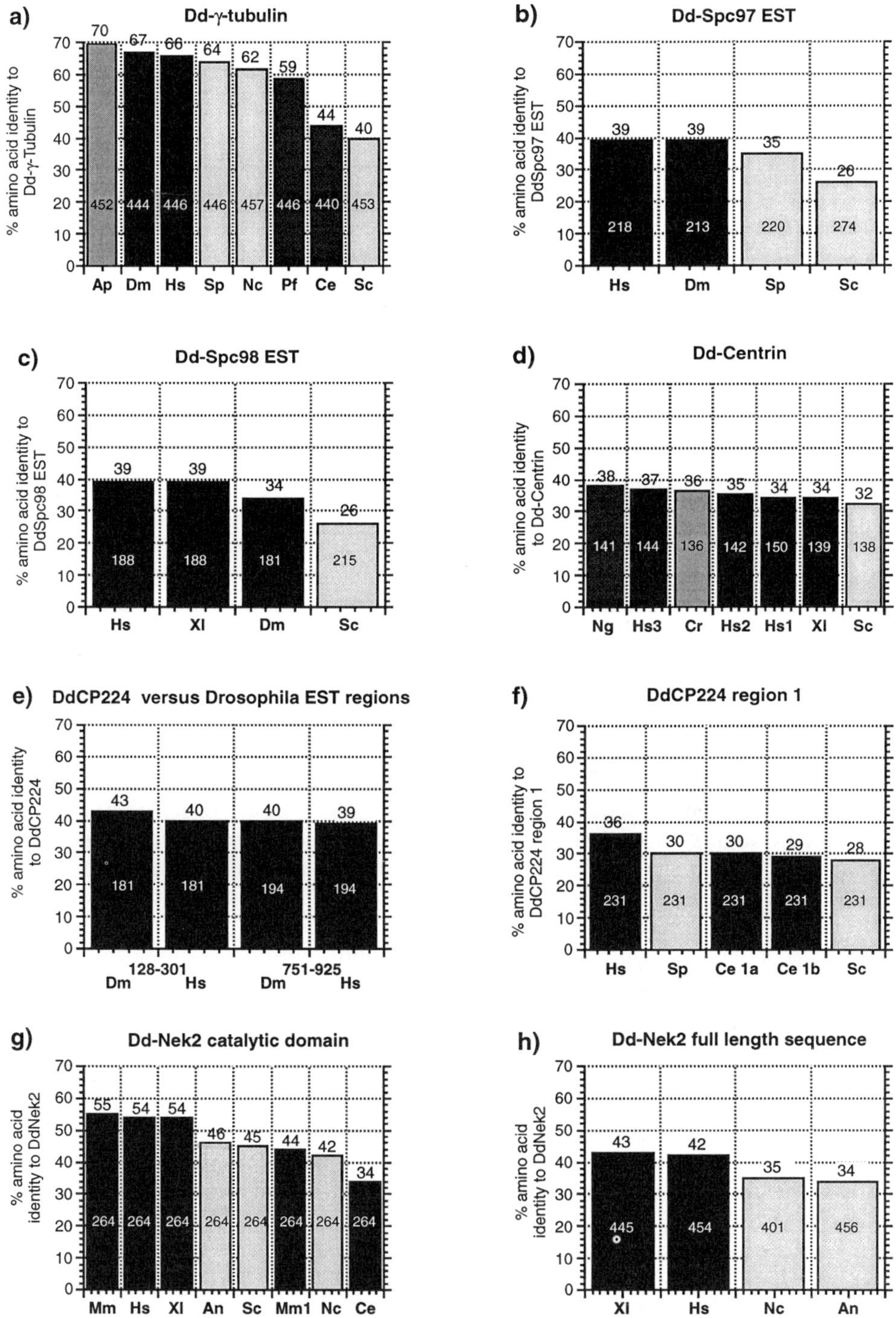
a)
Dd-γ-tubulin
% amino acid identity to Dd-γ-Tubulin
70 67 66 64 62 59 44 40
452 444 446 446 457 446 440 453
Ap Dm Hs Sp Nc Pf Ce Sc
b)
Dd-Spc97 EST
% amino acid identity to DdSpc97 EST
39 39 35 26
218 213 220 274
Hs Dm Sp Sc
c)
Dd-Spc98 EST
% amino acid identity to DdSpc98 EST
39 39 34 26
188 188 181 215
Hs Xl Dm Sc
d)
Dd-Centrin
% amino acid identity to Dd-Centrin
38 37 36 35 34 34 32
141 144 136 142 150 139 138
Ng Hs3 Cr Hs2 Hs1 Xl Sc
e)
DdCP224 versus Drosophila EST regions
% amino acid identity to DdCP224
43 40 40 39
181 181 194 194
128-301
Dm Hs
751-925
Dm Hs
f)
DdCP224 region 1
% amino acid identity to DdCP224 region 1
36 30 30 29 28
231 231 231 231 231
Hs Sp Ce 1a Ce 1b Sc
g)
Dd-Nek2 catalytic domain
% amino acid identity to DdNek2
55 54 54 46 45 44 42 34
264 264 264 264 264 264 264 264
Mm Hs Xl An Sc Mm1 Nc Ce
h)
Dd-Nek2 full length sequence
% amino acid identity to DdNek2
43 42 35 34
445 454 401 456
Xl Hs Nc An

B. Components of the *Dictyostelium* Centrosome

In addition to γ-tubulin, we now have complete cDNA sequences or at least ESTs of five more centrosomal components from *Dictyostelium*. These are (i) DdCP224, the Dictyostelium homolog of *S. cerevisiae* Stu2p (Wang and Huffaker, 1997); (ii) centrin; (iii) Nek2, a centrosomal NIMA-related kinase; (iv) Spc97p; and (v) Spc98p.

Spc97p and Spc98p are ubiquitous components of γ-tubulin complexes in all organisms analyzed so far. In *S. cerevisiae,* cytosolic γ-tubulin complexes seem to be composed only of one Spc97p/Spc98 and two γ-tubulin subunits (Knop and Schiebel, 1997). Such small γ-tubulin complexes also exist in higher organisms such as *Drosophila* and are called γ-TuSC (γ-tubulin small complex), but at least four additional components are required for the assembly of functional γ-tubulin complexes (γTuRC) (Oegema *et al.,* 1999). This leads to the question of the organization of γ-tubulin complexes in *Dictyostelium,* which is phylogenetically distant from yeasts and higher animals. Although several experimental attempts for biochemical analysis of γ-tubulin complexes employing a double-tagged γ-tubulin-myc/his mutant cell line (C. Daunderer, R. Gräf; unpublished) have been made, this cannot be answered to date, but the molecular data allow us to speculate on more similarity to γ-TuRCs than to the simple γ-TuSC-like complexes of yeast. Thus, *Dictyostelium* γ-tubulin is approximately 70% identical to human γ-tubulin (Zheng *et al.,* 1991) but only 40% identical to yeast Tub4p (Fig. 3a) (Marschall *et al.,* 1996; Spang *et al.,* 1996). The expressed sequence tags of Dd-Spc97p and Dd-Spc98p, each encoding approximately 200 amino acids, display about 40% identity to the human and *Drosophila* sequences (Murphy *et al.,* 1998; Tassin *et al.,* 1998) but only 26% identity to the *S. cerevisiae* sequences (Figs. 3b, 3c) (Geissler *et al.,* 1996; Knop *et al.,* 1997). Likewise, amino acid sequence comparison of the three other known centrosomal *Dictyostelium* proteins (centrin, Nek2, and DdCP244) to their respective homologs suggests a closer relationship of the *Dictyostelium* centrosome to centriolar centrosomes than to the SPB of *S. cerevisiae* (Figs. 3d–3h). This is in agreement with the morphological and centrosome-duplication data discussed previously.

As already mentioned, Dictyostelium centrin is more similar to animal centrins (Schiebel and Bornens, 1995) than to the yeast homolog Cdc31p (Baum *et al.,* 1986). Centrins are usually highly conserved members of the EF-hand family of Ca^{2+}-binding proteins. However, the sequence identity of Dictyostelium centrin compared to its closest homolog in *Naegleria gruberi* (Levy *et al.,* 1996) is only 38%, demonstrating considerable sequence divergence of some of the *Dictyostelium* centrosome components in comparison to other organisms as well (Fig. 3d).

DdCP224 was recently cloned in our lab with the aid of monoclonal antibodies raised against purified *Dictyostelium* centrosomes (Gräf *et al.,*

1999; Gräf *et al.,* 2000). Sequence comparisons revealed that DdCP224 belongs to a protein family including Stu2p, an essential SPB component in yeast (Wang and Huffaker, 1997). Conservation of DdCP224 is difficult to discuss, because to date full-length sequences of homologous proteins have been published only from humans (ch-TOG-protein; Charrasse *et al.,* 1998), *Caenorhabditis elegans* (ZYG-9; Matthews *et al.,* 1998), *S. pombe* (p93^{dis1}; Nabeshima *et al.,* 1995), and *S. cerevisiae* (Stu2p; Wang and Huffaker, 1997). Peptide sequences or ESTs are available for *Xenopus* XMAP215 (Charrasse *et al.,* 1998) and *Drosophila* (Fig. 3e), respectively. The size of DdCP224-like proteins varies markedly: Only the *Dictyostelium,* human, and *Xenopus* proteins have a similar molecular mass of ~220 kDa and display a moderate, but significant amino acid identity throughout the whole sequence (28% identity and 40% similarity between DdCP224 and the ch-TOG protein). The length of the *Drosophila* protein is unknown, but the available ESTs suggest that it is more closely related to DdCP224 than is the ch-TOG protein (Fig. 3e). The other proteins of this family are considerably smaller than the *Dictyostelium* and human proteins: The two yeast proteins have only ~90 kDa and correspond to the N-terminal half of the ch-TOG protein and DdCP224, respectively. ZYG-9 has a molecular mass of ~150 kDa, and its sequence exhibits only two conserved ~250 amino acids regions (regions 1 and 2; Matthews *et al.,* 1998) compared to the ch-TOG protein and DdCP224. Region 2 is missing in yeasts. Thus, when all five sequences at hand are compared, only region 1 (231 amino acids) is conserved in all of them (Fig. 3f). The role of this region, which is even duplicated in ZYG-9 (designated Ce1a, Ce1b in Fig. 3f), is not known, but it is conceivable that it is involved in an ubiquitous function that might be binding to microtubules or mitotic spindle poles, respectively. All members of this protein family are microtubule-binding proteins, but some of them exhibit cell-cycle-dependent centrosomal localization as well. The ch-TOG protein, ZYG-9, and p93^{dis1} are components of mitotic spindle poles but seem to be absent from the centrosome during interphase, whereas DdCP224 and Stu2p are permanent residents at the centrosome and SPB, respectively. Thus, members of this protein family may fulfill, at least in part, different tasks within the cell, which might explain the considerable sequence differences.

By contrast, Nek2-like NIMA-related kinases (Schultz *et al.,* 1994) are highly conserved and have a similar molecular mass of approx. 50 kDa. The serine/threonine kinase NIMA was first characterized in *A. nidulans.* Temperature-sensitive mutants in the *nimA* gene revealed a second NIMA-dependent kinase pathway (*nim* stands for "never in mitosis") in addition to the Cdc2/cyclin B pathway that is required to drive cells into mitosis, since these mutants failed to enter mitosis at the restrictive temperature (Osmani *et al.,* 1991). Overexpression of wild-type NIMA or dominant negative NIMA mutants in several heterologous species resulted in severe cell cycle disruption, suggesting that NIMA pathways play a role in cell cycle regulation in other organisms (*Xenopus,* humans, *S. pombe*) as well (Lu and Hunter, 1995; O'Connell *et al.,* 1994). Kinases related, but not necessarily functionally homologous, to NIMA, have now been characterized in a variety of organisms (for review, see Fry and Nigg, 1995, 1997). Among the human NIMA-related kinases, Nek2 is of special interest within this context because, in addition to its cytosolic presence, it is localized at the centrosome throughout the entire cell cycle and seems to play an important role in centrosome duplication (Fry *et al.,* 1998a,b). The *Dictyostelium* homolog (Dd-Nek2) is more similar to human, mouse, and *Xenopus* Nek2 than to any other NIMA-related kinase (Figs. 3g,3h). The catalytic domain extending over 2/3 of the protein chain displays ~55% identity and the whole sequence ~43% amino acid identity to human, mouse, and *Xenopus* Nek2. Thus, Dd-Nek2 is the first nonvertebrate member closely related to the Nek2-kinase subfamily. It is tempting to speculate that Dd-Nek2, as its vertebrate counterpart, is a centrosomal component and a regulator of centrosome duplication as well.

C. Regulation of Centrosome Duplication

In humans, a Nek2-binding partner, called C-Nap1, was isolated in a two-hybrid screen (Fry *et al.,* 1998a). Independently, C-Nap1 was also cloned using a human autoantiserum (Mack *et al.,* 1998). C-Nap1 turned out to be a 281-kDa centrosomal component that was localized to the proximal ends of mother and daughter centrioles during interphase. C-Nap1 can be phoshorylated by Nek2 *in vitro* and *in vivo,* and it dissociates from the centrosome at the onset of mitosis. Since earlier electron microscopic studies revealed that the proximal ends of mother and daughter centrioles are connected by fibers (Paintrand *et al.,* (1992), Fry *et al.* (1998a) hypothesized that these fibers contain, or consist, of C-Nap1 and mediate centriole cohesion during interphase. At the onset of mitosis C-Nap1 phosphorylation by Nek2 could trigger C-Nap1 dissociation or degradation, leading to a loss of centriole cohesion, which in turn is necessary for centriole separation during mitosis. Such a role of Nek2 is also supported by the observation that its overexpression in cultured cells resulted in centrosome splitting (Fry *et al.,* 1998b). A prospective binding partner of Nek2 also exists in *Dictyostelium:* Recent results indicated that the "350-kDa antigen" characterized by the monoclonal antibody NAB350 (Kalt and Schliwa, 1996) migrates as a ~280 kDa protein in high-resolution SDS-gels and is absent from the centrosome during mitosis (R. Gräf, unpublished). In addition, it is recognized by C-Nap1-specific antibodies in immunoblots and immunofluorescence. These properties make it a good candidate for a C-Napl homolog in *Dictyostelium* (R. Gräf, T. Mayor, E. A. Nigg; unpublished results). Since the "350-kDa antigen" was localized to the outer surface of the corona (Gräf *et al.,* 1998; Kalt and Schliwa, 1996), it may help to stabilize the corona during interphase. In early prophase, phosphorylation by Dd-Nek2 would trigger dissociation of the "350-kDa antigen" from the centrosom, which would destabilize the corona, causing it to dissipate (see Section III). Currently we are investigating the localization of Dd-Nek2, and whether the "350-kDa antigen" serves as a substrate for recombinant Dd-Nek2.

Centrosome duplication is certainly not regulated by Nek2 alone; rather, the data just reviewed suggest that Nek2 is placed near the end of a regulatory cascade. Despite its ability to undergo autophosphorylation (Fry *et al.,* 1995), Nek2 may require additional phosphorylation by other kinases for full activity, as does NIMA, which depends on phosphorylation by cdk1-cyclin B (also known as $p34^{cdc2}$-cyclin B) for full activation (Ye *et al.,* 1995). The regulation of human Nek2 is not clear, but experiments with S-phase arrested sea urchin zygotes undergoing repeated centrosome duplication have shown that cdkl/cyclin B does not play a role in the control of this process (Hinchcliffe *et al.,* 1998). Moreover, Nek2 lacks consensus sequences for cdk1-cyclin B phosphorylation. Thus, Nek2, unlike NIMA, may not be a target for cdk1-cyclin B regulation. However, other cyclin dependent kinases (cdk) might be involved. Cdk2-cyclin E would be an especially good candidate for a key regulator of centrosome duplication since it controls the onset of DNA replication (Jackson *et al.,* 1995; Strausfeld *et al.,* 1994). It is an attractive hypothesis that both processes, which take place coordinately at the G1/S transition, are coupled by the same effector. Recently, four independent reports indicated that centrosome duplication is indeed regulated by cdk2-cyclin complexes (Hinchcliffe *et al.,* 1999; Lacey *et al.,* 1999; Matsumoto *et al.,* 1999; Meraldi *et al.,* 1999). Microinjection of the cdk2 inhibitors into *Xenopus* early embryo cells prevented centrosome duplication, but coinjection of an excess of cyclin E alleviated the block. Furthermore, if isolated centrosomes were incubated with cdk2-depleted extracts from fertilized *Xenopus* eggs, centrosome duplication was inhibited, but it was restored if recombinant cdk2-cyclin E was added. By contrast, in somatic cells, cyclin A rather than cyclin E seems to be the primary partner of Cdk2. Meraldi *et al.* (1999) investigated the effect of various effectors of the Rb (retinoblastom protein) pathway on repeated centrosome cycles which occur in CHO cells after hydroxyurea treatment. The authors showed that centrosome duplication in these somatic cells requires Cdk2-cyclin A and release of transcription factors of the E2F family. However, Cdk2-cyclin E may play a role in centrosome duplication in these cells as well since it is, in addition to Cdk4/6-cyclin D, involved in Rb phosphorylation, which in turn is required for activation of E2F. In both experimental systems, *Xenopus* and CHO cells, the targets for the Cdk2/cyclins remain to be identified.

In addition to cyclin-dependent kinases, members of two further kinase families have been implicated in the process of centrosome duplication: kinases related to the polo kinase and the aurora kinase in *Drosophila.* Polo-like kinases (PLKs) are an evolutionarily conserved family of kinases regulating diverse cell-cycle functions (Glover *et al.,* 1998; Nigg, 1998). Among other functions, PLKs seem to act upstream of cyclin-dependent kinases in centrosome duplication by regulating the recruitment of pericentriolar material prior to centriole separation in mitosis. Mammalian Plk1 was localized to spindle poles during prophase and metaphase, and HeLa cells injected with anti-PLK-antibody fail to accumulate PCM proteins at the centrosome in late G2 (Lane and Nigg, 1996). Thus, these cells often arrest as pseudomitotic cells with monoastral spindles. Other players in centrosome duplication are kinases related to the *Drosophila* aurora kinase. They were localized to mitotic spindle poles (Bischoff *et al.,* 1998; Kimura *et al.,* 1997; Schumacher *et al.,* 1998), and ectopic expression of *STK15* (also known as *aurora2* and *BTAK*) in NIH 3T3 cells leads to abnormal centrosome amplification and aneuploidy (Zhou *et al.,* 1998). The complex control of centrosome duplication by a variety of, possibly interlinked, kinase pathways seems to be necessary to ensure the coordination of this crucial process with other mitotic events. Misregulation of centrosome duplication often leads to supernumerary centrosomes, which in turn seem to be a frequent cause of aneuploidy in numerous cancer cell lines (Brinkley and Goepfert, 1998).

In yeast, the role of most of the proteins involved in SPB duplication was discovered by the analysis of mutant strains that displayed an arrest at various stages of SPB duplication under restrictive conditions (reviewed by Winey and Byers, 1993, and Sobel, 1997). Mutated genes encoded SPB components such as the half-bridge proteins Cdc31p and Kar1p, or non-SPB components such as Ndc1p, Mps2p, and the dual specificity kinase Mps1p. Cyclin-dependent kinases play a similar role as in mammalian cells; i.e., Cdc28/Cln, the yeast Cdk/cyclin complex, promotes DNA synthesis and is required for progression through START. However, a comparable role of NIMA-, polo, or aurora-related kinases, as discussed earlier, has not been shown in *S. cerevisiae* so far, raising the possibility that some of the SPB duplication events are regulated by yeast-specific pathways. For example, the only polo-like kinase in yeast, Cdc5p, is known as a regulator of the anaphase-promoting complex (APC) and is required for exit from mitosis (Shirayama *et al.,* 1998), but does not seem to be involved in SPB duplication.

Thus far, Dd-Nek2 is the only candidate involved in the regulation of centrosome duplication in *Dictyostelium,* but it is likely that the genome project will soon reveal more potential regulators in *Dictyostelium* belonging to one of the kinase families discussed earlier. However, duplication may also involve novel regulatory pathways compared to animal cells and yeast, because in *Dictyostelium* centrosome duplication takes place in M (see Section III), and thus, it cannot be coupled to the onset of DNA replication in S-phase by the action of cdk2-cyclin E.

V. Conclusions

Some aspects of the *Dictyostelium* centrosome, such as its morphology and mode of duplication, are unique compared to other model systems, although the data presented in this review suggest that the centrosome of *Dictyostelium* shares more similarities with animal centriolar centrosomes than with *S. cerevisiae* SPBs. Recent progress justifies a role of *Dictyostelium* as a model system that may help elucidate the nature and function of this enigmatic cell organelle. Future studies on the analysis of proteins directly implicated in centrosome duplication, including Dd-Nek2 and its binding partners, and proteins involved in the regulation of the highly dynamic microtubule cytoskeleton, such as DdCP224 or γ-tubulin complexes, should contribute to our understanding of these processes in this fascinating organelle.

References

Adams, I. R., and Kilmartin, J. V. (1999). Localization of core spindle pole body (SPB) components during SPB duplication in *Saccharomyces cerevisiae. J. Cell Biol.* **145,** 809–823.

Balczon, R. (1996). The centrosome in animal cells and its functional homologs in plant and yeast cells. *Int. Rev. Cytol.* **169,** 25–82.

Baldauf, S. L., and Doolittle W. F. (1997). Origin and evolution of the slime molds (Mycetozoa). *Proc. Natl. Acad. Sci. USA* **94,** 12007–12012.

Baum, P., Furlong, C., and Byers, B. (1986). Yeast gene required for spindle pole body duplication: homology of its product with Ca^{2+}-binding proteins. *Proc. Natl. Acad. Sci. USA* **83,** 5512–6.

Bischoff, J. R., Anderson, L., Zhu, Y., Mossie, K., Ng, L., Souza, B., Schryver, B., Flanagan, P., Clairvoyant, F., Ginther, C., Chan, C. S. M., Novotny, M., Slamon, D. J., and Plowman, G. D. (1998). A homologue of *Drosophila* aurora kinase is oncogenic and amplified in human colorectal cancers. *EMBO J.* **17,** 3052–3065.

Brinkley, B. R., and Goepfert, T. M. (1998). Supernumerary centrosomes and cancer: Boveri's hypothesis resurrected. *Cell Motil. Cytoskeleton* **41,** 281–288.

Burns, R. G. (1996). Highly divergent *Caenorhabditis* and *Saccharomyces* tubulins evolved recently from genes encoding γ-tubulin. *Trends Cell Biol.* **6,** 376.

Byers, B., and Goetsch, L. (1975). Behavior of spindles and spindle plaques in the cell cycle and conjugation of *Saccharomyces cerevisiae. J. Bacteriol.* **124,** 511–523.

Calarco Gillam, P. D., Siebert, M. C., Hubble, R., Mitchison, T., and Kirschner, M. (1983). Centrosome development in early mouse embryos as defined by an autoantibody against pericentriolar material. *Cell* **35,** 621–9.

Charrasse, S., Schroeder, M., Gauthier Rouvieve, C., Ango, F., Cassimeris, L., Gard, D. L., and Larroque, C. (1998). The TOGp protein is a new human microtubule-associated protein homologous to the *Xenopus* XMAP215. *J. Cell Sci.* **111,** 1371–1383.

Chien, C.-T., Bartel, P. L., Stenrglanz, R., and Fields, S. (1991). The two-hybrid system: A method to identify and clone genes for proteins that interact with a protein of interest. *Proc. Natl. Acad. Sci. USA* **88,** 9578–9582.

Chretien, D., Buendia, B., Fuller, S. D., and Karsenti, E. (1997). Reconstruction of the centrosome cycle from cryoelectron micrographs. *J. Struct. Biol.* **120,** 117–133.

Corthesy Theulaz, I., Pauloin, A., and Rfeffer, S. R. (1992). Cytoplasmic dynein participates in the centrosomal localization of the Golgi complex. *J. Cell Biol.* **118,** 1333–45.

Daunderer, C., Schliwa, M., and Gräf, R. (1999). *Dictyostelium discoideum:* a promising centrosome model system. *Biol. Cell* **91,** 313–320.

Debec, A., and Abbadie, C. (1989). The acentriolar state of the *Drosophila* cell lines 1182. *Biol. Cell* **67,** 307–311.

Dictenberg, J. B., Zimmerman, W., Sparks, C. A., Young, A., Vidair, C., Zheng, Y. X., Carrington, W., Fay, F. S., and Doxsey, S. J. (1998). Pericentrin and γ-tubulin form a protein complex and are organized into a novel lattice at the centrosome. *J. Cell Biol.* **141,** 163–174.

Ding, R., West, R. R., Morphew, M., Oakley, B. R., and Mclntosh, J. R. (1997). The spindle pole body of *Schizosaccharomyces pombe* enters and leaves the nuclear envelope as the cell cycle proceeds. *Mol. Biol. Cell* **8,** 1461–1479.

Doxsey, S. J., Stein, P., Evans, L., Calarco, P. D., and Kirschner, M. (1994). Pericentrin, a highly conserved centrosome protein involved in microtubule organization. *Cell* **76,** 639–50.

Elliott, S., Knop, M., Schlenstedt, G., and Schiebel, E. (1999). Spc29p is a component of the Spc110p subcomplex and is essential for spindle pole body duplication. *Proc. Natl. Acad. Sci. USA* **96,** 6205–6210.

Euteneur, U., Gräf, R., Kube-Granderath, E., and Schliwa, M. (1998). *Dictyostelium* γ-tubulin: molecular characterization and ultrastructural localization. *J. Cell Sci.* **111,** 405–412.

Felix, M. A., antony, C., Wright, M., and Maro, B. (1994). Centrosome assembly *in vitro:* role of γ-tubulin recruitment in *Xenopus* sperm aster formation. *J. Cell Biol.* **124,** 19–31.

Fry, A. M., and Nigg, E. A. (1995). Cell cycle. The NIMA kinase joins forces with Cdc2. *Curr. Biol.* **5,** 1122–1125.

Fry, A. M., and Nigg, E. A. (1997). Characterization of mammalian NIMA-related kinases. *Methods Enzymol.* **283,** 270–282.

Fry, A. M., Mayor, T., Meraldi, P., Stierhof, Y. D., Tanaka, K., and Nigg, E. A. (1998a). C-Nap1, a novel centrosomal coiled-coil protein and candidate substrate of the cell cycle-regulated protein kinase Nek2. *J. Cell Biol.* **141,** 1563–1574.

Fry, A. M., Meraldi, P., and Nigg, E. A. (1998b). A centrosomal function for the human Nek2 protein kinase, a member of the NIMA family of cell cycle regulators. *EMBO J.* **17**(2), 470–481.

Fry, A. M., Schultz, S. J., Bartek, J., and Nigg, E. A. (1995). Substrate specificity and cell cycle regulation of the Nek2 protein kinase, a potential human homolog of the mitotic regulator NIMA of *Aspergillus nidulans. J. Biol. Chem.* **270,** 12899–12905.

Fukui, Y. (1993). Toward a new concept of cell motility: cytoskeletal dynamics in amoeboid movement and cell division. *Int. Rev. Cytol.* **144,** 85–127.

Geissler, S., Pereira, G., Spang, A., Knop, M., Soues, S., Kilmartin, J., and Schiebel, E. (1996). The spindle pole body component Spc98p interacts with the γ-tubulin-like Tub4p of *Saccharomyces cerevisiae* at the sites of microtubule attachment. *EMBO J.* **15,** 3899–3911.

Glover, D. M., Hagan, I. M., and Tavares, A. A. (1998). Polo-like kinases: a team that plays throughout mitosis. *Genes Dev.* **12,** 3777–3787.

Gräf, R., Euteneuer, U., Ueda, M., and Schliwa, M. (1998). Isolation of nucleation-competent centrosomes from *Dictyostelium discoideum. Eur. J. Cell Biol.* **76,** 167–175.

Gräf, R., Daunderer, C., and Schliwa, M. (1999). Cell cycle-dependent localization of monoclonal antibodies raised against isolated *Dictyostelium* centrosomes. *Biol. Cell* **91,** 471–477.

Gräf, R., Daunderer, C., and Schliwa, M. (2000). Dictyostelium DolCP224 is a microtubule-associated protein and a permanent centrosomal resident involved in centrosome duplication. *J. Cell Sci.* **113,** 1747–1758.

Hagan, I. M. (1998). The fission yeast microtubule cytoskeleton. *J. Cell Sci.* **12,** 1603–1612.

Hinchcliffe, E. H., Cassels, G. O., Rieder, C. L., and Sluder, G. (1998). The coordination of centrosome reproduction with nuclear events of the cell cycle in the sea urchin zygote. *J. Cell Biol.* **140,** 1417–1426.

Hinchcliffe, E. H., Li, C., Thompson, E. A., Maller, J. L., and Sluder, G. (1999). Requirement of Cdk2-cyclin E activity for repeated centrosome reproduction in *Xenopus* egg extracts. *Science* **283,** 851–4.

Hoyt, M. A., and Geiser, J. R. (1996). Genetic analysis of the mitotic spindle. *Annu. Rev. Genet.* **30,** 7–33.

Infante, C., Ramos-Morales, F., Fedriani, C., Bornens, M., and Rios, R. M. (1999). GMAP-210, a cis-Golgi network-associated protein, is a minus end microtubule-binding protein. *J. Cell Biol.* **145,** 83–98.

Jackson, P. K., Chevalier, S., Philippe, M., and Kirschner, M. W. (1995). Early events in DNA replication require cyclin E and are blocked by p21CIP1. *J. Cell Biol.* **130,** 755–769.

Jensen, C. G., Bollard, S. M., and Roos, U. P. (1991). Analysis of microtubule arms and bridges in mitotic and interphase amebae of *Dictyostelium discoideum. Eur. J. Cell Biol.* **54,** 121–131.

Kalnins, V. I. (1992). "The Centrosome." Academic Press, San Diego.

Kalt, A., and Schliwa, M. (1993). Molecular components of the centrosome. *Trends Cell Biol.* **3,** 118–128.

Kalt, A., and Schliwa, M. (1996). A novel structural component of the *Dictyostelium* centrosome. *J. Cell Sci.* **109,** 3103–12.

Kellogg, D. R., Moritz, M., and Alberts, B. M. (1994). The centrosome and cellular organization. *Annu. Rev. Biochem.* **63,** 639–74.

Kimura, M., Kotani, S., Hattori, T., Sumi, N., Yoshioka, T., Todokoro, K., and Okano, Y. (1997). Cell cycle-dependent expression and spindle pole localization of a novel human protein kinase, Aik, related to Aurora of *Drosophila* and yeast Ipl1.*J. Biol. Chem.* **272,** 13766–71.

Kitanishi, T., Shibaoka, H., and Fukui, Y. (1984). Disruption of microtubules and retardation of development of *Dictyostelium* with ethyl *N*-phenylcarbamate and thiabendazole. *Protoplasma* **120,** 185–196.

Kitanishi-Yumura, T., and Fukui, Y. (1987). Reorganization of microtubules during mitosis in *Dictyostelium:* dissociation from MTOC and selective assembly/disassembly in situ. *Cell Motil. Cytoskel.* **8,** 106–117.

Knop, M., Pereira, G., Geissler, S., Grein, K., and Schiebel, E. (1997). The spindle pole body component Spc97p interacts with the γ-tubulin of *Saccharomyces cerevisiae* and functions in microtubule organization and spindle pole body duplication. *EMBO J.* **16,** 1550–1564.

Knop, M., and Schiebel, E. (1997). Spc98p and Spc97p of the yeast γ-tubulin complex mediate binding to the spindle pole body via their interaction with Spc110p. *EMBO J.* **16,** 6985–6995.

Kochanski, R. S., and Borisy, G. G. (1990). Mode of centriole duplication and distribution. *J. Cell Biol.* **110,** 1599–1605.

Lacey, K. R., Jackson, P. K., and Stearns, T. (1999). Cyclin-dependent kinase control of centrosome duplication. *Proc. Natl. Acad. Sci. USA* **96,** 2817–2822.

Lane, H. A., and Nigg, E. A. (1996). Antibody microinjection reveals an essential role for human polo-like kinase 1 (Plk1) in the functional maturation of mitotic centrosomes. *J. Cell Biol.* **135,** 1701–1713.

Lange, B. M., and Gull, K. (1995). A molecular marker for centriole maturation in the mammalian cell cycle. *J. Cell Biol.* **130,** 919–927.

Lange, B. M. H., and Gull, K. (1996). Structure and function of the centriole in animal cells: Progress and questions. *Trends Cell Biol.* **6,** 348–352.

Levy, Y. Y., Lai, E. Y., Remillard, S. P., Heintzelman, M. B., and Fulton, C. (1996). Centrin is a conserved protein that forms diverse associations with centrioles and MTOCs in *Naegleria* and other organisms. *Cell Motil. Cytoskel.* **33,** 298–323.

Lu, K. P., and Hunter, T. (1995). Evidence for a NIMA-like mitotic pathway in vertebrate cells. *Cell* **81,** 413–24.

Mack, G. J., Rees, J., Sandblom, O., Balczon, R., Fritzler, M. J., and Rattner, J. B. (1998). Autoantibodies to a group of centrosomal proteins in human autoimmune sera reactive with the centrosome. *Arthritis Rheum.* **41,** 551–558.

Maeda, Y., Inouye, K., Takeuchi, I., and (eds.). (1997). "*Dictyostelium:* A Model System for Cell and Developmental Biology." Universal Academic Press, Inc., Tokyo.

Maniak, M., Rauchenberger, R., Albrecht, R., Murphy, J., and Gerisch, G. (1995). Coronin involved in phagocytosis: Dynamics of particle-induced relocalization visualized by a green fluorescent protein tag. *Cell* **83,** 915–924.

Maniotis, A., and Schliwa, M. (1991). Microsurgical removal of centrosomes blocks cell reproduction and centriole generation in BSC-1 cells. *Cell* **67,** 495–504.

Mann, S. K. O., Devreotes, P. N., Eliott, S., Jermyn, K., Kuspa, A., Fechheimer, M., Furukawa, R., Parent, C. A., Segall, J., Shaulsky, G., Vardy, P. H., Williams, J., Williams, K. L., and Firtel, R. A. (1998). Cell biological, molecular genetic, and biochemical methods used to examine *Dictyostelium. In* "Cell Biology: A Laboratory Handbook" (J. E. Celis, Ed.), Vol. 1, pp. 431–465. Academic Press, San Diego.

Marschall, L. G., Jeng, R. L., Mulholland, J., and Steams, T. (1996). Analysis of Tub4p, a yeast γ-tubulin-like protein: implications for microtubule-organizing center function. *J. Cell Biol.* **134,** 443–54.

Matsumoto, Y., Hayashi, K., and Nishida, E. (1999). Cyclin-dependent kinase 2 (Cdk2) is required for centrosome duplication in mammalian cells. *Curr. Biol.* **9,** 425–428.

Matthews, L. R., Carter, P., Thierry Mieg, D., and Kemphues, K. (1998). ZYG-9, a *Caenorhabditis elegans* protein required for microtubule organization and function, is a component of meiotic and mitotic spindle poles. *J. Cell Biol.* **141,** 1159–1168.

McCully, E. K., and Robinow, C. F. (1971). Mitosis in the fission yeast *Schizosaccharomyces pombe:* a comparative study with light and electron microscopy. *J. Cell Sci.* **9,** 475–507.

Meraldi, P., Lukas, J., Fry, A. M., Bartek, J., and Nigg, E. A. (1999). Centrosome duplication in mammalian somatic cells requires E2F and Cdk2-Cyclin A. *Nat. Cell Biol.* **1,** 88–93.

Mignot, J. P. (1996). The centrosomal big bang: From a unique central organelle towards a constellation of MTOCs. *Biol. Cell* **86,** 81–91.

Moens, P. B. (1976). Spindle and kinetochore morphology of *Dictyostelium discoideum. J. Cell Biol.* **68,** 113–122.

Morio, T., Urushihara, H., Saito, T., Ugawa, Y., Mizuno, H., Yoshida, M., Yoshino, R., Mitra, B. N., Pi, M., Sato, T., Takemoto, K., Yasukawa, H., Williams, J., Maeda, M., Takeuchi, I., Ochiai, H., and Tanaka, Y. (1998). The *Dictyostelium* developmental cDNA project: generation and analysis of expressed sequence tags from the first-finger stage of development. *DNA Res.* **5,** 335–340.

Moritz, M., Braunfeld, M. B., Sedat, J. W., Alberts, B., and Agard, D. A. (1995). Microtubule nucleation by γ-tubulin-containing rings in the centrosome. *Nature* **378,** 638–640.

Murphy, S. M., Urbani, L., and Stearns, T. (1998). The mammalian γ-tubulin complex contains homologues of the yeast spindle pole body components Spc97p and Spc98p. *J. Cell Biol.* **141,** 663–674.

Nabeshima, K., Kurooka, H., Takeuchi, M., Kinoshita, K., Nakaseko, Y., and Yanagida, M. (1995). p93dis1, which is required for sister chromatid separation, is a novel microtubule and spindle pole body-associating protein phosphorylated at the Cdc2 target sites. *Genes Dev.* **9,** 1572–85.

Neujahr, R., Albrecht, R., Kohler, J., Matzner, M., Schwartz, J. M., Westphal, M., and Gerisch, G. (1998). Microtubule-mediated centrosome motility and the positioning of cleavage furrows in multinucleate myosin II-null cells. *J. Cell Sci.* **111,** 1227–1240.

Nigg, E. A. (1998). Polo-like kinases: positive regulators of cell division from start to finish. *Curr. Opin. Cell Biol.* **10,** 776–783.

Noegel, A. A., and Luna, J. E. (1995). The *Dictyostelium* cytoskeleton. *Experientia* **51,** 1135–1143.

O'Connell, M. J., Norbury, C., and Nurse, P. (1994). Premature chromatin condensation upon accumulation of NIMA. *EMBO J.* **13,** 4926–4937.

Oakley, B. R., Oakley, C. E., Yoon, Y., and Jung, M. K. (1990). γ-Tubulin is a component of the spindle pole body that is essential for microtubule function in *Aspergillus nidulans. Cell* **61,** 1289–1301.

Oakley, C. E., and Oakley, B. R. (1989). Identification of γ-tubulin, a new member of the tubulin superfamily encoded by mipA gene of *Aspergillus nidulans. Nature* **338,** 662–664.

Oegema, K., Wiese, C., Martin, O. C., Milligan, R. A., Iwamatsu, A., Mitchison, T. J., and Zheng, Y. (1999). Characterization of two related *Drosophila* γ-tubulin complexes that differ in their ability to nucleate microtubules. *J. Cell Biol.* **144,** 721–733.

Omura, F., and Fukui, Y. (1985). *Dictyostelium* MTOC: Structure and linkage to the nucleus. *Protoplasma* **127,** 212–221.

Osmani, A. H., McGuire, S. L., and Osmani, S. A. (1991). Parallel activation of the NIMA and p34cdc2 cell cycle-regulated protein kinases is required to initiate mitosis in *A. nidulans. Cell* **67,** 283–91.

Paintrand, M., Moudjou, M., Delacroix, H., and Bornens, M. (1992). Centrosome organization and centriole architecture: their sensitivity to divalent cations. *J. Struct. Biol.* **108,** 107–128.

Pereira, G., and Schiebel, E. (1997). Centrosome-microtubule nucleation. *J. Cell Sci.* **110,** 295–300.

Raff *et al.* (1993). Drosophila γ-tubulin is part of a complex containing two previously identified MAPs. *J. Cell Biol.* **121,** 823–835.

Roos, U. P. (1975). Fine structure of an organelle associated with the nucleus and cytoplasmic microtubules in the cellular slime mould *Polysphondylium violaceum. J. Cell Sci.* **18,** 315–326.

Roos, U. P., and Camenzind, R. (1981). Spindle dynamics during mitosis in *Dictyostelium discoideum. Eur. J. Cell Biol.* **25,** 248–257.

Roos, U. P., De Brabander, M., and De Mey, J. (1984). Indirect immunofluorescence of microtubules in *Dictyostelium discoideum.* A study with polyclonal and monoclonal antibodies to tubulins. *Exp. Cell Res.* **151,** 183–193.

Roos, U. P., De Brabander, M., and Nuydens, R. (1986). Cell shape and organization of F-actin and microtubules in randomly moving and stationary amebae of *Dictyostelium discoideum. Cell Motil. Cytoskel.* **6,** 176–185.

Rubino, S., Fighetti, M., Unger, E., and Cappuccinelli, P. (1984). Location of actin, myosin and microtubular structures during directed locomotion of *Dictyostelium* amebae. *J. Cell Biol.* **98,** 382–390.

Sawin, K. E., and Endow, S. A. (1993). Meiosis, mitosis and microtubule motors. *Bioessays* **15,** 399–407.

Schiebel, E., and Bornens, M. (1995). In search of a function for centrins. *Trends Cell Biol.* **5,** 197–201.

Schliwa, M., Euteneuer, U., Gräf, R., and Ueda, M. (1998). Centrosomes, microtubules, and cell migration. *In* "Cell Behavior: Control and Mechanism of Motility" (J. M. Lackie, G. A. Dunn, and G. E. Jones, Eds.), pp. 223–231. Portland Press, London.

Schultz, S. J., Fry, A. M., Sutterlin, C., Ried, T., and Nigg, E. A. (1994). Cell cycle-dependent expression of Nek2, a novel human protein kinase related to the NIMA mitotic regulator of *Aspergillus nidulans. Cell Growth Differ.* **5,** 625–635.

Schumacher, J. M., Ashcroft, N., Donovan, P. J., and Golden, A. (1998). A highly conserved centrosomal kinase, AIR-1, is required for accurate cell cycle progression and segregation of developmental factors in *Caenorhabditis elegans* embryos. *Development* **125,** 4391–402.

Shirayama, M., Zachariae, W., Ciosk, R., and Nasmyth, K. (1998). The Polo-like kinase Cdc5p and the WD-repeat protein Cdc20p/fizzy are regulators and substrates of the anaphase promoting complex in *Saccharomyces cerevisiae. EMBO J.* **17,** 1336–49.

Singer, S. J., and Kupfer, A. (1986). The directed migration of eukaryotic cells. *Annu. Rev. Cell Biol.* **2,** 337–365.

Sobel, S. G. (1997). Mitosis and the spindle pole body in *Saccharomyces cerevisiae. J. Exp. Zool.* **277,** 120–138.

Spang, A., Geissler, S., Grein, K., and Schiebel, E. (1996). γ-Tubulin-like Tub4p of *Saccharomyces cerevisiae* is associated with the spindle pole body substructures that organize microtubules and is required for mitotic spindle formation. *J. Cell Biol.* **134,** 429–441.

Stearns, T., and Kirschner, M. (1994). In vitro reconstitution of centrosome assembly and function: the central role of γ-tubulin. *Cell* **76,** 623–37.

Strausfeld, U. P., Howell, M., Rempel, R., Maller, J. L., Hunt, T., and Blow, J. J. (1994). Cip1 blocks the initiation of DNA replication in *Xenopus* extracts by inhibition of cyclin-dependent kinases. *Curr. Biol.* **4,** 876–883.

Takeuchi, I., Noce, T., and Tasaka, M. (1986). Prestalk and prespore differentiation during development of *Dictyostelium discoideum. Curr. Top. Dev. Biol.* **20,** 243–256.

Tanaka, K., and Kanbe, T. (1986). Mitosis in the fission yeast *Schizosaccharomyces pombe* as revealed by freeze-substitution electron microscopy. *J. Cell Sci.* **80,** 253–268.

Tassin, A. M., Celati, C., Paintrand, M., and Bornens, M. (1997). Identification of an Spc110p-related protein in vertebrates. *J. Cell Sci.* **110,** 2533–2545.

Tassin, A. M., Celati, C., Moudjou, M., and Bornens, M. (1998). Characterization of the human homologue of the yeast Spc98p and its association with γ-tubulin. *J. Cell Biol.* **141,** 689–701.

Trivinos-Lagos, L., Ohmachi, T., Albrightson, C., Burns, R. G., Ennis, H. L., and Chisholm, R. L. (1993). The highly divergent α-tubulins and β-tubulins from *Dictyostelium discoideum* are encoded by single genes. *J. Cell Sci.* **105,** 903–911.

Ueda, M., Gräf, R., MacWilliams, H. K., Schliwa, M., and Euteneuer, U. (1997). Centrosome positioning and directionality of cell movements. *Proc. Natl. Acad. Sci. USA* **94,** 9674–9678.

Ueda, M., Schliwa, M., and Euteneuer, U. (1999). Unusual centrosome cycle in *Dictyostelium:* correlation of dynamic behavior and structural changes. *Mol. Biol. Cell* **10,** 151–160.

Unger, E., Hadrich, H., Rubino, S., Cappuccinelli, P., Muhlig, P., and Jelke, E. (1992). The microtubule system and the reduplication of microtubule organizing centres in *Dictyostelium discoideum. Acta Histochem.* **Suppl. 41,** 19–28.

Vogel, J. M., Stearns, T., Rieder, C. L., and Palazzo, R. E. (1997). Centrosomes isolated from *Spisula solidissima* oocytes contain rings and an unusual stoichiometric ratio of α/β tubulin. *J. Cell Biol.* **137,** 193–202.

Vorobjev, I. A., and Chentsov-Yu, S. (1982). Centrioles in the cell cycle. I. Epithelial cells. *J. Cell Biol.* **93,** 938–949.

Vorobjev, I. A., and Nadezhdina, E. S. (1987). The centrosome and its role in the organization of microtubules. *Int. Rev. Cytol.* **106,** 227–293.

Wang, P. J., and Huffaker, T. C. (1997). Stu2p: A microtubule-binding protein that is an essential component of the yeast spindle pole body. *J. Cell Biol.* **139,** 1271–1280.

Weijer, C. J., Duschl, G., and David, C. N. (1984). A revision of the *Dictyostelium discoideum* cell cycle. *J. Cell Sci.* **70,** 111–131.

Welker, D. L., and Williams, K. L. (1980). Mitotic arrest and chromosome doubling using thiabendazole, cambendazole, nocodazole, and ben late in the slime mould *Dictyostelium discoideum. J. Gen. Microbiol.* **116,** 397–407.

White, E., Tolbert, E. M., and Katz, E. R. (1983). Identification of tubulin in *Dictyostelium discoideum:* characterization of some unique properties. *J. Cell. Biol.* **97,** 1011–1019.

Wigge, P. A., Jensen, O. N., Holmes, S., Soues, S., Mann, M., and Kilmartin, J. V. (1998). Analysis of the *Saccharomyces* spindle pole by matrix-assisted laser desorption/ ionization (MALDI) mass spectrometry. *J. Cell Biol.* **141,** 967–977.

Winey, M., and Byers, B. (1992). Spindle pole body of *Saccharomyces cerevisiae*: A model for genetic analysis of the centrosome cycle. *In* "The Centrosome" (V. I. Kalnins, Ed.), pp. 197–218. Academic Press, San Diego.

Winey, M., and Byers, B. (1993). Assembly and functions of the spindle pole body in budding yeast. *Trends Genet.* **9,** 300–304.

Winsor, B., and Schiebel, E. (1997). Review: An overview of the *Saccharomyces cerevisiae* microtubule and microfilament cytoskeleton. *Yeast* **13,** 399–434.

Ye, X. S., Xu, G., Pu, R. T., Fincher, R. R., McGuire, S. L., Osmani, A. H., and Osmani, S. A. (1995). The NIMA protein kinase is hyperphosphorylated and activated downstream of p34cdc2/cyclin B: coordination of two mitosis promoting kinases. *EMBO J.* **14,** 986–994.

Zheng, Y., Jung, M. K., and Oakley, B. R. (1991). γ-Tubulin is present in *Drosophila melanogaster* and *Homo sapiens* and is associated with the centrosome. *Cell* **65,** 817–823.

Zheng, Y. X., Wong, M. L., Alberts, B., and Mitchison, T. (1995). Nucleation of microtubule assembly by a γ-tubulin-containing ring complex. *Nature* **378,** 578–583.

Zhou, H., Kuang, J., Zhong, L., Kuo, W. L., Gray, J. W., Sahin, A., Brinkley, B. R., and Sen, S. (1998). Tumour amplified kinase STK15/BTAK induces centrosome amplification, aneuploidy and transformation. *Nat. Genet.* **20,** 189–193.

8

Are There Nucleic Acids in the Centrosome?

*Wallace F. Marshall** and *Joel L. Rosenbaum*
Department of Molecular, Cellular, and Developmental Biology
Yale University
New Haven, Connecticut 06520

Many a long dispute among divines may be thus abridged,
It is so: It is not so; It is so: It is not so.
—Benjamin Franklin

I. Introduction

A surprisingly long controversy has hinged on the question: Do centrosomes contain nucleic acids? Centriole duplication, suggesting as it does a replication process, prompted speculation that centrioles might contain their own genome. Early biochemical reports that centrioles contained DNA began a 30-year-long debate. It has now been definitively shown that centrioles do not in fact contain any DNA.

At the same time, tantalizing hints emerged that centrosomes contain RNA and might, like ribosomes, require RNA for their function. This also

* Author to whom correspondence should be addressed.

Current Topics in Developmental Biology, Vol. 49

0070-2153/00 $35.00

sparked a long series of conflicting reports. A preponderance of evidence now suggests that centrosome-localized RNA is unlikely to play a role in centrosome assembly or function. Instead, this localization is more likely to be important for intracellular transport of specific mRNA species or the protein they encode.

II. Do Centrioles or Centrosomes Contain DNA?

A. Hints of a Centriole Genome

Centriole duplication is a remarkable process, in which new centrioles form adjacent to, and at right angles with, preexisting centrioles. The preexisting centriole specifies the position in which the new centrioles will form, as elegantly demonstrated by Beisson and Sonneborn (1965), who showed that if rows of basal bodies (modified centrioles that nucleate cilia) on the surface of *Paramecium* become reoriented, these rearranged centrioles cause subsequent centrioles to form in the altered pattern during cell division.

The precise geometric relation between old and new centrioles, and the ability of rearranged arrays of centrioles to propagate through cytoplasmic inheritance, indicates that centrioles somehow direct their own synthesis. Perhaps centrioles, like mitochondria and chloroplasts, might contain their own DNA-based genome. This idea sparked a controversy that has raged for decades, with strong assertions being made both for and against centriolar DNA. In order to make sense of these contradictory studies, we will divide them according to the experimental system, which fall into four classes: centrosomes, sperm centrioles, ciliate basal bodies, and *Chlamydomonas* basal bodies.

B. DNA in Centrosomes

Perhaps the most obvious place to look for centriolar DNA is in the centrosomes of mammalian cells. Early work exploited DNA binding compounds. McGill *et al.* (1976) found that when CHO cells were treated with ethidium bromide or propidium iodide, centriole ultrastructure was perturbed. Structural defects included abnormally narrow microtubules, centriole blades that had four microtubules instead of three, and large gaps between centriole blades. During division these cells exhibited an increased frequency of multipolar spindles, consistent with a defect in centrosome function. However, both ethidium bromide and propidium iodide can bind RNA as well as DNA, so although these results may suggest the presence of nucleic acid in centrosomes, it could well be RNA. Moreover, these compounds could interfere with DNA and RNA in other cellular structures such as the nucleus and produce indirect effects.

To circumvent such difficulties, Peterson and Berns (1978) treated PtK_2 cells with psoralen derivatives selected for their high binding specificity for DNA over RNA, and then irradiated the centrosome region of the cells with a laser microbeam. Because psoralens only bind DNA when activated by light, drug binding was restricted to just the centrosome region of the cells. DNA-specific psoralens had no effect on mitosis, thus failing to support a role for DNA in centrosome function.

The enzyme DNase is a more specific reagent to test for DNA in centrosomes. DNase digestion of centrosomes has been conducted by Zackroff *et al.* (1976) on Spisula centrosomes, by Byers *et al.* (1978) and Hyams and Borisy (1978) on yeast spindle pole bodies, by Pepper and Brinkley (1980) and Snyder (1980) on permeabilized tissue culture cells, by Roobol *et al.* (1982) on centrosomes isolated from *Physarum,* by Kuriyama *et al.* (1982) on *Dictyostelium* microtubule organizing centers, and by Kuriyama (1984) using centrosomes isolated from CHO cells. In no case did DNase treatment reduce the microtubule nucleation activity of these centrosomes, supporting the argument that either centrosomes do not contain any DNA, or if they do, it is not functionally relevant for microtubule nucleation.

Klotz *et al.* (1990) exploited a different functional assay in which isolated centrosomes were used to induce parthenogenetic development in unfertilized *Xenopus* oocytes. They showed that treatment of these centrosomes with micrococcal nuclease, which digests both RNA and DNA, had no effect on the ability of these centrosomes to support parthenogenesis, again implying that DNA is either not present, or not important for centrosome function.

Finally, centrosomes can undergo multiple rounds of duplication when DNA synthesis is inhibited (Rattner and Philips, 1973), implying that centrosomes do not contain DNA essential for replication.

C. DNA in Sperm Centrioles

Although DNA thus appears irrelevant for centrosome microtubule nucleation, these studies did not address the primary proposed role of DNA as a centriole-specific genome, a function that need not be required for microtubule nucleation. Without knowing what function a centriolar genome might play, it might seem prudent to avoid functional studies and attempt to test for DNA more directly. However, it is difficult to test for the presence of nucleic acids directly in centrosomes because they are large and sticky and can easily become contaminated during isolation, especially if the nucleus is disrupted during cell lysis. A major problem is the quantity of pericentriolar material (PCM) in which the centrioles are buried. Sperm provide a convenient system to examine the nucleic acid content of centrioles because sperm centrioles are free of PCM.

Dadoune *et al.* (1987) used colloidal-gold labeled DNase as a stain to detect DNA in monkey sperm by electron microscopy, and reported staining of the chromatoid body, a structure intimately associated with the sperm centriole. This suggested that this was a form of centriole-associated DNA. However, DNase binds not only DNA, but also RNA and actin, and indeed, preincubation with actin abolished staining of the chromatoid bodies. Moreover, Walt and Armbruster (1984) showed that the chromatoid bodies of sperm contain large quantities of actin, which suggests that Dadoune's results were artifactual and reflected DNase binding to actin.

Friedländer and Salet (1971) used a laser microbeam to irradiate sperm of the insect *Chrysopa carnea.* These sperm have a giant centriole visible by phase contrast microscopy. After treating cells with the DNA binding fluorophore acridine orange to sensitize any DNA-containing structures, they irradiated different parts of these sperm with intense light. Irradiation destroyed both the nucleus and mitochondria, but the centriole remained intact, suggesting that it does not contain DNA.

Chemes *et al.* (1987) identified three sterile patients whose sperm were headless, missing their nuclei but still retaining centrioles. These headless sperm did not contain any DNA as judged by Feulgen staining.

Most directly, Johnson and Rosenbaum (1990) performed immuno–electron microscopy on sea urchin sperm using monoclonal antibodies specific for DNA, and found heavy labeling of the nucleus but no labeling of the centriole.

Thus, sperm centrioles do not contain DNA.

D. DNA in Ciliate Basal Bodies

Ciliated protists provide an even more favorable situation in which to search for centriole DNA. The cortex (known as the pellicle) of these organisms contains arrays of basal bodies, centrioles that, like those of sperm, are unencumbered by PCM. Because each cell contains hundreds or thousands of basal bodies, the abundance of centriole DNA, if it exists, should be hundreds of times greater per cell than for organisms having only two centrioles.

The first published evidence for centriolar DNA was a biochemical study by Seaman (1960), who isolated basal bodies from pellicles of *Tetrahymena.* This enriched preparation of basal bodies was roughly 3% DNA by dry weight. These preparations, however, were only characterized by light microscopy, and given that mitochondria are associated with the pellicle in intact cells, much of this DNA was probably mitochondrial. Indeed, Seaman tested the enzymatic activities in his preparation and found that oxidative phosphorylation was enriched more than 12-fold, demonstrating that his purification method also enriched mitochondria. Seaman also found increased specific activity of succinic dehydrogenase and fumarase, enzymes located in the mitochondrial inner membrane and matrix, respectively. Clearly, better isolation procedures were needed to reduce mitochondrial contamination.

Argetsinger (1965) and Hoffman (1965) carried out modified versions of Seaman's isolation procedure in *Tetrahymena.* Importantly, Argetsinger (1965) used electron microscopy to guide her purification. These improved preparations contained roughly 10 times less DNA than that reported by Seaman, for a similar yield of basal bodies, suggesting the majority of DNA detected by Seaman was a contaminant. Hoffman (1965) also found significantly less DNA than reported by Seaman. Similarly, Hufnagel (1969) isolated pellicles from *Paramecium* and found that only 0.1% of the cellular DNA copurified with the pellicles. She further showed that isolated pellicles have a high affinity for exogenous DNA, raising the possibility that the small amounts of DNA found in these preparations could be DNA released from other sources, such as the nucleus, during cell lysis. Finally, Flavell and Jones (1971) isolated oral plates, a specialized array of basal bodies used for feeding in *Tetrahymena.* These preparations were completely free of mitochondria and were found to contain no detectable DNA. Overall, then, biochemical evidence does not support the presence of DNA in ciliate basal bodies.

A major limitation with the biochemical approach is that one cannot assign the DNA to a particular structure. Histochemical approaches were used to avoid this problem. Randall and Disbrey (1965) stained *Tetrahymena* pellicles with the nucleic acid binding dye acridine orange (AO). *In vitro,* the spectral properties of AO depend on whether it is bound to DNA or RNA. AO fluoresces green when bound to DNA and orange-red when bound to RNA. When Randall and Disbrey (1965) stained *Tetrahymena* cell ghosts containing basal bodies, they saw an array of green-fluorescing spots matching the pattern of the basal bodies. This implied that basal bodies contained DNA. The fluorescence was eliminated by DNase treatment, further strengthening the case. Smith-Sonneborn and Plaut (1967) attempted to duplicate these results in *Paramecium,* and succeeded in obtaining the same DNase-sensitive green fluorescence at the basal bodies, but found that RNase also eliminated the fluorescence, raising the possibility that the fluorescence might not be due to DNA at all. Finally, Hartman and co-workers (1974) reported that in their hands the AO fluorescence of *Tetrahymena* pellicles was sensitive to RNase but not DNase. Thus, the results of enzymatic treatment do not consistently support the presence of DNA in centrioles. The other piece of evidence supporting DNA, the green color of the AO fluorescence, is not convincing, either. Although AO emission wavelength can distinguish DNA from RNA in aqueous solution, it is sensitive to local environment and can be affected by bound proteins. For instance, Mayor and Diwan (1961) found that tobacco mosaic virus, which contains only RNA and no DNA, fluoresces green when stained with AO. Thus, the green fluorescence of AO staining cannot distinguish DNA from RNA. Histochemical studies, therefore, do not support the presence of DNA in ciliate basal bodies.

A more specific test for DNA is autoradiography using tritiated thymidine. This is the most definitive approach because it allows high spatial resolution if coupled with electron microscopy, and is more specific than acridine orange because RNA does not incorporate thymidine. Randall and Disbrey (1965) and Smith-Sonneborn and Plaut (1967) carried out autoradiography using tritiated thymidine in *Tetrahymena* and *Paramecium,* respectively, and observed rows of silver grains that resembled the rows of basal bodies on the cell surface. These studies were only done using light microscopy and could not distinguish incorporation in basal bodies versus incorporation in cortical mitochondrial DNA. Stone and Miller (1965) carried out a similar autoradiographic labeling of *Tetrahymena,* but using electron microscopy to localize the silver grains at much higher resolution. They found that all the label incorporated into mitochondria. Unfortunately, in their published images, very few basal bodies are visible, making it difficult to verify the absence of basal body incorporation. Pyne (1968) also used electron-microscopy autoradiography with tritiated thymidine in *Tetrahymena,* but presented better images showing both basal bodies and mitochondria in the same section, and all the label incorporated into the mitochondria, with no labeling of basal bodies. Pyne (1968) labeled cells during multiple cell divisions, during which all basal bodies replicated several times, so if they did contain DNA, it would have been expected to incorporate label. Thus, based on autoradiography, ciliate basal bodies appear not to contain any DNA.

Two other studies touch on the question of DNA in ciliate basal bodies. Dippell (1976) showed that whereas RNase treatment could cause subtle defects in basal body ultrastructure, DNase had no such effect, implying that if DNA was present, it was not required to maintain basal body morphology. Younger *et al.* (1972) showed that basal body production in regenerating *Stentor* was not affected by drugs that blocked DNA synthesis. The ciliate *Stentor* can be induced to shed its membranellar band of cilia, which then regenerates within 10 hr. Regeneration of the membranellar band involves the synchronous production of 20,000 basal bodies, and Younger *et al.* showed that this process was unaffected by DNA synthesis inhibitors, under conditions where they could confirm a complete absence of DNA synthesis. These studies imply that basal bodies do not contain any DNA that is essential for their assembly.

In summary, biochemical, histochemical, and autoradiographic studies indicate that ciliate basal bodies do not contain any DNA, and functional studies show that DNA is not required for basal body assembly. Before leaving the question of centriole DNA, however, we must recount one final case, in which it seemed that the Holy Grail had been attained: the identification of a specific centriole-associated DNA genome in *Chlamydomonas.*

E. DNA in *Chlamydomonas* Basal Bodies

Chlamydomonas is a unicellular green alga with genetics similar to yeast. *Chlamydomonas* swims using two flagella that are nucleated by basal bodies morphologically identical to centrioles in other organisms. Because of the excellent genetics of *Chlamydomonas,* many genes required for flagellar assembly have been described. One such gene, *UNI1,* is necessary for the maturation of centrioles into fully functional basal bodies. In *uni1* mutants, more than one cell cycle is required before a centriole can nucleate a flagellum. When the *UNI1* gene was first mapped, it was not linked to any known genes, and thus defined its own linkage group (Huang *et al.,* 1982), initially called the *UNI* linkage group, or ULG, but now more commonly called Linkage Group XIX (LGXIX). Ramanis and Luck (1986) later found that nine additional genes related to flagellar function also mapped to LGXIX. They pointed out that of the 10 genes found to reside on LGXIX, all were related to flagellar function. This clustering of flagellar genes suggested that perhaps LGXIX was a basal body genome, encoding only genes necessary for basal body function or assembly. Ramanis and Luck (1986) made an even more remarkable observation: that the genetic map of LGXIX appeared to be circular, a result confirmed by Dutcher (1986). This circular map is unlike that of any other chromosome in *Chlamydomonas,* or in fact any naturally occurring eukaryotic nuclear chromosome, which are always linear. Because all known organellar genomes are circular, the circularity of LGXIX made a strong case that it might in fact represent the long-sought-after centriole genome.

Direct evidence that LGXIX was centriole-associated came from fluorescence *in situ* hybridization (FISH) experiments. Hall *et al.* (1989) cloned a fragment of LGXIX and used it to generate FISH probes specific for this chromosome. FISH staining of pronase-treated cell spreads using probes from LGXIX gave, in many cells, a pair of dots that appeared to overlap the basal bodies. Control hybridizations with chloroplast or nuclear rDNA probes gave signals in the chloroplast nucleoids or the nucleus, respectively, whereas the LGXIX probes only stained the basal bodies and not the nucleus or chloroplast. Therefore, LGXIX appeared to localize to the basal bodies instead of the nucleus. By comparing intensities on Southern blots of pure cloned LGXIX fragment versus total cellular DNA probed with LGXIX-specific probes, Hall *et al.* found that twice as many genome equivalents of cloned LGXIX were needed in a lane to give the same hybridization intensity as seen with cellular DNA, implying that there are two copies of LGXIX per cell. Because these *Chlamydomonas* cells are haploid, the presence of two copies per cell is hard to explain if LGXIX is a nuclear chromosome, but is exactly what one would predict for a centriole-associated genome, because each cell has two centrioles. Taken together, the FISH data, as well as the clustering of flagellar genes and the circular genetic map, provided strong evidence that LGXIX was in fact the centriole genome.

Unfortunately, this work did not hold up under more intense scrutiny. First, as more genes were mapped in *Chlamydomonas,* it became clear that not all genes on LGXIX were related to flagellar function. Several genes involved in tryptophan metabolism were mapped to LGXIX (Dutcher *et al.,* 1992), as well as a histone gene cluster and a glycine tRNA gene (Walther and Hall, 1995). Second, by no means did all genes related to basal body function map to LGXIX. In fact, of three genes with phenotypes like mutations in *UNI1,* only *UNI1* maps to LGXIX; the other two, *UNI2* and *UNI3,* map to LGIX and LGIII, respectively (Dutcher and Trabuco, 1998). Other basal-body-related genes not located on LGXIX include genes involved in basal body segregation and regulation (Wright *et al.,* 1983) and the gene encoding gamma tubulin (Silflow *et al.,* 1999). Finally, two flagellar-related genes, *SUN1* and *PF29,* which were initially mapped to LGXIX (Ramanis and Luck, 1986), were found to have been mismapped, and in fact they are located on LGII and LGXI, respectively (Holmes *et al.,* 1993). Strains containing mutations in a third flagellar gene initially used to support the notion of a flagellar gene cluster on LGXIX, *FLA20,* were later found to have been mislabeled (Holmes *et al.,* 1993). Thus, the reported clustering of flagellar or basal body related genes on LGXIX, the feature that first drew attention to this chromosome, was incorrect.

The original assertion that LGXIX is circular, the other major genetic argument in favor of it being a centriole genome, was also incorrect. As noted earlier, several genes originally mapped to LGXIX (*SUN1, PF29,* and possibly *FLA20*) were later found to have been mismapped. These genes were particularly critical in assigning a circular linkage map to LGXIX, and when a corrected genetic map of LGXIX was made using only genes legitimately on LGXIX, the map was found to be linear (Holmes *et al.,* 1993). Thus, LGXIX is not circular after all, but linear just like any other *Chlamydomonas* chromosome.

A third piece of evidence supporting LGXIX as a centriole genome was its copy number (Hall *et al.,* 1989). The method used by Hall *et al.* to establish copy number is potentially sensitive to errors in estimating genome equivalents and genome size. Johnson and Dutcher (1991) developed an alternative method in which Southern blots were probed for transposon insertions found in both LGXIX and other chromosomes. By comparing, in a single lane, the ratio of intensity of LGXIX-specific bands to bands from other chromosomes, they found that this ratio was not significantly different from 1, that is, the copy number of LGXIX in haploid cells was 1, the same as that of other *Chlamydomonas* chromosomes. Because haploid cells have two basal bodies, each of which would be expected to contain a copy of the putative centriole genome, this result argues that LGXIX is not a centriole genome. In diploids, the copy number increased from 1 to 2, even though the number of centrioles remained the same (Johnson and Dutcher, 1991), again arguing that LGXIX is not a centriole genome. Finally, in *bld2* mutants, which lack basal bodies in greater than 95% of the cells, the copy number is unchanged. Thus, the copy number of LGXIX correlates exactly with the copy number of other nuclear chromosomes, and does not correlate with the number of basal bodies. Additional molecular characterization of LGXIX genes indicated that they have the same chromatin structure (as judged by micrococcal nuclease digestion), codon usage, and intron junction sequences as nuclear genes (Schloss and Croom, 1991; Walther and Hall, 1995), consistent with a nuclear location for LGXIX.

To summarize these results, the three pieces of genetic evidence initially suggesting that LGXIX was the basal body genome were all subsequently found to be incorrect. The apparent clustering of flagellar genes was not borne out as more genes were identified, the circular map resulted from mapping errors, and the chromosome copy number of 2 in haploid cells was not reproducible when a more careful method was used to assess it. Careful microscopic analyses of *Chlamydomonas* cells provided the final nails in the coffin by showing that *Chlamydomonas* basal bodies do not contain any detectable DNA.

Kuroiwa *et al.* (1990) isolated a complex from *Chlamydomonas* containing basal bodies and flagella, and stained this purified preparation with the DNA dye DAPI. Using highly sensitive video-intensified photon-counting fluorescence microscopy, they were able to measure weak DAPI fluorescence in these samples. This method had previously been shown capable of detecting as little as 50 kb of DNA. Because LGXIX was supposedly 6–9 Mb in size, it would have been expected to show up in this analysis. The fluorescence of the basal bodies was the same as that of the flagella, and corresponded to at least 100 times less DNA than *Escherichia coli* cells, whose total genome is about the size of LGXIX, clearly inconsistent with LGXIX being located at basal bodies. This weak fluorescence was yellow, not blue like DAPI emission, and was DNase insensitive, and thus it probably represented weak autofluorescence. Finally, Kuroiwa *et al.* (1990) showed that if cells were treated with pronase, as was done by Hall *et al.* during their FISH procedure, part of the nucleus was seen to bleb out and overlay the basal body region, possibly explaining the results of Hall *et al.* Together these experiments argued that *Chlamydomonas* basal bodies do not contain any DNA and that the staining seen by Hall *et al.* was an artifact of pronase treatment.

Johnson and Rosenbaum (1990) looked for basal body DNA in *Chlamydomonas* using anti-DNA monoclonal antibodies for immuno–electron microscopy. The nucleus stained heavily with the gold-labeled antibodies, and they could also detect the nucleoids of the chloroplasts and mitochondria. However, in hundreds of thin sections containing basal bodies, immunogold staining was never seen in the basal body region. Because a mitochondrial nucleoid contains approximately 200 kb of DNA in *Chlamydomonas,* this method should easily have been sensitive enough to pick up the 6–9 Mb LGXIX. These results thus provide a high-resolution complement to the work of Kuroiwa *et al.,* and together argue rather conclusively that basal bodies do not contain any DNA.

How, then, can one explain the FISH results of Hall *et al.*? Hall and Luck (1995) repeated the FISH experiments, this time using cells that had not been blasted apart by pronase. They used three different probes derived from LGXIX. In interphase, all three probes indicated that LGXIX was inside the nucleus after all, thus completing the proof that LGXIX is not a basal body genome. However, Hall, and Luck still claimed that the hybridization signal, while nuclear, was found in a region of the nucleus close to the basal bodies, and thus they maintained that LGXIX still had a basal-body-related localization. In the course of these FISH experiments, they also carried out FISH on metaphase cells, and showed that all three probes hybridized near the centromere of the chromosome. It is well known that, in many cell types, centromeric DNA clusters at one end of the nucleus as a remnant of the anaphase configuration of chromosomes, an arrangement called the "Rabl Configuration" after its discoverer, Carl Rabl (for review, see Marshall *et al.,* 1997). The Rabl configuration means that centromeres, after having been pulled to the spindle poles, will remain nonrandomly close to the centrosomes during interphase. Because in *Chlamydomonas* the basal bodies are associated with the spindle poles during mitosis, it follows that centromeric sequences from any chromosome will be found nonrandomly close to the basal bodies. Thus, there is nothing special about the position of the LGXIX centromere. The localization of these probes probably just reflects the role of anaphase in setting up nuclear architecture, rather than a special relation between LGXIX and basal bodies.

Thus, one is forced to conclude that LGXIX is not a special basal body chromosome. LGXIX does not contain a unique cluster of strictly flagellar genes; LGXIX does not have a circular linkage map; the copy number of LGXIX does not correspond with the number of basal bodies; and finally, LGXIX does not localize to basal bodies.

F. Summary: Centrioles and Centrosomes Do Not Contain DNA

In summary, there is no evidence to support DNA in centrosomes, sperm centrioles, ciliate basal bodies, or *Chlamydomonas* basal bodies. One must conclude that centrioles and centrosomes do not contain DNA.

III. Do Centrosomes Contain RNA?

A. Thinking about Centrosomal RNA

Several studies (summarized later) have suggested that centrosomes might contain RNA and require it for centrosome assembly or for microtubule nucleation. This is theoretically possible, since RNA is found throughout the cell and plays a variety of roles. Several important multiprotein complexes involve RNA as either a structural or catalytic component, including the signal sequence recognition particle, the ribosome, and the spliceosome. Many viruses also use their RNA genome as a structural scaffold around which to assemble. Might RNA play similar scaffolding roles in centrosome assembly or function?

The first possibility is that RNA might be required for microtubule nucleation by centrosomes, by acting as a scaffold to organize microtubule nucleating proteins. A second possibility is that RNA might be required to assemble a functional centrosome or centriole, in which case RNase might have no effect on intact centrosomes but might block centrosome assembly around centrioles. The third possibility is that RNA is localized to centrosomes in order to position it within the cell or to accumulate it at a discrete site as part of a more general intracellular trafficking pathway. In this last case, centrosomal RNA would not be required either for microtubule nucleation or for centrosome assembly.

We will first review evidence that centrosomes contain RNA, and then consider evidence regarding these three possible functions.

B. Direct Evidence for RNA in Centrosomes

Early work on centrosomal/centriolar RNA exploited the abundance of centrioles in the form of basal bodies in the cortex of ciliated protists.

Argetsinger (1965) and Hoffman (1965) found that RNA coenriched with isolated basal bodies. Hartman *et al.* (1974) showed that the acridine orange staining of *Tetrahymena* basal bodies was RNase sensitive. RNA has also been detected biochemically in basal bodies isolated from both *Chlamydomonas* and *Tetrahymena* (Heidemann *et al.* 1977). Of course, this RNA might be contributed by contaminating ribosomes or mitochondria, or cytoplasmic mRNA nonspecifically adhering to the basal bodies. Nevertheless, the possibility of RNA in centrioles or centrosomes is raised by these studies.

RNA has also been detected in isolated mitotic spindles by Goldman and Rebhun (1969) using the histochemical stain azure B. Peterson and Berns (1978) provided evidence for RNA in centrosomes within intact cells by showing that light-activated psoralen binding to centrosomes caused mitotic abnormalities when psoralens reactive with both DNA and RNA were used, while they saw no effect with DNA-specific psoralen derivatives. Histochemistry also indicates the presence of RNA in centrosomes. Rieder (1979) using uranyl-EDTA-lead stain, which stains ribonucleoprotein for electron microscopy, found staining of centrioles, and this staining was RNase sensitive, suggesting that centrioles contain RNA.

The most dramatic demonstration of centrosome-associated RNA is that of Klotz *et al.* (1990), who extracted RNA from isolated centrosomes and analyzed it by gel electrophoresis. They detected five distinct bands at 6, 4, 2, 0.8. and 0.4 kb, all of which were eliminated by prior RNase digestion of the centrosomes. This established that centrosomes probably contain specific RNA species and that these are accessible to RNase digestion. It would, presumably, be straightforward to prepare and sequence cDNA clones from these RNA species, but this has not yet been reported. Of course, even if centrosomes do contain specific RNA molecules, they need not necessarily have anything to do with centrosome function. The key question is whether the RNA that is bound by centrosomes has any centrosome-related function.

C. Role of RNA in Microtubule Nucleation by Centrosomes

Could RNA be directly involved in microtubule nucleation? RNA is adept as a scaffolding molecule and is able to arrange large numbers of proteins into precise configurations, such as in the ribosome. It is thus not unreasonable to ask whether RNA is involved in initiating microtubule assembly.

The role of RNA in microtubule nucleation has been tested by digesting centrosomes with RNase. RNase digestion had no effect on the microtubule nucleation capacity of yeast spindle pole bodies (Byers *et al.*, 1978; Hyams and Borisy, 1978), *Dictyostelium* microtubule organizing centers

(Kuriyama *et al.*, 1982), or on the more structurally typical centrosomes of *Spisula* (Zackroff *et al.*, 1977), *Physarum* (Roobol *et al.*, 1982), CHO cells (Kuriyama, 1984), and a human lymphocyte cell line (Klotz *et al.*, 1990). On the other hand, RNase digestion did reduce microtubule nucleation in lysed PtK_1 cells (Snyder, 1980) and in permeabilized PtK_2 cells (Pepper and Brinkley, 1980). It may be worth noting that the only cell types in which RNase had an effect on microtubule nucleation were those derived from rat kangaroo (*Potorous tridactylus*) cells.

Because RNase digestion has no effect on microtubule nucleation in many studies, it seems likely that RNA is not generally required for microtubule nucleation. According to Pepper and Brinkley (1980) the reduction in microtubule nucleation caused by RNase digestion was accompanied by a depletion of pericentriolar material, suggesting that the effect on nucleation was an indirect consequence of the effect on the PCM. But why did RNase remove the PCM and decrease microtubule nucleation in some studies, but not in others?

One possibility is that nonspecific RNA adhering to centrosomes might stabilize them *in vitro*. Centrosome-associated RNA could also bind basic proteins nonspecifically, and since many basic proteins can enhance microtubule nucleation (Erickson, 1976), this could create an RNase-sensitive enhancement of microtubule nucleation. Such a mechanism could potentially be dependent on cell type or preparation method. Preparations involving lysed or permeabilized cells should be the most susceptible to this type of artifact, and indeed, it is exactly these kinds of preparations in which the most dramatic RNase effects were seen. In principle, RNA may be involved in microtubule nucleation, but its sensitivity to RNase digestion may be variable because of protection by bound proteins whose abundance varies between different cell types. This was ruled out by Klotz *et al.* (1990), who directly demonstrated that in their hands RNase digestion did in fact remove all centrosome-associated RNA but had no effect on microtubule nucleation. It thus seems probable that the variable effect of RNase on microtubule nucleation reflects either a species- or cell-type-specific artifact, or a variable requirement for RNA in different cell types, such that in at least some species, RNA is not required for microtubule nucleation.

D. Role of RNA in Centrosome Assembly

RNA might be involved in maintaining centriole structure itself. Dippell (1976) found that RNase treatment of ciliary basal bodies caused an apparent emptying of the basal body lumen, suggesting RNA might normally be found there. Likewise, McGill *et al.* (1976) have shown that ethidium bromide causes dramatic alterations in centriole structure, such as missing triplets, and since centrioles apparently do not contain DNA, this could be a result of ethidium bromide binding centriolar RNA.

Several studies have tested the effect of RNase digestion on centrosome structure. Zackroff *et al.* (1977) found that RNase treatment of *Spisula* centrosomes causes a change in the appearance of the pericentriolar material by phase contrast microscopy, and causes the centrioles to dissociate from the rest of the centrosome. However, the RNase treatment did not affect the number of microtubules emanating from the centrosome. This suggests that perhaps RNA might link the PCM to the centrioles, so that removal of RNA breaks this link and the centrioles diffuse away. Similarly, Pepper and Brinkley (1980) found that RNase treatment of permeabilized PtK_2 cells decreased the quantity of PCM surrounding the centrioles. These results are difficult to find fault with, although it could perhaps be argued that nonspecific centrosome-associated RNA might help to hold the centrosome together after cell lysis, but might not have this function in living cells.

One of the most interesting reports implicating RNA in centrosome assembly is that of Heidemann *et al.* (1977), who injected isolated centrioles (basal bodies) from *Chlamydomonas* or *Tetrahymena* into *Xenopus* oocytes. These injected centrioles nucleated microtubule asters, and this was blocked by prior treatment of the centrioles with RNase. Because the centrioles at the time of treatment do not contain PCM, this approach suggests that centriole-associated RNA is required to recruit PCM around the centrioles, consistent with the results of Zackroff *et al.* (1977) and Pepper and Brinkley (1980). An alternative interpretation of these results has, however, been published (Sluder and Hinchcliffe, 1999). The *Chlamydomonas* basal body isolation procedure used in this study (Snell *et al.,* 1974) involves high-speed centrifugation, which aggregates the basal bodies into large clumps that cannot be fully dissociated even by vigorous homogenization. *In vitro* using purified brain tubulin, basal bodies isolated by this procedure nucleate microtubules largely from the ends of the centriolar microtubule triplets and also from the basal body walls (Snell *et al.,* 1974), even without the addition of any PCM components. Large clusters of randomly oriented centrioles all nucleating microtubules in this manner will be indistinguishable from an aster at the light-microscopy level. Indeed, repetition of the procedure of Heidemann *et al.* followed by electron microscopy analysis shows that each "aster" in fact contains many basal bodies (J. Kilmartin and J. L. Rosenbaum, unpublished). Thus, the aster assembly observed by Heidemann *et al.* probably has nothing to do with the normal mechanism of aster nucleation by centrosomes. Further experiments will be needed to absolutely rule out this possibility.

E. Role of Centrosomes in RNA Localization

Centrosomes might localize RNA to a particular part of the cell. Many mRNAs in cells are asymmetrically localized, and in some cases this localization requires microtubule-based motor proteins (reviewed in Wilhelm and Vale, 1993). If an mRNA were transported by a minus-end directed motor, it could potentially accumulate at the centrosome. Because centrosomes are able to position themselves within a cell, this could be a way to accumulate message in defined cellular locations. For example, Han *et al.* (1997) have found that mRNA encoding flagellar proteins accumulates specifically around basal bodies in *Naegleria* from which the translated proteins can directly enter the flagella. We must also consider the possibility that the centrosome might serve as a temporary staging area for RNA that will ultimately be exported to other cellular regions along microtubules. The key point is that even if highly specific centrosome-associated RNA species are eventually identified, this will not in itself prove that these RNAs are related to centrosome function.

F. Summary: Centrosomal RNA Is Still an Open Question

We are forced to conclude that the question of centrosomal RNA has not yet been answered decisively. Detection of centrosome-associated RNA is inconclusive because it could just reflect nonspecific adhesion, or else localization, via binding to centrosomes, of mRNA molecules unrelated to centrosome function. *In vitro* RNase digestion experiments have produced a Babel of inconsistent reports, which suggest that RNA almost certainly has no role in microtubule nucleation, but may be required for centrosome assembly (Heidemann *et al.,* 1977). However, the abundance and ubiquity of cytoplasmic RNA, together with the adhesive tendencies of centrosomes, means that any *in vitro* experiment, no matter how carefully done, will always be haunted by the specter of artifact.

This dilemma can only be resolved by identifying specific centrosome-associated RNA species, knocking out the corresponding genes, and observing their function *in vivo.* The identification of specific centrosome-associated RNA by Klotz *et al.* (1990) provides a starting point for this type of analysis. At any rate, until a functionally relevant centrosome-specific RNA, if such exists, is thus identified, the matter will remain unresolved.

IV. Conclusions

Centrioles and centrosomes do not contain any DNA. Although centrosomes clearly contain RNA, it is not required for centrosome function as judged by microtubule nucleation activity. Evidence for RNA playing a role in centrosome assembly remains inconclusive and will await more detailed characterization of the specific RNA species located at centrosomes.

We suggest that instead of RNA playing a role in centrosome function or assembly, perhaps centrosomes bind specific mRNA molecules as a step in the intracellular trafficking of the mRNAs or the proteins they encode. Given that many mRNAs and proteins are transported to their final destinations by kinesin motors, it might make sense to accumulate the RNA at the centrosome as a platform from which to load the mRNA or protein onto microtubules. Perhaps then we should ask not what RNA can do for the centrosome, but rather what the centrosome can do for RNA.

Acknowledgments

This work was supported by a postdoctoral fellowship from the Helen Hay Whitney Foundation (W.F.M.) and by a grant from the National Institutes of Health (J.L.R.).

References

Argetsinger, J. (1965). The isolation of ciliary basal bodies (kinetosomes) from *Tetrahymena pyriformis*. *J. Cell Biol.* **24,** 154–157.

Beisson, J., and Sonneborn, T. M. (1965). Cytoplasmic inheritance of the organization of the cell cortex in *Paramecium aurelia*. *Proc. Natl. Acad. Sci. USA* **53,** 275–282.

Byers, B., Shriver, K., and Goetsch, L. (1978). The role of spindle pole bodies and modified microtubule ends in the initiation of microtubule assembly in *Saccharomyces cerevisiae*. *J. Cell Sci.* **30,** 331–352.

Chemes, H. E., Carizza, C., Scarinci, F., Brugo, S., Neuspiller, N., and Schwarzstein, L. (1987). Lack of a head in human spermatozoa from sterile patients: a syndrome associated with impaired fertilization. *Fertil. Steril.* **47,** 310–316.

Dadoune, J. P., Alfonsi, M.-F., and Fain-Maurel, M.-A. (1987). Marquage ultrastructural du corps chromatoïde et du corps associé au centriole à l'aide du complexe DNase-or colloïdal dans les spermatides de singe. *C. R. Acad. Sci. Paris ser. III* **305,** 135–140.

Dippell, R. V. (1976). Effects of nuclease and protease digestion on the ultrastructure of *Paramecium* basal bodies. *J. Cell Biol.* **69,** 622–637.

Dutcher, S. K. (1986). Genetic properties of linkage group XIX in *Chlamydomonas reinhardtii*. *In* "Extrachromosomal Elements in Lower Eukaryotes" (R. B. Wickner, A. Hinnebusch, A. M. Lambowitz, I. Gunsalus, and A. Hollaender, Eds.), pp. 303–325. Plenum Press, New York.

Dutcher, S. K., and Trabuco, E. C. (1998). The UNI3 gene is required for assembly of basal bodies of *Chlamydomonas* and encodes delta-tubulin, a new member of the tubulin superfamily. *Mol. Biol. Cell* **9,** 1293–1308.

Dutcher, S. K., Galloway, R. E., Barclay, W. R., and Poortinga, G. (1992). Tryptophan analog resistance mutations in *Chlamydomonas reinhardtii. Genetics* **131,** 593–607.

Erickson, H. P. (1976). Facilitation of microtubule assembly by polycations. *In* "Cell Motility" (R. Goldman, T. Pollard, and J. Rosenbaum, Eds.), pp. 1069–1080. Cold Spring Harbor Laboratory Press, Cold Spring Harbor.

Flavell, R. A., and Jones, I. G. (1971). DNA from isolated pellicles of *Tetrahymena. J. Cell Sci.* **9,** 719–726.

Friedländer, M., and Salet, C. (1971). Absence of DNA in giant centrioles of the neuropteran insect Chrysopa carnea. A laser microbeam irradiation study. *Cytobios* **4,** 171–176.

Goldman, R. D., and Rebhun, L. I. (1969). The structure and some properties of the isolated mitotic apparatus. *J. Cell Sci.* **4,** 179–209.

Hall, J. L., and Luck, D. J. L. (1995). Basal body-associated DNA: in situ studies in *Chlamydomonas reinhardtii. Proc. Natl. Acad. Sci. USA* **92,** 5129–5133.

Hall, J. L., Ramanis, Z., and Luck, D. J. L. (1989). Basal body/centriolar DNA: molecular genetic studies in *Chlamydomonas. Cell* **59,** 121–132.

Han, J. W., Park, J. H., Kim, M., and Lee, J. H. (1997). mRNAs for microtubule proteins are specifically colocalized during the sequential formation of basal body, flagella, and cytoskeletal microtubules in the differentiation of *Naegleria gruberi. J. Cell Biol.* **137,** 871–879.

Hartman, H., Puma, J. P., and Gurney, T. (1974). Evidence for the association of RNA with the ciliary basal bodies of *Tetrahymena. J. Cell Sci.* **16,** 241–259.

Heidemann, S. R., Sander, G., and Kirschner, M. W. (1977). Evidence for a functional role of RNA in centrioles. *Cell* **10,** 337–350.

Hoffman, E. J. (1965). The nucleic acids of basal bodies isolated from *Tetrahymena pyriformis. J. Cell Biol.* **25,** 217–228.

Holmes, J. A., Johnson, D. E., and Dutcher, S. K. (1993). Linkage group XIX of *Chlamydomonas reinhardtii* has a linear map. *Genetics* **133,** 865–874.

Huang, B., Ramanis, Z., Dutcher, S. K., and Luck, D. J. L. (1982). Uniflagellar mutants of *Chlamydomonas:* evidence for the role of basal bodies in transmission of positional information. *Cell* **29,** 745–753.

Hufnagel, L. A. (1969). Properties of DNA associated with raffinose-isolated pellicles of *Paramecium aurelia. J. Cell Sci.* **5,** 561–573.

Hyams, J. S., and Borisy, G. G. (1978). Nucleation of microtubules in vitro by isolated spindle pole bodies of the yeast *Saccharomyces cerevisiae. J. Cell Biol.* **78,** 401–414.

Johnson, D. E., and Dutcher, S. K. (1991). Molecular studies of linkage group XIX of *Chlamydomonas reinhardtii:* evidence against a basal body location. *J. Cell Biol.* **113,** 339–346.

Johnson, K. A., and Rosenbaum, J. L. (1990). The basal bodies of *Chlamydomonas reinhardtii* do not contain immunologically detectable DNA. *Cell* **62,** 615–619.

Klotz, C., Dabauvalle, M.-C., Paintrand, M., Weber, T., Bornens, M., and Karsenti, E. (1990). Parthenogenesis in *Xenopus* eggs requires centrosomal integrity. *J. Cell Biol.* **110,** 405–415.

Kuriyama, R. (1984). Activity and stability of centrosomes in Chinese hamster ovary cells in nucleation of microtubules in vitro. *J. Cell Sci.* **66,** 277–295.

Kuriyama, R., Sato, C., Fukui, Y., and Nishibayashi, S. (1982). In vitro nucleation of microtubules from microtubule-organizing center prepared from cellular slime mold. *Cell Motil.* **2,** 257–272.

Kuroiwa, T., Yorihuzi, T., Yabe, N., Ohta, T., and Uchida, H. (1990). Absence of DNA in the basal body of *Chlamydomonas reinhardtii* by fluorimetry using a video-intensified microscope photon-counting system. *Protoplasma* **158,** 155–164.

Marshall, W. F., Fung, J. C., and Sedat, J. W. (1997). Deconstructing the nucleus: global architecture from local interactions. *Curr. Opin. Gen. Dev.* **7,** 259–263.

Mayor, H. D., and Diwan, A. R. (1961). Studies on the acridine orange staining of two purified RNA viruses: poliovirus and tobacco mosaic virus. *Virology* **14,** 74–82.

McGill, M., Highfield, D. P., Monahan, T. M., and Brinkley, B. R. (1976). Effects of nucleic acid specific dyes on centrioles of mammalian cells. *J. Ultrastruc. Research* **57,** 43–53.

Pepper, D. A., and Brinkley, B. R. (1980). Tubulin nucleation and assembly in mitotic cells: evidence for nucleic acids in kinetochores and centrosomes. *Cell Motil.* **1,** 1–15.

Peterson, S. P., and Berns, M. W. (1978). Evidence for centriolar region RNA functioning in spindle formation in dividing PtK2 cells. *J. Cell Sci.* **34,** 289–301.

Pyne, C. K. (1968). Sur l'absence d'incorporation de la thymidine tritée dans les cinétosomes de *Tetrahymena pyriformis* (Ciliés Holotriches). *C. R. Acad. Sci. Paris D* **267,** 755–757.

Ramanis, Z., and Luck, D. J. L. (1986). Loci affecting flagellar assembly and function map to an unusual linkage group in *Chlamydomonas reinhardtii. Proc. Natl. Acad. Sci. USA* **83,** 423–426.

Randall, J., and Disbrey, C. (1965). Evidence for the presence of DNA at basal body sites in *Tetrahymena pyriformis. Proc. Roy. Soc. B* **162,** 473–491.

Rattner, J. B., and Phillips, S. G. (1973). Independence of centriole formation and DNA synthesis. *J. Cell Biol.* **57,** 359–372.

Rieder, C. L. (1979). Ribonucleoprotein staining of centrioles and kinetochores in newt lung cell spindles. *J. Cell Biol.* **80,** 1–9.

Roobol, A., Havercroft, J. C., and Gull, K. (1982). Microtubule nucleation by the isolated microtubule-organizing centre of *Physarum polycephalum* myxamoebae. *J. Cell Sci.* **55,** 365–381.

Schloss, J. L., and Croom, H. B. (1991). Normal *Chlamydomonas* nuclear gene structure on linkage group XIX. *J. Cell Sci.* **100,** 877–881.

Seaman, G. R. (1960). Large-scale isolation of kinetosomes from the ciliated protozoan *Tetrahymena pyriformis. Exp. Cell Res.* **21,** 292–302.

Silflow, C. D., Liu, B., LaVoie, M., Richardson, E. A., and Palevitz, B. A. (1999). γ-Tubulin in *Chlamydomonas:* characterization of the gene and localization of the gene product in cells. *Cell Motil. Cyt.* **42,** 285–297.

Sluder, G., and Hinchcliffe, E. H. (1999). Control of centrosome reproduction: the right number at the right time. *Biol. Cell* **91,** 413–427.

Smith-Sonneborn, J., and Plaut, W. (1967). Evidence for the presence of DNA in the pellicle of *Paramecium. J. Cell Sci.* **2,** 225–234.

Snell, W. J., Dentler, W. L., Haimo, L. T., Binder, L. I., and Rosenbaum, J. L. (1974). Assembly of chick brain tubulin onto isolated basal bodies of *Chlamydomonas reinhardtii. Science* **185,** 357–360.

Snyder, J. A. (1980). Evidence for a ribonucleoprotein complex as a template for microtubule initiation in vivo. *Cell Biol. Intl. Rep.* **4,** 859–868.

Stone, G. E., and Miller, O. L. (1965). A stable mitochondrial DNA in *Tetrahymena pyriformis. J. Exp. Zool.* **159,** 33–38.

Walt, H., and Armbruster, B. L. (1984). Actin and RNA are components of the chromatoid bodies in spermatids of the rat. *Cell Tiss. Res.* **236,** 487–490.

Walther, Z., and Hall, J. L. (1995). The uni chromosome of *Chlamydomonas:* histone genes and nucleosome structure. *Nuc. Acid. Res.* **23,** 3756–3763.

Wilhelm, J. E., and Vale, R. D. (1993). RNA on the move: the mRNA localization pathway. *J. Cell Biol.* **123,** 269–274.

Wright, R. L., Chojnacki, B., and Jarvik, J. W. (1983). Abnormal basal-body number, location, and orientation in a striated fiber-defective mutant of *Chlamydomonas reinhardtii. J. Cell Biol.* **96,** 1697–1707.

Younger, K. B., Banerjee, S., Kelleher, J. K., Winston, M., and Margulis, L. (1972). Evidence that the synchronized production of new basal bodies is not associated with DNA synthesis in *Stentor coeruleus*. *J. Cell Sci.* **11,** 621–637.

Zackroff, R. V., Rosenfeld, A. C., and Weisenberg, R. C. (1977). Effects of RNase and RNA on in vitro aster assembly. *J. Supramol. Struct.* **5,** 577–589.

9

Basal Bodies and Centrioles: Their Function and Structure

Andrea M. Preble,[1] *Thomas M. Giddings, Jr.,*[2] and *Susan K. Dutcher*[3]
[1] Department of Biological Sciences
Stanford University
Stanford, California 94305-5020

[2] Molecular, Cellular, and Developmental Biology
University of Colorado
Boulder, Colorado 80309-0347

[3] Department of Genetics
Washington University School of Medicine
St. Louis, Missouri 63110

I. Introduction to Centrioles and Basal Bodies

Centrioles were first described in 1887 by Van Beneden and Nyet and extensively studied in roundworm eggs by Boveri. The centriole was a darkly staining dot at the center of each mitotic aster and near the nucleus in interphase cells. The staining was observed at the center of the differenti-

Current Topics in Developmental Biology, Vol. 49

0070-2153/00 $35.00

ated cytoplasm that was termed the centrosome (Boveri, 1901). In addition, Boveri suggested that the centriole was supplied to the egg by the sperm, since no similar staining pattern was observed in the unfertilized egg. Meves in 1903 observed the interconversion of centrioles and basal bodies, in which the centriole staining of the meiotic spindle of the moth *Pygaera* became the basal bodies used for flagellar assembly in the sperm (cited in Gall, 1961). Fifty years later using electron microscopy, several groups showed that the ultrastructure of these organelles was similar. Both centrioles and basal bodies are cylindrical structures that contained nine sets of microtubules (Manton and Clarke, 1952; Fawcett and Porter, 1954; Bernhard and deHarven, 1960).

Although Boveri was careful to distinguish the centriole from the centrosome, this distinction is not always maintained. A microtubule organizing center (MTOC) is a region or structure that organizes and nucleates microtubules (Pickett-Heaps, 1969). A centrosome is one type of MTOC; it consists of a pair of centrioles and the differentiated cytoplasm as described by Boveri, now known as pericentriolar material. A centriole is a cylinder consisting of nine sets of radially arrayed triplet microtubules and its associated structures. A basal body is also a cylinder with nine sets of radially arrayed triplet microtubules that can serve to template the assembly of a flagellar or ciliary axoneme.

A. The Role of Microtubule Organizing Centers

Microtubule organizing centers are responsible for the nucleation and organization of microtubules. MTOCs organize microtubule structures that play a central role in many cellular processes. For example, microtubules are important for certain types of cell motility; they are the primary cytoskeletal element of cilia and flagella. Microtubules play a role in cell morphology. Microtubules and microtubule-associated proteins are involved in the formation of nerve axons. Microtubules are also important for the organization of organelles; drug studies have demonstrated that microtubules are required for the proper formation and positioning of the mitochondria, Golgi, and ER (reviewed in Balczon, 1996). Some of the most striking roles of MTOCs and microtubules are in cell division. The mitotic spindle, which is responsible for chromosome segregation, requires microtubules to function. Spindle positioning is also microtubule-based. MTOCs and microtubules are also important for cleavage furrow placement during cytokinesis. When the asters of two spindle poles are brought in close proximity to one another, cleavage furrow formation can be induced, and this induction does not require chromosomes (Rappaport, 1961; Rieder *et al.*, 1997).

1. The Role of Pericentriolar Material

The roles of the two morphological components of the centrosome have been studied. Pericentriolar material was originally observed as an electron-dense cloud surrounding centrioles. Experiments performed by Gould and Borisy (1977) demonstrated that pericentriolar material nucleates microtubules. Centrosomes isolated from colcemid-blocked Chinese hamster ovary cells nucleated microtubules when incubated with purified tubulin *in vitro.* When centrioles and pericentriolar material were separated, the pericentriolar material nucleated arrays of microtubules. In recent years, many components of the pericentriolar material have been isolated. It is now clear that the pericentriolar material is not amorphous, but instead has an organized structure. Pericentrin and γ-tubulin are present in a lattice that surrounds the centrioles (Dictenberg *et al.,* 1998). Multiple centrosomal components, including γ-tubulin, are found in ring-shaped structures that are found in pericentriolar material (Moritz *et al.,* 1995; Zheng *et al.,* 1995; Vogel *et al.,* 1997). These structures have a similar diameter to microtubules and are likely to nucleate microtubules.

2. The Role of the Centriole

Although it is clear that the pericentriolar material nucleates microtubules, the role of centrioles remains ambiguous. Certainly, centrioles can interconvert to basal bodies, which are needed for the assembly of cilia or flagella. Several examples demonstrate that basal bodies and centrioles are interchangeable. In many organisms, including humans, basal bodies take on the role of centrioles during fertilization (reviewed in Schatten, 1994), as mentioned earlier for *Pygaera.* In these organisms, it is the sperm basal bodies that serve as centrioles in the zygotic MTOC. In *Chlamydomonas,* the basal bodies serve as centrioles during mitosis (Coss, 1974). The role that centrioles play in cell division is murkier. Several lines of evidence exist that suggest that centrioles are not required for mitosis. Evolutionarily, centrioles are not essential for proper centrosome function. Higher plants, fungi, and diatoms do not contain centrioles, but are able to establish functional spindles and undergo mitosis (Pickett-Heaps, 1971). The lack of centrioles in these organisms, however, does not rule out centrioles having a function. These organisms may have other structures, such as spindle pole bodies in fungi, that carry out analogous functions. Laser ablation experiments also suggest that centrioles serve no function. In a study conducted by Berns and Richardson (1977), centrioles in prophase cells were laser ablated, but these cells underwent a phenotypically normal mitosis. These experiments do not completely dismiss centrioles from having a function. The effect of centriole ablation on subsequent divisions was not examined, and some remnants of the centriole may have remained after the ablation. A *Drosophila* cell line also points to centrioles not being essential. The haploid *Drosophila* cell line 1182-4 lacks centrioles and undergoes growth and mitotic divisions (Debec *et al.,* 1982; Szöllösi *et al.,* 1986). However, it is difficult to tell whether or not the divisions are affected since 1182-4 and all related haploid cell lines spontaneously diploidize, have high levels of aneuploidy, and contain large numbers of multinucleate cells (Debec *et al.,* 1982; Debec, 1984).

In contrast, a number of experiments indicate that centrioles play an essential role in mitosis. Maniotis and Schliwa (1991) removed centrioles from cells by micromanipulation. Despite extensive growth, reformation of the microtubule aster, and DNA replication, these cells failed to undergo mitosis. This arrest suggests that centrioles are required for mitosis. Several experiments using sea urchin eggs have demonstrated that there is a correlation between the presence of centrioles and the ability of a centrosome to duplicate. In an experiment in which centriole duplication was blocked by mercaptoethanol, centrosomes with a single centriole did not double before the next mitosis, producing monopolar spindles (Sluder and Rieder, 1985). In contrast, centrosomes with two centrioles were able to duplicate before the next mitosis, producing bipolar spindles. In a separate experiment, fragments of fertilized sea urchin eggs had distinct phenotypes correlated with the presence or absence of centrioles (Sluder *el al.,* 1989). Centrioles are donated by the sperm and are associated with the male pronucleus in the initial stages of fertilization. Egg fragments with a male pronucleus and centrioles showed centrosomal duplication and formation of a bipolar spindle in each round of mitosis. However, egg fragments with a female pronucleus and no centrioles produced monasters that failed to duplicate through multiple rounds of division. Thus, although material capable of nucleating microtubules was present in cell fragments that lacked centrioles, these fragments could not duplicate centrosomes. A recent experiment by Bobinnec and co-workers (1998a) demonstrated that centrioles are likely to play a key role in centrosomal organization. Antibodies specific for glutamylated tubulin primarily stain centrioles and faintly stain spindle microtubules (Bobinnec *et al.,* 1998b). Injection of these antibodies into HeLa cells results in the disassembly of centrioles and a corresponding disruption in the organization of the pericentriolar material (Bobinnec *et al.,* 1998a), indicating that centrioles may be required for the proper organization of the centrosome. *bld2-1,* a mutation in *Chlamydomonas reinhardtii,* results in nearly the complete loss of centrioles and basal bodies (Goodenough and St. Clair, 1975; Ehler *et al.,* 1995). The phenotype of *bld2-1* cells indicates that, although centrioles may not be essential, they play a role in mitotic and meiotic cell division in *Chlamydomonas* (see Section III,C).

B. Unresolved Questions about Basal Bodies and Centrioles

Just as a genetic analysis of the spindle pole body, the MTOC in yeast, has been very fruitful (Wigge *et al.*, 1998), a genetic approach in the green alga *Chlamydomonas reinhardtii* has yielded information about the structure and function of centrioles and basal bodies. Many questions remain to be addressed and answered: (1) How is basal body duplication regulated? Basal body duplication, like chromosome duplication, occurs once per cell cycle and the new basal bodies form at right angles to the preexisting basal body. Molecular genetic analysis of mutants that lose this control (Section IV,A) may help to address this issue. (2) What are the molecular cues that account for morphological and functional differences between the old and new basal bodies? Differences exist between the old (mother) and new (daughter) basal bodies (Fig. 1). These differences include additional distal appendages on the mother basal body as well as immunological differences (Lange and Gull, 1995). Functional differences in the two basal bodies are observed in the scaly green algae in which mother basal bodies template flagella with different elaborations (Melkonian *et al.*, 1987; Beech *et al.*,

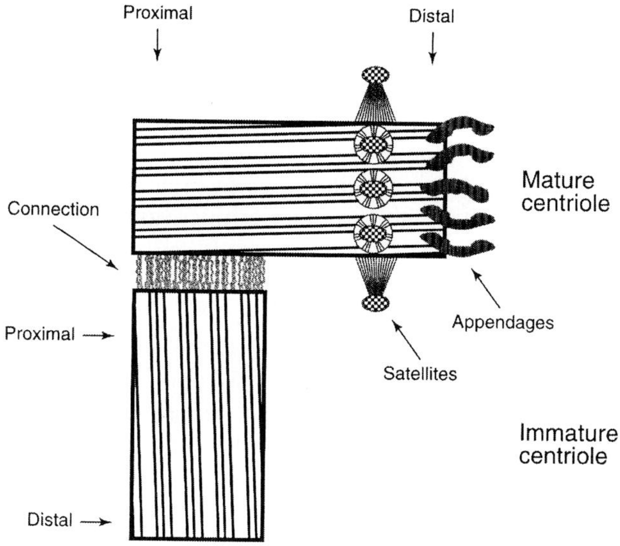

Figure 1 Schematic diagram of a typical pair of centrioles. Centrioles are arranged at right angles with the daughter (immature) centriole intersecting the mother (mature) centriole. The centrioles are connected at their proximal ends by fibrous material. The mother centriole has appendages and satellites at its distal end, whereas the daughter centriole does not. The proximal and distal ends are indicated. (Adapted from Lange and Gull, *Trends in Cell Biology*, 1996a, pp. 348–352, with permission from Elsevier Science.)

1988). The *uni* mutants in *Chlamydomonas* have flagellar assembly defects that are more severe for the daughter basal body than for the mother basal body. (3) How are triplet and doublet microtubules templated? The mechanism for assembling triplet microtubules is not yet known. The mutant phenotype associated with the loss of delta tubulin provides some clues (Section III,A,2). (4) What is the basis for the attachment of different fibers onto different triplets around the circumference and on different positions on daughter and mother basal bodies? Little molecular information is available (Section II,E).

II. The Structure of Centrioles and Basal Bodies

A. The Microtubule Cylinder

A pair of centrioles is found in an interphase centrosome as well as at each spindle pole during mitosis. Centrioles are cylindrical microtubule structures that are 200 nm wide and 400 nm long and consist of nine sets of triplet microtubules known as microtubule blades (Fig. 2). The tubules in each blade are termed A, B, and C, with the innermost tubule being the A tubule. In most mammalian cells, the centrioles are arranged at right angles to one another and are connected at their proximal ends. The microtubules are positioned at a slight angle with respect to the circumference of the cylinder.

The structure of basal bodies is similar to that of centrioles. Both structures are highly conserved in a wide range of organisms (Table I). A- and B-tubules are continuous with the doublet microtubules of the flagellum, and the C-tubule terminates. Like cytoplasmic microtubules, the A-tubule consists of 13 protofilaments, whereas B- and C-tubules consist of 10–11 protofilaments (Tilney *et al.,* 1973).

Although triplet microtubules are a hallmark of basal bodies, several exceptions have been observed. *Mastigoameoba,* a pleobiont, has basal bodies with only doublet microtubules, and these basal bodies are competent to assemble flagella. In *Drosophila* embryos, centrioles with singlet and doublet microtubules are observed (McDonald and Morphew, 1993; Callaini *et al.,* 1997). In both the testis and ovary of *Drosophila,* the centrioles have triplet microtubules (Mahowald and Strassheim, 1970). In *Caenorhabditis,* the meiotic spindle in the testes has a pair of centrioles that have singlet microtubules (Wolf *et al.,*1968; Albertson and Thomson, 1993). The sensory cilia of *Caenorhabditis* have no obvious basal body associated with them (Perkins *et al.,* 1986). A tabulation of centriole structure in other organism is shown in Table I.

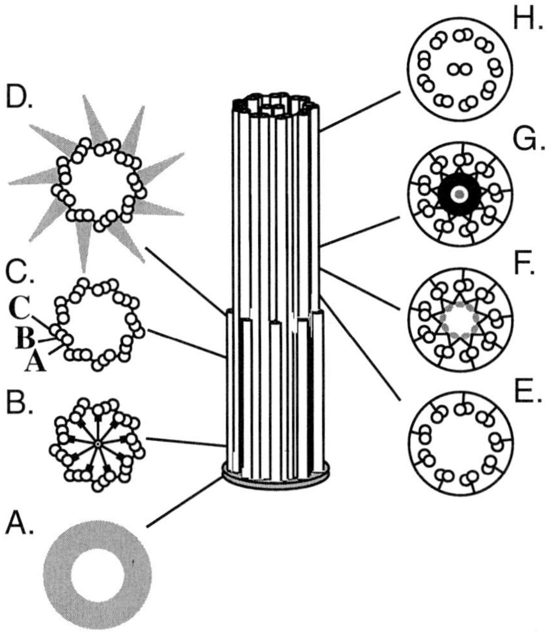

Figure 2 Diagram of a *Chlamydomonas* basal body. The basal body contains triplet microtubules arranged cylindrically. The tubules are designated A, B, and C, with the innermost tubule being the A-tubule. (A) An amorphous, electron-dense ring is found at the proximal end of the basal body. (B–D) Cross sections through the structures of a basal body shared by centrioles. (B) The proximal end contains a cartwheel structure in the center of the basal body with a central hub and spokes. Each spoke is connected to an A tubule. (C) The middle portion of the basal body lacks recognizable structures or appendages. (D) At the distal end of the triplet microtubules are transitional fibers that extend radially from each triplet (similar to the satellites found in Fig. 1). (E–H) Cross sections through the transition zone, which is unique to the basal body and contains doublet microtubules. (E) The doublets are attached to the plasma membrane. (F) A nine-pointed star with attachment to the A tubule is observed in the interior. Small electron-dense spots are observed around an inner circumference. (G) The nine-pointed star is still present, but a dense ring with a central hub is observed. (H) The flagellar axoneme is observed with a central pair, but no dynein arms.

B. Structures Shared between Basal Bodies and Centrioles

The beginning of the microtubular structure is often proceeded by an amorphous disc that remains associated with the basal body. Basal bodies and centrioles have several appendages that are associated with the microtubules. At the proximal end of the basal body, closest to the cell center, is a cartwheel structure, with a central hub and nine spokes radiating off of the hub (Fig. 2B). Each spoke connects to one of the nine A-tubules. These

Table I The Structure of Centrioles in Various Cell Types and Organisms

Cell type	Nine Sets of Triplet Microtubules	Orientation	Associated Structures	References
3T3 (mouse embryo)		Centrioles are present in pairs and are arranged orthogonally	Centrioles have distal appendages and primary cilia	Albrecht-Buehler and Bushnell (1979, 1980)
Anthoceros laevis L.		Centrioles are present in pairs, are arranged linearly or in parallel	Centrioles contain a cartwheel structure at their proximal end	Moser and Kreitner (1970)
Calf thymus	Yes	Centrioles are present in pairs, are arranged linearly, and are connected at their proximal ends	One centriole from each set has distal appendages	Komesli *et al.* (1989)
Caenorhabditis elegans	No, singlet microtubules in the male meiotic spindle	Centrioles are present in pairs and arranged orthogonally		Wolf *et al.* (1968); Albertson and Thomson (1993)
Caenorhabditis elegans	No basal bodies at the base of the sensory cilia			Perkins *et al.* (1986)
Chlamydomonas reinhardtii	Yes	Centrioles and basal bodies are present in pairs and are arranged orthogonally; two sets of fibers connect the basal bodies, one at the proximal end and one mid-cylinder	Basal bodies contain cartwheel structures at their proximal end and distal appendages (transitional fibers); a fourth tubule is associated with each triplet microtubule; it is present on the inside of the basal body cylinder	Ringo (1967); Johnson and Porter (1968); Cavalier-Smith (1974); Giddings and Dutcher, personal communication
Drosophila embryo	No, singlet microtubules	Centrioles are present in pairs and are arranged orthogonally		McDonald and Morphew (1993); Moritz *et al.* (1995)

HeLa (human)		Centrioles are present in pairs and are arranged orthogonally		Robbins *et al.* (1968)
Human sperm	The proximal centriole contains triplet microtubules	Centrioles are present in pairs and are arranged orthogonally		Zamboni and Stefanini (1971)
KE37 (human lymphoblastic cells)	Yes	Centrioles are present in pairs, and are connected by fibrous material; they are arranged orthogonally	Centrioles have material radiating inward from each triplet at the proximal end, and have internal ring at the distal end; the mother centriole has distal appendages and satellites	Bornens *et al.* (1987) Paintrand *et al.* (1992)
Lamb thymus	Yes	Centrioles are present in pairs; 65% are arranged linearly, 35% are arranged orthogonally; centrioles are connected by fibrous material at proximal end	The mother centriole has distal appendages and subdistal appendages and proximal projections	Lange and Gull (1996b)
Mastigamoeba schizophrenia	Basal bodies contain doublet microtubules		Basal body contains distal appendages and a partial cartwheel structure in the proximal portion of the basal body	Simpson *et al.* (1997)
Marchantia polymorpha L.		Centrioles are present in pairs are arranged linearly or in parallel	Centrioles contain a cartwheel structure at their proximal end	Moser and Kreitner (1970)
Paramecium	Yes	Basal bodies originate orthogonally from mother and then migrate to a parallel position	Basal bodies have an amorphous disk at their proximal end, and a cartwheel structure at their proximal end	Dippell (1968)

Table I (*Continued*)

Cell type	Nine Sets of Triplet Microtubules	Orientation	Associated Structures	References
PE (pig kidney embryo cells)	Yes	The centrioles are present in pairs and are arranged orthogonally	The mother centriole has distal appendages; the daughter centriole has a cartwheel structure at the proximal end	Vorobjec and Chentsov (1982)
Physarum		Basal bodies are present in pairs	A cartwheel structure is present at the proximal end of both basal bodies	Gely and Wright (1985)
Ptk_1 (rat kangaroo cells)	Centrioles are present in pairs and are arranged orthogonally	A primary cilium is associated with one of the two centrioles; centrioles have cartwheel structure		Jensen *et al.* (1979); Rieder *et al.* (1979)
Ptk_2 (rat kangaroo cells)		Centrioles are present in pairs and are arranged orthogonally	The mother centriole has distal appendages and a primary cilium	Rieder and Borisy (1982)
Starfish spermatazoa	Yes	Centrioles are present in pairs and are usually arranged at an angle to one another	Distal appendages are present on the basal body that templates the flagellum	Kuriyama and Kanatani (1981)

spokes end and are followed by a relatively long region that is devoid of any organized structure (Fig. 2C). At the distal end are distal appendages called transitional fibers that extend outward from each set of triplet microtubules. These structural features are shared with centrioles.

C. Structures Unique to Basal Bodies

The basal body undergoes a transition to double microtubules. The C-tubule ends, but the two central singlet microtubules of the flagellar axoneme are not yet present. This region is known as the transitional region or transition zone. The ultrastructure of this region differs among different organisms to some degree; the information presented is for *Chlamydomonas* (Fig. 3).

Fibers attaching the microtubules to the plasma membrane are first observed (Fig. 2E). These are followed by stellate fibers, which are one of the most striking structures in basal bodies (Fig. 2F). The stellate fibers form a nine-pointed star from a central region with nine distinct spots of staining. Each point of the star ends on an A-tubule. The spots of staining in the central region change into a ring of osmiophilic material with a small centrally stained hub (Fig. 2G). In longitudinal sections of basal bodies, an

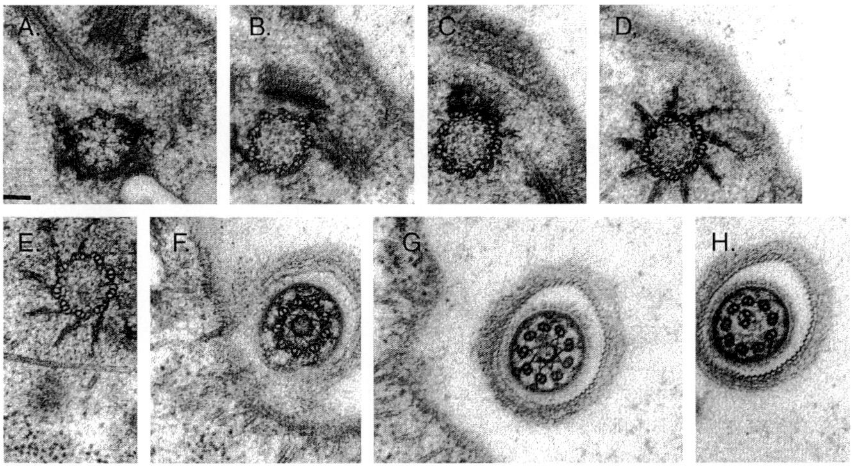

Figure 3 Electron microscopic images taken of one basal body. (A) The proximal portion of the basal body contains a cartwheel structure in the center of the basal body with a central hub and spokes. Each spoke is connected to an A tubule. (B, C) The middle portion of the basal body lacks appendages. (D, E) At the distal end of the basal body are transitional fibers. (F, G) A nine-pointed star with the points ending near the A tubule. (H) Beginning of the axoneme. Central pair microtubules are present, but the dynein arms are not yet present. The bar in panel A is 10 nm.

H-shaped structure is observed. The central pair of microtubules, characteristic of the ciliary or flagellar axoneme, appears before the assembly of the dynein arms or radial spokes (Fig. 2H). The role of most of these structures remains unresolved.

D. Associated Microtubule Structures

Several microtubule structures are associated with the basal bodies. For the purpose of this discussion, the structures present in *Chlamydomonas* will be discussed. These structures include the flagella, basal bodies, microtubule rootlets, and cytoplasmic microtubules (Fig. 4). The two mature basal bodies lie at 90° angles to each other and template the two flagella. Each of the mature basal bodies is associated with an immature basal body or probasal body. The basal bodies are connected to one another by distal and proximal striated fibers, and are connected to the nucleus by nucleus–basal body connectors. The basal bodies are also connected to specialized, stable microtubule structures called microtubule rootlets. An interphase cell contains four microtubule rootlets arranged in a cruciate array originating at the basal bodies and radiating outward, just under the plasma membrane (Ringo, 1967). Two of these rootlets are composed of two microtubules, and the other two are composed of four microtubules. The rootlets alternate to form a 4-2-4-2 pattern (Goodenough and Weiss, 1978). Different green algae have different numbers of microtubules in their rootlets (Moestrup, 1978). These microtubules contain α-tubulin that has been acetylated, which makes observations of them straightforward (LeDizet and Piperno, 1986). Cytoplasmic microtubules originate at the MTOC as well (Doonan and Grief, 1987; Holmes and Dutcher, 1989).

E. Other Associated Structures

1. Centrin

The microtubule structures at the anterior of the cell are associated with other cytoskeletal structures that are composed of several different proteins. Centrin is an EF-hand protein that forms calcium-sensitive, contractile fibers (Salisbury *et al.*, 1984; Huang *et al.*, 1988). In interphase cells, it is found in the stellate fibers of the transition zone, inner dynein arms of the flagella, nucleus–basal body connectors, and distal striated fibers (Salisbury *et al.*, 1988). The distal striated fibers, as the name implies, are striated and appear as bundles of oriented fibers. They connect the two mature basal bodies. They are attached to seven of the nine triplets and are preferentially

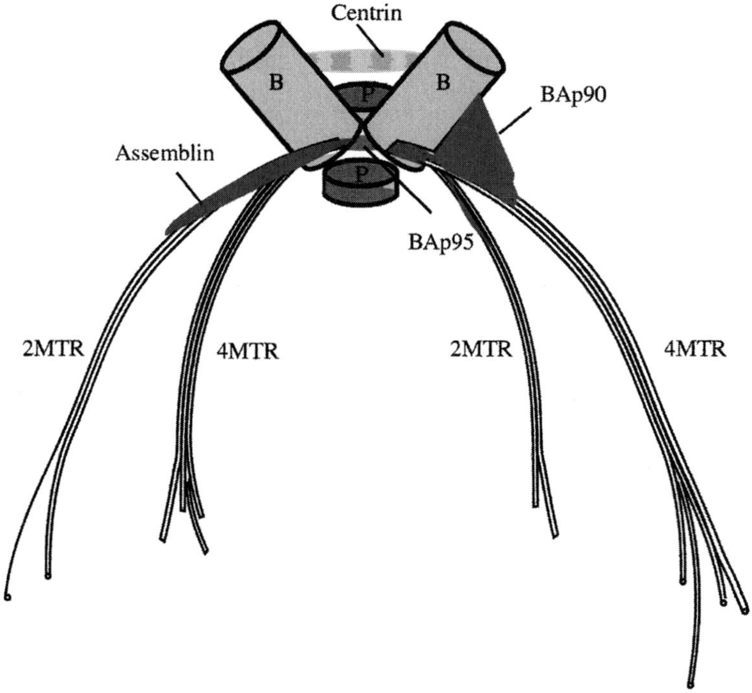

Figure 4 Diagram of the microtubule structures of an interphase *Chlamydomonas* cell. The two basal bodies lie at approximate 90° angles to one another. They are connected by distal striated fibers that contain centrin and proximal striated fibers that contain BAp95. Each mature basal body (B) is associated with an immature or probasal body (P). The four microtubule rootlets are closely associated with the basal bodies. They are arranged to form a cruciate pattern with two 2-membered microtubule rootlets and two 4-membered microtubule rootlets. Striated microtubule associated fibers are present at the proximal ends of all four microtubule rootlets and contain assemblin. The Bap90 polypeptide is found on the lateral side of the basal bodies. There are many cytoplasmic microtubules that also originate at the anterior of the cell near the basal bodies. These microtubules are not included in the diagram for clarity. B = mature basal body, P = probasal body, 2MTR = 2-membered microtubule rootlets, 4MTR = 4-membered microtubule rootlets.

associated with the C-tubule of the triplets. The nucleus basal body connector makes contact with only two of the triplets near the proximal end of the basal bodies (Geimer *et al.*, 1998b). Centrin is thought to be important for orienting the two basal bodies with respect to one another and with respect to the nucleus. The *VFL2* gene encodes centrin (Taillon *et al.*, 1992), and a deletion of *VFL2* is lethal (K. Mills, J. Rogers and S. K. Dutcher, personal communication). The *vfl2-1* mutation is a missense mutation and results in cells with reduced amounts of assembled centrin. The mutant cells are viable but have variable numbers of MTOCs and flagella. They lack nucleus–basal body connectors, distal striated fibers, and stellate fibers (Kuchka and Jarvik, 1982; Wright *et al.,* 1985; Jarvik and Suhan, 1991). It has been suggested that centrin plays an important role in the placement of the nucleus with respect to the MTOCs during mitosis (Kirk, 1998).

2. Assemblin

Studies of the MTOCs of *Spermatozopsis similis,* an alga that is closely related to *Chlamydomonas,* have resulted in the identification of several other proteins that are associated with basal bodies and microtubule rootlets (Geimer *et al.,* 1997). Assemblin is the major component of the striated microtubule associated fibers (SMAFs). These fibers are found at the base of all four microtubule rootlets in interphase cells, in close proximity to the basal bodies (Goodenough and Weiss, 1978; Weiss, 1984; Lechtreck and Silflow, 1997). Assemblin is a 30-kDa coiled-coil protein (Lechtreck *et al.,* 1991). Assemblin is thought to be important for the assembly and stability of microtubule rootlets and may play a role in basal body orientation as well.

3. BAp95

Immunoelectron microscopic studies in *Spermatozopsis similis* with antibodies to this 95-kDa polypeptide suggest that the protein assembles into additional fibers that connect microtubule rootlets to basal bodies (Geimer *et al.,* 1998a, b). These fibers associate with only three of nine triplet microtubules at the proximal end of the basal bodies. These antibodies also stain the basal body region of *Chlamydomonas* cells.

4. BAp90

Immunological studies suggest that this protein, which is predicted to form alpha helices, also assembles into fibers. These fibers associate with the four of the triplets in daughter basal bodies and five of the triplets in mother basal bodies. This protein may serve as a marker of possible lateral heterogeneity around the basal body.

F. Asymmetry around the Basal Body

The triplet microtubules in the basal body are highly differentiated. Different proteins associate with different triplets as noted above (Fig. 5, Color insert). The proteins responsible for targeting these different fibers to different triplets remain unknown. However, this asymmetry can be imagined to play an important role in generating asymmetry in the cell. *Chlamydomonas* employs the asymmetry of the microtubule rootlets to position the eyespot (Huang *et al.,* 1982; Holmes and Dutcher, 1989).

III. The Function of Basal Bodies and Centrioles as Revealed by Mutational Analysis

A. The Role of Basal Bodies in Flagellar Assembly

1. Templating of Flagellar Axonemes

Basal bodies function in two ways to promote flagellar assembly. The nine triplet microtubules in a basal body are hypothesized to template the nine doublet microtubules of the flagellum, so that flagella have the same nine-fold rotational symmetry as basal bodies. Evidence for the role of templating is provided by the analysis of the double mutant *bld2-2 rgn1-1.* A small percentage of basal bodies in this strain are missing one or two of the microtubule blades in the basal body. Instead of having nine microtubule blades, they have seven or eight blades. There are gaps in the ring of microtubules (Figs. 6A, 6B). Flagella assembled onto these basal bodies also have only seven or eight microtubules (Fig. 6C). This analysis is the first example of an altered basal body producing an altered flagellar axoneme (T. Giddings, A. Preble, and S. K. Dutcher, unpublished observations) and provides support for direct templating of the flagellar axoneme.

2. Basal Body Structural Defects Affect Flagellar Assembly

a. Delta Tubulin. *tud1-1* (previously designated *uni3-1),* a deletion of the gene encoding δ-tubulin in *Chlamydomonas reinhardtii,* affects basal body morphology (Dutcher and Trabuco, 1998). The basal bodies of *tudl-1* cells contain doublet instead of triplet microtubules; the C-tubule is missing. Delta tubulin appears to play a role in the formation of the C-tubule, perhaps by providing assembly information. This basal body defect results in a flagellar phenotype. One-fourth of *tud1-1* cells are biflagellate, one-fourth are uniflagellate, and one-half are aflagellate. The ability of *tud1-1* cells to build a flagellum corresponds to basal body age. Pedigree analysis suggests that a cell with a basal body that has been through one division can template a single flagellum, and a cell with a basal body that has been through two divisions can template two flagella.

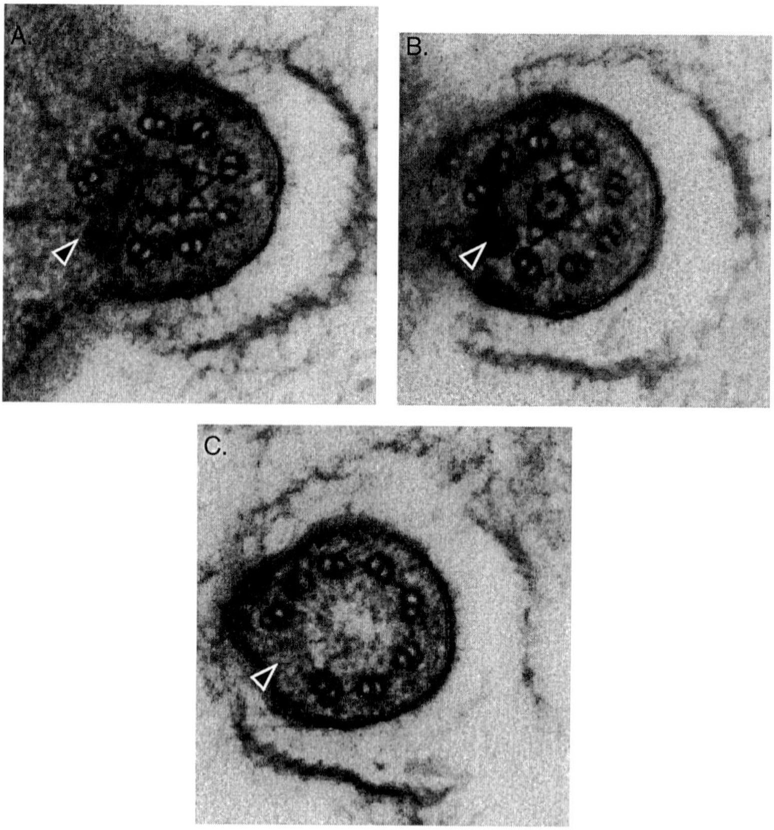

Figure 6 Electron micrographs of a basal body from *bld2-1 rgn1-1* cells. Serial sections of a basal body were examined. In both the transition zone and in the early stages of the flagellum, one set of the doublet microtubules is missing. (A) The star found in the transition zone is shown. Points of the star end on the A-tubule; the star is interrupted at the region at about 8:00 to 9:00 on a clock face. (B) A more distal section through the same basal body. The hub is present in the center, but the star is abnormal at the position of the missing doublet microtubule. (C) A section through the base of an axoneme, which still lacks a central pair and dynein arms. No doublet microtubule is present above the region lacking the doublet microtubule in the basal body.

This possible difference between mother and daughter basal bodies that is revealed in the *tud1-1* mutant strain can be modified genetically. We have identified mutations that in combination with *tud1-1* allow both basal bodies to assemble flagella (*tua2-6*) or make neither basal body able to assemble flagella (*fla10-1*). Basal bodies are believed to be docking sites for the machinery that transports flagellar components. Flagellar extension occurs at the distal end of a flagellum (Johnson and Rosenbaum, 1992).

Many components are required for flagellar assembly, and transport of these components from the cell body to the tip of a flagellum involves microtubule motor proteins (Cole *et al.,* 1998; Pazour *et al.,* 1998, 1999; Porter *et al.,* 1999). One of the components of a heterotrimeric kinesin, p85 (FLA10), which is required for this transport, is concentrated in a horseshoe-shaped pattern at the basal bodies as well as being found along the length of flagella (Cole *et al.,* 1998). This distribution is consistent with basal bodies serving as a docking site for the flagellar transport machinery. The homolog of FLA10 is also required for flagellar transport in the sensory organelles in worms (Collet *et al.,* 1998) and in embryonic cilia in mice (Nonaka *et al.,* 1998).

fla10-1 is a temperature-sensitive mutation in the gene encoding the p85 subunit of the heterotrimeric kinesin. Cells appear wild-type at 21° but aflagellate at 32°. Strains with both the *tud1-1* and *fla10-1* mutations fail to assemble flagella at 21°, which suggests that the C-tubule may be important for binding of this motor complex. It is intriguing to note that in *Mastigoameoba,* which has only doublet microtubules, there is a striking horseshoe-shaped structure visible by electron microscopy (Simpson *et al.,* 1997). It would interesting if this structure contained kinesin II and was exaggerated as a means to increase the docking of the kinesin II complex.

tua2-6, a mutation in one of two genes encoding α-tubulin in *Chlamydomonas reinhardtii,* restores flagellar assembly in the absence of delta tubulin (S. Fromherz, T. Giddings, N. Gomez-Ospina, N. Lai, and S. K. Dutcher, manuscript in preparation). The *tua2-6* allele is one of 42 alleles isolated as suppressors of *tud1-1. tua2-6; tud1-1* cells have a wild-type number of flagella but regain triplet microtubules in only a small subset of the basal bodies.

b. UNI1. The *uni1* mutation was identified by its unique flagellar phenotype. Cells assemble only one flagellum rather than two (Huang *et al.,* 1982). The single flagellum is assembled on the older basal body. The older basal body is found in the quadrant of the cell opposite the eyespot (Fig. 7, color insert). This mutant phenotype reveals a distinct difference between the two basal bodies that is likely to result from a difference in cell cycle age. Increased cell cycle times fail to rescue the mutant phenotype.

B. The Role of Basal Bodies in the Organization of Interphase *Chlamydomonas* Cells

In *Chlamydomonas,* the organization of the interphase cytoskeleton is highly stereotyped (Fig. 7). The flagella and the MTOC of *Chlamydomonas* are located at the anterior of the cell. The nucleus also lies at the anterior of the cell, just beneath the MTOC, and a cup-shaped chloroplast lies at the posterior of the cell. The eyespot, which is used to detect light, is located to one side of the cell, always at a 45° angle with respect to the plane of the flagella (Holmes and Dutcher, 1989). This organization is likely to be important for phototaxis, because the location of this eyespot needs to be coordinated with the flagella in order for cells to process light signals properly (reviewed in Kirk, 1998).

In addition to the highly organized arrangement of the cytoskeleton and organelles during interphase, *Chlamydomonas* also has a highly stereotyped organization of structures during cell division (Fig. 8). This organization ensures proper segregation of the chromosomes, chloroplast, mitochondria, basal bodies and microtubule rootlets. One of the first events of division is the disassembly and resorption of flagella (Johnson and Porter, 1968; Cavalier-Smith, 1974). During prophase, basal bodies migrate away from the plasma membrane and separate from one another (Johnson and Porter, 1968; Gaffal, 1988). At metaphase, the basal bodies lie at the spindle poles (Coss, 1974). The nuclear membrane remains intact throughout mitosis, except for large fenestrae that form at the spindle poles (Johnson and Porter, 1968; Coss, 1974). By anaphase, a microtubule structure called the phycoplast forms perpendicular to the plane of the spindle, between the two nuclei (Johnson and Porter, 1968, Gaffal and el-Gammal, 1990). At cytokinesis, the basal bodies migrate to the anterior of the cell, and a

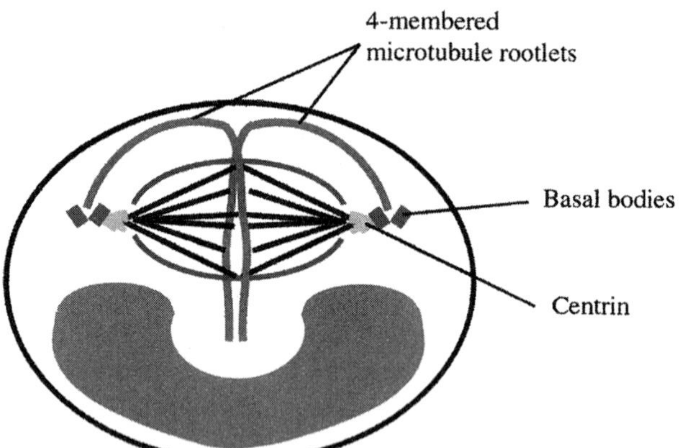

Figure 8 Diagram of a *Chlamydomonas* cell during metaphase. Metaphase arrangement of organelles and microtubule structures. During cell division, the flagella retract, and basal bodies migrate to opposite ends of the cell, where they are associated with the spindle poles. Centrin connects the basal bodies to the spindle poles. The four-membered microtubule rootlets are arranged such that their proximal ends are parallel with the mitotic spindle, and their distal ends are perpendicular to the spindle. The distal ends of these microtubule rootlets lie in the same plane as the cleavage furrow.

cleavage furrow forms coincidentally with the phycoplast (Johnson and Porter, 1968).

During mitosis the proximal portions of the microtubule rootlets remain associated with the basal bodies. A two-membered and a four-membered microtubule rootlet move to each spindle pole with each set of basal bodies. The two-membered microtubule rootlets partially disassemble, and the four-membered microtubule rootlets elongate (Doonan and Grief, 1987). At metaphase, the proximal portions of the four-membered microtubule rootlets arc in the same plane as the spindle; they extend from the basal bodies to the midline of the spindle. The distal portions of the four-membered microtubule rootlets extend perpendicularly to the spindle, so that they are over the spindle in the plane of the future cleavage furrow (Johnson and Porter, 1968; Gaffal and el-Gammal, 1990). The examination of mutant strains has provided evidence for the idea that microtubule rootlets are responsible for the location of the cleavage furrow (Ehler *et al.,* 1995).

Like the microtubule rootlets, centrin and assemblin remain associated with the basal bodies throughout mitosis. Centrin remains associated with the nucleus and connects the spindle poles to basal bodies and microtubule rootlets (Salisbury *et al.,* 1988). Assemblin is also associated with basal bodies and microtubule rootlets throughout cell division, but its pattern changes through the different stages of mitosis (Lechtreck and Silflow, 1997). In interphase, assemblin is associated with the proximal ends of all four microtubule rootlets, forming a cruciate pattern. Assemblin disassembles during prophase. By metaphase, assemblin is present at two dots at each spindle pole. Reassembly of assemblin fibers starts in anaphase and is complete by cytokinesis.

C. *BLD2* Is Required for the Positioning of the Spindle and Cleavage Furrow

Goodenough and St. Clair (1975) isolated the *bld2-1* mutation as an aflagellate strain and observed that few cells had basal bodies. Those cells that did have basal body structures had singlet instead of triplet microtubules. This observation led these authors to propose that the BLD2 gene product is required for the formation of doublet and triplet microtubules.

Ehler and co-workers (1995) found that, in addition to the basal body defect, *bld2-1* strains have coordination defects during cell division. *bld2-1* strains show defects in timing; nearly half of *bld2-1* cells initiate cytokinesis prematurely with respect to karyokinesis. *bld2-1* cells also have positioning defects. Seventy-five percent of *bld2-1* cells show defective positioning of either the spindle or the cleavage furrow. The position of the spindle is random with respect to the position of the organelles. Furthermore, the position of the cleavage furrow is random with respect to the position of both the spindle and organelles. The microtubule rootlets are also abnormal in *bld2-1* cells. They are often abnormal in number and placement within cells. Despite this mispositioning, the position of the cleavage furrow is still correlated with the position of one of the microtubule rootlets. This indicates that the microtubule rootlets play a central role in designating the position of the cleavage furrow. The coordination defects that are manifested in the mispositioning of the spindle, cleavage furrow, and microtubule rootlets are likely to occur because *bld2-1* cells lack basal bodies/centrioles, which lie at the center of the MTOC. The placement defects observed in *bld2* cells can be phenocopied by treatment with compounds that promote microtubule depolymerization (Ehler and Dutcher, 1998).

Few mutations have been identified that affect the assembly of basal bodies. The *bld2-1* mutation is one of these. However, it is striking that we have been unable to identify additional alleles at this locus or at other loci that lack basal bodies among large collections of aflagellate strains. Recently, a new allele *bld2* allele was isolated that has a lethal phenotype (Preble, Giddings, and Dutcher, manuscript in preparation). This result suggests that basal bodies and centrioles have an essential function or that gene products like BLD2 are required both for basal body and centiole assembly as well as a second essential function.

IV. Basal Body/Centriole Replication

Centriole duplication in animal cells occurs during G_1 or S (Robbins *et al.*, 1968; Rieder and Borisy, 1982; Vorobjev and Chentsov, 1982). During centriole duplication, the new (daughter) centriole emerges at a right angle from the side of the old (mother) centriole. The centrioles maintain this orientation until the next round of duplication, when the two separate. Developing daughter centrioles have smaller diameters than mother centrioles do, which suggests that the inner tubules are formed before the outer tubules (Chrétien *et al.*, 1997). Kochanski and Borisy (1990) elegantly demonstrated that duplication is semiconservative. Biotinylated tubulin was injected into G_1 cells. This tubulin was incorporated into new centrioles, but not old centrioles, and each daughter cell received an old and a new centriole.

Dippell (1968) performed a detailed examination of *Paramecium* basal body duplication. As in centrioles, duplication occurs at a right angle to the mother basal body. The first sign of basal body genesis is the appearance of a flat, amorphous disk. Then singlet tubules appear that correspond to the A-tubule. Before all the singlets have formed, the formation of doublets begins, and before all doublets have formed, the formation of triplets begins. Thus, in a given cross section, it is possible to observe singlet, doublet, and triplet microtubules simultaneously. The cartwheel structure is complete by the time the triplets have formed. Finally, the daughter basal body migrates to lie parallel to the mother.

In *Chlamydomonas,* basal body replication begins in metaphase (Gaffal, 1988). The daughter basal body remains immature throughout the following interphase, and the basal bodies elongate to their full length by the following prophase (Gaffal, 1988; Johnson and Porter, 1968). It appears that *Chlamydomonas* basal bodies form by the same mechanism as *Paramecium* basal bodies, with A-tubule formation followed by that of B- and C-tubules. *Chlamydomonas* basal bodies with singlet microtubules or mixtures of singlet, doublet, and triplet microtubules have been described (Cavalier-Smith, 1974; Johnson and Porter, 1968; Adams *et al.,* 1985).

A. A Mutation That Affects Basal Body Duplication

The *vfl1-1* mutation is defective in the spatial and temporal regulation of basal body synthesis (Adams *et al.,* 1985). *vfl1-1* cells have abnormal numbers of flagella, ranging from 0 to 10 flagella per cell. Instead of 2 flagella being assembled at the beginning of the cell cycle, new flagella are constructed throughout G_1 of the cell cycle. *vfl1-1* cells also have abnormal numbers of basal body precursors and immature basal bodies; amorphous disks and basal bodies with doublet microtubules were observed. It is likely that the unregulated new synthesis of basal bodies results in abnormal numbers of flagella being templated. The morphology of *vfl1-1* basal bodies is normal, but often the basal bodies are placed abnormally within the cell. These basal bodies can be very misaligned; in extreme cases, the distal end of the basal body points towards the cell interior instead of toward the cell surface. Electron microscopic analysis revealed that *vfl1-1* basal bodies lack an association with the distal striated fibers or microtubule rootlets. Thus, misregulation of basal body synthesis in *vfl1-1* cells results in mistiming of basal body synthesis, abnormal placement within cells, and a loss of association of basal bodies with cytoskeletal elements. The product of the *VFL1* gene has been described. It is a novel polypeptide that has no significant homology to other polypeptides. Epitope-tagged protein is found localized to the basal bodies (Silflow *et al.,* 1998).

The role of gamma tubulin in basal body assembly is unknown, although gamma tubulin has been observed in basal bodies (Silflow *et al.,* 1999). It is not known if gamma tubulin or delta tubulin may serve as the initiator for the triplet microtubules of the basal bodies and centrioles.

B. Centrosome Duplication and Genetic Stability

One of the characteristics of malignant tumors is genetic instability as shown by the presence of aneuploid or polyploid genomes. The role of the centrosome and possibly the centrioles has been suggested as a means of generating abnormal numbers of chromosomes. Several groups (Lingle *et al.,* 1998; Pihan *et al.,* 1998) have shown that various tumors have abnormal centrosomes, which includes increased numbers of centrosomes, excess pericentriolar material, increased numbers of centrioles, and inappropriate phosphorylation of centrosomal components. In addition, centrosome amplification has been observed in the absence of the tumor suppressor gene product p53 (Cross *et al.,* 1995; Fukasawa *et al.,* 1996) or in the presence of overexpression of Mdm2 (Carroll *et al.,* 1999). The mutations in *Chlamydomonas* that alter the replication and duplication of basal bodies and centrioles may be targets for checkpoint genes such as *p53.*

References

Adams, G. M. W., Wright, R., and Jarvik J. W. (1985). Defective temproal and spatial control of flagellar assembly in a mutant of *Chlamydomonas reinhardtii* with variable flagellar number. *J. Cell Biol.* **100,** 955–964.

Albertson, D. G., and Thomson, J. N. (1993). Segregation of holocentric chromosomes at meiosis in the nematode, *Caenorhabditis elegans. Chromo. Res.* **1,** 15–26.

Albrecht-Buehler, G., and Bushnell A. (1979). The orientation of centrioles in migrating 3T3 cells. *Exp. Cell Res.* **120,** 111–118.

Albrecht-Buehler, G., and A. Bushnell (1980). The ultrastructure of primary cilia in quiescent 3T3 cells. Exp. Cell Res. **126,** 427–437.

Balczon, R. (1996). The centrosome in animal cells and its functional homologs in plants and yeast cells. *Inter. Rev. Cytol.* **169,** 25–82.

Beech, P. L., Wetherbee, R., and Pickett-Heaps, J. D. (1988). Transformation of the flagella and associated flagellar components during cell division in the Coccolithophorid *Pleurochysis carterie. Protoplasma* **145,** 37–46.

Bernhard, W., and de Harven, E. (1960). L'ultrastructure de centriole et d'autres éléments de l'appareil achromatique. Proc. 4th Intern. Congr. Electron Microscopy (Ed. Bargmann, W.) Vol. 2, pp. 217–227. Berlin-Göttingen-Heidelberg, Springer.

Berns, M. W., and Richardson, S. M. (1977). Continuation of mitosis after selective laser microbeam destruction of the centriolar region. *J. Cell Biol.* **75,** 977–982.

Bobinnec, Y., Khodjakov, A., Mir, L. M., Rieder, C. L., Eddé, B., and Bornens, M. (1998a). Centriole disassembly in vivo and its effect on centrosome structure and function in vertebrate cells. *J. Cell Biol.* **143,** 1575–1589.

Bobinnec, Y., Moudjou, M., Fouqet, J. P., Desbruyères, E., Eddé, B., and Bornens, M. (1998b). Glutamylation of centriole and cytoplasmic tubulin in proliferating non-neuronal cells. *Cell Motil. Cytoskeleton* **39,** 223–232.

Bornens, M., Paintrand, M., Berges, J., Marty, M.-C., and Karsenti, E. (1987). Structural and chemical characterization of isolated centrosomes. *Cell Motil. Cytoskeleton* **8,** 238–249.

Boveri, T. (1901). Zellen-Studien IV. Über die Natur der Centrosomen. *Jena. Z. Med. Naturw.* **28,** 1–220.

Callaini, G., Whitfield, W. G., and Riparbelli, M. G. (1997). Centriole and centrosome dynamics during the embryonic cell cycles that follow the formation of the cellular blastoderm in *Drosophila. Exp. Cell Res.* **234,** 183–190.

Carroll, P. E., Okuda, M., Horn, H. F., Biddinger, P., Stambrook, P. J., Gleich, L. L., Li, Y. Q., Tarapore, P., and Fukasawa, K. (1999). Centrosome hyperamplification in human cancer: chromosome instability induced by p53 mutation and/or Mdm2 overexpression. *Oncogene* **18,** 1935–1944.

Cavalier-Smith, T. (1974). Basal bodies and flagellar development during the vegetative cell cycle and the sexual cycle of *Chlamydomonas reinhardii. J. Cell Sci.* **16,** 529–556.

Chrétien, D., Buendia, B., Fuller, S. D., and Karsenti, E. (1997). Reconstruction of the centrosome cycle from cryoelectron micrographs. *J. Struct. Biol.* **120,** 117–133.

Cole, D. G., Diener, D. R., Himelblau, A. L., Beech, P. L., Fuster, J. C., and Rosenbaum, J. L. (1998). *Chlamydomonas* kinesin-II-dependent intraflagellar transport (IFT): IFT particles contain proteins required for ciliary assembly in *Caenorhabditis elegans* sensory neurons. *J. Cell Biol.* **141,** 993–1008.

Collet J., Spike, C. A., Lundquist, E. A., Shaw, J. E., and Herman, R. K. (1998). Analysis of *osm-6,* a gene that affects sensory cilium structure and sensory neuron function in *Caenorhabditis elegans. Genetics* **148,** 187–200.

Coss, R. A. (1974). Mitosis in *Chlamydomonas reinhardtii* basal bodies and the mitotic apparatus. *J. Cell Biol.* **63,** 325–329.

Cross, S. M., Sanchez, C. A., Morgan, C. A., Schimke, M. K., Ramel, S., Idzerda, R. L., Raskind, W. H., and Reid, B. J. (1995). A p53-dependent mouse spindle checkpoint. *Science* **267,** 1353–1356.

Debec, A. (1984). Evolution of karyotype in haploid cell lines of *Drosophila melanogaster. Exp. Cell Res.* **151,** 236–246.

Debec, A., Szöllösi, A., and Szöllösi, D. (1982). A *Drosophila melanogaster* cell line lacking centriole. *Biol. Cell.* **44,** 133–138.

Dictenberg, J. B., Zimmerman, W., Sparks, C. A., Young, A., Vidair, C., Zheng, Y., Carrington, W., Fay, F. S., and Doxsey, S. J. (1998). Pericentrin and μ-tubulin form a protein complex and are organized into a novel lattice at the centrosome. *J. Cell Biol.* **141,** 163–174.

Dippell, R. V. (1968). The development of basal bodies in *Paramecium. Proc. Natl. Acad. Sci. USA* **61,** 461–468.

Doonan, J. H., and Grief, C. (1987). Microtubule cycle in *Chlamydomonas reinhardtii:* an immunofluorescence study. *Cell Motil. Cytoskeleton* **7,** 381–392.

Dutcher, S. K., and Trabuco, E. C. (1998). The *UNI3* gene is required for assembly of basal bodies of *Chlamydomonas* and encodes δ-tubulin, a new member of the tubulin superfamily. *Mol. Biol. Cell* **9,** 1293–1308.

Ehler, L. L., and Dutcher, S. K. (1998). Pharmacological and genetic evidence for a role of rootlet and phycoplast microtubules in the positioning and assembly of cleavage furrows in *Chlamydomonas reinhardtii. Cell Motil. Cytoskeleton* **40,** 193–207.

Ehler, L. L., Holmes, J. A., and Dutcher, S. K. (1995). Loss of spatial control of the mitotic spindle apparatus in a *Chlamydomonas reinhardtii* mutant strain lacking basal bodies. *Genetics* 141, 945–960. Corrigendum: *Genetics* **142,** 651–654.

Fawcett, D. W., and K. R. Porter (1954). A study of the fine structure of ciliated epithelia. *J. Morphol.* **94,** 221–282.

Fukasawa, K., Choi, T., Kuriyama, R., Rulong, S., and Vande Woude, G. F. (1996). Abnormal centrosome amplification in the absence of p53. *Science* **271,** 1744–1747.

Gaffal, K. P. (1988). The basal body-root complex of *Chlamydomonas reinhardtii* during mitosis. *Protoplasma* **143,** 118–129.

Gaffal, K. P., and el-Gammal, S. (1990). Elucidation of the enigma of the metaphase band of *Chlamydomonas reinhardtii. Protoplasma* **156,** 139–148.

Gall, J. G. (1961). Centriole replication: A study of spermatogenesis in the snail *Viviparus. J. Biophys. Biochem. Cytol.* **10,** 163–193.

Geimer, S., Teltenkötter, A., Plessmann, U., Weber, K., and Lechtreck, K.-F. (1997). Purification and characterization of basal apparatuses from a flagellate green alga. *Cell Motil. Cytoskeleton* **37,** 72–85.

Geimer, S., Clees, J., Melkonian, M., and Lechtreck, K.-F. (1998a). A novel 95-kD protein is located in a linker between cytoplasmic microtubules and basal bodies in a green flagellate and forms striated filaments in vitro. *J. Cell Biol.* **140,** 1149–1158.

Geimer, S., Lechtreck, K.-F., and Melkonian, M. (1998b). A novel basal apparatus protein of 90 kD (Bap90) from the flagellate green alga *Spermatozopsis similis* is a component of the proximal plates and identifies the d- (dexter) surface of the basal body. *Protist* **149,** 173–184.

Gely, C., and Wright, M. (1985). Centriole size modifications during the cell cycle of the amoebae of the myxomycete *Physarum polycephalum. J. Ultrastr. Res.* **91,** 127–137.

Goodenough, U. W., and St. Clair, H. S. (1975). *BALD-2:* a mutation affecting the formation of doublet and triplet sets of microtubules in *Chlamydomonas reinhardtii. J. Cell Biol.* **66,** 480–491.

Goodenough, U. W., and Weiss, R. L. (1978). Interrelationships between microtubules, a striated fiber, and the gametic mating structure of *Chlamydomonas reinhardi. J. Cell Biol.* **76,** 430–438.

Gould, R. R., and Borisy, G. G. (1977). The pericentriolar material in Chinese hamster ovary cells nucleates microtubule formation. *J. Cell Biol.* **73,** 601–615.

Holmes, J. A., and Dutcher, S. K. (1989). Cellular asymmetry in *Chlamydomonas reinhardtii. J. Cell Sci.* **94,** 273–285.

Huang, B., Ramanis, Z., Dutcher, S. K., and Luck, D. J. L. (1982). Uniflagellar mutant of *Chlamydomonas:* evidence for the role of basal bodies in the transmission of positional information. Cell **29,** 745–753.

Huang, B., Mengersen, A., and Lee, V. D. (1988). Molecular cloning of cDNA for caltractin, a basal body associated Ca^{2+}-binding protein: homology in its protein sequence with calmodulin and the yeast *CDC31* gene product. *J. Cell Biol.* **107,** 133–140.

Jarvik, J. W., and Suhan, J. P. (1991). The role of the flagellar transition region: inferences from the analysis of a *Chlamydomonas* mutant with defective transition region structures. *J. Cell Sci.* **99,** 731–740.

Jensen, C. G., Jensen, L. C. W., and Rieder, C. L. (1979). The occurrence and structure of primary cilia in a subline of *Potorous tridactylus. Exp. Cell Res.* **123,** 444–449.

Johnson, U. G., and Porter, K. R. (1968). Fine structure of cell division in *Chlamydomonas reinhardi J. Cell Biol.* **38,** 403–425.

Johnson, K. A., and Rosenbaum, J. L. (1992). Polarity of flagellar assembly in *Chlamydomonas. J. Cell Biol.* **119,** 1605–1611.

Kirk, D. L. (1998). "Volvox: Molecular–Genetic Origins of Multicellularity and Cellular Differentiation." Cambridge University Press.

Kochanski, R. S., and Borisy, G. G. (1990). Mode of centriole duplication and distribution. *J. Cell Biol.* **110,** 1599–1605.

Komesli, S., Tournier, F., Painstrand, M., Margolis, R. L., Job, D., and Bornens, M. (1989). Mass isolation of calf thymus centrosomes: Identification of a specific configuration. *J. Cell Biol.* **109,** 2869–2878.

Kuchka, M. R., and Jarvik, J. W. (1982). Analysis of flagellar size control using a mutant of *Chlamydomonas reinhardtii* with a variable number of flagella. *J. Cell Biol.* **92,** 170–175.

Kuriyama, R., and Kanatani, H. (1981). The centriolar complex isolated from starfish spermatozoa. *J. Cell Sci.* **49,** 33–49.

Lange, B. M., and Gull, K. (1995). A molecular marker for centriole maturation in the mammalian cell cycle. *J. Cell Biol.* **130,** 919–927.

Lange, B. M. and Gull, K. (1996a). Structure and function of the centriole in animal cells: progress and questions. *Trends Cell Biol.* **6,** 348–352.

Lange, B. M. H., and Gull, K. (1996b). A structural study of isolated mammalian centrioles using negative staining electron microscopy. *J. Struct. Biol.* **117,** 222–226.

Lechtreck, K.-F., and Silflow, C. D. (1997). SF-assemblin in *Chlamydomonas:* sequence conservation and localization during the cell cycle. *Cell Motil. Cytoskeleton* **36,** 190–201.

Lechtreck, K.-F., and Melkonian, M. (1991). Striated microtubule-associated fibers: Identification of of assemblin, a novel 34-kD protein that forms paracrystals of 2-nm filaments *in vitro. J. Cell Biol.* **115,** 705–716.

LeDizet, M., and Piperno, G. (1986). Cytoplasmic microtubules containing acetylated α-tubulin in *Chlamydomonas reinhardtii:* spatial arrangement and properties. *J. Cell Biol.* **103,** 13–22.

Lingle, W. L., Lutz, W. H., Ingle, J. N., Maihle, N. J., and Salisbury, J. L. (1998). Centrosome hypertrophy in human breast tumors: Implications for genomic stability and cell polarity. *Proc. Natl. Acad. Sci. USA* **95,** 2950–2955.

Mahowald, A. P., and Strassheim, J. M. (1970). Intercellular migration of centrioles in the germarium of *Drosophila melanogaster.* An electron microscopic study. *J. Cell Biol.* **45,** 306–320.

Maniotis, A., and Schliwa, M. (1991). Microsurgical removal of centrosomes blocks cell reproduction and centriole generation in BSC-1 cells. *Cell.* **67,** 495–504.

Manton, I., and Clarke, B. (1952). An electron microscope study on the spermatozoid of *Sphagnum. J. Exp. Bot.* **3,** 265–275.

McDonald, K., and Morphew, M. K. (1993). Improved preservation of ultrastructure in difficult-to-fix organisms by high pressure freezing and freeze substitution: I. *Drosophila melanogaster* and *Strongylocentrotus purpuratus* embryos. *Microsc. Res. Tech.* **24,** 465–473.

Melkonian, M., Reize, I. B., and Preising, H. R. (1987). Maturation of a flagellum/basal body requires more than one cell cycle in algal flagellates: studies on *Nephroselmis olivacea* (Prasinphyceae). *In* "Algal Development" (Ed., Wiessner, W., Robinson, D. G., and Starr, R. C.) Berlin: Springer-Verlag, pp. 102–113.

Moestrup, Ø. (1978). On the phylogenetic validity of the flagellar apparatus in green algae and other chlorophyll A and B containing plants. *BioSystems* **10,** 117–144.

Moritz, M., Braunfeld, M. B., Fung, J. C., Sedat, J. W., Alberts, B. M., and Agard, D. A. (1995). Three-dimensional structural characterization of centrosomes from early *Drosophila* embryos. *J. Cell Biol.* **130,** 1149–1159.

Moser, J. W., and Kreitner, G. L. (1970). Centrosome structure in *Anthoceros laevis* and *Marchantla polymorpha. J. Cell Biol.* **44,** 454–458.

Nonaka, S., Tanaka, Y., Okada, Y., Takeda, S., Harada, A., Kanai, Y., Kido, M., and Hirokawa, N. (1998). Randomization of left-right asymmetry due to due of nodal cilia generating leftward flow of extraembryonic fluid in mice lacking Kif3B motor protein. *Cell* **95,** 829–838.

Paintrand, M., Moudjou, M., Delacroix, H., and Bornens, M. (1992). Centrosome organization and centriole architecture: their sensitivity to divalent cations. *J. Struct. Biol.* **108,** 107–128.

Pazour, G. J., Wilkerson, C. G., and Witman, G. B. (1998). A dynein light chain is essential for the retrograde particle movement of intraflagellar transport (IFT). *J. Cell Biol.* **141,** 979–992.

Pazour, G. J., Dickert, B. L., and Witman, G. B. (1999). The DHC1b (DHC2) isoform of cytoplasmic dynein is required for flagellar assembly. *J. Cell Biol.* **144,** 473–481.

Perkins, L. A., Hedgecock, E. M., Thomson, J. N., and Culotti, J. G. (1986). Mutant sensory cilia in the nematode *Caenorhabditis elegans. Dev. Biol.* **117,** 456–487.

Pickett-Heaps, J. (1969). The evolution of the mitotic spindle: an attempt at comparative ultrastructural cytology in dividing plant cells. *Cytobios.* **3,** 257–280.

Pickett-Heaps, J. (1971). The autonomy of the centriole: fact or fallacy? *Cytobios* **3,** 205–214.

Pihan, G. A., Purohit, A., Wallace, J., Knecht, H., Woda, B., Quesenberry, P., and Doxsey, S. J. (1998). Centrosome defects and genetic instability in malignant tumors. *Cancer Res.* **58,** 3974–3985.

Porter, M. E., Bower, R., Knott, J. A., Byrd, P., and Dentler, W. (1999). Cytoplasmic dynein heavy chain 1b is required for flagellar assembly in *Chlamydomonas. Mol Biol. Cell* **10,** 693–712.

Rappaport, R. (1961). Experiments concerning the cleavage stimulus in sand dollar eggs. *J. Exp. Zool.* **148,** 81–89.

Rieder, C. L., and Borisy, G. G. (1982). The centrosome cycle in PtK_2 cells: asymmetric distribution and structural changes in pericentriolar materiel. *Biol. Cell.* **44,** 117–132.

Rieder, C. L., Jensen, C. G., and Jensen, L. C. W. (1979). The resorption of primary cilia during mitosis in a vertebrate (PtK_1) cell line. *J. Ultrastruc. Res.* **68,** 173–185.

Rieder, C. L., Khodjakov, A., Paliulis, L. V., Fortier, T. M., Cole, R. W., and Sluder, G. (1997). Mitosis in vertebrate somatic cells with two spindles: implications for the metaphase/anaphase transition checkpoint and cleavage. *Proc. Natl. Acad. Sci. USA* **94,** 5107–5112.

Ringo, D. L. (1967). Flagellar motion and fine structure of the flagellar apparatus in *Chlamydomonas. J. Cell Biol.* **33,** 543–571.

Robbins, E., Jentzsch, G., and Micali, A. (1968). The centriole cycle in synchronized HeLa cells. *J. Cell Biol.* **36,** 329–339.

Salisbury, J. L., Baron, A. T., and Sanders, M. A. (1988). The centrin-based cytoskeleton of *Chlamydomonas reinhardtii:* distribution in interphase and mitotic cells. *J. Cell Biol.* **107,** 635–641.

Salisbury, J. L., Baron, A. T., Surek, B., and Melkonian, M. (1984). Striated flagellar roots: isolation and partial characterization of a calcium-modulated contractile organelle. *J. Cell Biol.* **99,** 962–970.

Schatten, G. (1994). The centrosome and its mode of inheritance: the reduction of the centrosome during gametogenesis and its restoration during fertilization. *Dev. Biol.* **165,** 299–335.

Silflow, C. D., La Voie, B., Wu, W., Tam, L.-W. (1998). The product of the *VFL1* gene in *Chlamydomonas* localizes to the basal bodies. *Mol. Biol. Cell* **9,** 398a.

Silflow, C. D., Liu, B., LaVoie, M., Richardson, E. A., and Palevitz, B. A. (1999). γ-Tubulin in *Chlamydomonas:* Characterization of the gene and localization of the gene product in cells. *Cell Motil. Cytoskeleton* **42,** 285–297.

Simpson, A. G. B., Bernard, C., Fenchel, T., and Patterson, D. J. (1997). The organisation of *Mastigamoeba schizophrenia* n. sp.: More evidence of ultrastructural idiosyncrasy and simplicity in pelobiont protists. *Europ. J. Protistol.* **33,** 87–98.

Sluder, G., Miller, F. J., and Rieder, C. L. (1989). Reproductive capacity of sea urchin centrosomes without centrioles. *Cell Motil. Cytoskeleton* **13,** 264–273.

Sluder, G., and Rieder, C. L. (1985). Centriole number and the reproductive capacity of spindle poles. *J. Cell Biol.* **100,** 887–896.

Szöllösi, A., Ris, H., Szöllösi, D., and Debec A. (1986). A centriole-free *Drosophila* cell line: a high voltage EM study. *Eur. J. Cell Biol.* **40,** 100–104.

Tailon, B. E., Adler, S. A., Suhan, J. P., and Jarvik, J. W. (1992). Mutational analysis of centrin, an EF-hand protein associated with three distinct contractile fibers in the basal body apparatus of *Chlamydomonas. J. Cell Biol.* **119,** 1613–1624.

Tilney, L. G., Bryan, J., Bush, D. J., Fujiwara, K., Mooseker, M. S., Murphy, D. B., and Snyder, D. H. (1973). Microtubules: evidence for 13 protofilaments. *J. Cell Biol.* **59,** 267–275.

Van Beneden, E., and Neyt, A. (1887). Nouvelles recherches sur la fecondation et la division mitosique chez l'*Ascaridae megalocephale, Burx. Ac. Bull.* **14,** 215–295.

Vogerl, J. M., Stearns, T., Rieder, C. L., and Palazzo, R. E. (1997). Centrosomes isolated from *Spisula solidissima* oocytes contain rings and unusual stoichiometric ratio of alpha/beta tubulin. *J. Cell Biol.* **137,** 193–202.

Vorobjev, I. A., and Chentsov, Y. S. (1982). Centrioles in the cell cycle. I. Epithelial cells. *J. Cell Biol.* **93,** 938–949.

Weiss, R. L. (1984). Ultrastructure of the flagellar roots in *Chlamydomonas* gametes. *J. Cell Sci.* **67,** 133–143.

Wigge, P. A., Jensen, O. N., Holmes, S., Soues, S., Mann, M., and Kilmartin, J. V. (1998). Analysis of the *Saccharomyces* spindle pole by matrix-assisted laser desorption/ionization (MALDI) mass spectrometry. *J. Cell Biol.* **141,** 967–977.

Wolf, N., Hirsh, D., and McIntosh, J. R. (1968). Spermatogenesis in males of the free-living nematode, *Caenorhabditis elegans. J. Ultrastruc. Res.* **6,** 155–169.

Wright, R. L., Salisbury, J., and Jarvik, J. W. (1985). A nucleus-basal body connector in *Chlamydomonas reinhardtii* that may function in basal body localization or segregation. *J. Cell Biol.* **101,** 1903–1912.

Zamboni, L., and Stefanini, M. (1971). The fine structure of the neck of mammalian spermatozoa. *Anat. Rec.* **169,** 155–172.

Zheng, Y., Wong, M. L., Alberts, B. M., and Mitchison, T. (1995). Nucleation of microtubule assembly by a γ-tubulin-containing ring complex. *Nature* **378,** 578–583.

10

Centriole Duplication and Maturation in Animal Cells

B. M. H. Lange,[1] *A. J. Faragher,*[2] *P. March,*[2] and *K. Gull*[2]
[1] EMBL
D-69012 Heidelberg, Germany

[2] School of Biological Sciences
University of Manchester
Manchester, M13 9PT, United Kingdom

I. Overview

The centriole has been an object of study in cell biology for the past 100 years or so, and Boveri provided original and modified definitions of the centrosome and centriole more than a century ago (Boveri, 1888, 1895, 1901). Although we now know much about its structure in many different organisms because of excellent electron microscopy studies (for reviews to such classical studies, see Wheatley, 1982; Vorobjev and Nadezhdina, 1987), less is known about the centriole's precise functional role, molecular composition, duplication, and maturation. Only recently have new centriolar and basal body components been identified, shedding light on the functional role of the centriole and on the molecular events underlying centriolar duplication and maturation. This review mainly concentrates on recently published data, with emphasis on the centriole of mammalian cells. However, it also highlights and rehearses important insights obtained from studies of microbial systems and some invertebrates.

Current Topics in Developmental Biology, Vol. 49

0070-2153/00 $35.00

II. Centriole Lineage Patterns

A number of early workers drew attention to the fact that new centrioles appeared at distinct points in cell cycle or life cycle stages, and that these new centrioles then later changed into "mature" centrioles. These events were soon recognized as defining milestones in the ontogeny of the centriole/basal body. It was also recognized that a distinct pattern of inheritance existed—a lineage—characterizing centriole production and distribution to daughter cells. In algal cells these lineage's have been described by particular nomenclatures (Heimann *et al.,* 1989; Moestrup and Hori, 1989). Beech *et al.* (1991) reviewed much of the basal body lineage patterns of unicellular algae, and centriole/basal body lineages have been rehearsed for many organisms, including ameboflagellate slime molds such as *Physarum* (Wright *et al.,* 1980) and kinetoplastid flagellates such as *Trypanosoma brucei* (Sherwin and Gull, 1989).

However, one of the clearest demonstrations of this duplication, maturation cycle, and lineage pattern came from the classic work of Cleveland on flagellated protozoa. Cleveland (1957, 1963) chose to study the giant centrioles of flagellates such as *Barbulanympha* and *Trichonympha* species. The centrioles of these organisms are so large that they are easily visualized by light microscopy. Cleveland described the relationship of the centrioles to the centrosome (surrounding the distal end of *Barbulanympha* centrioles) and the existence in *Trichonympha* of two centrioles, one long and one short. At prophase the short centriole began to elongate and new centrioles were subsequently observed to form. These remained short however, and only elongated at prophase of the next generation. Thus, in retrospect we might suggest that studies such as Cleveland's light microscopy defined five important principles of centriole biology:

- Cells often possess centrioles of different structural form and age
- New centrioles form at particular points in the cell or life cycle
- This point may vary between organisms
- In most cases, the centriole requires at least two "generations" for "formation and completion of its growth" (Cleveland's terminology)
- Centrioles are inherited by daughter cells in a particular lineage pattern

These principles apply to many evolutionary diverse organisms and form some of the unifying concepts at the heart of centriole biology.

III. Structural Features of Centriole Maturation and Duplication

Beech *et al.* pointed out in their 1991 algal review that "the details of centriole/basal body ontogeny have been best studied in ciliate and animal

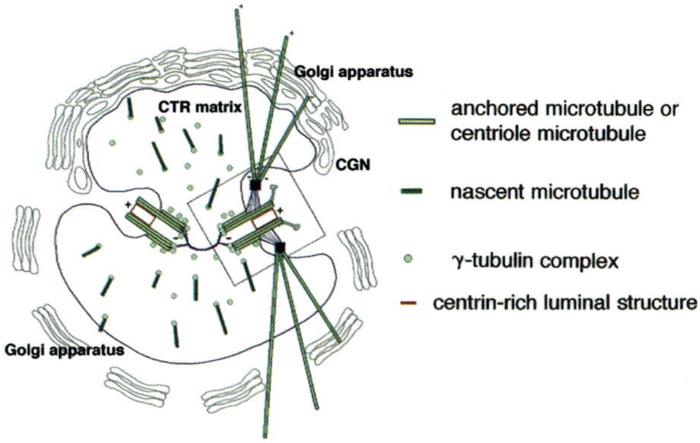

Chapter 1, Figure 1. Cartoon of the overall organization of the animal centrosome. The centriole pair, represented in green with the mother centriole on the right and the daughter centiole on the left, is held together through multiple interactions between the centrosome matrix, in blue, and the proximal part of each centriole (-). This matrix is compsed of several large coil-coiled proteins, some of them interacting with the end, others with the wall, of the triplet-containing proximal side of both centrioles. In addition, subdistal appendages (see Paintrand *et al.*, 1992, for the nomenclature), which can be considered as specialized domains of the matrix, are associated with the doublet-containing distal side of the mother centriole and have microtubules anchored at their tips (blue squares), but possibly nucleated elsewhere. Microtubules are supposed to be nucleated throughout the matrix from γ-tublin-containing complexes, which are, however, more abundant in the immediate vicinity of each centriole. Appendages at the distal ends of the mother centriole (+), the function of which is unknown, are represented in gray, as is the Golgi apparatus that surrounds the centrosome. The rectangle that frames the mother centriole corresponds to the part that is tentatively represented at higher resolution in Fig. 2 below.

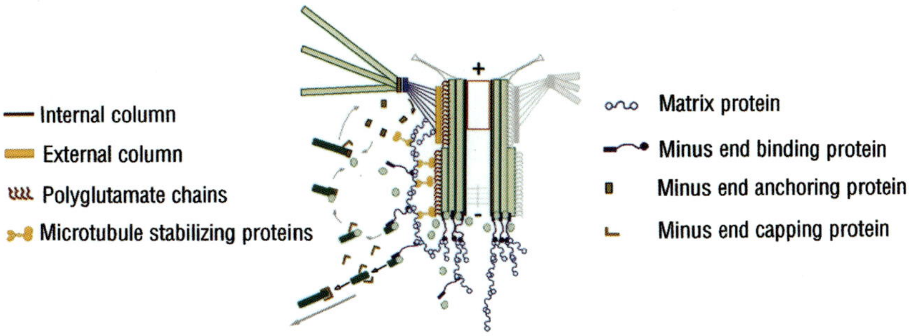

Chapter 1, Figure 2. Hypothetical model of how centrioles organize the centrosome matrix. Minus end binding proteins, including γ-tubulin-containing complexes, are concentrated at the proximal end of centrioles (-), whereas the heavy polyglutamylation of the centriole tubulin controls the lateral binding of other proteins acting as MAPs or microtubule stablizing proteins. The two sets of centriole-interacting proteins are in turn able to bind and accumulate matrix proteins in a specific manner. Matrix proteins, through homotypic or heterotypic interactions, are progressively assembling a matrix that is itself capable of binding and concentrating γ-tubulin-containing complexes, as well as regulatory activities (see text). A release and capture mechanism (see text) could control the overall microtubule nucleating activity of the centrosome. Once nucleated, microtubules could be further capped at their minus ends and either released away from the centrosome or anchored at some sites. The tips of the subdistal appendages of the mother centriole, which are represented as interacting with the distal wall of the centriole through polyglutamylation-dependent MAPs, are candidate structures for anchoring microtubules (see text). The other symbols are those of Fig. 1, above, and the nomenclature is that of Paintrand *et al.*, (1992).

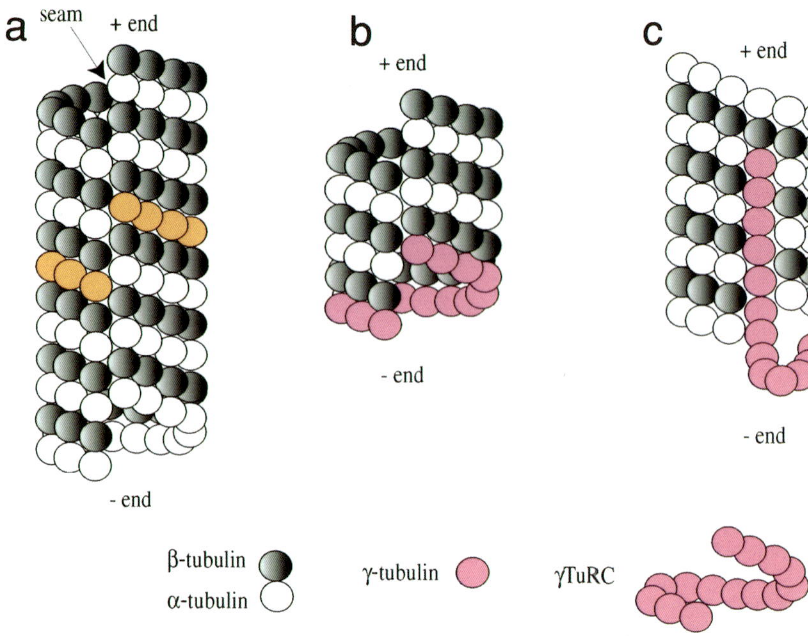

Chapter 3, Figure 5. The three-start helix of a microtubule and the two models for microtubule nucleation. The three-start helix refers to a complete helical path of tubulin monomers up the microtubule lattice that leads three tubulin monomers above the starting point (highlighted in orange in a). The adjacent tubulins in a three-start helix are either both α-tubulin or both β-tubulin. The exception to this is at the seam that spans the length of the MT lattice, where an α-tubulin will interact with an adjacent β-tubulin and vice versa. (b) The "seeded nucleation" model for microtubule nucleation where γTuRC is thought to act as a template for nucleation. According to this model, γ-tubulins in γTuRC would form direct longitudinal interactions with the tubulin heterodimers to nucleate a microtubule. In the "protofilament" model (c), γ-tubulins in γTuRC are thought to form the first "protofilament," which interacts laterally with the tubulin heterodimers to form a microtubule. Of the γTuRC subunits, only γ-tubulin is shown (pink circles). The plus and minus signs denote the polarity of the microtubule.

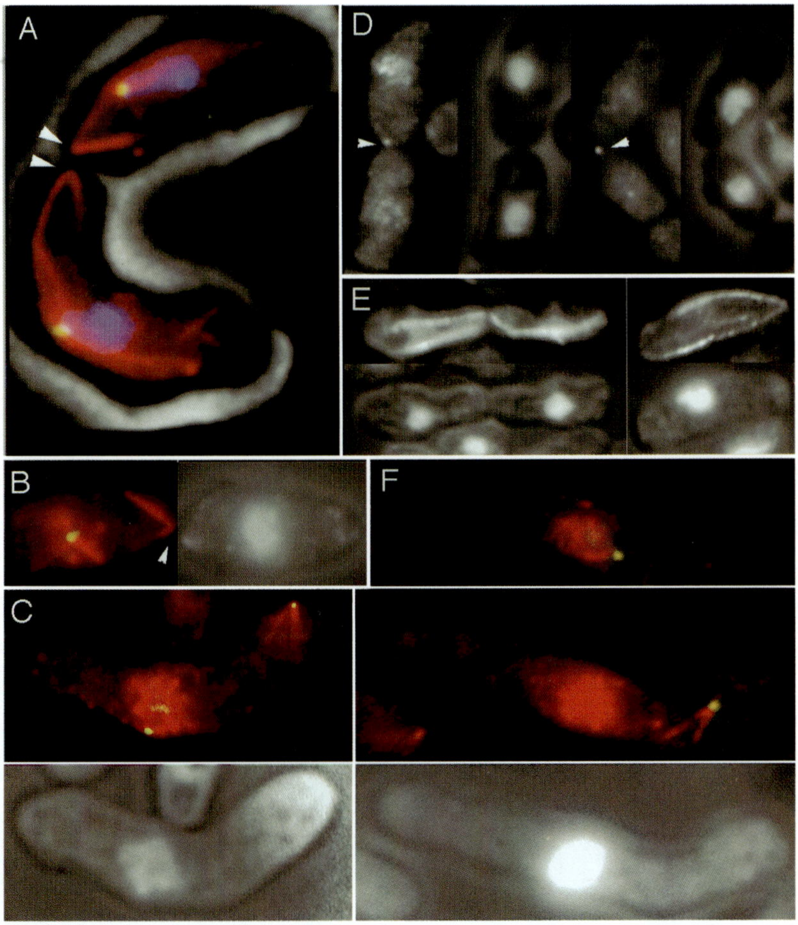

Chapter 6, Figure 2. The TAM forms at the touching projection tips of mating cells. The *fus1.B20* mutation blocks cell fusion during mating, thus increasing the frequency with which the TAM can be seen. (A) A *fus1.B20* h^{90} cell stained with DAPI (blue) and a comination of anti-tubulin (red) and anti-Sad1 antibodies (green). The microtubules clearly come together at a point of the broad tip of each of the two touching cells. (B) Transient depolymerization of the microtubules by temporary cold shock in a *fus1.B20* h^{90} strain revealed that microtubules regrew from both the Sad1 positive SPB and a point at the cell tip. Enlarging the *fus1.B20* h^{90} cell by temporary arrest at the *cdc10.v50* arrest point before the cold shock clearly showed that the microtubules do not come from the SPB (red staining in C and F). Furthermore, staining for γ–tubulin (green) revealed two points in these cells, both points from which the cytoplasmic microtubules polymerized, indicating that it is indeed an MTOC. The TAM was occasionally seen to disassociate from the cell tip following incubation on ice for 30 min and rewarming to 30°C for 2.5 min (F). To determine whether the F-actin cytoskeleton was required for the association of the TAM with the cell tip, 200μM Latrunculin A was added 5 h after nitrogen starvation of *fus1.B20* h^{90}. Staining cells for γ and α-tubulin 15 min later revealed that the association of the TAM with the tip (D) and the distribution of microtubules was unaffected by this treatment (E). Cells were cultured and prepared as described in Petersen *et al.* (1998a). Some panels in the figure are reproduced with permission from *Current Biology* (Petersen *et al.*, 1998a).

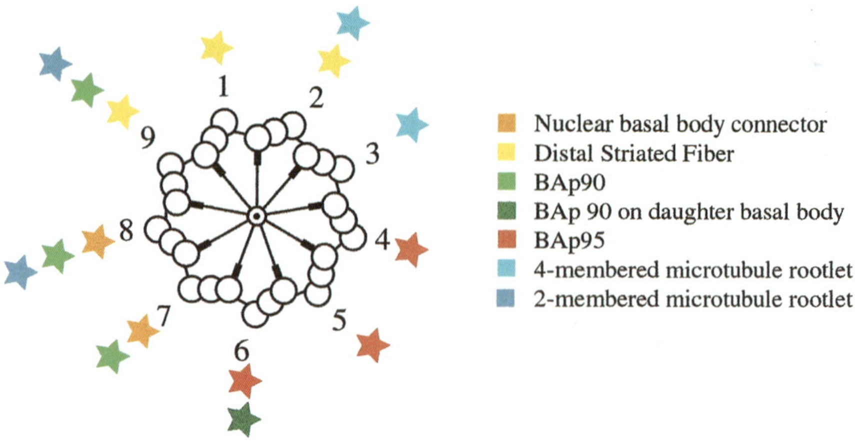

Chapter 9, Figure 5. Diagram of the radial placement on the basal body of different structures. The triplet microtubules are labeled from 1 to 9. The triplet microtubule that is associated with a particular structure of protein is indicated by the colored stars. Color key indicates the structure or protein involved.

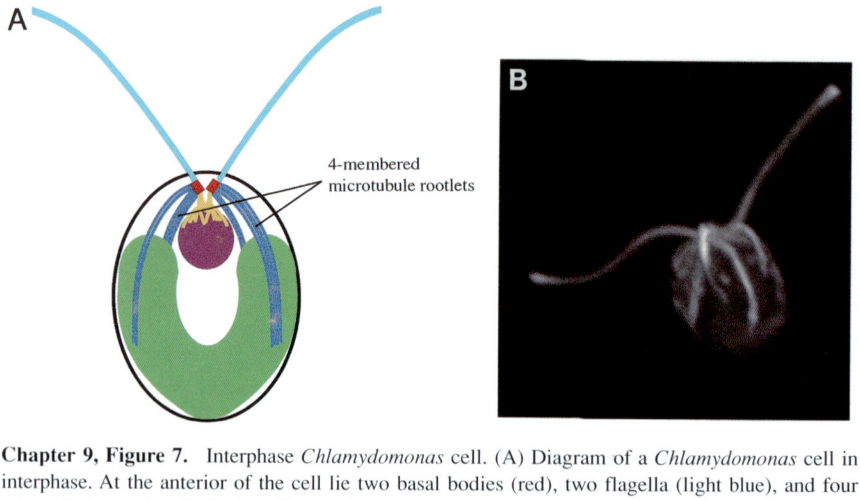

Chapter 9, Figure 7. Interphase *Chlamydomonas* cell. (A) Diagram of a *Chlamydomonas* cell in interphase. At the anterior of the cell lie two basal bodies (red), two flagella (light blue), and four microtubule rootlets (dark blue). Two of the microtubule rootlets consist of two microtubules, and the other two consist of four microtubules. The nucleus–basal body connectors connect the nucleus to the basal bodies and contain centrin (yellow). The cup-shaped chloroplast is found at the posterior of the cell (green). (B) Immunofluorescence image of an interphase Chlamydomanas cell stained with an antibody that recognizes acetylated α-tubulin. The flagella and the rootlet microtubules are visible. This image is a complication of images taken with a deconvoluting microscope.

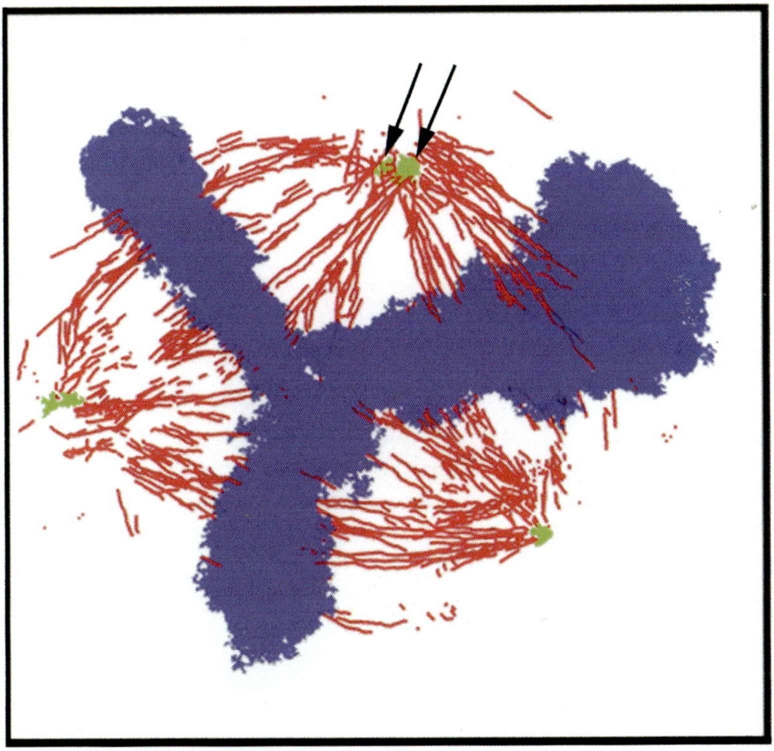

Chapter 14, Figure 1. Reconstruction of multipolar mitosis. This reconstruction is from six transmission electron micrographs of nonadjacent serial sections. The condensed chromosomes (blue) are arranged in a triskelion with three spindle poles (green) generating three sets of spindle microtubules (red). The spindle pole near the 12 o'clock position is actually a cluster of two groups of centrioles generating two slightly separate microtubule foci that function as one spindle pole. Even though some of the supernumerary centrioles were clustered at one spindle pole, this cell has three spindle poles. Daughter cells from mitoses such as these will be aneuploid. (From Lingle and Salisbury, *American Journal of Pathology*, 1999.)

Chapter 16, Figure 1. γ-Tubulin and centrin loss during mouse spermiogenesis. A, B, D, and E are triple immunofluorescent images of spermatids labeled with γ-Tubulin (red), β-tubulin (green), and Hoechst (blue). (A) The round spermatid displaying random and reticulate microtubules in the cortex. γ-Tubulin is located as pundtate spots in the juxtanuclear region (arrow) without any microtubule nucleation. (B) An elongating spermatid showing manchette (arrowheads) around the nucleus. The γ-Tubulin spots are located in the neck region, well below the manchette nucleation site. (C) A schematic diagram of an elongating spermatid drawn on the basis of immunofluorescent, electron microscopic, and immunogold electron microscopic observations. The manchette microtubules (a cone of green lines) nucleate from a perinuclear ring, which is obliquely located at this stage. The neck region has two centrioles (filled yellow cylinders) in an orthogonal orientation. The proximal centriole produces an adjunct (open yellow cylinder) and the distal centriole extends an axoneme below (yellow tube). γ-Tubulin is found around the centrioles and the adjunct (red spots). The latter nucleates the microtubule aster (radiating green lines). (D) A spermiation stage spermatid with a residual body attached. The residual body removes the microtubule bundles (green, arrowheads) and γ-tubulin (red, arrows) from the spermatid. The head region of this spermatid shows two small γ-tubulin spots (small arrows). Inset, phase contrast image of the spermatid. (E) Mature and immature testicular sperm

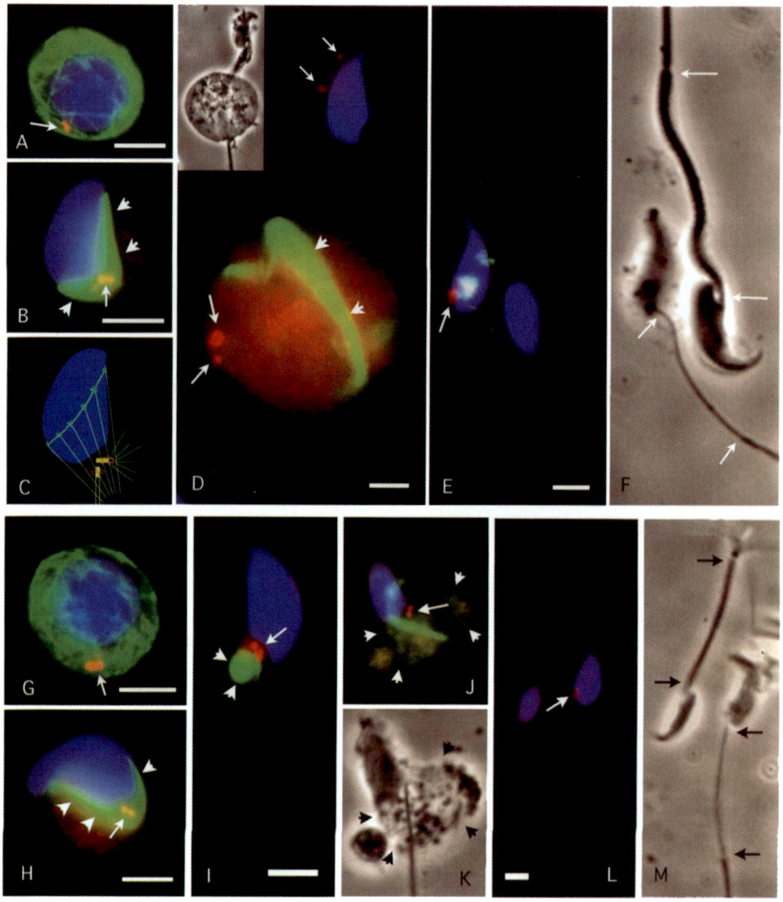

Continued from Chapter 16, Figure 1

compared; (F) phase-contrast image of E. The mature testicular sperm (the right one) has a thick midpiece (arrows in F) due to the formation of mitochondrial sheath around the axoneme that is lacking in the immature sperm. The immature sperm possesses γ-tubulin spots (red, arrow in E) in the neck region and a microtubule aster (green and white due to overlapping with the red signal γ-tubulin) radiating from them. Mature testicular sperm lacks γ-tubulin or microtubule. Bar, 5 μm. (A–F, from Manandhar *et al.*, 1998, *Dev. Biol.* **203**, 424–434.) (G) A mouse round spermatid showing cortical microtubule network (green) and two centrin spots (red, arrow) near the nucleus (blue). (H) An elongating spermatid displaying manchette (green, arrowheads) and two centrin spots (red, arrow). (I) A spermiation stage spermatid with two centrin spots (arrow) and a microtubule aster (arrowheads) in the neck region. (J) A spermiation stage spermatid with a residual body attached (arrowheads). The residual body possesses microtububle bundles (green) but the centrin spots are located in the neck region. (J) A spermiation stage spermatid with a residual body attached (arrowheads). The residual body possesses microtubule bundles (green) but the centrin spots are located in the neck region of the spermatid (arrow). (K) Phase contrast image of (J). (L, M) Immature testicular sperm (right) and mature epididymal (left) sperm compared. Aliquots of testicular sperm and epidiymal sperm were mixed together before labeling with the antibodies. The midpiece is thick in the epididymal sperm but thin in the immature testicular sperm (arrows in M). The testicular sperm has two inconspicuous centrin spots (arrow in L), while the epididymal sperm lacks them. Bar=5μm. (G–M, from Manandhar *et al.*, 1999, *Cell Motil. Cytoskel.* **43**, 137–144, with permission.)

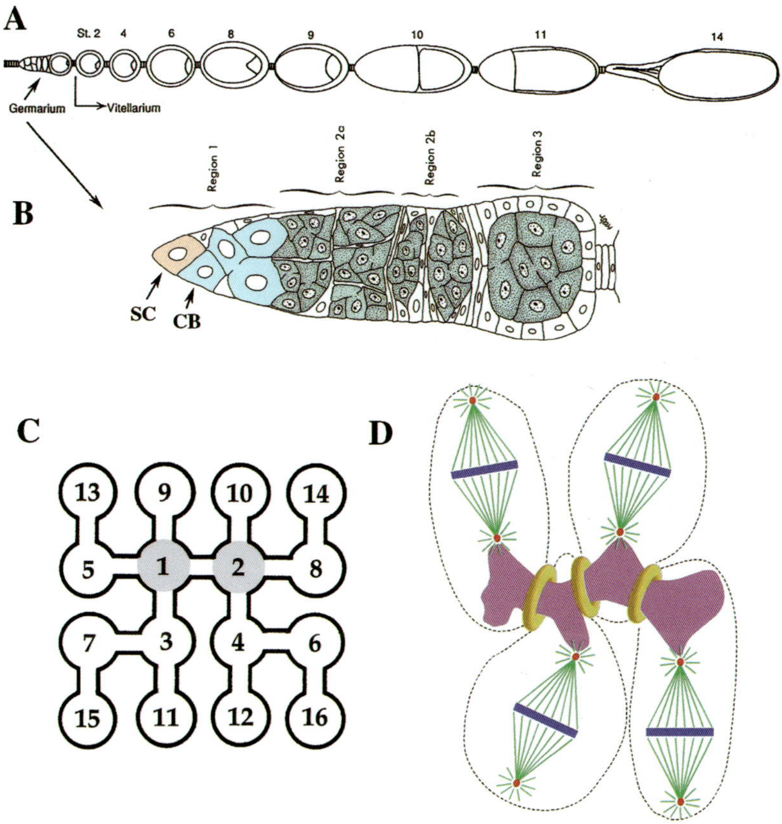

Chapter 18, Figure 1. The ovariole and the germarium. The ovariole comprises the germarium and the vitellarium (A). The germarium lies at the distal end of the ovariole and contains the stem cells, cystoblasts, mature cysts, and stage 1 egg chambers. The vitellarium contains the stage 2–14 egg chambers. The germarium is shown in schematic detail in (B) with its four regions indicated. Region 1 contains the stem cells (SC, shown in orange), the cystoblast cells (CB), and the dividing cystocytes (shown in light blue). Region 2a contains 16-cell cysts (mature cystocytes shown in green). In region 2b the cysts are flattened into a lens shape that spans the width of the germarium. At this stage the oocyte has already been determined and is positioned in the middle of the cyst. Region 3 contains a stage 1 egg chamber with the oocyte positioned at the posterior. The anterior is to the left in (A) and (B). The configuration of the mature cyst is illustrated in (C). The products of the first cystoblast division and the most branched of the cystocytes (cells 1 and 2) are shaded. One of these two cells will become the oocyte. During the cystocyte divisions the spindles are attached to the fusome. The fusome connects all four cells through the ring canals and the mitotic spindles are attached to the fusome through the centrosome. This is illustrated in (D); which shows a four-cell stage cyst in mitosis. The four cells are outlined with a dotted line, the fusome is shown in purple, the ring canals in yellow, the centrosomes in red, the spindle and aster microtubules in green, and the chromosomal DNA in blue. The drawing in (A) was used with permission from Drs. S. Mahajan-Miklos and L. Cooley (Mahajan-Miklos and Cooley, 1994). The drawing in (B) was modified and used with permission from Dr. A. Mahowald (Mahowald and Strassheim, 1970).

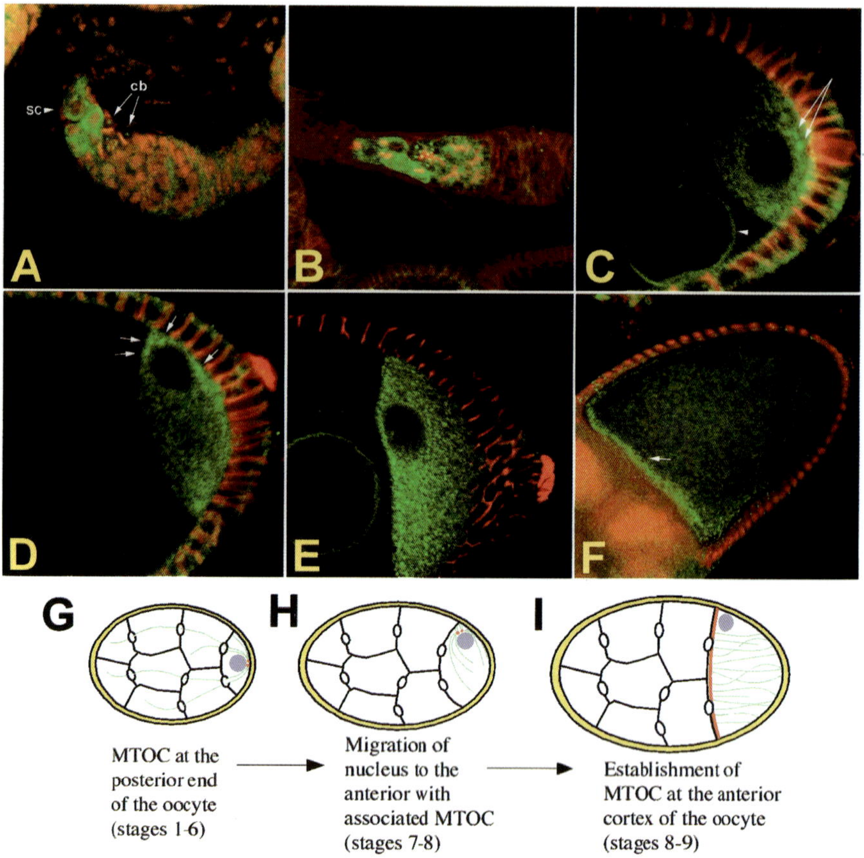

Chapter 18, Figure 2. Centrosomin is a marker for the MTOC throughout oogenesis. A–F show the localization of centrosomin with anti-centrosomin antibodies (in green) and DNA in red (A and F), HTS in red (B), and actin in red (C–E). Centrosomin localizes to the centrosomes in dividing stem cells (SC) and cystoblasts (CB) at mitosis. In the germarium (A) a four-cell cyst can be seen in metaphase with the DNA stained with propidium iodide (red) and the centrosomes labeled with anti-centrosomin antibodies (green). In (B) the germarium has been counterstained with anti-HTS antibodies to show the relative positions of the centrosomes to the fusome in an eight-cell cyst in mitosis. In a stage 6 oocyte (C) centrosomin accumulates in one or more bodies (arrows in C) at the posterior of the oocyte in the space between the nucleus and the follicle cells (the germinal vesicle appears as a dark hole amidst the CNN staining in C–E). Also, centrosomin accumulates on the apical face of the posterior follicle cells at this stage. The arrowhead in (C) indicates the perinuclear localization of centrosomin in nurse cells. In a stage 7 oocyte (D), when the microtubule polarity is shifting and the nucleus is migrating to the anterior, centrosomin accumulates in several bodies on the nuclear periphery (arrows in D). After the nucleus has migrated to its anterior cortical position, which marks the anterior/dorsal coordinate of the oocyte, centrosomin disperses from the bodies and begins to accumulate at the anterior cortex (E). By stage 10, centrosomin accumulates to a significant level at the anterior of the oocyte (arrow in F). Centrosomin remains at the anterior cortex after cytoplasmic streaming begins at stage 10b (not shown). Ovaries were fixed and stained according to a published protocol (Peifer *et al.*, 1993). In G–I the dynamics of microtubule polarity and MTOC localization at midoogenesis are illustrated. The microtubules are shown in green, the nucleus in blue, and the MTOCs in red.

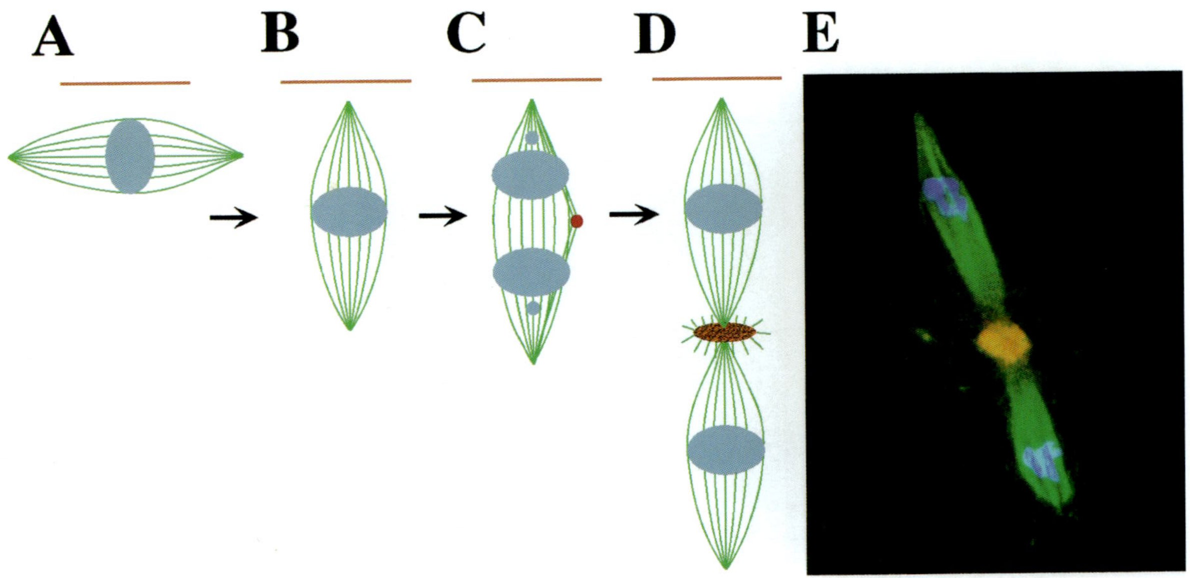

Chapter 18, Figure 3. The meiotic spindles are acentriolar. The poles of the meiosis I spindle contain no centrosomes. Shown is a schematic drawing of the dynamics of meiotic spindles upon activation and the assembly of the meiosis II spindle from the meiosis I spindle. In nonactivated oocytes the meiosis I spindle is arrested at metaphase (A) and the spindle lies in a parallel register to the cortex (shown as a red line). The chromosomes are shown in blue and the microtubules in green. The mieosis I spindle contains no centrosomes at its poles and is anastral. Upon activation the spindle pivots 90° , orienting the axis of the spindle perpendicular to the cortex (B). The meiosis II spindles have a shared pole that forms within the meiosis I spindle at anaphase (shown as a red dot in (C)). The small blue dots in (C) represent the smaller fourth chromosomes, which move toward the poles in advance of the exchange chromosomes. The meiosis II spindles share an astral pole that has a ring structure (D) and contains centrosomal proteins (shown in red). A real image of meiosis II spindles from a wild-type oocyte that shows the localization of centrosomin (red) at the central spindle pole is shown in (E). The micrograph in (E) was provided courtesy of Drs. Salud Llamazares and Cayetano Gonzalez.

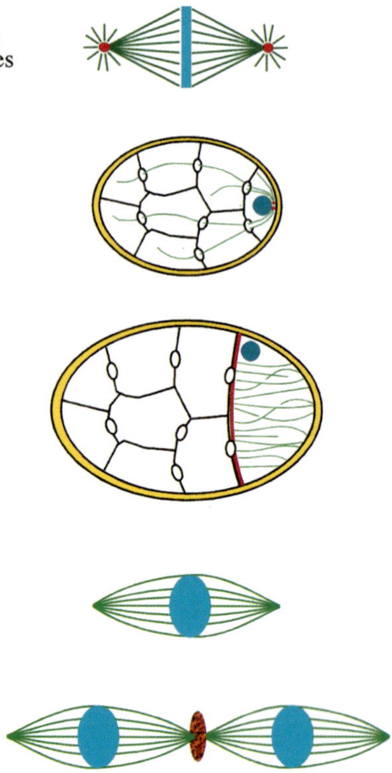

Stage	Appearance	Characteristics
Germarium stem cells and cystocytes		Centrosomes. Contains proteins found in mitotic centrosomes.
Early oocyte		Perinuclear discrete bodies in proximity to cluster of centrioles. Contains Centrosomin.
Late oocyte		Acentriolar. Associated with the anterior cortex. Contains Centrosomin.
Meiosis I		Acentriolar. No known centrosomal proteins associated with spindle poles.
Meiosis II		Acentriolar. The central MTOC is astral and contains proteins found in mitotic centrosomes.

Chapter 18, Figure 4. Summary of the MTOCs in oogenisis. The appearance of the MTOCs at different stages of oogenesis is illustrated in the middle column. These figures are taken from Figs 1–3 and are described in their legends. An abbreviated description of the characteristics of each MTOC is given in the right-hand column.

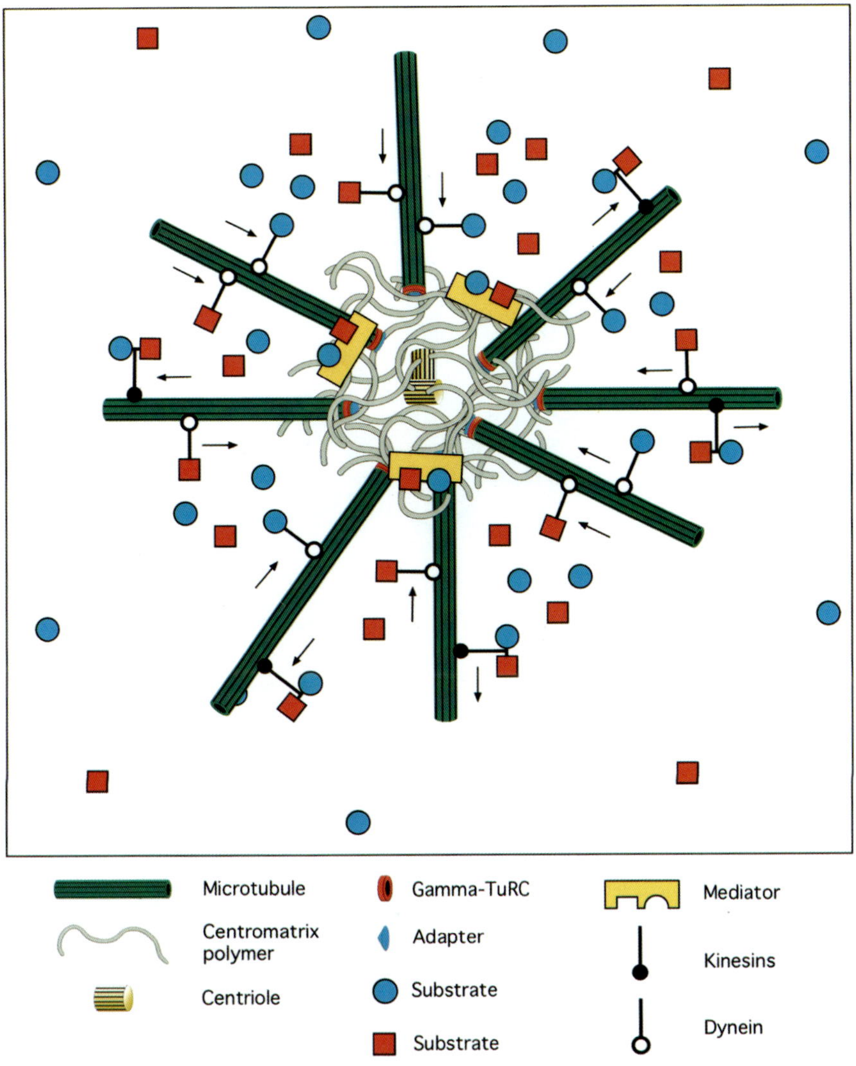

Chapter 20, Figure 1. A model for the centrosome as a central processing station. The centromatrix serves as a scaffold for the binding of specialized molecules that mediate interactions between substrates to yield higher-order structures. Delivery of substrates and dispatch of macromolecular assemblies is accomplished by microtubule-dependent motor proteins.

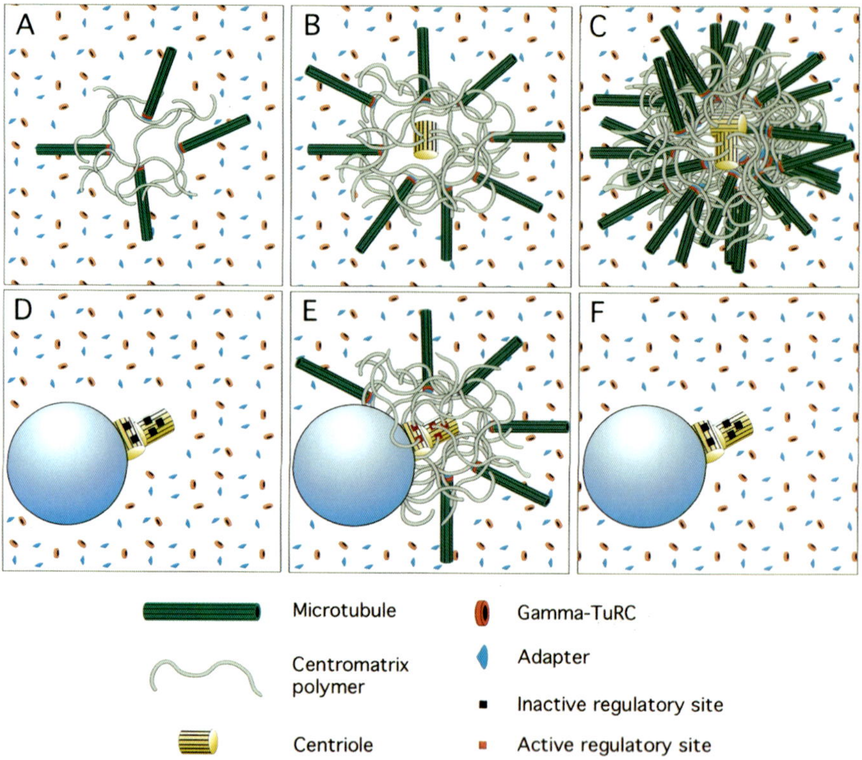

Chapter 20, Figure 2. A model for centrosome maturation in *Spisula* embryos. Unfertilized oocytes may contain maternal procentrosomes that, in the absence of centrioles, maintain integrity and are capable of nucleating microtubules (A). Artificial activation or fertilization induces maternal centrosome maturation such that by prometaphase of meiosis I, each maternal centrosome contains a single centriole (B). In addition, centromatrix polymer is added that allows the binding of additional γ-TuRCs, thus increasing the microtubule nucleation potential (B). Maternal centrosome maturation is completed with the assembly of a second centriole and additional PCM by metaphase of meiosis I (C). Sperm centrioles are not capable of recruiting maternal PCM components in unactivated oocyte cytoplasm (D). Shortly after fertilization, during prometaphase of meiosis I, sperm centriole-specific components are activated to initiate centromatrix assembly, resulting in paternal centrosome maturation (E). By metaphase of meiosis I, sperm centriole-specific regulatory sites are inactivated, resulting in a loss of centriole-centromatrix interactions, dispersal of the PCM, and loss of microtubule nucleation potential (F). As sperm centrioles undergo this transient maturation process (D–F), maternal centrosomes continue to mature and function (A–C).

cell systems." Electron microscopy and other molecular cell biology approaches have revealed novel aspects of centriole duplication and maturation and have been influential in consolidating our views of inheritance patterns (Vorobjev and Chentsov, 1982; Kochanski and Borisy, 1990; Lange and Gull, 1995; Andersen, 1999).

The centriolar or basal body duplication cycle has been followed in many systems such as somatic mammalian cells (Rieder and Borisy, 1982; Vorobjev and Chentsov, 1982; Robbins *et al.,* 1968), in developmental systems such as sea urchin eggs (Sluder and Rieder, 1985), *Spisula* oocytes (Palazzo *et al.,* 1992) and *Drosophila* (Callaini and Riparbelli, 1990; Callaini *et al.,* 1997, Gonzalez *et al.,* 1998), in algae (Beech *et al.,* 1988, 1991; Brugerolle, 1992; Gould, 1975; Holmes and Dutcher, 1989), and in ameboflagellates such as *Naegleria* (Fulton and Dingle, 1971) and *Physarum* (Wright *et al.,* 1980). In mammalian somatic cells the centriolar pair consists, at the beginning of the cell cycle, of a mature and an immature centriole. In G1/G0 the mature centriole often subtends a 9 + 0 immotile, primary cilium. Both centrioles duplicate around S-phase, but it is not until the beginning of mitosis that the immature centriole of only the original pair matures by acquiring specific structural features or precursors thereof (Rieder and Borisy, 1982; Vorobjev and Chentsov, 1982; Lange and Gull, 1996b; Sluder and Hinchcliffe, 1998; Andersen, 1999) (Fig. 1).

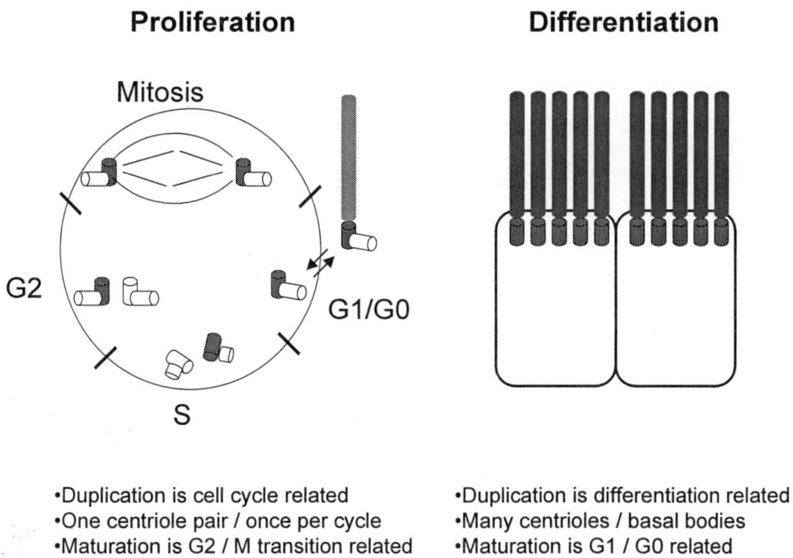

Figure 1 A diagram illustrating the major conceptual differences between centriole duplication and maturation in a proliferating somatic cell and during formation of differentiated ciliated epithelium. (Unshaded centrioles are immature. Shaded centrioles are cenexin-positive, mature centrioles. The proliferative cell cilium is a primary, immotile cilium. The ciliated epithelial cells bear motile cilia.)

The two main features marking the mature centriole of the pair in interphase cells are the subdistal appendages (also called satellite structures) and distal appendages. Classic electron microscopic studies have described these features (reviewed, e.g., by Vorobjev and Nadezhdina, 1987). Some more recent work gives additional detailed description of these markers of centriole maturity. Paintrand *et al.* (1992) undertook an extensive ultrastructural study of centrosomes isolated from a human lymphoid cell line (KE37). The authors described the distal appendages, which are orientated with a ninefold symmetry around the centriolar cylinder, as rigid sticks likely to be attached to the edges of the B-tubules of the centriole. The subdistal appendages (satellite structures) show more variability than the distal appendages in their number, thickness, and distribution along the centriole in these isolated centrosomes. The base of the subdistal appendages is wide enough to interact with two adjacent microtubule sets, whereas the tip is rather narrower and more electron dense. It remains ambiguous whether or not the satellites are composed of discrete filaments and, if so, how they are arranged (Paintrand *et al.*, 1992). Isolated centrioles have been viewed by negative stain electron microscopy (Lange and Gull, 1996a). Kenney *et al.* (1997) have analyzed centriole structure from tilt series using cryoelectron microscopy and drawn attention to the complexity of the area of the distal base.

Although these appendages assist in defining a structural description of the centriole, their roles are less clear. This uncertainty holds for both the distal and subdistal appendages. However, the tip of the subdistal appendages has been described as a binding site or nucleation site of microtubules, as, for example, in PE cells (Vorobjev and Chentsov, 1982) and PtK2 cells, in which large numbers of microtubules are seen attached to the subdistal appendages after release of cells from nocodazole treatment (de Brabander *et al.*, 1986). Microtubule regrowth from the subdistal appendages was also shown beautifully *in vitro* on isolated KE37 centrosomes by cryoelectron microscopy of vitrified centrosome specimens (Chretien *et al.*, 1997). The subdistal appendages or satellites are likely to be related to the striated wedge shaped fibers (sinister fibers) in algal cells, which connect basal bodies and cytoplasmic microtubules (Geimer *et al.*, 1998). One possible role of these appendages might be the positioning, in relation to other cytoplasmic organelles, of the centriole pair or the basal bodies via the microtubules.

In mammalian somatic cells the satellite structures disappear in the G2 period of the cell cycle; at the same time a fibrillar halo forms around both mother centrioles (Vorobjev and Chentsov, 1982; Rieder and Borisy, 1982). This halo formation is likely to reflect the increase of microtubule nucleation elements, such as γ-tubulin, occurring at the onset of mitosis (Lajoie-Mazenc *et al.*, 1994; Zheng *et al.*, 1991; Khodjakov and Rieder, 1999). This halo disappears at the beginning of the G1 phase simultaneously with the formation of the subdistal appendages on the mature centriole. Each daughter cell therefore receives one mature and one immature centriole, both of which were generated at least one cell cycle ago (Fig. 1).

Other systems, for example the early postblastodermic mitoses of the *Drosophila* embryo, are much different, with no obvious evidence for a structural maturation of the centriole (Callaini *et al.,* 1997). The centrioles in the *Drosophila* embryo are of a more minimal structure compared to mammalian somatic cell's centrioles. They are composed of a ring of either single (Moritz *et al.,* 1995) or doublet microtubules (Callaini *et al.,* 1997) connected to a central hub via spokes (see also Gonzalez *et al.,* 1998, for a review of this topic). This reduced form of the centriole structure might be explained by the need for rapid mitoses in the embryo, limiting the time available for building more complex centrioles (Moritz *et al.,* 1995).

IV. Centriolar Duplication, Maturation, and the Cell Cycle

The events of centriole duplication and maturation are clearly integrated into the cell cycle (reviewed by Karsenti, 1999; Sluder and Hinchcliffe, 1998). Some insights into key elements of the regulation of the centriolar duplication cycle have recently been provided (Meraldi *et al.,* 1999; Hinchcliffe *et al.,* 1999; Lacey *et al.,* 1999). We therefore discuss the regulation of centriolar duplication and maturation, the possible mechanisms involved, and the molecular components required in this process.

Both embryonic (Sluder and Rieder, 1985; Palazzo *et al.,* 1992; Hinchcliffe *et al.,* 1998; 1999; Lacey *et al.,* 1999) and somatic cells (Fukasawa *et al.,* 1996; Balczon *et al.,* 1995; Meraldi *et al.,* 1999) have been intensively used to study the molecular regulation events that are involved in the process of centriolar duplication. Recently, a centriole duplication assay was established using a Chinese hamster ovary (CHO) cell line (Balczon *et al.,* 1995; Meraldi *et al.,* 1999). The assay is based on the fact that in the CHO cell line centriole duplication can be uncoupled from DNA replication. This is accomplished by treating cells with the DNA synthesis inhibitor hydroxyurea, leading to extra rounds of centrosomal duplication while DNA synthesis is halted (Kuriyama and Borisy, 1981). This system was used to test the effect of a number of cell cycle regulators, which might have a potential role in centrosomal duplication. CHO cells were treated with hydroxyurea, transfected with complementary DNAs of cell-cycle regulators, and then analyzed by immunofluorescence and electron microscopy to determine the effect of the expressed gene products on centriole duplication. This elegant approach established that the cyclin dependent kinase 2 (Cdk2) is a major factor influencing centriole duplication. Moreover, phosphorylation of the retinoblastoma protein (Rb) and the presence of the transcription factor E2F (Meraldi *et al.,* 1999) are implicated in centriolar duplication. Essentially, progression toward S-phase requires the E2F factor, which is repressed by the Rb protein (Weinberg, 1995). Phosphorylation of Rb through the cyclin D–Cdk4/6 complex permits entry into S-phase by releasing E2F from Rb. The subsequent activation of the E2F transcription factor permits transcription of cyclin A and cyclin E and, in general, DNA replication, i.e., entry into S-phase. This means that the factors, which are essential for DNA replication, are also implicated in progression of centriole duplication in CHO cells. The same authors also established a more direct link between centriole duplication and Cyclin A/Cdk2 activity. Multiple rounds of centriolar duplication were supported by Cyclin A/Cdk2 even in cells where activation of E2F was inhibited by expressing a Rb mutant lacking CDK phosphorylation sites. This demonstrated that cyclin A/Cdk2 is likely to be downstream of RB/E2F and may be more directly involved in the regulation of centriolar duplication.

Embryonic systems often exhibit differences to somatic cells in their regulation of centriole cycles. For example, in the early *Xenopus* embryo, cyclin A does not associate with CdK2 before the mid-blastula transition, whereas cyclin E does (Meraldi *et al.,* 1999). Hence, in this system the concept exists that the Cdk2/cyclin E complex (rather than Cdk2/cyclin A) is required to promote centriolar duplication (Hinchcliffe *et al.,* 1999; Lacey *et al.,* 1999). Indeed, in cell-cycle-arrested *Xenopus* extracts where Cdk2/cyclin E was specifically inhibited with the truncated recombinant Cdk inhibitor Δ34Xic1, centrosome duplication can be rescued by addition of purified Cdk2/cyclin E complex (Hinchcliffe *et al.,* 1999). In further support of this concept, Lacey *et al.* (1999) showed that extra cycles of centrosomal duplication depend on cyclin/Cdk2 activity *in vivo.* The authors microinjected the Cdk inhibitor p21 into *Xenopus* embryos, and then the embryos were subsequently treated with cyclohexamide, which blocks cell division and DNA replication but not centrosome duplication. p21 indeed halted centrosome duplication, specifying the role of Cdk2 in centrosome duplication. However, it is still unclear (Hinchcliffe *et al.,* 1999; Lacey *et al.,* 1999) whether or not the centrosomes studied actually contained centrioles. Nevertheless, there is some background evidence to support the notion that they do, as it has been shown before that treatment of *Xenopus* or sea urchin embryos with protein synthesis inhibitors leads to multiple centrosomes that actually contain both "normal" centrioles and pericentriolar material (Gard *et al.,* 1990; Sluder *et al.,* 1990).

The separation of centrioles, an early step in centrosome duplication, is shown to be driven by the activity of cyclin E/Cdk2 in *in vitro* experiments with *Xenopus* extracts (Lacey *et al.,* 1999). In contrast, the presence of high levels of cyclin B and cdc2 kinase (as in mitotic extracts) in these *Xenopus* extracts inhibited centriolar separation (Lacey *et al.,* 1999). This implies a mechanism by which the periodic event of centriolar duplication is linked to the cell cycle. In this manner the destruction of the physical link between the centrioles and therefore also the event of centriolar duplication (to which centriolar separation might be a prerequisite) is prevented in mitosis. Yet, the direct regulatory link between destruction of the physical connection of the centrioles as a prerequisite to centrosomal duplication remains so far moot. Interestingly, a human kinase has been identified (Nek2) that, when overexpressed in mammalian U20S cells, leads to centrosomal splitting and ultimately, to centrosomal dispersion (Fry *et al.,* 1998a). One of the possible substrates of this kinase is C-Nap1, a centrosomal protein localized at the proximal ends of both the mother and the daughter centriole. This protein, together with Nek2, are likely to be directly involved in regulation of the connection between the two centrioles of the pair and therefore might have an important role in the control of the centriole cycle (Fry *et al.* 1998b). The upstream signals for the Nek2 kinase are as yet unidentified, although protein phosphorylation is a good candidate for regulating Nek2 kinase activity, as well as many other centrosomal and centriolar cell-cycle-related transitions (Mayor *et al.,* 1999).

While centriolar duplication is controlled via the Cdk2/cyclin E or Cdk2/cyclin B complex, less is known about centriolar maturation on the molecular level. The molecular marker cenexin has been identified. This is a protein component of the mature centriole of the pair in all mammalian cell lines tested (Lange and Gull, 1995). It is localized to the distal end of the mature centriole as identified by immunoelectron microscopy. At the moment it is unclear whether cenexin actually corresponds to one of the appendagelike structures at the distal end of the centriole, or whether its function is related to the capacity of the mature centriole of the pair to nucleate a primary cilium from this end. Nevertheless, acquisition of cenexin occurs exactly at the G2/M transition, and its addition may well be controlled by a cdc2-related pathway—although this still remains to be shown. The timing of cenexin acquisition at the onset of mitosis coincides with the increase of microtubule nucleation elements, such as γ-tubulin (see earlier discussion). This sudden increase in the level of centrosomal γ-tubulin (Khodjakov and Rieder, 1999) is again likely to be regulated by phosphorylation of centrosomal components through cell-cycle-related kinases. Indeed, it has been demonstrated in an elegant microinjection study in immortalized HeLa and nonimmortalized Hs68 cells (Lane and Nigg, 1996) that plk1, the human homolog of the *Drosophila* polo kinase (Glover *et al.* 1996), is involved in the functional maturation of mitotic centrosomes, i.e., the increase in centrosomal material at the G2/M border (reviewed by Kellogg *et al.,* 1994; Kalt and Schliwa, 1993). Microinjection of specific anti-plk1 antibodies leads to duplicated but nonseparated centrosomes. Moreover, the size of the centrosomes in injected cells was much reduced (as indicated by the diminished immunostaining with different centrosomal markers such as γ-tubulin or CTR453), providing evidence that the process of centrosomal maturation is severely affected (Lane and Nigg, 1996). In summary, it remains open which of the components of the cell cycle regulation machinery are real players for the process of centriolar maturation. However, the recent studies do certainly strongly suggest causal links between the two processes.

V. Centriolar Maturation and Centriolar Function

Centriolar maturation is marked by the acquisition of appendages and satellite structures and the protein cenexin (Lange and Gull, 1995, 1996b; Andersen, 1999). A widely accepted functional role of the mature centriole is nucleation of a cilium or flagellum. Other possible functional roles of the (mature) centriole such as the organization of the pericentriolar material are a matter of debate. However, new data on functional aspects of centrioles and basal bodies has recently appeared.

A serious caveat affecting the task of identifying centriolar functions is that we do not have a catalogue of proteins that are specific for the centriole. Recently, though, proteins have been identified that not only are part of the centriolar structure, but also are important for centriole/basal body assembly or duplication. These insights provide us with some clues to the functional roles of basal bodies and centrioles, other than their nucleation of a cilium or flagellum (review by Marshall and Rosenbaum, 1999). More specifically, γ-tubulin has been identified in the centriolar cylinder or basal body (Fuller *et al.,* 1995; Liang *et al.,* 1996) and is essential for basal body duplication in *Paramecium* (Ruiz *et al.,* 1999). Using a gene silencing approach by injection of coding sequence for either of the two γ-tubulin genes (γ-PT1 and γ-PT2) of *Paramecium* into the macronucleus resulted in individuals with only half the numbers of basal bodies. Normally, the cortex of *Paramecium* possesses units of double basal bodies (with a linked lineage of formation and maturation) arranged in a precise localization in the cortex. However, the inactivation of γ-tubulin produced rows of single basal bodies, indicating that γ-tubulin is indispensable for proper basal body duplication (Ruiz *et al.,* 1999). Although not shown by the authors, it seems reasonable to assume that the basal bodies produced in the microinjected cells were only the mature ones, subtending a flagellum. Hence, the lack of immature basal bodies probably indicates that γ-tubulin is possibly required directly or indirectly in an early stage of basal body assembly.

Another example of a recently identified basal body component is δ-tubulin (Dutcher and Trabuco, 1998). δ-Tubulin is the product of the

UNI3-gene in *Chlamydomonas reinhardtii* and is a member of the tubulin superfamily. Mutants for this gene produce a uniflagellelate phenotype. On the ultrastructural level all basal bodies in the *uni3* mutant showed abnormalities. Surprisingly, the C-tubule of the basal body was missing, resulting in basal bodies with only doublet microtubules and indicating a role of Uni3 in basal body assembly. Moreover, cell division defects involving the proper placement of the mitotic spindles and, in addition, misplacement of the eyespot were also revealed. Similar defects had been described before for the *bld2* mutant in *Chlamydomonas,* which lacks the basal bodies (Ehler *et al.,* 1995). Importantly, these mutations are revealing a role for the basal body not only in flagella assembly but in the orchestration of spatial organization of the cell and placement of the mitotic spindle in *Chlamydomonas.* Moreover, in these uniflagellate cells the flagellum grows from the basal body located opposite to the eyespot which is, at least in wild-type cells, indicative of the older of the two basal bodies. Dutcher and Trabuco (1998) suggest that the *uni3* mutants need to go through at least three generations before they can generate a flagellum, because their basal bodies need to age from daughter basal body to parent basal body and finally to a mature basal body that can then produce a flagellum.

Hence, it is possible that flagella assembly requires mature basal bodies with triplet microtubules. An analogy can be drawn with *Drosophila,* where, for example, in spermatogenesis only the mature, triplet-containing centrioles subtend a flagellum, while other immature "neotenic" centrioles, for example in the early embryo, do not (Callaini *et al.,* 1997). In summary these results indicate that functional maturation of centrioles or basal bodies is a prerequisite for this organelle to produce a flagellum, for proper placement of the mitotic spindles and at least some spatial organization of the cell.

Recently, two additional lines of evidence have shown that in flies and in mammalian cells the centriole has a function in addition to its role as basal body. This work suggests a role in maintaining proper centrosome structure and function. Microinjection of antibodies against polyglutamated tubulin (a posttranslationally modified tubulin isoform predominantly restricted to centrioles) into HeLa cells lead to the disappearance of centrioles after about 12 hr postinjection (Bobinnec *et al.,* 1998). Moreover, the pericentriolar material (PCM) became scattered and a large proportion of monopolar spindles and spindles with abnormal PCM distribution was detected. Although these effects were only transient, they indicate that centrioles are likely to have a role in maintaining proper centrosome structure, duplication, and centrosome function in mammalian somatic cells. Similar conclusions can be drawn from work in unfertilized *Sciara* embryos, which can develop parthenogenetically without centrosomes and centrioles, yet form functional mitotic spindles. When compared to fertilized embryos, the spindles of unfertilized embryos are anastral. Furthermore, the distance between the daughter nuclei in the subsequent interphase is reduced (de Saint Phalle and Sullivan, 1998). These data argue again for a possible role of centrioles in centrosome congression and again a role in the spatial organization of the cell.

However, there are many systems that do not require a centriole for proper mitosis and spatial cell organization (Debec and Abbadie, 1989; Szöllösi *et al.,* 1972, 1986; Vaughn and Harper, 1998). The cell biology of the system studied is critical to the interpretation of the result, the main focus being, what is the function of centrioles in those cells that possess them (Lange and Gull, 1996b)?

VI. Ciliogenesis and Cell Cycle

Much of the preceding discussion of centriole replication and lineage relates to the situation where a proliferative cell faces the task of endowing its daughters with a similar component set of centrioles/basal bodies. Hence, in cells as diverse as flagellates, ciliates, and mammalian cells this form of duplication and maturation of centrioles/basal bodies is intimately linked to progression through the cell cycle. Number control and other phenomena are related to the existing centriole complement.

In somatic animal cells one can recognize three main characteristics of this process. First, the duplication of centrioles is cell cycle related. There is precise number control, with one centriole pair duplicated once per cell cycle. Finally, the maturation event is related to the G2/M phase transition.

However, there are situations where centriole replication and maturation occur, yet are unrelated to the existing centriolar status of the cell or to progression through the cell cycle. Perhaps the best examples of these events are spermiogenesis in the fern *Marsilea* (Hepler, 1976) and ciliogenesis in the respiratory and reproductive epithelia of animals (Dirksen, 1991). There is a strong contrast between the process of centriole formation during ciliogenesis and that involved in the proliferative cell cycle as rehearsed earlier. During ciliogenesis many hundreds of centrioles are formed. Ciliated epithelial cells are terminally differentiated and the 200–300 cilia are produced in nondividing cells blocked in G1/G0 (Conti *et al.,* 1981).

The major route for production of these cilia involves the coincident assembly of many hundreds of procentrioles in the cytoplasm around amorphous clouds of variously termed fibrillar/granular material (Dirksen, 1991). The centrioles develop and migrate to the apex of the cell where axoneme formation occurs. The differences of this centriole morphogenesis process from that described in proliferating cells are striking. Duplication of the centrioles is differentiation related rather than cell cycle related, since the ciliogenic cell does not divide and is in G1/G0. Second, many centrioles are formed—and by a novel morphogenetic route. Finally, in contrast to the proliferative cell situation, maturation of the centrioles is not linked to the G1/M phase transition, but rather to the G1/G0 stage of the cell cycle (Fig. 1).

Recent studies described here have started to unravel the regulatory links between cell cycle control and centriole duplication/maturation. Ultimately, however, we will need to understand how in ciliogenesis these regulatory links can be modulated to facilitate the uncoupling of centriolar duplication, number control, and maturation from progression through the cell cycle.

VII. Future Developments and Directions

Finally, we outline some of the possible future developments in the study of centriolar cell biology. There are at least three main avenues driving future developments. First, there are developments that will expand our knowledge of the molecular components of the centrosome/centriole complex. This approach will be aided by new techniques such as MALDI mass spectroscopy, which have been applied successfully in the analysis of the yeast spindle pole body and the *Drosophila* centrosome components (Wigge *et al.,* 1998; Lange *et al.,* 2000). These techniques will probably be combined with specific salt-extraction and centrosome reconstruction protocols (Schnackenberg *et al.,* 1998; Moritz *et al.,* 1998). Secondly, this cellular organelle's high-resolution ultrastructure can now be resolved using modern electron microscopy techniques such as cryoelectron microscopy of isolated vitrified specimens (Chretien *et al.,* 1997) or automated electron tomography for three-dimensional structural analysis (Moritz *et al.,* 1995). Finally, new techniques combining high-resolution fluorescence imaging, laser microsurgery, and molecular biology are now more widely available. Such techniques have already been used in the study of live cells in which centrosomal components have been labeled with GFP (Khodjakov *et al.,* 1997; Khodjakov and Rieder, 1999; Endow and Komma, 1996; Arnaud *et al.,* 1998; Tsakraklides *et al.,* 1999). These and other new and remarkable experimental approaches will allow us to answer at least some of the questions that make this cryptic cell organelle so fascinating.

Acknowledgments

Our work has been funded by Wellcome Trust Programme and Equipment Grants and BBSRC Project and Equipment Grants.

References

Andersen, S. S. L. (1999). Molecular characteristics of the centrosome. *Int. Rev. Cytol.* **187,** 51–109.

Arnaud, L., Pines, J., and Nigg E. A. (1998). GFP tagging reveals human Polo-like kinase 1 at the kinetochore/centromere region of mitotic chromosomes. *Chromosoma* **107,** 424–429.

Balczon, R., Bao, L., Zimmer, W. E., Brown, K., Zinkowski, R. P., and Brinkley, B. R. (1995). Dissociation of centrosome replication events from cycles of DNA synthesis and mitotic division in hydroxyurea-arrested Chinese hamster ovary cells. *J. Cell Biol.* **130,** 105–115.

Beech, P. L., Wetherbee, R., and Pickett-Heaps, J. D. (1988). Transformation of the flagella and associated flagellar components during cell division in the Coccolithophorid *Pleurochysis carterie. Protoplasma* **145,** 37–46.

Beech, P. L., Heinman, K., and Melkonian, M. (1991). Development of the flagellar apparatus during the cell cycle in unicellular algae. *Protoplasma* **164,** 23–37.

Bobinnec, Y., Moudjou, M., Fouquet, J. P., Desbruyeres, E., Edde, B. and Bornens, M. (1998). Glutamylation of centriole and cytoplasmic tubulin in proliferating non-neuronal cells. *Cell Motil. Cytoskeleton* **39,** 223–232.

Boveri, T. (1888). "Zellen Studien," Vol. 2. Fischer, Jena.

Boveri, T. (1895). *Verhandl. physik-med. Ges. Wurzburg* [NF] **29,** 1.

Boveri, T. (1901). Über die Natur der Zentrosomen. *Jena. Z. Med. Naturwiss.* **28,** 1–220.

Brugerolle, G. (1992). Flagellar apparatus duplication and partition, flagellar transformation during division in *Entosiphon sulcatum. Biosystems* **28,** 203–209.

Callaini, G., and Riparbelli, M. G. (1990). Centriole and centrosome cycle in the early *Drosophila* embryo. J. Cell Sci. **97,** 539–543.

Callaini, G., Whitfield, W. G., and Riparbelli, M. G. (1997). Centriole and centrosome dynamics during the embryonic cell cycles that follow the formation of the cellular blastoderm in *Drosophila. Exp. Cell Res.* **234,** 183–190.

Chretien, D., Buendia, B., Fuller, S. D. and Karsenti, E. (1997). Reconstruction of the centrosome cycle from cryoelectron micrographs. *J. Struct. Biol.* **120,** 117–133.

Cleveland, L. R. (1957). Types and life cycles of centrioles in flagellates. *J. Protozool.* **4,** 230–241.

Cleveland, L. R. (1963). Functions of flagellate and other centrioles in cell reproduction. *In* "The Cell in Mitosis" (L. Levine, Ed.), pp. 3–53. Academic Press, New York.

Conti, C. J., Conner, E. A., Gimenez-Conti, I. B., Silverberg, S. G., and Gerschensen, L. E. (1981). Regulation of ciliogenesis and proliferation of uterine epithelium by 20-hydroxypregn-4-en-3-one administration and withdrawal in overiectomized rabbits. *Biol. Reprod.* **24,** 903–921.

Debec, A., and Abbadie, C. (1989). The acentriolar state of the *Drosophila* cell lines 1182. *Biol. Cell* **67,** 307–311.

de Brabander, M., Geuens, G., Nuydens, R., Willebrords, R., Aerts, F., De Mey, J., with the participation of J. R. Macintosh (1986). Microtubule dynamics during the cell cycle: the effects of taxol and nocodazole on the microtubule system of Pt K2 cells at different stages of the mitotic cycle. *Int. Rev. Cytol.* **101,** 215–274.

de Saint Phalle, B., and Sullivan, W. (1998). Spindle assembly and mitosis without centrosomes in parthenogenetic *Sciara* embryos. *J. Cell Biol.* **141,** 1383–1391.

Dirksen, E. R. (1991). Centriole and basal body formation during ciliogenesis revisited. *Biol. Cell* **72,** 31–38.

Dutcher, S. K., and Trabuco, E. C. (1998). The UNI3 gene is required for assembly of basal bodies of *Chlamydomonas* and encodes delta-tubulin, a new member of the tubulin superfamily. *Mol. Biol. Cell* **9,** 1293–1308.

Ehler, L. L., Holmes, J. A., and Dutcher, S. K. (1995). Loss of spatial control of the mitotic spindle apparatus in a *Chlamydomonas reinhardtii* mutant strain lacking basal bodies. *Genetics* **141,** 945–960. Published erratum appeared in *Genetics* (1996). **142,** 650.

Endow, S. A., and Komma D. J. (1996). Centrosome and spindle function of the *Drosophila* Ncd microtubule motor visualized in live embryos using Ncd-GFP fusion proteins. *J. Cell Sci.* **109,** 2429–2442.

Fry, A. M., Meraldi, P., and Nigg, E. A. (1998a). A centrosomal function for the human Nek2 protein kinase, a member of the NIMA family of cell cycle regulators. *EMBO J.* **17,** 470–481.

Fry, A. M., Mayor, T., Meraldi, P., Stierhof, Y. D., Tanaka, K., and Nigg, E. A. (1998b). C-Nap 1, a novel centrosomal coiled-coil protein and candidate substrate of the cell cycle-regulated protein kinase Nek2. *J. Cell Biol.* **141,** 1563–1574.

Fukasawa, K., Choi, T., Kuriyama, R., Rulong, S. and Vande-Woude, G. F. (1996). Abnormal centrosome amplification in the absence of p53. *Science* **271,** 1744–1747.

Fuller, S. D., Gowen, B. E., Reinsch, S., Sawyer, A., Buendia, B., Wepf, R., and Karsenti, E. (1995). The core of the mammalian centriole contains gamma-tubulin. *Curr. Biol.* **5,** 1384–1393.

Fulton, C., and Dingle, A. D. (1971). Basal bodies, but not centrioles, in *Naegleria. Cell Biol.* **51,** 826–836.

Gard, D. L., Hafezi, S., Zhang, T., and Doxsey, S. J. (1990). Centrosome duplication continues in cycloheximide-treated *Xenopus* blastulae in the absence of a detectable cell cycle. *J. Cell Biol.* **110,** 2033–2042.

Geimer, S., Clees, J., Melkonian, M., and Lechtreck, K. F. (1998). A novel 95-kD protein is located in a linker between cytoplasmic microtubules and basal bodies in a green flagellate and forms striated filaments *in vitro. J. Cell Biol.* **140,** 1149–1158.

Glover, D. M., Okhura, H., and Tavares, A. (1996). Polo kinase: the choreographer of the mitotic stage? *J. Cell Biol.* **135,** 1681–1684.

Gonzalez, C., Tavosanis, G., and Mollinari, C. (1998). Centrosomes and microtubule organisation during *Drosophila* development. *J. Cell Sci.* **111,** 2697–2706.

Gould, R. R. (1975). The basal bodies of *Chlamydomonas reinhardtii.* Formation from probasal bodies, isolation, and partial characterization. *J. Cell Biol.* **65,** 65–74.

Heimann, K., Benting, J., Timmerman, S., and Melkonian, M. (1989). The flagellar developmental cycle in algae; two types of flagella development in unicellular algae. *Protoplasma* **153,** 14–23.

Hepler, P. (1976). The blepharoplast of *Marsilea;* its *de novo* formation and spindle association. *J. Cell. Sci.* **21,** 361–390.

Hinchcliffe, E. H., Cassels, G. O., Rieder, C. L., and Sluder, G. (1998). The coordination of centrosome reproduction with nuclear events of the cell cycle in the sea urchin zygote. *J. Cell Biol.* **140,** 1417–1426.

Hinchcliffe, E. H., Li, C., Thompson, E. A., Maller, J. L., and Sluder, G. (1999). Requirement of Cdk2-cyclin E activity for repeated centrosome reproduction in *Xenopus* egg extracts. *Science* **283,** 851–854.

Holmes, J. A., and Dutcher, S. K. (1989). Cellular asymmetry in *Chlamydomonas reinhardtii. J. Cell Sci.* **94,** 273–285.

Kalt, A., and Schliwa, M. (1993). Molecular components of the centrosome. *Trends Cell Biol.* **3,** 118–128.

Karsenti, E. (1999). Centrioles reveal their secrets. *Nature Cell Biol.* **1,** 62–64.

Kellogg, D. R., Moritz, M., and Alberts, B. M. (1994). The centrosome and cellular organization. *Ann. Rev. Biochem.* **63,** 639–674.

Kenney, J., Karsenti, E., Gowen, B., and Fuller, S. D. (1997). Three-dimensional reconstruction of the mammalian centriole from cryoelectron micrographs: The use of common lines for orientation and alignment. *J. Struct. Biol.* **120,** 320–328.

Khodjakov, A., and Rieder, C. L. (1999). The sudden recruitment of gamma-tubulin to the centrosome at the onset of mitosis and its dynamic exchange throughout the cell cycle, do not require microtubules. *Cell Biol.* **146,** 585–596.

Khodjakov, A., Cole, R. W., and Rieder, C. L. (1997). A synergy of technologies: combining laser microsurgery with green fluorescent protein tagging. *Cell Motil. Cytoskel.* **38,** 311–317.

Kochanski, R. S., and Borisy, G. G. (1990). Mode of centriole duplication and distribution. *J. Cell Biol.* **110,** 1599–1605.

Kuriyama, R. and Borisy, G. G. (1981). Microtubule-nucleating activity of centrosomes in Chinese hamster ovary cells is independent of the centriole cycle but coupled to the mitotic cycle. *J. Cell Biol.* **91,** 822–826.

Lacey, K. R., Jackson, P. K., and Stearns, T. (1999). Cyclin-dependent kinase control of centrosome duplication. *Proc. Natl. Acad. Sci. USA* **96,** 2817–2822.

Lajoie-Mazenc, I., Tollon, Y., Detraves, C., Julian, M., Moisand, A., Geuth-Hallonet, C., Debec, A., Salles-Passador, I., Puget, A., Mazarguil, H., Raynaud-Messina, B., and Wright, M. (1994). Recruitment of antigenic gamma-tubulin during mitosis in animal cells: presence of gamma-tubulin in the mitotic spindle. *J. Cell Sci.* **107,** 2825–2837.

Lane, H. A., and Nigg, E. A. (1996). Antibody microinjection reveals an essential role for human polo-like kinase 1 (Plk1) in the functional maturation of mitotic centrosomes. *J. Cell Biol.* **135,** 1701–1713.

Lange, B. M. H., and Gull, K. (1995). A molecular marker for centriole maturation in the mammalian cell cycle. *J. Cell Biol.* **130,** 919–927.

Lange, B. M. H., and Gull, K. (1996a). A structural study of isolated mammalian centrioles using negative staining electron microscopy. *J. Struct. Biol.* **117,** 222–226.

Lange, B. M. H., and Gull, K. (1996b). Structure and function of the centriole in animal cells: progress and questions. *Trends Cell Biol.* **6,** 348–352.

Lange, B. M. H., Bachi, A., Wilm, M., and Gonzalez, C. (2000). Hsp90 is a core centrosomal component and is required at different stages of the centrosome cycle in Drosophila and vertebrates. *EMBO J.* **19,** 1252–1262.

Liang, A., Ruiz, F., Heckmann, K., Klotz, C., Tollon, Y., Beisson, J., Wright, M. (1996). Gamma-tubulin is permanently associated with basal bodies in ciliates. *Eur. J. Cell. Biol.* **70,** 331–338.

Marshall, W. F., and Rosenbaum, J. L. (1999). Cell division: The renaissance of the centriole. *Curr. Biol.* **9,** 218–220.

Mayor, T., Meraldi, P., Stierhof, Y. D., Nigg, E. A., and Fry, A. M. (1999). Protein kinases in control of the centrosome cycle. *FEBS Lett.* **452,** 92–95.

Meraldi, P., Lukas, J., Fry, A. M., Bartek, J., and Nigg, E. A. (1999). Centrosome duplication in mammalian somatic cells requires E2F and Cdk2-Cyclin A. *Nature Cell Biol.* **1,** 88–93.

Moestrup, O., and Hori, T. (1989). Ultrastructure of the flagellar apparatus in *Pyramimonas octopus* (Prasinophyceae) II. Flagellar roots, connecting fibres, and the numbering of flagella in green algae. *Protoplasma* **148,** 41–56.

Moritz, M., Braunfeld, M. B., Fung, J. C., Sedat, J. W., Alberts, B. M., and Agard, D. A. (1995). Three-dimensional structural characterization of centrosomes from early *Drosophila* embryos. *J. Cell Biol.* **130,** 1149–1159.

Moritz, M., Zheng, Y., Alberts, B. M., and Oegema, K. (1998). Recruitment of the gamma-tubulin ring complex to *Drosophila* salt-stripped centrosome scaffolds. *J. Cell Biol.* **142,** 775–786.

Paintrand, M., Moudjou, M., Delacroix, H., and Bornens, M. (1992). Centrosome organization and centriole architecture: their sensitivity to divalent cations. *J. Struct. Biol.* **108,** 107–128.

Palazzo, R. E., Vaisberg, E., Cole, R. W. and Reider, C. L. (1992). Centriole duplication in lysates of *Spisula solidissima* oocytes. *Science* **256,** 219–221.

Rieder, C. L., and Borisy, G. G. (1982). The centrosome cycle in in PtK2 cells: asymmetric distribution and structural changes in the pericentriolar material. *Biol. Cell* **44,** 117–132.

Robbins, E. G., Jentzsch, G., and Micali, A. (1968). The centriole cycle in synchronized HeLa cells. *J. Cell Biol.* **36,** 329–339.

Ruiz, F., Beisson, J., Rossier, J., and Dupuis-Williams, P. (1999). Basal body duplication in *Paramecium* requires gamma-tubulin. *Curr. Biol.* **9,** 43–46.

Schnackenberg, B. J., Khodjakov, A., Rieder, C. L., and Palazzo, R. E. (1998). The disassembly and reassembly of functional centrosomes *in vitro. Proc. Natl. Acad. Sci. USA* **95,** 9295–9300.

Sherwin, T., and Gull, K. (1989). The cell division cycle of *Trypanosoma brucei brucei:* Timing of events markers and cytoskeletal modulations. *Phil. Trans. Roy. Soc,* **323,** 573–588.

Sluder, G., and Hinchcliffe, E. H. (1998). The apparent linkage between centriole replication and the S phase of the cell cycle. *Cell Biol. Int* **22,** 3–5.

Sluder, G., and Rieder, C. L. (1985). Centriole number and the reproductive capacity of spindle poles. *J. Cell Biol.* **100,** 887–896.

Sluder, G., Miller, F. J., Cole, R., and Rieder, C. L. (1990). Protein synthesis and the cell cycle: centrosome reproduction in sea urchin eggs is not under translational control. *J. Cell Biol.* **110,** 2025–2032.

Szöllösi, D., Calarco, P., and Donahue, R. P. (1972). Absence of centrioles in the first and second meiotic spindles of mouse oocytes. *J. Cell Sci.* **11,** 521–541.

Szöllösi, A., Ris, H., Szöllösi, D., and Debec, A. (1986). A centriole-free *Drosophila* cell line: A high voltage EM study. *Eur. J. Cell Biol.* **40,** 100–104.

Tsakraklides, V., Krogh, K., Wang, L., Bizario, J. C., Larson, R. E., Espreafico, E. M., and Wolenski, J. S. (1999). Subcellular localization of GFP-myosin-V in live mouse melanocytes. *J. Cell. Sci.* **112,** 2853–2865.

Vaughn, K. C., and Harper, J. D. (1998). Microtubule-organizing centers and nucleating sites in land plants. *Int. Rev. Cytol.* **181,** 75–149.

Vorobjev, I. A., and Chentsov, Y. S. (1982). Centrioles in the cell cycle. I. Epithelial cells. *J. Cell. Biol.* **93,** 938–949.

Vorobjev, I. A., and Nadezhdina, E. S. (1987). The centrosome and its role in the organization of microtubules. *Int. Rev. Cytol.* **106,** 227–293.

Weinberg R. A. (1995). The retinoblastoma protein and cell cycle control. *Cell* **81,** 323–330.

Wheatley, D. N. (1982). "The Centriole: A Central Enigma of Cell Biology." Elsevier Biomedical Press, Amsterdam.

Wigge, P. A., Jensen, O. N., Holmes, S., Soues, S., Mann, M., and Kilmartin, J. V. (1998). Analysis of the *Saccharomyces* spindle pole by matrix-assisted laser desorption/ionization (MALDI) mass spectrometry. *J. Cell. Biol.* **141,** 967–977.

Wright, M., Mir, L., and Moisand, A. (1980). The structure of the proflagellar apparatus of the amoebae of *Physarum polycephalum;* relationship to the flagellar apparatus. *Protoplasma* **103,** 69–81.

Zheng, Y., Jung, M. K., and Oakley, B. R. (1991). Gamma-tubulin is present in *Drosophila melanogaster* and *Homo sapiens* and is associated with the centrosome. *Cell* **5,** 817–823.

11

Centrosome Replication in Somatic Cells: The Significance of G_1 Phase

Ron Balczon
Department of Cell Biology and Neuroscience
The University of South Alabama
Mobile, Alabama 36688

I. Introduction

The centrosome, which is composed of a pair of centrioles surrounded by an osmiophilic substance termed pericentriolar material (PCM), is the principal microtubule organizing center in mammalian cells (Balczon, 1996). Most somatic mammalian cells contain a single centrosome, although exceptions to this general rule exist. To ensure that each of the progeny cells that results from a mitotic division receives a centrosome, it is imperative that the centrosome be replicated once, and only once, during each cell cycle. The cellular processes that regulate the timing and fidelity of centrosome doubling have not been completely defined. However, a variety of cellular systems have been developed that are allowing the events of centrosome duplication to be investigated, and several recent observations have begun to solidify our thinking about the centrosome replication cycle. Before describing these critical observations, it first would be useful to consider the complexity of the mammalian cell cycle.

II. The Cell Cycle in Somatic Mammalian Cells

The somatic cell cycle in multicellular organisms is the most complex of the cell cycles that have been described. Unlike the well-characterized cell cycles of *Xenopus, Drosophila,* and sea urchin embryos, which consist

Current Topics in Developmental Biology, Vol. 49

0070-2153/00 $35.00

exclusively of rapidly repeating cycles of S and M phases, cell cycles in somatic cells include intervening G_1 and G_2 stages. These additional cell cycle phases allow somatic cells to double their size during each cell cycle and also permit cells to perform various functions that ensure the fidelity of genome replication prior to the cell progressing into mitosis. Examples of regulatory networks that are present for controlling cell cycle progression in somatic cells include the various DNA damage checkpoints, the replication checkpoint, the rereplication block, and the spindle checkpoint (Elledge, 1996; Gorbsky, 1997). The functions of these regulatory systems are to monitor the accuracy and completeness of DNA synthesis, to ensure that events occur in the proper sequence during the cell cycle, and then to guarantee that chromosomes are segregated appropriately during division events.

Additional levels of regulatory complexity are placed on the cell cycle in somatic cells in multicellular organisms. This is a result of the important role played by the environment in driving cell cycle progression in multicellular organisms. In animals, such as humans, several different signals are received at the cell surface, including growth factor input, information from the extracellular matrix, and signals due to cell–cell contact. A somatic cell must interpret all of the growth stimulatory and inhibitory regulatory information received at the cell surface and decide whether to progress through the cell cycle or to arrest. This critical decision is made during the G_1 phase, and advancement through this junction in the cell cycle defines the restriction point.

As a somatic cell is progressing through interphase, the cell must coordinate centrosome replication with other cell cycle events. For example, the machinery responsible for replicating the centrosome must complete its task prior to progression to M phase; otherwise, the cell will be unable to form a bipolar spindle. However, the cell must also be able to suppress the overproduction of centrosomes that might occur during prolonged arrest while DNA damage is repaired; otherwise, a multipolar spindle would form during M phase. The formation of a multipolar spindle would result in the inappropriate segregation of chromosomes, an event that is proposed to contribute to genomic instability and cell transformation (Brinkley and Goepfert, 1998). Our understanding of the processes that allow somatic cells to integrate centrosome replication with the various other cell cycle progression events is only beginning to come into focus.

III. Electron Microscopic Descriptions of Centrosome Doubling, and the Cellular Events of Centrosome Replication

Initial studies that defined centrosome replication in somatic mammalian cells were performed more than 30 years ago (Brinkley and Stubblefield,

1970; Robbins *et al.,* 1968). These studies, and several subsequent investigations that utilized formation of progeny centrioles as a morphological landmark for centrosome doubling (Rattner and Phillips, 1973; Kuriyama and Borisy, 1981; Vorobjev and Chentsov, 1982; Alvey, 1985), provided a time frame for the process of centrosome doubling. These morphological studies determined that the first identifiable event of centrosome doubling occurs during G_1 phase, when the orthogonally arranged centrioles become disoriented and slightly separated (reviewed in Vandre and Borisy, 1989 and Hinchcliffe *et al.,* 1998). Although not definitively established, this event may occur as a result of growth factor signaling and passage through the restriction point (Sherline and Mascardo, 1982, 1984) and may ultimately set in motion the events that will allow centrosome doubling. By early S phase, a small structure, termed a procentriole, becomes apparent at the proximal end of each parent centriole. These procentrioles, which are oriented at right angles to the original parental centrioles, continue to elongate through S phase and early G_2 phase. By the transition from G_2 into M phase, centriole maturation is complete. Centriole maturation can be demonstrated using a monoclonal antibody that recognizes a relatively uncharacterized centriole component that is only associated with fully developed centrioles (Lange and Gull, 1995). This antibody reacts with parental centriole cylinders during the majority of interphase, and is not observed associated with elongating procentrioles; staining of the progeny centrioles only occurs during the later stages of G_2 when procentriole elongation is completed.

The pattern of semiconservation centriole inheritance that was proposed from ultrastructural observations was elegantly proven in cytological studies performed by Kochansky and Borisy (1990). These investigators microinjected a derivitized tubulin probe into cultured cells, and then analyzed the pattern of tubulin incorporation into the replicating centrioles. Microscopic examination demonstrated that as the centrosome doubled, each of the progeny centrosomes received a centriole pair comprising one parental centriole and one daughter centriole. These studies by Kochansky and Borisy (1990), as well as the numerous electron microscopy studies that were described previously, suggest that the parental centriole may direct the nucleation and/or assembly of the progeny centriole. However, the nature of this putative nucleating template remains to be determined, and the details concerning the events that lead to the activation of this putative structure at the G_1/S boundary of the cell cycle remain to be elucidated.

Although the morphological events of centriole duplication have been described, relatively little is known about the processes that result in doubling of the PCM during interphase. In one set of studies from this lab, Northern blot analysis demonstrated that the mRNA encoding the PCM autoantigen PCM-1 was synthesized in an oscillatory manner during the cell cycle, with highest levels of PCM-1 mRNA being detected in late G_1,

S, and early G_2 phases of the cell cycle (Balczon *et al.*, 1995). The levels of PCM-1 mRNA were undetectable during other cell cycle stages. The pattern of PCM-1 mRNA synthesis and degradation overlapped identically with the reported events for centriole duplication, suggesting that replication of the PCM components is coordinated with centriole doubling. A critical question that remains to be addressed is the nature of the events that occur to allow activation of the genes encoding PCM and centriole components at the appropriate times during the cell cycle so that the sequential events of centrosome replication indicated by centriole separation, procentriole assembly, procentriole elongation, and centriole/centrosome maturation occur in proper order. For example, it is possible that the genes encoding centrosome proteins may contain conserved elements that allow the genes for the centrosome subunits to be activated in unison during G_1 so that the individual protein subunits will be available for incorporation into the progeny centrosome as replication occurs. Alternatively, elements may exist that permit gene activation in sequence during different cell cycle stages so that centrosome proteins will only be synthesized as needed as the centrosome is duplicated. The most likely scenario is that the genes encoding centrosome subunits are activated in unison. The reason for this conclusion is that centrosome overproduction occurs in cells that are arrested near the G_1/S boundary of the cell cycle by treatment with either hydroxyurea or aphidicolin (Balczon *et al.*, 1995), indicating that all of the genes necessary to form centrosomes are active at this cell cycle stage. However, few of the genes encoding components of the mammalian centrosome have been cloned and analyzed, and this possibility has not been tested. It is worth mentioning that the genes encoding several of the protein subunits of the yeast spindle pole body, the functional homolog of the centrosome in yeast cells, have been cloned and characterized. The yeast genes that have been analyzed contain a conserved element that allows their activation during G_1 phase of the cell cycle (Winey and Byers, 1993). It will be interesting to see if similar types of regulatory sequences can be identified in the genes encoding mammalian centrosome components.

Once synthesized, centrosome subunits must be targeted to the site of the parental centrosome and then assembled into a functional organelle. The mechanism of centrosome assembly is unknown. However, recent studies from this lab have provided insight into the procedure used by somatic cells for targeting PCM components to the centrosome during centrosome replication (Balczon *et al.*, 1999). Studies by Kuriyama (1984) had suggested a role for microtubules in centrosome replication processes; this observation was reinvestigated by this lab, and our studies identified a role for microtubules in targeting of PCM subunits to the centrosome during centrosome replication. These studies demonstrated that once synthesized, PCM subunits colocalized along cellular microtubules with dynein/

dynactin motor complexes (Balczon *et al.,* 1999). Moreover, microtubule pelleting studies determined that dynein/dynactin and the centrosome autoantigen PCM-1 copelleted with microtubules from cytosolic extracts prepared from CHO cells, and the pelleting of PCM-1 with microtubules was inhibited by immunodepleting dynactin from the extracts. These observations lead to a model for centrosome subunit targeting where centrosome components are transported along cellular microtubules by the dynein/dynactin motor complex to the parental centrosome during centrosome doubling events. Once the subunits reach the parental centrosome, they then are assembled into a functional organelle. These observations are in contrast to those reported for centrosome assembly following fertilization (Schatten, 1994) and in cell extracts prepared from unfertilized eggs (Schnackenberg *et al.,* 1998) where centrosome assembly processes have been demonstrated to occur in the absence of microtubules. Presumably, unfertilized eggs contain large pools of centrosome subunits that are free to diffuse within the cytosol during centrosome assembly. However, centrosome subunits must be synthesized during each cell cycle in somatic cells (Phillips and Rattner, 1976), and these proteins are produced in relatively low quantities in mammalian cells. Therefore, diffusion would not appear to be an efficient mechanism for targeting subunits to the centrosome during centrosome replication in somatic cell types. Coupling centrosome components to a minus-end-directed motor complex might help to ensure that the subunits are transported to the centrosome. Moreover, this process presumably would allow a localized increase in the concentration of centrosome subunits that might be critical for any assembly processes that need to occur during centrosome doubling. Although this model does not eliminate other possible roles for microtubules in centrosome doubling (e.g., microtubule assembly most certainly must occur for centriole doubling events), this scheme does provide a potential explanation for the observation by Kuriyama (1984) that microtubule inhibitors block centrosome replication. Clearly, much more remains to be learned about the events of centrosome subunit targeting and the mechanism of centrosome assembly during interphase.

IV. Spindle Pole Body Doubling in Yeast Cells

Much of what is known about the replication of centrosomes in somatic cells has been derived from studies performed in unicellular organisms, such as yeasts. These organisms are useful for investigating regulation of doubling of the microtubule organizing center for a variety of reasons. First, unlike embryonic cells, the yeast cell cycle contains G_1 and G_2 phases, which makes yeasts reliable indicators of cell cycle regulatory events that occur in somatic mammalian cells. Second, although yeasts do not contain a centrosome, they do contain a functionally homologous organelle termed the spindle pole body (SPB) (Snyder, 1994). Third, yeasts are amenable to genetic analyses that allow the identification of useful mutants that are disrupted in the ability to complete SPB duplication accurately. In addition, although cell division in yeasts can be regulated by environmental factors, such as nutrient availability, there is no input from environmental signals that occur due to cell–cell and cell–ECM interactions (as occurs in somatic cells in multicellular organisms), which removes a layer of regulatory complexity from the process of SPB replication. Finally, the recent characterization of the protein composition of the core SPB (Adams and Kilmartin, 1999) and the completion of the yeast genome sequencing project should allow rapid advancements in our understanding of SPB assembly processes. For these reasons, the analysis of SPB replication in yeasts will likely provide critical insights into the process of centrosome doubling in somatic cells.

The SPB has a morphological organization distinct from the animal cell centrosome, and, although it is discussed in more detail elsewhere in this volume, a brief summary of SPB architecture will be useful. The SPB is a trilaminar structure that traverses the nuclear envelope in the yeast *S. cerevisiae.* The dense central plaque, which is embedded in the nuclear envelope, is abutted on the cytoplasmic and intranuclear surfaces by additional plaques. Functionally, the cytoplasmic plaque nucleates the cytoplasmic array of microtubules while the intranuclear plaque of the SPB assembles the mitotic spindle microtubules. A structure termed the half-bridge lies adjacent to the central plaque within the plane of the nuclear envelope. Several different proteins have been localized to specific domains of the SPB, and the roles of each of these proteins are being defined (Balczon, 1996; Winey and Byers, 1993; Snyder 1994; see also chapter by Francis and Davis in this volume).

The initial events of SPB replication occur early during G_1 phase in *S. cerevisiae.* Ultrastructural analyses have determined that a structure termed the "satellite" appears in the cytoplasm directly adjacent to the bridge structure (Winey and Byers, 1993; Snyder, 1994; Adams and Kilmartin, 1999). Biochemical studies have demonstrated that formation of the satellite occurs in the absence of cdk activity and in cells arrested early in G_1 by treatment with the mating pheromone α-factor (Snyder, 1994; Adams and Kilmartin, 1999). Soon after START, the cell cycle point in yeast that is analagous to the restriction point in higher eukaryotic cells and the point when cdk activity is detectable, SPB assembly is completed. By the G_1/S transition, the duplicated SPB is inserted into the nuclear envelope at the distal end of the bridge structure (Snyder, 1994; Francis and Davis, this volume).

The observations highlight several important events that occur to the SPB during G_1 phase of the cell cycle in *S. cerevisiae.* First, the initial events of SPB duplication, which are apparent in the form of an assembled satellite structure, occur in the absence of cdk activity, whereas the later events of the process occur when this kinase is active. Similarly, recent studies have also demonstrated that cdk activity is necessary for centrosome doubling in higher eukaryotes (Hinchcliffe *et al.,* 1999; Lacey *et al.,* 1999; Matsumoto *et al.,* 1999), and elucidating the role of cdk in SPB replication should provide critical information concerning the role of cdks in centrosome doubling. Second, there does not appear to be a mixing of parts between the parental and progeny MTOC in yeasts, although the half-bridge may play a key role in initiating SPB assembly (Adams and Kilmartin, 1999). As mentioned previously, centrosome replication in mammalian cells occurs by a semiconservative mechanism with each of the progeny centrosomes formed during doubling events receiving a parental centriole and a daughter centriole. In *S. cerevisiae,* there does not appear to be a semiconservative mixing of parts between the parental SPB and the assembled SPB. Moreover, studies have demonstrated that, in budding yeast, the progeny SPB preferentially enters the bud while the parental SPB remains in the parental cell (Vallen *et al.,* 1992). Finally, these observations concerning the formation of the satellite and maturation to an SPB suggest that a nucleating component for SPB formation may be associated with the half-bridge (Adams and Kilmartin, 1999). If this is correct, then identifying the SPB nucleating activity within the half-bridge may have important implications for understanding centrosome replication in somatic mammalian cells. The demonstration that centrosomes originate from preexisting centrosomes in higher eukaryotes suggests the existence of a nucleating template of some type that directs the assembly of the progeny centrosome during doubling events. Identifying the factor that organizes the SPB satellite may provide important clues that will be useful for understanding centrosome replication and may allow the identification of a homologous structure in higher eukaryotes.

Mutation analyses have identified several proteins whose activities appear to be critical for the early events of SPB replication (Byers and Goetsch, 1975). The first mutation that was identified as having a defect in SPB doubling was the Cdc31 mutant, with studies determining that the CDC31 gene encodes a calcium binding protein with homology to centrin. Immunoelectron microscopy has localized the Cdc31 protein to the half-bridge structure (Spang *et al.,* 1993), and Cdc31 mutants are unable to form satellites (Snyder, 1994). These results demonstrate that Cdc31 activity is essential for early G_1 SPB replication events. A second mutant, Kar1, shows a similar phenotype, and genetic analyses suggest that Kar1 and Cdc31 act in a common pathway (Winey and Byers, 1993). An additional gene that appears to be involved in the early events of SPB duplication is MPS1. The MPS1 gene product encodes a protein kinase whose activity is critical for SPB replication, and the MPS1 kinase also appears to have a role in maintaining the spindle checkpoint (Weiss and Winey, 1996). Studies have shown that the MPS1 gene product is involved in maintaining satellite integrity, and shifting Mps1 cells to restrictive temperature causes loss of the satellite and failure of SPB doubling (Winey *et al.,* 1991). Therefore, Mps1 kinase activity is probably required for the maturation of the satellite to a SPB. Finally, the activities of heat shock proteins/chaperones appear to be essential for early stages of SPB doubling. Mutant cells that were unable to produce heat shock proteins appropriately arrested with a single SPB containing an elongated half-bridge structure (Zarzov *et al.,* 1997). Interestingly, the half-bridge nucleated microtubules in these mutant cells, suggesting that proteins involved in microtubule nucleation processes associated with the half-bridge in these mutants. The role of heat shock proteins and other chaperones in the early events of SPB doubling remains to be established, although it is possible that these proteins are required either for the synthesis of the SPB proteins or for the assembly of the SPB subunits into a functional organelle.

Mutants also have been identified whose gene products appear to be essential for latter events of SPB replication during G_1 phase. An important aspect of these additional mutants is that they have allowed observation of intermediates in the SPB doubling process. For example, the Ndc1 and Mps2 mutants progress much further with SPB replication than the previously described mutants, but fail to form a mature SPB. Specifically, these two mutants form an incomplete SPB that fails to be inserted into the nuclear envelope (Winey *et al.,* 1991, 1993). Morphological studies have demonstrated that SPBs in Mps2 and Ndc1 mutant cells consist of the central and outer plaques only, with the outer plaque capable of nucleating microtubules. As stated earlier, this incomplete SPB is not inserted into the nuclear envelope, and, as a result, intranuclear microtubules do not form off of this SPB during mitosis. These results demonstrate that Ndc1 and Mps2 are required for appropriate assembly of the SPB. Whether the activity of either, or both, of these gene products is required for insertion of the SPB into the nuclear envelope remains to be determined. However, in support of this possibility, the NDC1 gene has been cloned and it has been determined that Ndc1 protein contains six or seven potential membrane-spanning domains (Winey and Byers, 1993; Winey *et al.,* 1993).

An interesting yeast mutant is the strain mutated in the gene ESP1. Cells mutated in this gene exhibit a phenotypic overproduction of SPBs (Winey and Byers, 1993); hence the name Esp mutants (for "extra spindle poles"). Each of the SPBs that forms in these mutants appears morphologically and functionally normal, and recent studies (Ciosk *et al.,* 1998) suggest the mechanism of SPB overproduction in Esp1 mutants. The Esp1 protein appears to be involved in regulating the proteolytic destruction of proteins at the metaphase-to-anaphase transition and, as a result, the mutant cells are unable to segregate chromosomes during M phase. The mutant cells, however, do advance to the following interphase and produce additional SPBs. Although these studies argue against a direct role for Esp1 in SPB replication, the results do suggest that yeast cells lack an endogenous counting mechanism that limits the number of SPBs in a cell.

An important observation from the preceding studies is that cells will transit from interphase to mitosis even in the absence of appropriate SPB doubling. For example, Esp1 mutants form multipolar spindles while many of the mutants that were described earlier enter M phase in the absence of SPB replication and form an aberrant spindle. These results argue against an essential role for accurate SPB doubling in the completion of interphase. It is worth mentioning that similar observations concerning centrosome replication have been made in higher eukaryotic cells, and the formation of multipolar spindles is a relatively common occurrence in various cultured cell lines and tumor tissues (Brinkley and Goepfert, 1998). Taken together, these observations argue against the presence of a cell cycle checkpoint system that monitors either SPB or centrosome number and replication. Moreover, these results demonstrate that accurate SPB and centrosome doubling are not an essential prerequisite for completion of interphase and transition from interphase to mitosis.

The role of the cdc2/cdc28 kinase (or cdk1) in SPB doubling events remains to be defined completely. As mentioned previously, yeast cells only advance as far as the "satellite" stage of SPB replication prior to passage through START, which corresponds to activation of cdk. Later events in SPB maturation occur after activation of the kinase and prior to the onset of DNA synthesis. These cytological observations suggest that active cdk1 is required for completion of SPB doubling, and this supposition is supported by the recent demonstrations that a member of the cdk family, cdk2 kinase, is essential for centrosome doubling in vertebrate cells (Hinchcliffe *et al.*, 1999; Lacey *et al.*, 1999; Matsumoto *et al.*, 1999). Provided cdk1 activity is required for SPB doubling, it might be possible to identify a cdc2/cdc28 mutation that would permit cell cycle progression while inhibiting SPB replication. Isolation of such a putative mutant might permit the identification of downstream targets that are regulated by cdk1 during SPB doubling.

In summary, yeast cells represent a simple example of a somatic cell cycle. During G_1 phase, the SPB, which serves as a functional homolog of the centrosome, undergoes a conservative replication process. Understanding the genetic regulation of SPB doubling should provide critical information for elucidating the process of centrosome replication in somatic cells in higher eukaryotes. The recent completion of the sequencing of the yeast genome combined with the successful identification of the core proteins that comprise the SPB (Adams and Kilmartin, 1999; Francis and Davis, this volume) should lead to rapid advances in our understanding of the genetic and biochemical processes that occur during G_1 phase to direct SPB replication events.

V. Events in G_1 That Trigger Centrosome Replication in Somatic Mammalian Cells

The earliest recognizable events of centrosome replication occur during G_1 phase in somatic mammalian cells, and it is presumed that these events are set in motion following activation of the machinery that is responsible for regulating centrosome replication. Whole mount EM studies by Kuriyama and Borisy (1981) demonstrated that the centrioles undergo a subtle change in orientation during G_1 phase. The most likely candidates for initiating these processes are extracellular signals. Although it has not been shown experimentally, it seems reasonable to assume that the environmental signals responsible for activating centrosome doubling in somatic cells would vary from cell to cell depending upon the phenotypes of individual cells. However, to date, the only extracellular ligand that has been shown to affect centrosome doubling is EGF. The role of EGF in activating centrosome replication was demonstrated initially by Sherline and Mascardo (1982, 1984), who cultured various mammalian cell lines in serum-free medium and then added several different growth factors to the cultures. Sherline and Mascardo (1982, 1984) observed that EGF treatment induced centriole separation in several cell types, which may have been an indicator of the earliest events of centrosome doubling. These observations subsequently were extended by Balczon *et al.* (1995), who demonstrated that EGF drove multiple rounds of centrosome replication in CHO cells that were arrested in early S phase by treatment with either hydroxyurea or aphidicolin. These results suggest not only that EGF triggers early centriole separation, but that EGF has the capacity to activate the entire repertoire of events necessary for centrosome doubling in many cell types.

The cellular pathways that are activated following receptor stimulation remain to be defined. However, recent studies suggest that the ultimate intracellular target may be the cyclin/cdk2 complexes (Hinchcliffe *et al.*, 1999; Lacey *et al.*, 1999; Matsumoto *et al.*, 1999; Meraldi *et al.*, 1999). Using *Xenopus* egg extracts, cultured *Xenopus* cells, and cultured mammalian cells, it has been demonstrated that the early events of centrosome replication, such as centriole separation (Lacey *et al.*, 1999), as well as actual centrosome doubling and maturation (Hinchcliffe *et al.*, 1999; Matsumoto *et al.*, 1999; Meraldi *et al.*, 1999), depend upon the activation of cdk2. The major difference between the studies utilizing embryonic cells and somatic cells is the demonstration the cyclin E is the cdk2 partner in embryonic cells (Hinchcliffe *et al.,* 1999) while cyclin A is the cdk2 activating subunit in somatic cells (Lacey *et al.,* 1999; Matsumoto *et al.,* 1999; Meraldi *et al.,* 1999). This may simply indicate differences between cell cycles in embryonic and somatic cells. For example, in embryonic cells, cyclin E is responsible for activating DNA synthesis, whereas cyclin A fulfills that role in somatic cell types (Hartley *et al.,* 1997). In addition, in embryonic cells cyclin E levels do not oscillate during the cell cycles (Knoblich *et al.,* 1994). This observation fits nicely with the demonstration that rounds of centrosome replication will occur in the absence of protein synthesis in embryonic cells (Gard *et al.,* 1990; Sluder *et al.,* 1990). However, a few questions need to be addressed before the role of cdk2 in centrosome replication can be appreciated fully. First, the targets of the activated complex need to be identified. Elucidating the targets of the active cyclin/cdk2 complex will provide essential information concerning the events that allow formation of the progeny centrioles and the duplication of the entire centrosome complex. In addition, an inconsistency concerning the intracellular location of cyclin A and E must be resolved. Specifically, the events of centrosome assembly occur in the cytoplasm, whereas cyclins A and E are thought to be intranuclear proteins. For example, Pines and Hunter (1991) have demonstrated that the cdk associated form of cyclin A is nuclear. Likewise, although Hinchcliffe *et al.* (1999) have presented immunofluorescence data demonstrating an enrichment of cyclin E at centrosomes, others have shown cyclin E to be an intranuclear protein (Knoblich *et al.,* 1994; Ohtsubo *et al.,* 1995; Moore *et al.,* 1999). In order to define the role of cyclin/cdk2 complexes in centrosome replication, it must be firmly established whether the complexes formed between cdk2 and cyclins A/E are intranuclear, cytoplasmic, or both.

In somatic cells, the intracellular pathway that triggers centrosome doubling following receptor activation remains to be established. To be precise, what sequence of events occurs in G_1 between the moment when EGF (or some other ligand) binds to its cell surface receptor and the time when cyclin A/cdk2 is activated? Most likely, several different kinases must be turned on during G_1 to allow the early events needed for centrosome doubling. This possibility is supported by the observations that a number of kinases, phosphatases, and other signaling molecules localize to centrosomes (Brinkley and Goepfert, 1998). Moreover although the actual target of the signal transduction pathway that extends from the cell surface to the centrosome to set in motion centrosome replication remains to be identified, an argument can be made for a kinase called Nek2. Nek2 is a serine–threonine kinase that induces centrosome splitting when overexpressed in mammalian cells (Fry *et al.,* 1998). Although these results argue that Nek2 is the target of extracellular signals that induce centrosome separation and the earliest events of centrosome doubling during G_1, biochemical analyses have shown that Nek2 levels and activity are lowest during G_1 phase and peak during G_2 phase (Fry *et al.*, 1995). A possible explanation for these apparently contradictory results is that the low levels of Nek2 in G_1 cells may be sufficient to drive centriole separation events, with elevated levels later in the cell cycle being required for an additional function of Nek2 such as triggering centrosome separation during spindle formation. Additional studies will be required before the role of Nek2 in centrosome doubling can be defined.

Additional intracellular kinases, besides Nek2 and cdk2, appear to be involved in the decision to replicate centrosomes by somatic cells. For example, the BTAK/aurora 2 kinase localizes to centrosomes. Like Nek2, BTAK aurora 2 activity appears to be critical for centrosome separation during spindle formation (Glover *et al.*, 1995), and protein and activity profiles for aurora kinase show low levels in G_1 and high levels at G_2/M (see Fry *et al.*, this volume). However, overexpression of BTAK/aurora 2 leads to amplification of centrosomes in cultured cells (Zhou *et al.*, 1998), suggesting a role for this enzyme in regulating commitment to induce centrosome doubling processes in cells. As with Nek2, it remains to be established whether BTAK makes up part of the G_1 signaling network that leads to the activation of cyclin A/cdk2 in somatic cells. Finally, the localization of other kinases to the centrosome region (Brinkley and Goepfert, 1998) suggests that these additional kinases will play a critical role in regulating centrosome number. However, the significance of these localization data remain to be established. One of the reasons for confusion is that the extracellular signaling networks that control cell cycle progression are only beginning to be elucidated. Until we gain an increased understanding of how input received from cell–cell adhesion, cell–matrix attachment, and growth factors is interpreted by a cell to induce progression through the cell cycle, it will be difficult to establish the role of various kinases in regulating the decision of a cell to commence the centrosome replication process.

A simple prediction that can be made from the previously outlined studies demonstrating the importance of kinases in controlling centrosome replication is that phosphatase activity should be equally important in regulating centrosome doubling events. For example, if kinases are not inactivated once they complete their respective functions in the centrosome replication process, one might predict that centrosome overproduction would occur. In fact, recent studies from Cheng *et al.* (1998) have demonstrated that treatment of various types of cultured cells with phosphatase inhibitors resulted in the overproduction of centrosomes and the formation of multipolar spindles at mitosis. These studies further highlight the importance of intracellular signaling networks in regulating centrosome number.

Because there does not appear to be a block or a checkpoint mechanism to limit centrosome doubling to a single event during the cell cycle, appropriate activation and repression of the intracellular machinery responsible for interpreting extracellular stimuli must be critical for controlling centrosome number.

The final cellular component that deserves consideration as a potential regulator of centrosome replication during G_1 phase in multicellular animals is the p53 tumor suppressor protein (see Lingle and Salisbury, this volume). Studies have shown that p53 is a critical regulator of the G_1 DNA damage checkpoint, although a role for p53 in G_2 checkpoint activity cannot be discounted (Levine, 1997). In addition to this cellular role, analyses of primary cultures of fibroblasts generated from tissues obtained from p53 −/− transgenic mice (Fukasawa *et al.*, 1996) and studies of papillomas that arose in transgenic mice expressing a mutant form of p53 (Wang *et al.*, 1998) determined that these cells overproduced centrosomes. It is not clear how p53 might be involved in regulating centrosome number, although one possibility is that p53 participates in centrosome replication directly. Alternatively, the absence of functional p53 may allow the buildup of mutations in genes encoding proteins whose activity is important for regulating centrosome doubling. Finally, the absence of a functional p53-mediated G_1 DNA damage checkpoint may result in cells arresting for prolonged periods at other points in the cell cycle while they attempt to repair DNA damage. As mentioned previously, the synthesis of the centrosome protein PCM-1 occurs in an oscillatory manner during the cell cycle, and arresting CHO cells for prolonged periods at a cell cycle stage when PCM-1 (and presumably other centrosome subunits) synthesis is occurring results in the overproduction of centrosomes. The absence of an intact G_1 checkpoint in p53 null cells may allow the cells to progress further into the cell cycle before being arrested in an effort to repair DNA damage. This arrest stage may be at a point in the cell cycle that would permit overproduction of centrosome subunits and assembly of those subunits into functional centrosomes. Clearly, considerable additional work will be necessary before the role of p53 in the centrosome replication process can be established.

VI. Summary

Proper cell division requires that the cell be able to form a bipolar spindle during mitosis. To achieve this, the centrosome must be replicated accurately during interphase. Our understanding of the mechanisms that allow centrosome doubling to be coordinated with other cell cycle progression processes is advancing at a rapid pace. Several different experimental systems have been developed that are allowing detailed studies of centrosome replication. For example, the identification of mutants in yeast that are unable to duplicate the SPB accurately during interphase has provided important insights concerning centrosome duplication. In addition, intact embryonic cells and extracts prepared from unfertilized eggs are powerful tools for investigating the molecular regulation of centrosome doubling during the cell cycle. Many of the observations from these embryonic systems are directly applicable to understanding centrosome doubling in somatic cells. Finally, transgenic mouse models and cultured mammalian cell systems have been developed for analyzing the regulation of centrosome doubling in cells with more complex cell cycles. As our knowledge of the cell cycle advances, particularly our understanding of the intricate series of events that must occur for somatic cells to traverse G_1 phase, it should be possible to use the systems that have been developed to determine how the replication of the centrosome is coordinated with other cell cycle progression processes. The next few years should see rapid advances in our understanding of this critical cell biological process.

References

Adams, I. R., and Kilmartin, J. V. (1999). Localization of core spindle pole body components during SPB duplication in *Saccharomyces cerevisiae*. *J. Cell Biol.* **145,** 809–823

Alvey, P. L. (1985). An investigation of the centriole cycle using 3T3 and CHO cells. *J. Cell Sci.* **78,** 147–162.

Balczon, R. (1996). The centrosome in animal cells and its functional homologs in plants and yeast cells. *Int. Rev. Cytol.* **169,** 25–82.

Balczon, R., Bao, L., Zimmer, W. E., Brown, K. Zinkowski, R. P., and Brinkley, B. R. (1995). Dissociation of centrosome replication events from DNA synthesis and mitotic division in hydroxyurea arrested Chinese hamster ovary cells. *J. Cell Biol.* **130,** 105:105–115.

Balczon, R., Varden, C., and Schroer, T. (1999). A role for microtubules in centrosome doubling in Chinese hamster ovary cells. *Cell Motil. Cytoskel.* **20,** 121–135.

Brinkley, B. R., and Goepfert, T. M. (1998). Supernumary centrosomes and cancer: Boveri's hypothesis resurrected. *Cell Motil. Cytoskel.* **41,** 281–288.

Brinkley, B. R., and Stubblefield, E. (1970). Ultrastructure and interaction of the kinetochore and the centriole in mitosis and meiosis. *Adv. Cell Biol.* **1,** 119–185.

Byers, B., and Goetsch, L. (1975). Behavior of the spindle plaques in the cell cycle and conjugation of *Saccharomyces cerevisiae*. *J. Bacteriol.* **124,** 511–523.

Ciosk, R., Zachariae, W., Michaelis, C., Shevchenko, A., Mann, M., and Nasmyth, K. (1998). An ESP1/PDS1 complex regulates loss of sister chromatid cohesion at the metaphase to anaphase transition in yeast. *Cell,* **93,** 1067–1076.

Cheng, A., Balczon, R., Zuo, Z., Koons, J. S., Walsh, A. H., and Honkanen, R. E. (1998). Fostriecin-mediated G_2-M phase growth arrest correlates with abnormal centrosome replication, the formation of aberrant mitotic spindles, and the inhibition of serine/threonine protein phosphatase activity. *Canc. Res.* **58,** 3611–3619.

Elledge, S. J. (1996). Cell cycle checkpoints: preventing an identity crisis. *Science* **274,** 1664–1672.

Fry, A. M., Meraldi, P., and Nigg, E. A. (1998). A centrosomal function for the human Nek2 protein kinase, a member of the NIMA family of cell cycle regulations. *EMBO J.* **17,** 470–481.

Fry, A. M., Schultz, S. J., Barrek, J., and Nigg, E. A. (1995). Substrate specificity and cell cycle regulation of the Nek2 protein kinase, a potential human homolog of the meiotic regulator NIMA of *Aspergillus nidulans. J. Biol. Chem.* **2705,** 12899–12905.

Fukasawa, K., Choi, T., Kuriyama, R., Rulong, S., and Vande Woude, G. F. (1996). Abnormal centrosome amplification in the absence of p53. *Science* **271,** 1744–1747.

Gard, D. L., Hafezi, S., Zhang, T., and Doxsey, S. J. (1990). Centrosome duplication continues in cycloheximide-treated *Xenopus* blastulae in the absence of a detectable cell cycle. *J. Cell Biol.* **110,** 2033–2042.

Glover, D. M., Leibovitz, M. H., McLean, D. A., and Perry, H. (1995). Mutations in aurora prevent centrosome separation leading to the formation of monopolar spindles. *Cell* **81,** 95–105.

Gorbsky, G. J. (1997). Cell cycle checkpoints: Arresting progress in mitosis. *Bioessays* **19,** 193–197

Hartley, R. S., Sible, J. C., Lewellyn, and Maller, J. L. (1997). A role for cyclin E/cdk2 in the timing of the midblastula transition in *Xenopus* embryos. *Dev. Biol.* **188,** 312–321.

Hinchcliffe, E. H., Cassels, G. O., Rieder, C. L., and Sluder, G. (1998). The coordination of centrosome reproduction with nuclear events of the cell cycle in the sea urchin zygote. *J. Cell Biol.* **140,** 1417–1426.

Hinchliffe, E. H., Li, C., Thompson, E. A., Maller, J. L., and Sluder, G. (1999). Requirement of cdk2-cyclin E activity for repeated centrosome reproduction in *Xenopus* egg extracts. *Science* **283,** 851–854.

Knoblich, J. A., Sauer, K., Jones, L., Richardson, H., Saint, R., and Lehner, C. F. (1994). Cyclin E controls S phase progression and its down-regulation during *Drosophila* embyogenesis is required for the arrest of cell proliferation. *Cell* **77,** 107–120.

Kochansky, R., and Borisy, G. G. (1990). Mode of centriole ducplication and distribution. *J. Cell Biol.* **110,** 1599–1605.

Kuriyama, R. (1984). Activity and stability of centrosomes in Chinese hamster ovary cells in nucleation of microtubules in vitro. *J. Cell Sci.* **66,** 277–295.

Kuriyama, R., and Borisy, G. G. (1981). Centriole cycle in Chinese hamster ovary cells as determined by whole mount electron microscopy. *J. Cell Biol.* **91,** 814–821.

Lacey, K. R., Jackson, P. K., and Stearns, T. (1999). Cyclin-dependent kinase control of centrosome duplication. *Proc. Natl. Acad. Sci. USA* **96,** 2817–2822.

Lange, B. M. H., and Gull, K. (1995). A molecular marker for centriole maturation in the mammalian cell cycle. *J. Cell Biol.* **130,** 919–927.

Levine, A. J. (1997). P53, the cellular gatekeeper for growth and division. *Cell* **88,** 323–331.

Matsumoto, Y., Hayashi, K., and Nishida, E. (1999). Cyclin-dependent kinase 2 (Cdk2) is required for centrosome duplication in mammalian cells. *Curr. Biol.* **9,** 429–432.

Meraldi, P., Lukas, J., Fry, A. M., Bartek, J., and Nigg, E. A. (1999). Centrosome duplication in mammalian somatic cells requires E2F and cdk2/cyclin A. *Nature Cell Biol.* **1,** 88–93.

Moore, J. D., Yang, J., Truant, R., and Kornbluth, S. (1999). Nuclear import of cdk/cyclin complexes: Identification of distinct mechanisms for import of cdk2/cyclin E and cdc2/ cyclin B1. *J. Cell. Biol.* **144,** 213–224.

Ohtsubo, M., Theodoras, A. M., Schumacher, J., Roberts, J. M., and Pagano, M. (1995). Human cyclin E, a nuclear protein essential for the G_1 to S phase transition. *Mol. Cell. Biol.* **15,** 2612–2624.

Phillips, S. G., and Rattner, J. B. (1976). Dependence of centriole formation on protein synthesis. *J. Cell Biol.* **70,** 9–19.

Pines, J., and Hunter, T. (1991). Human cyclins A and B1 are differentially located in the cell and undergo cell cycle–dependent nuclear transport. *J. Cell Biol.* **115,** 1–17.

Rattner, J. B., and Phillips, S. G. (1973). Independence of centriole formation and DNA synthesis. *J. Cell Biol.* **57,** 359–372.

Robbins, E., Jentzsch, G., and Micali, A. (1968). The centriole cycle in synchronized HeLa cells. *J. Cell Biol.* **36,** 329–339.

Schatten, G. (1994). The centrosome and its mode of inheritance—the reduction of the centrosome during gametogenesis and its restoration during fertilization. *Dev. Biol.* **165,** 299–335

Schnackenberg, B. J., Khodjakov, A., Rieder, C. L., and Palazzo, R. E. (1998). The disassembly and reassembly of functional centrosomes in vitro. *Proc. Natl. Acad. Sci. USA* **95,** 9295–9300.

Sherline, P., and Mascardo, R. N. (1982). Epidermal growth factor-induced centrosomal separation: Mechanism and relationship to mitogenesis. *J. Cell Biol.* **95,** 316–322.

Sherline, P., and Mascardo, R. N. (1984). Co-ordinate control of centrosomal separation and DNA synthesis by growth regulators. *Exp. Cell Res.* **153,** 109–120.

Sluder, G., Miller, F. J., Cole, R., and Rieder, C. L. (1990). Protein synthesis and the cell cycle: Centrosome reproduction in sea urchin eggs is not under translational control. *J. Cell Biol.* **110,** 2025–2032

Snyder, M. (1994). The spindle pole body of yeast. *Chromosoma* **103,** 369–380.

Spang, A., Courtney, I., Fackler, U., Matzner, M., and Schiebel, E. (1993). The calcium-binding protein Cdc31 of *Saccharomyces cerevisiae* is a component of the half bridge of the spindle pole body. *J. Cell Biol.* **123,** 405–416.

Vallen, E. A., Schernson, T. Y., Roberts, T., Van Zee, K., and Rose, M. D. (1992). Asymetric mitotic segregation of the yeast spindle pole body. *Cell* **69,** 505–515.

Vandre, D. D., and Borisy, G. G. (1989). The centrosome cycle in animal cells. *In* "Mitosis: Molecules and Mechanisms" (J. S. Hyams and B. R. Brinkley, Eds.), pp. 39–75. Academic Press, San Diego

Vorobjev, I. A., and Chentsov, Y. S. (1982). Centrioles in the cell cycle. I. Epithelial cells. *J. Cell Biol.* **93,** 938–949.

Wang, X-J., Greenhalgh, D. A., Jiang, A., He, D., Zhong, L., Medina, D., Brinkley, B. R., and Roop, D. R. (1998). Expression of a p53 mutant in the epidermis of transgenic mice accelerates chemical carcinogenesis. *Oncogene* **17,** 35–45.

Weiss, E., and Winey, M. (1996). The *Saccharomyces cerevisiae* spindle pole body duplication gene MPS1 is part of the mitotic checkpoint. *J. Cell Biol.* **132,** 111–123.

Winey, M., and Byers, B. (1993). Assembly and functions of the spindle pole body in budding yeast. *Trends Genet.* **9,** 300–304.

Winey, M., Goetsch, L., Baum, P., and Byers, B. (1991). MPS1 and MPS2: Novel yeast genes defining distinct steps of spindle pole body duplication. *J. Cell Biol.* **114,** 745–754.

Winey, M., Hoyt, A., Chan, C., Goetsch, L., Zho, Z., Botstein, D., and Byers, B. (1993). Nuclear envelope localized Ndc1 protein is required for yeast spindle pole body duplication. *J. Cell Biol.* **122,** 743–751

Zarzov, P., Boucherie, H., and Mann, C. (1997). A yeast heat shock transcription factor (Hsf1) mutant is defective in both Hsc82/Hsp82 synthesis and spindle pole body duplication. *J. Cell Sci.* **110,** 1879–1891.

Zhou, H., Kuang, J., Zhang, L., Kuo, W-L., Gray, J. W., Sahin, A., Brinkley, B. R., and Sen, S. (1998). Tumor amplified kinase STK15/BTAK induces centrosome amplification, aneuploidy, and transformation, *Nature Genet.* **20,** 189–193.

12

The Coordination of Centrosome Reproduction with Nuclear Events during the Cell Cycle

Greenfield Sluder and *Edward H. Hinchcliffe*
Department of Cell Biology
University of Massachusetts Medical School
Worcester, Massachusetts 01605

I. Introduction

The centrosome is the ensemble of structures that nucleates the interphase array of cytoplasmic microtubles and assembles the poles of the mitotic spindle (for comprehensive reviews on centrosome structure, composition, and function, see Brinkley, 1985, and Kellogg *et al.,* 1994). Since the equal segregation of daughter chromosomes depends upon the spindle being strictly bipolar, it is of obvious importance for the cell to coordinate the events of centrosome doubling with nuclear events during the cell cycle, thereby limiting centrosome copy number at the start of mitosis to two and only two. If the centrosome fails to fully duplicate before the onset of mitosis, the cell may eventually return to interphase without dividing and becomes polyploid. If the centrosome duplicates more than once in a cell cycle, a multipolar spindle may be assembled, and the chromosomes will be unequally distributed to the daughter cells (Fig. 1). Although many daughter cells with abnormal chromosome number will be inviable, progeny that accumulate chromosomes with growth promoting genes and/or lose chromosomes containing tumor suppressor genes will produce a population of cells with aggressive growth characteristics (reviewed in Orr-Weaver and Weinberg, 1998; Brinkley and Goepfert, 1998). Indeed, many human

Current Topics in Developmental Biology, Vol. 49

0070-2153/00 $35.00

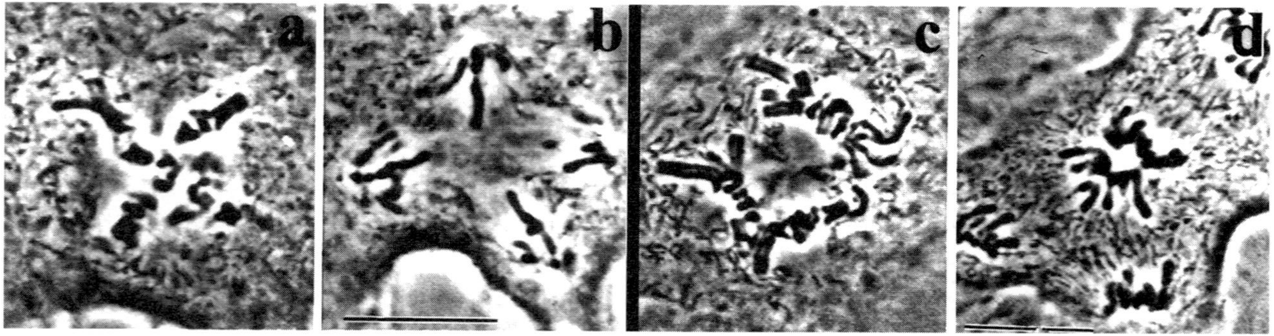

Figure 1 Random chromosome segregation by tetrapolar spindles in PtK cells. (a-b) First cell: This tetrapolar cell has all four poles located at the cell periphery. (a) Metaphase, all chromosomes bioriented to pairs of spindle poles. (b) Late anaphase, the normal chromosome compliment is randomly distributed to four spindle poles. (c-d) Second cell: This cell has one spindle pole located at the cell center and three poles at the cell periphery. (c) Anaphase onset. (d) Telophase. Phase contrast optics. Bar in (b) 20 μm.

tumor cells, which are genetically unstable, have abnormally high numbers of centrosomes (Pihan *et al.,* 1998; Lingle *et al.,* 1998). This is a real problem because neither somatic or embryonic cells appear to have a checkpoint for the metaphase–anaphase transition that monitors spindle bipolarity, as long as all chromosomes establish bipolar attachments to the spindle (Sluder *et al.,* 1997).

II. The Events of Centrosome Reproduction

At the end of mitosis each daughter cell inherits a single centrosome, and by the onset of the next mitosis it contains just two centrosomes. This precise doubling of the interphase centrosome in preparation for mitosis is called centrosome reproduction. At the functional level centrosome reproduction in higher animal cells consists of at least four morphological events that are distinct and experimentally separable (Mazia *et al.,* 1960; Sluder and Begg, 1985). Figure 2 illustrates the sequence of these events. The first event is the *splitting* apart of the pair of inherited mother–daughter centrioles, which coincides with the loss of their orthogonal relationship within the centrosome (also referred to as centriole disorientation; see Kuriyama and Borisy, 1981). The splitting event is followed by *duplication,* in which the centrosome goes from a single unit capable of organizing one centrosome to two units that have the ability to form two centrosomes. Duplication is correlated with the formation of daughter centrioles at right angles to each mother or mature centriole. The third event entails the splitting apart or *disjunction* of the two sister centrosomes, each containing a pair of mother–daughter centrioles. The fourth and final step is the physical *separation* of the now disjoined sister centrosomes through the action of microtubule-based motor proteins.

For normal cells, be they embryonic or somatic, the steps of centrosome reproduction are correlated with distinct stages of the cell cycle (Fig. 2). In mammalian somatic cells the start of centrosome reproduction is commonly thought to begin in late G_1, when the inherited centriole pair loses its orthogonal arrangement (Kuriyama and Borisy, 1981). Depending upon the cell type, daughter centrioles are first seen between late G_1 and mid-S phase with the appearance of short, annular centrioles (often called procentrioles) at right angles to and separated slightly from the proximal end of each mature centriole (Robbins *et al.,* 1968; reviewed in Hinchcliffe and Sluder, 1998). These daughter centrioles elongate during S and G_2, and depending on the cell type, may reach their mature length as late as the following G_1 (Kuriyama and Borisy, 1981; Lange and Gull, 1995).

The completion of centrosome reproduction typically occurs when the daughter centrosomes disjoin and separate at a variable time in G_2, with

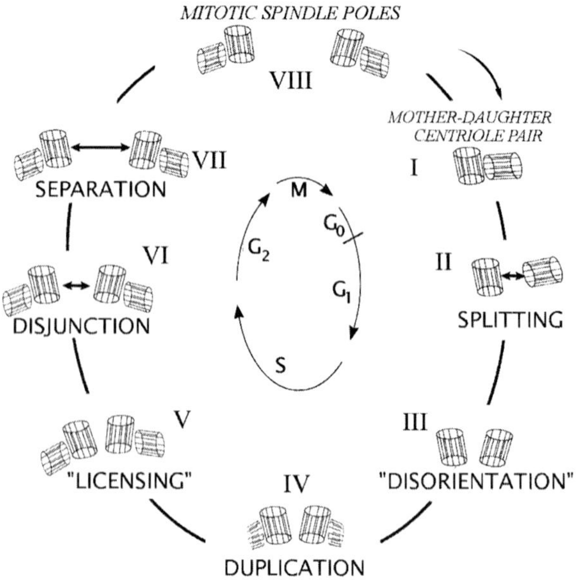

Figure 2 Schematic illustrating the centrosome reproduction cycle. At each stage of the cycle the centrioles are represented as barrels. Clockwise from the top right: (I) The single centrosome inherited at the end of mitosis; the mother–daughter pair of centrioles are arranged at right angles to and in close proximity with each other. (II) During late G_1 or early S phase, the mother–daughter centriole pair splits apart, and (III) concurrently, the centriole pair loses its orthogonal orientation (termed "disorientation"). (IV) The centrosome then duplicates once, as seen by the appearance of short, annular daughter centrioles, called procentrioles, at right angles to the preexisting centrioles. The duplicated centrosome now consists of two mother–daughter pairs of centrioles. (V) During S phase the reproductive capacity of the duplicated centrosome is restored (termed "licensing"). (VI) The duplicated centrosome disjoins during G_2 with pairs of mother–daughter centrioles in each sister centrosome. (VII) The sister centrosomes physically separate from each other through the action of microtubule-based motor proteins. (VIII) At the time of the G_2/M transition the sister centrosomes assemble the poles of the mitotic spindle. As the cell completes mitosis and divides into two, each daughter cell inherits exactly one centrosome containing a mother–daughter pair of centrioles.

pairs of mother–daughter centrioles going to each daughter centrosome (Aubin *et al.,* 1980; Kochanski and Borisy, 1990). However, the extent to which daughter asters have separated during prophase can vary greatly between cells in the same population. In some cases the two centrosomes remain close together until nuclear envelope breakdown, while in others both asters are well separated around the nucleus before the end of prophase (reviewed in Rieder, 1990).

III. Controls for Centrosome Reproduction

Control of centrosome reproduction is exercised at two levels: (i) by limits that are intrinsic to the centrosome itself; and (ii) by extrinsic controls imposed by changing cytoplasmic conditions. Limits intrinsic to the centrosome determine centrosome copy number at the onset of mitosis; cytoplasmic controls determine when the centrosome reproduces in relation to the progression of nuclear events, such as DNA synthesis and the onset of mitosis.

A. Intrinsic Controls

Functional analyses of centrosome reproduction in sea urchin zygotes indicate that there is a "counting" mechanism within each centrosome that limits the number of daughters that can arise from the parent centrosome. The key finding is that it is possible to experimentally manipulate the reproductive capacity of centrosomes. When mitosis is prolonged by any of several independent methods (Fig. 3), the two spindle poles split during mitosis to yield four functional poles that will not further subdivide even when mitosis is prolonged to 20 times its normal duration (Mazia *et al.,* 1960; Sluder and Begg, 1985; Hinchcliffe *et al.,* 1998). Ultrastructural analysis of such tetrapolar spindles reveals that each pole contains only one centriole, confirming that the centrosomes have split, not duplicated (Fig. 3). In effect, the two mitotic centrosomes with normal reproductive capacity have subdivided into four centrosomes with half the normal reproductive capacity (and half the complement of centrioles). After the cell divides into four and as the daughter cells go through interphase, these half centrosomes each assemble a daughter centriole, thus becoming normal centrosomes with full reproductive capacity (Sluder and Rieder, 1985). Since they do not split during interphase, each daughter cell assembles a monopolar spindle at next mitosis (Fig. 4a). If a daughter cell with a monopolar spindle remains in mitosis longer than normal, as they often do, the centrosome of the monopolar spindle will split to give two functional spindle poles with one centriole apiece (Fig. 4b). These poles duplicate but do not double in interphase and monopolar spindles are once again formed at the following mitosis (Fig. 4b). Thus, the splitting of the mother–daughter centrioles and their reproduction are distinct events that can be experimentally uncoupled by prolonging mitosis (Sluder and Begg, 1985; Sluder and Rieder, 1985). The reproduction of a spindle pole cannot simply be the subdivision of the centrosomal microtubule organizing center (MTOC), because such a fission mechanism should always give two smaller centrosomes, each of which should have the same reproductive capacity as the parent pole.

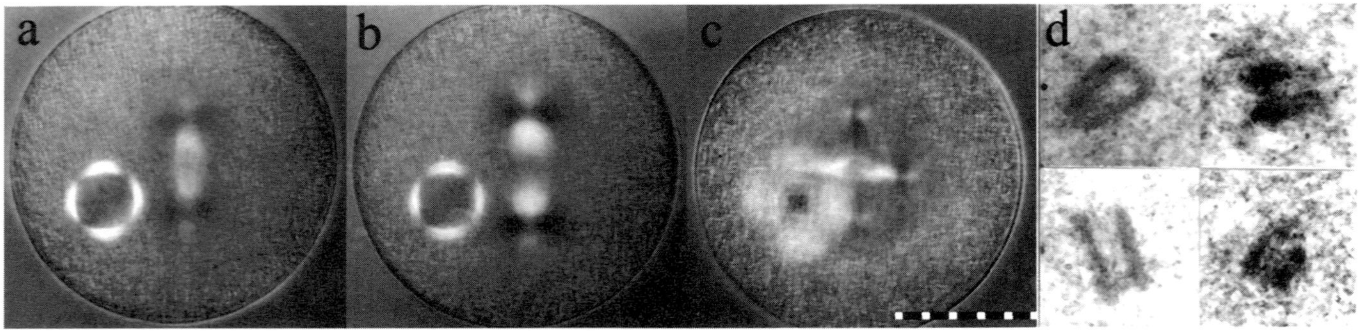

Figure 3 Microinjection of cyclin B Δ-90 mRNA at first prophase arrests zygotes in mitosis. (a) The zygote enters mitosis and forms a normal bipolar spindle. The refractile sphere in the lower left portion of the zygote is a drop of oil used to cap the micropipet. (b) Anaphase onset occurs at the normal time. (c) This zygote shortly before fixation for serial section ultrastructural analysis. Both of the spindle poles have split and a tetrapolar spindle forms. (e) Each spindle pole of this zygote contains only a single centriole. All four centrioles in this zygote are shown in this frame. (a-c) Polarization optics, 10 microns per scale division. (From Journal of Cell Biology (1986), with copyright permission of the Rockefeller University Press, **103,** 1873–1881).

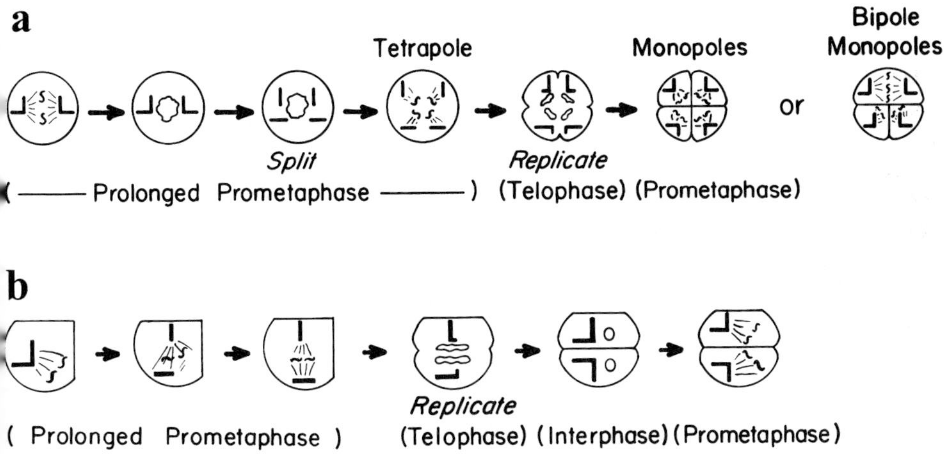

Figure 4 Diagrammatic representation of the experimental manipulation of the reproductive capacity of spindle poles in zygotes. (a) The first division spindle has a pair of orthogonally arranged centrioles at each spindle pole. During prolonged prometaphase the centriole pairs (and centrosomes) split without duplicating, and the four spindle poles separate from each other, each containing a single centriole. In telophase, as the cell divides into four, the singlet centrioles replicate but do not separate. The result is the formation of monopolar spindles at the next mitosis. Each centrosome has the normal complement of two centrioles. When a cleavage furrow fails to form, two monopolar spindle come together to assemble a functional bipolar spindle with poles that reproduce in a normal fashion. (b) Centriole behavior during prolonged prometaphase in a cell containing a monopolar spindle. The single pole splits and a bipolar spindle forms when prometaphase is prolonged, as it often is in such cells. Each sister aster contains only one centriole. After anaphase, the cell divides and the singlet centrioles duplicate but do not split during interphase. At the next mitosis monopolar spindles are again assembled.

The mechanism that limits the number of daughters that can be formed from the parent centrosome is a mystery. One possibility is that centrosome duplication depends upon the formation of a template assembly that seeds or nucleates the daughter centrosome in association with an existing centrosome. Not all centrosomal structures need be templated, only the new entities around which the familiar structures of the centrosome are later elaborated (discussed in Sluder and Reider, 1996). Possibly this "seed" represents the molecular assemblies that pattern the daughter centrioles, if centrioles in fact organize the centrosome in higher animal cells as a number of studies suggest (Sluder and Rieder, 1985; Sluder *et al.,* 1989; reviewed in Marshall and Rosenbaum, 1999). Even if the nine triplet microtubules of centrioles per se do not do not participate in the organization of the centrosome in animal cells, some specific activity associated with the centriole is required for the assembly of a complete daughter centrosome from nonlimiting subunit pools.

Evidence for templating activity in the centrosome can be found at both the functional and morphological levels. Even though before fertilization, sea urchin, clam, and *Xenopus* eggs all contain enough centrosomal subunits to assemble many complete centrosomes (Sluder *et al.,* 1986, 1990; Gard *et al.,* 1990; Palazzo *et al.,* 1992), they normally assemble only one new centrosome in close spatial association with each preexisting centrosome at each cell cycle. Somatic cells, after microsurgical removal of the centrosome, reform an asterlike array of microtubules, yet this "aster" does not reassemble centrioles or double even though the cell grows to a larger than normal size and enters S phase, the expected time of centrosome reproduction (Maniotis and Schliwa, 1991). Structural studies reveal that procentrioles form only at the proximal ends on parenting centrioles, and in some cases specific precursor structures have been described (Dippel, 1968; Gould, 1975).

B. Extrinsic Controls

The formation of a new centrosome through a templating mechanism is not sufficient, by itself, to fully explain the tight temporal coordination between the nuclear and centrosome cycles. Tests of the possibility that the coordination of centrosome reproduction with nuclear events in the cell cycle is determined by nuclear activities started with an examination of centrosome reproduction in enucleate sea urchin zygotes (Lorch, 1952; Sluder *et al.,* 1986). The approach was to remove the nucleus and one centrosome from individual prophase zygotes with a micropipet and determine if the centrosome left behind could reproduce. After the nucleus was removed, the cell cycle appeared to continue, as seen by the cyclical rise and fall of the astral microtubule assembly in normal coordination with the initiation of cleavage furrows. Importantly, the single centrosome reproduced in a precise 1-2-4-8 fashion in proper coordination with cycles of astral microtubule assembly/disassembly and cleavage furrow initiation (Fig. 5). The fact that all daughter centrosomes were found to contain two centrioles indicated that centrosome reproduction was normal and complete. These observations reveal that nuclear activities, such as the synthesis of DNA and/or the timed transcription of limiting RNAs, do not control when the centrosome reproduces. The finding that all the centrosomes reproduced in synchrony within an enucleated zygote suggests that the temporal control of their reproduction is exercised by a cyclical change in the state of the cytoplasm.

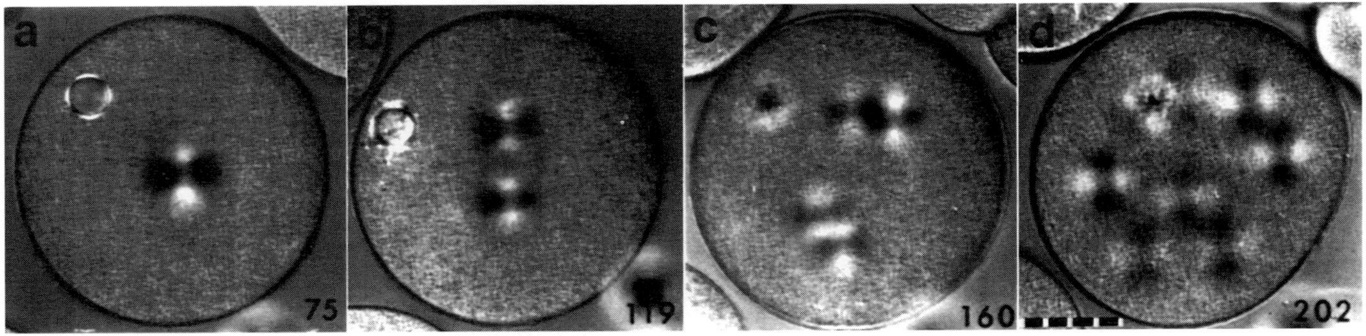

Figure 5 Repeated centrosome reproduction in an enucleated sea urchin zygote. (a) Mitotic zygote after the nucleus and one centrosome were removed with a micropipet. The remaining centrosome organizes a birefringent aster in the center of the cell; the oil drop expelled from the micropipet is seen in the upper left portion of the egg. (b) During second mitosis two birefringent daughter asters are visible. (c) Third mitosis; the two centrosomes have reproduced to four. (d) Fourth mitosis; the zygote contains eight asters, seven of which are visible at this plane of focus. All centrosomes contain a pair of centrioles indicating that centrosome reproduction is complete. Polarization microscopy; 10 microns per scale division. (From Journal of Cell Biology (1990), with copyright permission of the Rockefeller University Press, **110,** 2025–2032).

These results were obtained with zygotes, which are admittedly specialized cells that store relatively large pools of proteins to be used for the rapid divisions of early development. At a minimum, these findings reveal that control of centrosome reproduction by limitations on the availability of centrosomal subunits is not a conserved control strategy utilized by all cells. Nevertheless, it is clear that centrosome reproduction in somatic cells ultimately does require the transcription and translation of messages for centrosomal components, because these cells must grow between divisions. Evidence for the possible role of subunit synthesis in the control of centrosome reproduction in somatic cells comes from the finding that when CHO cells are arrested in S phase, centrosomes continue to reproduce (Balczon *et al.,* 1995). This is also the cell cycle phase in which the levels of PCM-1 mRNA, which codes for a protein associated with the pericentriolar material (Balczon *et al.,* 1994), are normally maximal. Importantly, during the prolonged S phase block the levels of this mRNA remained elevated, suggesting that there is continued synthesis of this centrosome-associated protein. Control experiments revealed that the increase in centrosome number was not simply the consequence of blocking the cell cycle, because cells arrested in G_2 contained only two centrosomes, the normal number for this cell cycle phase. G_2 is also a time that the levels of PCM-1 mRNA levels are normally minimal. At face value these findings suggest that the more centrosomal subunits that are synthesized, the greater the number of centrosomes that are assembled. Whether or not limits on the synthesis of new subunits at each cell cycle per se could act as a control that has sufficient finesse to ensure the formation of just one daughter centrosome per cell cycle is an important question (discussed in Sluder and Rieder, 1996). To limit centrosome number by this mechanism, the cell would have to precisely control transcript and protein pools within very tight tolerances.

IV. Coordination of Centrosome Reproduction with Nuclear Events in the Cell Cycle

At first glance the coordination of the centrosome and nuclear cycles might seem to be straightforward; the events of centrosome reproduction are linked to particular cell cycle stages and coordination is thereby assured. However, the appealing simplicity of this notion is clouded by findings that the cycle of centrosome reproduction can occur repeatedly in the absence of cell cycle progression. For example, when somatic or embryonic cells are arrested in S phase by inhibitors of DNA synthesis, the interphase centrosome undergoes multiple rounds of duplication (Sluder and Lewis, 1987; Raff and Glover, 1988; Balczon *et al.,* 1995). In addition, the centrosomes of sea urchin and *Xenopus* zygotes undergo repeated rounds of complete reproduction when protein synthesis is completely blocked and the cell cycle is arrested before mitosis due to the lack of cyclin A or B synthesis (Fig. 6a; and Sluder *et al.,* 1990; Gard *et al.,* 1990). Such findings

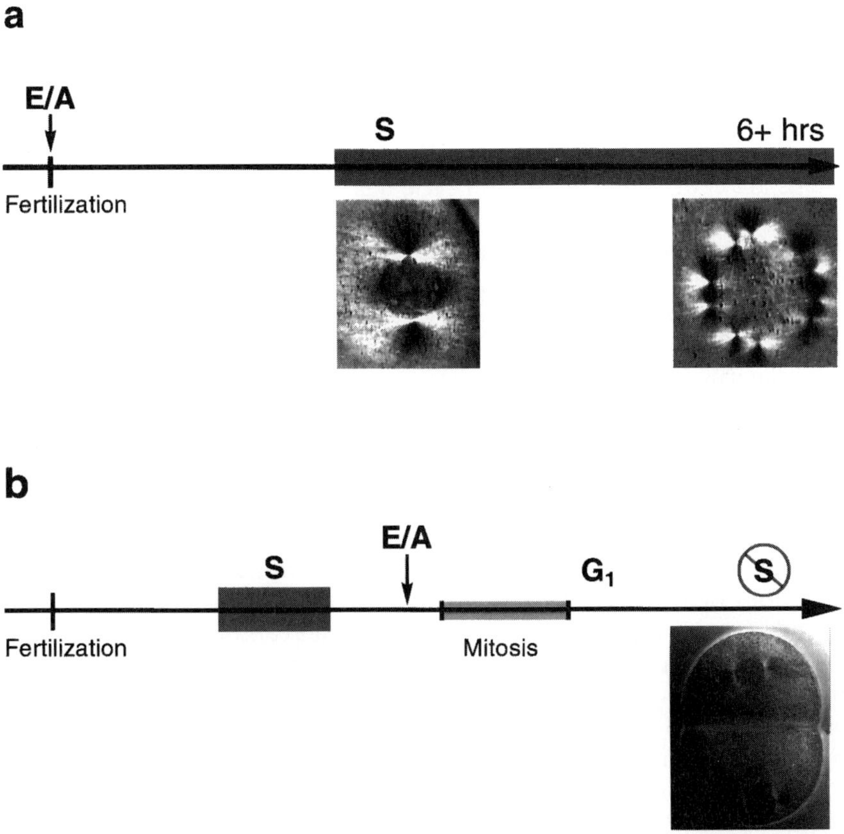

Figure 6 Patterns of centrosome reproduction in sea urchin zygotes treated with translation inhibitors before fertilization (a) or at first prophase (b). Diagrammatic representation of the experiment and its results with photographs of cells at corresponding stages. (a) Translation inhibitors (emetine plus anisomycin: E/A) were added to the zygotes at the time of fertilization. The zygotes enter S phase and contain two asters closely associated with the nucleus. After 6 hr, the zygotes are still in S phase, and some zygotes contain 10 or more asters. These zygotes were extracted with microtubule-stabilizing buffer, and astral birefringence was augmented by hexylene glycol. (b) E/A is added to zygotes at prophase of first mitosis, and they complete M phase and cleave at the normal time. These zygotes arrest at the two-cell stage prior to the onset of second S phase; each cell of the embryo contains exactly two asters positioned on either side of the reformed nuclear envelope. Each aster contains a pair of centrioles. The number of asters or centrioles never increased in these zygotes, even after 6 hr. The astral birefringence in this zygote was augmented by hexylene glycol. Polarization microscopy. (From Journal of Cell Biology (1998), with copyright permission of the Rockefeller University Press, **140,** 1417–1426).

raised the possibility that the nuclear and centrosomal cycles are regulated by independent pathways (Sluder *et al.,* 1990). Nevertheless, the normal tight linkage between centrosomal events and nuclear events in the cell cycle forces a search for the mechanisms that coordinate these important preparations for mitosis.

A. Cell Cycle Stage Dependency of Centrosome Reproduction

One avenue used to investigate how centrosome reproduction and nuclear events are coordinated in the cell cycle has been to characterize which cell cycle stages support centrosome reproduction and which do not. In general, the approach used has been to arrest cells in a particular phase of the cell cycle and then determine if the centrosome will reproduce one or more times without further experimental intervention.

The results of several studies have clearly demonstrated that the centrosome inherited at the end of the previous mitosis does not reproduce during prolonged G_0 (Tucker *et al.,* 1979; reviewed in Wheatley, 1982). However, it is not clear to what extent the centrosome will begin to reproduce during G_1, once the cell has passed the restriction point and is committed to prepare for division, because of the difficulty of credibly arresting cells in the G_1 phase of the cell cycle. The conventional view is that the reproduction of centrioles, and hence the centrosome, is controlled by the cytoplasmic conditions of S phase. This notion comes from well-documented observations that the disorientation of mother–daughter centrioles is not seen until late in G_1 and procentrioles are not evident until S phase in a wide variety of cultured cells (Robbins *et al.,* 1968; Kuriyama and Borisy, 1981; Vorobjev and Chentsov, 1982; Alvey, 1985). Also, treatment of cultured cells with mimosine, a plant amino acid, blocks the cell cycle in G_1, but the centrosome does not appear to double, as seen by the number of gamma tubulin immunoreactive spots (Matsumoto *et al.,* 1999). However, the temporal correlation between procentriole formation and onset of S phase does not appear to be universal. For example, a high proportion of L929 cells in culture form procentrioles 4 hr before the onset of DNA synthesis (Rattner and Phillips, 1976). Also, in sea urchin zygotes arrested in interphase before the onset of DNA synthesis (a cell cycle phase presumably equivalent to G_1), the centrosome inherited by each blastomere reproduces just once to form two centrosomes, each containing a pair of centrioles (Hinchcliffe *et al.,* 1998; also see Fig. 6b). In addition, Fukasawa *et al.* (1996) reported that $p53^{-/-}$ mouse embryonic fibroblasts (MEFs) may also support centrosome duplication in G_1. For MEFs arrested in G_0 by serum starvation, 17% of G_0 $p53^{-/-}$ MEFs had greater than two centrosomes, whereas 100% of $p53^{+/+}$ MEFs had only one centrosome. After addition of serum to release the cell cycle arrest, none of the $p53^{+/+}$ MEFs had more than two centrosomes. However, multiple centrosomes were observed in 49% of the $p53^{-/-}$ MEFs, at times well before the cell cycle had progressed into S phase.

Thus, it may be premature to eliminate the possibility that the early steps of centriole duplication, such as the formation of precursor structures, occur well before DNA synthesis begins; the assembly of procentrioles to the point when they are visible in the electron microscope may mark the end of the initial reproductive processes that began in G_1. This issue is important because any effort to experimentally identify cell cycle regulatory pathways that control centrosome duplication will be facilitated by an understanding of when the reproductive processes actually begin.

In contrast, it is clear that the cytoplasmic conditions of S phase are permissive for the complete cycle of centrosome reproduction. In fact, prolongation of this phase in both zygotes and somatic cells through the inhibition of DNA synthesis allows *multiple* rounds of centrosome duplication to occur (Hinchcliffe *et al.,* 1998; discussed in Winey, 1999)

Although the daughter centrosomes normally disjoin and separate during G_2, this phase of the cell cycle does not appear to support centrosome duplication. This notion comes from the observation that CHO cells arrested in G_2 with the topoisomerase inhibitor etopiside contain only two sister centrosomes, whereas the same cells arrested in S assemble multiple centrosomes (Balczon *et al.,* 1995). Importantly, the disjunction of the sister centrosomes is correlated with the rise in activity of Nek2 (Fry *et al.,* 1998a,b), the mammalian homolog of the *Aspergillus* protein NIMA—a cell cycle regulatory kinase that contributes to driving entry into mitosis (Osmani *et al.,* 1988). This suggests that the activity of Nek2 may function to coordinate sister centrosome disjunction with progression of the cell cycle into mitosis.

Finally, several studies have demonstrated that the cytoplasmic conditions of mitosis do not support centrosome reproduction. When in sea urchin zygotes are microinjected with mRNA coding for a nondegradable form of sea urchin cyclin B (cyclin B Δ90; Glotzer *et al.,* 1991; Holloway *et al.,* 1993), the cell cycle is arrested in M phase (Hinchcliffe *et al.,* 1998). Anaphase onset occurs at the normal time, indicating that proteolysis at the metaphase–anaphase transition occurred, yet the cells do not exit mitosis or undergo cytokinesis. Shortly after anaphase onset the two spindle poles split apart, but no further doubling of the spindle poles occurs regardless of how long the zygotes remain in mitosis. Since each of the four poles contains only one centriole, this doubling represents a splitting of the mother–daughter centrioles, not a duplication of the centrosomes themselves (Fig. 3). Although ubiquitin-mediated proteolysis at anaphase onset happened, this does not appear to signal when centrosome reproduction begins.

Similar results were obtained with somatic cells. Transfection of a nondegradable form of cyclin B into cultured cells arrests the cell cycle in mitosis and leads to the formation of a tetrapolar spindle (Gallant and Nigg, 1992). However, since ultrastructural analyses of these spindles were not performed, we do not know if this doubling of the spindle poles represents a splitting or full duplication of the centrosomes.

At this point the simplest explanation for normal tight coordination of centrosome reproduction with other cell cycle events would be to propose that centrosomes can start to reproduce only when the cell cycle reaches or enters S phase. However, a recent study with sea urchin zygotes indicates that the situation is somewhat more complex (Hinchcliffe *et al.,* 1998). The key observation came from an experiment in which zygotes were treated with translation inhibitors in first prophase, after they had synthesized all the necessary proteins to support first mitosis (Fig. 6b). As expected, they completed mitosis and arrested before second nuclear envelope breakdown because no new cyclin A or B was synthesized. Importantly, the two daughter blastomeres did not initiate DNA synthesis and were thus arrested in a cell cycle phase equivalent to G_1 (sea urchin zygotes in the rapid early divisions do not normally have a detectable G_1). The surprise was that the centrosome inherited by each daughter blastomere at the end of mitosis completely reproduced during G_1 arrest. Each daughter cell contained just two centrosomes, one on each side of the nucleus, and each centrosome contained a pair of centrioles (Fig. 6b). Importantly, neither the number of centrosomes nor the number of centrioles increased thereafter. These results indicate that the G_1 phase of the cell cycle supports the morphological aspects of centrosome reproduction, such as the assembly of daughter centrioles and the splitting/separation of the duplicated centrosomes. However, the finding that the sister centrosomes do not reproduce again under such supportive conditions indicates that they are lacking a component or activity necessary for reproduction.

This finding, coupled with the observation that centrosomes repeatedly reproduce in second division blastomeres when they are arrested in S phase, suggests that there is an event in S phase that restores this component to the daughter centrosomes, thereby preparing them to reproduce in the next cell cycle. Although we have no information on the nature of this event, possibilities include a "licensing" of the centrosome for reproduction through the posttranslational modifications of key centrosomal components and/or the assembly of essential precursor structures needed for the formation of the next generation of daughter centrosomes (see Fig. 2).

Taken together, these observations have led to the following working model for how centrosome reproduction is coordinated with nuclear events in the cell cycle. The centrosome inherited by a daughter cell in telophase is competent to reproduce (i.e., it was "licensed" during the previous S phase), and there is a G_1 signal, conceivably passage through the restriction point, that triggers the expression of the morphological events of centrosome reproduction. These events include the splitting of the mother and daughter centrioles, as well as the assembly of procentrioles. If progress through the remainder of G_1 is fast enough, the procentrioles would not be detectable until the onset of S or shortly thereafter. In addition, the centrosome cycle is linked to the nuclear cycle by a "licensing" event during S phase that prepares the daughter centrosomes for morphological duplication in the following G_1. Disjunction and separation or the sister centrosomes are normally associated with the conditions during G_2 (possibly the activity of the NIMA kinase homolog Nek2), which could couple bipolarization of the spindle with entry in M phase (Fry *et al.*, 1998a,b). Finally, as the cell cycle enters mitosis, centrosome duplication is prevented, and can only resume after the cell has divided. Thus, the perceived independence of centrosome reproduction from progression of the cell cycle suggested by a number of previous studies (Sluder *et al.*, 1990; Gard *et al.*, 1990; Balczon *et al.*, 1995) is simply due to greatly prolonging the cytoplasmic conditions of S phase, which allows both the expression of centrosome reproduction and the acquisition of reproductive capacity.

Observations that S-phase arrest supports multiple rounds of centrosome reproduction raises the important question of why centrosomes do not reduplicate during S phase in the normal cell cycle. A possible answer comes from observations on sea urchin zygotes and CHO cells that the period of centrosome reduplication during S phase arrest is on average more than twice as long as the entire cell cycle (Hinchcliffe *et al.*, 1998; Balczon *et al.*, 1995). Under normal circumstances, therefore, S phase does not last long enough for centrosomes to reduplicate. Thus, when the cell cycle proceeds at the normal rate, progress into late G_1 and entry into S are rate limiting for centrosome reproduction, and the essential coordination between centrosomal and nuclear events is assured.

B. Cyclin-Dependent Kinases in the Control of Centrosome Reproduction

The link between cell cycle stage and centrosome reproduction raises the question of what control pathways establish this link. In principle, the sequential activation and inactivation of one or more of the cyclin-dependent kinases (Cdks) could function as the cytoplasmic mechanism that coordinately drives centrosome duplication and nuclear events in the cell cycle. Although logical, this was initially a problematic proposal because centrosomes repeatedly reproduce when progression of the cell cycle is blocked before mitosis by complete inhibition of protein synthesis (Sluder *et al.,* 1990; Gard *et al.,* 1990). Thus, the reproduction of the centrosome cannot be driven by the cyclic rise and fall of Cdk1–cyclin A or Cdk1-B activities, which depend upon the synthesis of new cyclin proteins during each cell cycle. Further investigation of this issue has also established that the absolute value of Cdk1-B activity does not establish cytoplasmic conditions that favor or inhibit centrosome reproduction (Hinchcliffe *et al.,* 1998).

Since mitotic Cdk1-cyclin combinations did not appear to control the events of centrosome reproduction, attention then shifted to whether or not Cdk2 activity plays a role in this process. Cdk2, bound to cyclin A and cyclin E, is required for the G_1/S transition and the maintenance of S phase progression (Strausfeld *et al.,* 1996; Stillman, 1996), points in the cell cycle that are correlated with the assembly of procentrioles (Kuriyama and Borisy, 1981). In addition, the levels of both the Cdk2 and cyclin E proteins remain constant from the time of fertilization until the mid-blastula transition in early *Xenopus* embryos (Rempel *et al.,* 1995; Hartley *et al.,* 1997), thus providing a possible explanation for why repeated centrosome reproduction in *Xenopus* zygotes is not dependent upon cyclin protein synthesis (Gard *et al.,* 1990).

Several recent studies have demonstrated that Cdk2 activity is required for centrosome reproduction in both early zygotes and somatic cells. First, Hinchcliffe *et al.* (1999) developed a *Xenopus* egg extract system (arrested in S phase with aphidicolin) that supports multiple rounds of centrosome duplication (Fig. 7). Asters organized by added sperm centrosomes doubled three times in 1-2 fashion, which is characteristic of centrosome duplication and inconsistent with either centrosome splitting or the fragmentation of the centrosomal MTOC (see Sluder and Begg, 1985). The fact that the extract is already in S means that the role of Cdk-E in the control of centrosome reproduction could be tested without the concern, applicable to live cells and cycling extracts, that inhibition of Cdk2–cyclin E activity could potentially arrest the cell cycle at a point before centrosomes are normally scheduled to reproduce.

In these extracts, Cdk2-E activity was selectively inhibited by adding purified recombinant Δ34Xic-1. This is the NH_3-terminal truncated variant of Xic-1^{p27}, the *Xenopus* cyclin-dependent kinase inhibitor (CKI) that inhibits Cdk2-E activity but not Cdk1-cyclin A or B at the concentration used (Su *et al.,* 1995; Hartley *et al.,* 1997). Since Cdk2 does not complex with cyclin A until after the mid-blastula transition in *Xenopus* (Rempel *et al.,* 1995; Hartley *et al.,* 1997), Cdk2-cyclin A activity was not a factor in these experiments. Importantly, the inhibition of Cdk2-E should not drive the cell cycle out of S phase, since the majority of S phase promoting activity is provided by Cdk1-A activity (Strausfeld *et al.,* 1996), which is not inhibited by the concentrations of Δ34Xic-1 used.

Inhibiting Cdk2-E activity did not block the first round of aster doubling, but did prevent subsequent rounds over the duration of the experiment.

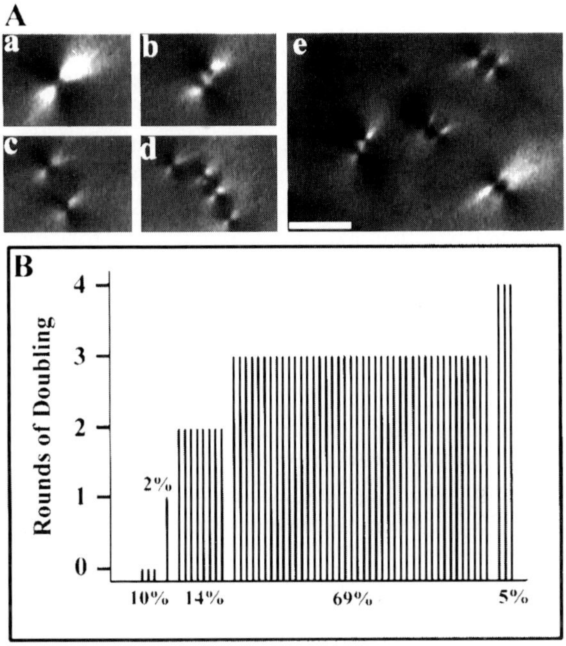

Figure 7 (A) Repeated rounds of aster doubling in an aphidicolin-treated extract (a–e) Frames from a time-lapse sequence showing an individual aster undergoing three rounds of doubling *in vitro*. Polarization optics. Bar in (e) = 10 μm. (B) Analysis of aster doubling in aphidicolin-treated extracts. Each vertical line illustrates the maximum number of rounds of duplication seen for the progeny of individual asters followed for the duration of the experiment. Ordinate: Total rounds of duplication. Abscissa: Percentage for each category shown. Data taken from three experiments (N = 59 asters).

When an excess of purified Cdk2-E was added to the Δ34Xic-1 treated extracts, multiple rounds of aster duplication took place. These observations reveal that Cdk2-E activity is required for repeated centrosome reproduction during prolonged S phase. Whether a single round of aster doubling in Δ34Xic-1 treated extracts reflects splitting of centriole pairs or complete centrosome reproduction is an important issue to address by serial section ultrastructural analysis.

Second, Lacey and colleagues (Lacey *et al.*, 1999) arrested living *Xenopus* embryos in S phase with protein synthesis inhibitors, which allows repeated centrosome duplication (Gard *et al.*, 1990). They then microinjected individual blastomeres in these arrested embryos with the CKIs p21 or p27, which bind to and block the activity of Cdk2-E (Elledge and Harper, 1994; Jackson *et al.*, 1995). The injected blastomeres did not show the repeated centrosome duplication found in the uninjected cells of the same zygote (Lacey *et al.*,

1999). In addition, these workers developed an *in vitro* assay that uses fluorescence light microscopy to monitor mother–daughter centriole splitting in *Xenopus* egg extracts treated with cylohexamide. They found that Cdk2-E activity is required to promote centriole splitting, thought to be the earliest event in the centrosome reproduction cycle (Kuriyama and Borisy, 1981). Their finding raises the question of why Hinchcliffe *et al.*, (1999) observed one round of astral doubling in Δ34Xic-1 treated extracts in which Cdk2-E activity was low. One explanation for this apparent discrepancy is that the Hinchcliffe *et al.* study used *Xenopus* sperm centrosomes, which may start with separated centrioles (see Felix *et al.*, 1994), whereas Lacey *et al.* used isolated centrosomes from cultured cell in which the mother–daughter centriole pairs may be functionally linked to each other.

Matsumoto *et al.* (1999) also provided evidence that repeated centrosome duplication in Chinese hamster ovary (CHO) cells that are arrested in S phase with HU is dependent upon Cdk2 activity, though their experiments did not differentiate between the importance of Cdk2-A versus Cdk2-E activity (reviewed in Winey, 1999).

That centrosome reproduction is driven solely by Cdk2 bound to cyclin E has been brought into question by a recent report that Cdk2-A is more effective than Cdk2-E in restoring multiple rounds of centrosome duplication in cultured mammalian cells (Meraldi *et al.*, 1999). For these studies, the authors used CHO cells arrested in S phase with hydroxyurea (HU). To test the role of cell cycle regulatory proteins, CHO cells (before or after addition of HU) were then transfected with plasmids encoding $p16^{ink4}$, to inhibit the Cdk4 and Cdk6 kinases, which in turn blocks phosphorylation of the retinoblastoma protein (Rb) and arrests the cell cycle in G_1 (Lukas *et al.*, 1997). They found that $p16^{ink4}$ transfection before HU addition blocked centrosome duplication, whereas transfection after S phase arrest did not, thus ruling out a direct role for either Cdk4 and Cdk6 activity in control of the centrosome cycle. Next, these workers tested the relative importance of Cdk2-cyclin A versus Cdk2-cyclin E activity in supporting centrosome reproduction. They coexpressed plasmids for $p16^{ink4}$ and either cyclin E and cyclin A. Not surprisingly, expression of these cyclin constructs could overcome the $p16^{ink4}$ block, probably by phosphorylation of Rb protein, allowing the cell cycle to progress into S phase. However, when they blocked the G_1/S transition by transfection with RbDCdk (a mutant form of Rb that lacks putative Cdk phosphorylation sites; Lukas *et al.*, 1997), only coexpression of cyclin A restored significant levels of repeated centrosome duplication; overexpression of cyclin E did not. The restoration of centrosome duplication by cyclin A was dependent strictly upon Cdk2 activity, as it could be inhibited by cotransfection with a dominant negative Cdk2 construct. Taken together, these findings suggest that Cdk2-A activity is required for centrosome duplication and this kinase acts downstream of its role in the Rb phosphorylation pathway needed to drive the G_1/S transition.

These findings are interesting, because they suggest that somatic cells and early cleavage stage zygotes use different Cdk2–cyclin complexes to regulate centrosome reproduction. However, this may not reflect fundamentally different control strategies; perhaps centrosomes, or the downstream pathways that regulate them, are responsive to both Cdk2-A and Cdk2-E. Since Cdk2 does not complex with cyclin A until the mid-blastula transition in *Xenopus* zygotes (Rempel *et al.,* 1995), Cdk2-E may be the only kinase complex available to do the job. In somatic cells that contain both Cdk2-A and Cdk2-E kinase complexes, perhaps Cdk2-A is more effective in promoting centrosome reproduction. Nevertheless, despite these early indications that the regulation of centrosome reproduction in various cell types may differ slightly in the details, it is clear that cell cycle progression into S phase and the establishment of conditions that support multiple rounds of centrosome reproduction are linked through a common pathway—Cdk2 activity.

Looking forward, there are a number of fascinating issues to explore. For example, when does centrosome reproduction first begin? Is the regulation of centrosome reproduction really a two-step process starting with the onset of conditions during G_1 that trigger the steps in the morphological completion of daughter centriole assembly followed by a "licensing" event specific to S phase (Hinchcliffe *et al.,* 1999)? Alternatively, does the whole reproductive cycle of the centrosome depend upon a rise in Cdk2 activity, be it based in Cdk2-A and/or Cdk2-E complexes? Also, are other control pathways involved? For example, do the activities of the polo kinases (reviewed in Glover *et al.,* 1998) or Aurora kinases (reviewed in Giet and Prigent, 1999) play a direct role in the events of centrosome reproduction, or do these kinases effect only the physical separation of duplicated centrosomes through their regulation of the activity of microtubule-based motor molecules? Is ubiquitin-mediated proteolysis, at times other than the metaphase–anaphase transition, necessary for the splitting and disorientation of the mother–daughter centrioles, as suggested by recent reports (Gstaiger *et al.,* 1999; Freed *et al.,* 1999); is this splitting event absolutely required for their reproduction? If proteolysis is important, is it signaled by the same activities that drive entry into S phase, or is it regulated independently? Finally, does centrosome reproduction require a microtubule network to transport subunit proteins to the cell center (Balczon *et al.,* 1999; Purohit *et al.,* 1999), and if so, by what mechanism does this transport occur? Although we may be overwhelmed by the emerging complexity of the events and control mechanisms involved in centrosome reproduction, we can be pleased that there are yet more intriguing avenues for further research. As Chandler Fulton (1982) aptly said of the centriole,

"Many of us have had a transient affair with centrioles, and retain a profound affection for them, but few have committed ourselves to a long-term relationship, perhaps because, in spite of our enchantment, we could not find a way to build a meaningful relationship—that is, to break through the enigmas." Thus, it is gratifying to find that, after being investigated for more than a hundred years, the centrosome still retains enough mysteries to continue attracting new generations of eager suitors.

Acknowledgments

We thank Jim Maller, Conly Rieder, and Bob Palazzo for stimulating conversations on the nature of centrosomes and the cell cycle. We especially appreciate the efforts of Rick Miller in preparing the figures for this article. This work was supported by grants from the NIH (GM 30758), the Cabot Family Charitable Trust, and the Trustees of the Worcester Foundation to GS. E.H.H. is supported by an NRSA Cell Biology of Development Post-Doctoral Training Grant (NIH HD 07312).

References

Alvey, P. L. (1985). An investigation of the centriole cycle using 3T3 and CHO cells. *J. Cell Sci.* **78,** 147–162.

Aubin, J. E., M. Osborn, and K. Weber (1980). Variations in the distribution and migration of centriole duplexes in mitotic PtK_2 cells studied by immunofluorescence microscopy. *J. Cell Sci.* **43,** 177–194.

Balczon, R., L. Bao, and W. E. Zimmer (1994). PCM-1, a 228 kDa centrosome autoantigen with a distinct cell cycle distribution. *J. Cell Biol.* **124,** 783–793.

Balczon, R., L. Bao, W. E., Zimmer, K. Brown, R. P. Zinkowski, and B. R. Brinkley (1995). Dissociation of centrosome replication events from cycles of DNA synthesis and mitotic division in hydroxyurea-arrested Chinese hamster ovary cells. *J. Cell Biol.* **130,** 105–115.

Balczon R, C. E. Varden, and T. A. Schroer (1999). Role for microtubules in centrosome doubling in Chinese hamster ovary cells. *Cell Motil. Cytoskeleton* **42,** 60–72.

Brinkley, B. R. (1985). Microtubule organizing centers. *Ann. Rev. Cell Biol.* **1,** 145–172.

Brinkley, B. R., and T. M. Goepfert (1998). Supernumerary centrosomes and cancer: Boveri's hypothesis resurrected. *Cell Motil. Cytoskeleton* **41,** 281–288.

Dippell, R. V. (1968). The development of basal bodies in paramecium. *Proc. Natl. Acad. Sci. USA* **61,** 461–468.

Elledge, S. J., and J. W. Harper (1994). Cdk inhibitors: on the threshold of checkpoints and development. *Curr. Opin. Cell Biol.* **6,** 847–52.

Freed E, K. R. Lacey, P. Huie, S. A. Lyapina, R. J. Deshaies, T. Stearns, and P. K. Jackson (1999). Components of an SCF ubiquitin ligase localize to the centrosome and regulate the centrosome duplication cycle. *Genes Dev.* **13,** 2242–2257.

Felix, M., C. Antony, M. Wright, and B. Maro (1994). Centrosome assembly in vitro: Role of γ-tubulin recruitment in *Xenopus* sperm aster formation. *J. Cell Biol.* **124,** 19–31.

Fry, A. M., T. Mayor, P. Meraldi, Y. D. Stierhof, K. Tanaka, and E. A. Nigg (1998a). C-Napl, a novel centrosomal coiled-coil protein and candidate substrate of the cell cycle-regulated protein kinase Nek2. *J. Cell Biol.* **141,** 1563–74.

Fry, A. M., P. Meraldi, and E. A. Nigg (1998b). A centrosomal function for the human Nek2 protein kinase, a member of the NIMA family of cell cycle regulators. *EMBO J.* **17,** 470–81.

Fukasawa, K., T. Choi, R. Kuriyama, S. Rulong, and G. F. Vande Woude (1996). Abnormal centrosome amplification in the absence of p53. *Science* **271,** 1744–1747.

Fulton, C. (1982). Pinwheels as cell's doodlebugs. *Cell* **30,** 341–343.

Gallant, P., and E. A. Nigg (1992). Cyclin B2 undergoes cell cycle-dependent nuclear translocation and, when expressed as a non-destructible mutant, causes mitotic arrest in HeLa cells. *J. Cell Biol.* **117,** 213–224.

Gard, D. L., S. Hafezi, T. Zhang, and S. J. Doxsey (1990). Centrosome duplication continues in cycloheximide-treated *Xenopus* blastulae in the absence of a detectable cell cycle. *J. Cell Biol.* **110,** 2033–2042.

Giet, R., and C. Prigent (1999). Aurora/Ipl1p-related kinases, a new oncogenic family of mitotic serine-threonine kinases. *J. Cell Sci.* **112,** 3591–3601.

Glotzer, M., A. W. Murray, and M. W. Kirschner (1991). Cyclin in degraded by the ubiquitin pathway. *Nature* **349,** 132–138.

Glover, D. M., I. M. Hagan, and A. A. Tavares (1998). Polo-like kinases: a team that plays throughout mitosis. *Genes Dev.* **12,** 3777–3787.

Gould, R. R. (1975). The basal bodies of *Chlamydomonas reinhardtii:* Formation from probasal bodies, isolation, and partial characterization. *J. Cell Biol.* **65,** 65–74.

Gstaiger M, A. Marti, and W. Krek (1999). Association of human SCF(SKP2) subunit p19(SKP1) with interphase centrosomes and mitotic spindle poles. *Exp. Cell Res.* **247,** 554–562.

Hartley, R. S., J. C. Sible, A. L. Lewellyn, and J. L. Maller (1997). A role for cyclin E/Cdk2 in the timing of the midblastula transition in *Xenopus* embryos. *Dev. Biol.* **188,** 312–321.

Hinchcliffe, E. H., and G. Sluder (1998). The apparent linkage between centriole replication and S phase of the cell cycle. *Cell Biol. Int.* **22,** 3–5.

Hinchcliffe, E. H., G. O. Cassels, C. L. Rieder, and G. Sluder (1998). The coordination of centrosome reproduction with nuclear events during the cell cycle in the sea urchin zygote. *J. Cell Biol.* **140,** 1417–1426.

Hinchcliffe, E. H., C. Li, E. A. Thompson, J. L. Maller, and G. Sluder (1999). Requirement of Cdk2–Cyclin E activity for repeated centrosome reproduction in *Xenopus* egg extracts. *Science* **283,** 851–854.

Holloway, S., M. Glotzer, R. W. King, and A. W. Murray (1993). Anaphase is initiated by proteolysis rather than by the inactivation of maturation-promoting factor. *Cell* **73,** 1393–1402.

Jackson, P. K., S. Chevalier, M. Philippe, and M. W. Kirschner (1995). Early events in DNA replication require cyclin E and are blocked by p21 CIP1. *J. Cell Biol.* **130,** 755–769.

Kellogg, D. R., M. Moritz, and B. M. Alberts (1994). The centrosome and cellular organization. *Annu. Rev. Biochem.* **63,** 639–674.

Kochanski, R. S., and G. G. Borisy (1990). Mode of centriole duplication and distribution. *J. Cell Biol.* **110,** 1599–1605.

Kuriyama, R., and G. G. Borisy (1981). Centriole cycle in Chinese hamster ovary cells as determined by whole-mount electron microscopy. *J. Cell Biol.* **91,** 814–821.

Lacey, K. R., P. K. Jackson, and T. Stearns (1999). Cyclin-dependent kinase control of centrosome duplication. *Proc. Natl. Acad. Sci. USA* **96,** 2817–2822.

Lange, B. M. H., and K. Gull (1995). A molecular marker for centriole maturation in the mammalian cell cycle. *J. Cell Biol.* **130,** 991–927.

Lingle, W. A., W. H. Lutz, J. N. Ingle, N. J. Maihle, and J. L. Salisbury (1998). Centrosome hypertrophy in human breast tumors: Implications for genomic stability and cell polarity. *Proc. Natl. Acad. Sci. USA* **95,** 2950–2955.

Lorch, I. J. (1952). Enucleation of sea urchin blastomeres with or without removal of asters. *Q. J. Microsc. Sci.* **93,** 475–486.

Lukas, J., T. Herzinger, K. Hansen, M. C. Moroni, D. Resnitzky, K. Helin, S. I. Reed, and J. Bartek (1997). Cyclin E-induced S phase without activation of the pRb/E2F pathway. *Genes Dev.* **11,** 1479–1492.

Maniotis, A., and M. Schliwa (1991). Microsurgical removal of centrosomes blocks cell reproduction and centriole generation in BSC-1 cells. *Cell* **67,** 495–504.

Marshall, W. F., and J. L. Rosenbaum (1999). Cell division: The renaissance of the centriole. *Curr. Biol.* **9,** R218–220.

Matsumoto, Y., K. Hayashi, and E. Nishida (1999). Cyclin-dependent kinase 2 (Cdk2) is required for centrosome duplication in mammalian cells. *Curr. Biol.* **9,** 429–432.

Mazia, D., P. Harris, and T. Bibring (1960). The multiplicity of the mitotic centers and the time-course of their duplication and separation. *Biophys. Biochem. Cytol.* **7,** 1–20.

Meraldi, P., J. Lukas, A. M. Fry, J. Bartek, and E. A. Nigg (1999). Centrosome duplication in mammalian somatic cells requires E2F and Cdk2-Cyclin A. *Nature Cell Biol.* **1,** 88–93.

Orr-Weaver, T. L., and R. A. Weinberg (1998). A checkpoint on the road to cancer. *Nature* **392,** 223–224.

Osmani, S. A., R. T. Pu, and N. R. Morris (1988). Mitotic induction and maintenance by overexpression of a G_2-specific gene that encodes a potential protein kinase. *Cell* **53,** 237–244.

Palazzo, R. E., E. Vaisberg, R. W. Cole, and C. L. Rieder (1992). Centriole duplication in lysates of *Spisula solidissima* oocytes. *Science* **256,** 219–221.

Pihan, G. A., A. Purohit, J. Wallace, H. Knecht, B. Woda, P. Quesenberry, and S. J. Doxsey (1998). Centrosome defects and genetic instability in malignant tumors. *Cancer Res.* **58,** 3974–3985.

Purohit, A., S. H. Tynan, R. Vallee, and S. J. Doxsey (1999). Direct interaction of pericentrin with cytoplasmic dynein light intermediate chain contributes to mitotic spindle organization. *J. Cell Biol.* **147,** 481–492.

Raff, J. W., and D. M. Glover (1988). Nuclear and cytoplasmic mitotic cycles continue in Drosophila embryos in which DNA synthesis is inhibited by aphidicolin. *J. Cell Biol.* **107,** 2009–2019.

Rattner, J. B., and S. A. Phillips (1976). Dependence of centriole formation on protein synthesis. *J. Cell Biol.* **70,** 9–19.

Rempel, R. E., S. B. Sleight, and J. L. Maller (1995). Maternal *Xenopus* Cdk2–cyclin E complexes function during meiotic and early embryonic cell cycles that lack a G_1 phase. *J. Biol. Chem.* **270,** 6843–6855.

Rieder, C. L. (1990). Formation of the astral mitotic spindle: ultrastructural basis for the centrosome–kinetochore interaction. *Electron Microsc. Rev.* **3,** 269–300.

Robbins, E. L., G. Jentzsch, and A. Micali (1968). The centriole cycle in synchronized HeLa cells. *J. Cell Biol.* **36,** 329–339.

Sluder, G., and D. A. Begg (1985). Experimental analysis of the reproduction of spindle poles. *J. Cell Sci.* **76,** 35–51.

Sluder, G., and K. Lewis (1987). Relationship between nuclear DNA synthesis and centrosome reproduction in sea urchin eggs. *J. Exp. Zoology* **244,** 89–100.

Sluder, G., and C. L. Rieder (1985). Centriole number and the reproductive capacity of spindle poles. *J. Cell Biol.* **100,** 887–896.

Sluder, G., and C. L. Rieder (1996). Controls for centrosome reproduction in animal cells: Issues and recent observations. *Cell Motil. Cyto.* **33,** 1–5.

Sluder, G., F. J. Miller, and C. L. Rieder (1986). The reproduction of centrosomes: Nuclear versus cytoplasmic controls. *J. Cell Biol.* **103,** 1873–1881.

Sluder, G., F. J. Miller, and C. L. Rieder (1989). Reproductive capacity of sea urchin centrosomes without centrioles. *Cell Motil. Cytoskeleton* **13,** 264–73.

Sluder, G., F. J. Miller, R. Cole, and C. L. Rieder (1990). Protein synthesis and the cell cycle: Centrosome reproduction in sea urchin eggs is not under translational control. *J. Cell Biol.* **110,** 2025–2032.

Sluder, G., E. A. Thompson, F. J. Miller, J. Hayes, and C. L. Rieder (1997). The checkpoint control for anaphase onset does not monitor excess numbers of spindle poles or bipolar spindle symmetry. *J. Cell Sci.* **110,** 421–429.

Stillman, B. (1996). Cell cycle control of DNA replication. *Science* **274,** 1659–1564.

Strausfeld, U. P., M. Howell, P. Descombes, S. Chevalier, R. Rempel, J. Adamczewski, J. L. Maller, T. Hunt, and J. J. Blow (1996). Both cyclin A and cyclin E have S-phase promoting (SPF) activity in *Xenopus* egg extracts. *J. Cell Sci.* **109,** 1555–1563.

Su, J.-Y., R. E. Rempel, E. Erikson, and J. L. Maller (1995). Cloning and characterization of the *Xenopus* cyclin-dependent kinase inhibitor $p27^{XIC1}$. *Proc. Natl. Acad. Sci. USA* **92,** 10187–10191.

Tucker, R. W., A. B. Pardee, and K. Fujiwara (1979). Centriole ciliation is related to quiescence and DNA synthesis in 3T3 cells. *Cell* **17,** 527–535.

Vorobjev, I. A., and Y. S. Chentsov (1982). Centrioles in the cell cycle. I. Epithelial cells. *J. Cell Biol.* **93,** 938–949.

Wheatley, D. N. (1982). *The Centriole: A Central Enigma of Cell Biology.* Elsevier Biomedical Press, Amsterdam.

Winey, M. (1999). Cell cycle: driving the centrosome cycle. *Curr. Biol.* **9,** R449–452.

13

Regulating Centrosomes by Protein Phosphorylation

Andrew M. Fry,[1,*] *Thibault Mayor*,[2] and *Erich A. Nigg*[2]
[1] Department of Biochemistry
University of Leicester
Leicester LE1 7RH, United Kingdom

[2] Department of Cell Biology
Max Planck Institute for Biochemistry
D-82152 Martinsried, Germany

I. Introduction

The centrosome is a highly dynamic structure that undergoes dramatic change in both organization and function as cells pass through their division cycle. During interphase, the centrosome nucleates the microtubule network that is responsible for intracellular transport, cell migration, and the establishment of cell shape and polarity. It also undergoes a single round of duplication in preparation for its role at mitosis, when, by constituting the two spindle poles, the centrosome orchestrates mitotic spindle formation. A major challenge in centrosome research today is to understand how these different activities are regulated at the molecular level. In this chapter, we discuss the role of reversible protein phosphorylation in controlling many different aspects of centrosome structure and function. In particular, we

* Corresponding author.

Current Topics in Developmental Biology, Vol. 49

0070-2153/00 $35.00

focus on the transitions in centrosome biology that are regulated in a cell-cycle-dependent manner and highlight the distinct roles of a number of different families of centrosome-associated protein kinases.

II. Protein Phosphorylation: A Major Theme in Cell Cycle Control

When protein phosphorylation was first described, it was not immediately obvious whether the regulation of proteins by covalent attachment of phosphate might extend beyond a subset of metabolic enzymes. However, it became rapidly apparent that many diverse biological processes are regulated by the protein kinases and phosphatases that catalyze reversible phosphorylation. Cell cycle control is a good example of this paradigm. Passage from one stage of the cell cycle to the next is largely controlled by the sequential activation and inactivation of specific protein kinases. Cyclin-dependent kinases (Cdks) are by far the best-characterized family of kinases controlling the cell cycle (Morgan, 1997; Nigg, 1995; Norbury and Nurse, 1992). They consist of a small catalytic subunit that is only active in the presence of a cyclin regulatory subunit. Mammalian cells have multiple Cdks that are active at different times in the cell cycle and control different transitions. Specifically, Cdk1 is activated at the G2/M transition and is necessary for cells to enter mitosis, Cdk2 is essential for passage through S-phase, and Cdk4/6 is required for transition from G1 to S. Substrate phosphorylation by Cdks is often directly responsible for the biochemical and morphological changes that characterize cell cycle transitions. For example, phosphorylation of the retinoblastoma protein is necessary for S-phase entry, whereas nuclear lamin phosphorylation leads to nuclear lamina disassembly and subsequent breakdown of the nuclear envelope at mitosis. Cdks, though, are not the only protein kinases essential for cell cycle progression, and during the past few years many structurally distinct protein kinases have been identified that are both regulated in a cell-cycle-dependent manner and essential for promoting cell cycle progression.

The centrosome is an organelle whose structure and activity clearly changes through the cell cycle, and these changes are critical not least for its important role in chromosome segregation (Kellogg *et al.,* 1994). Many of the protein components of the centrosome are therefore prime candidates for regulation by cell-cycle-dependent protein kinases. It was first demonstrated more than a decade ago that a significant increase occurs in the number of phosphorylated epitopes at the centrosome during mitosis (Vandre and Borisy, 1989; Vandre *et al.,* 1986). In the mean time, many kinases and phosphatases have been localized to the centrosome (see Table

I), and specific phosphorylation events are likely to take place at the centrosome throughout the cell cycle. Some kinases appear to be only transiently associated with the centrosome, usually coinciding with the time of their peak activity. Others are detected at the centrosome throughout the cell cycle. Many centrosome proteins, including kinases, exist in equilibrium between cytoplasmic and centrosomal pools. Activation of these kinases could therefore take place either at the centrosome itself or elsewhere in the cell before recruitment to the centrosome. It is possible that some cell-cycle-regulated changes are triggered by kinases that are constitutively active or active in response to extracellular stimuli but whose association with the centrosome is dependent upon cell cycle position.

It is worth stating early on in this review that the assignment of a protein kinase to the centrosome needs careful scrutiny. The frequent cross reactivity of animal sera with the centrosome means that information based on

Table I Centrosome-Associated Protein Kinases and Phosphatases

Centrosome Process	Kinase/Phosphatase	Substrates
Centrosome/SPB duplication	Cdk2 (cyclin E or A)	Unknown
	Mps1p	
Microtubule nucleation	Cdk1	CP60
	PKA	AKAP450[a]
	LK6	
	PP4 (phosphatase)	
Maturation	Cdk1	Eg5
	NIMA	Cdk1/cyclin B (recruitment)
	Polo-like kinases	Cdc25, Asp
Separation	Polo-like kinases	pavarotti
	Aurora-related kinases	Eg5
	Nek2	C-Nap1
Putative spindle function	Ca^{2+}/CaMKII	Centrin?
	PP1α (phosphatase)[b]	
Zygotic centrosome assembly	X-Nek2B	Unknown
Cell signaling	α-Casein Kinase I	Unknown
	Casein Kinase II	
	PI-3-kinase	
	Fyn	
	PKC-θ	
Cleavage furrow formation	Polo-like kinases	Unknown
	S.p. Cdc7p	

[a] AKAP450 has been characterized as a scaffold protein for PKA at the centrosome and as such is also likely to be phosphorylated by PKA.

[b] Protein phosphatase 1α has been shown to associate with mitotic spindle poles. In addition, although no direct function has yet been ascribed, mutations in PP1 genes of lower eukaryotes lead to complex mitotic phenotypes including abnormal spindles.

indirect immunofluorescence microscopy alone needs to be interpreted with caution. A more secure approach is to use a combination of methods, including observations with several different antibodies, localization of epitope- or green fluorescent protein–tagged kinases following ectopic expression, and immunoblot analysis of enriched centrosome preparations. If centrosomal localization can be confirmed using each of these methods, there is still the possibility that a kinase is accumulating in the region of the centrosome merely as a result of movement toward the minus ends of microtubules, through either microtubule flux or motor-driven poleward movement. Hence, it is often informative to examine centrosomal association following depolymerization of the microtubule network using drugs such as nocodazole or cold treatment. This should distinguish whether the kinase interacts directly with core centrosomal proteins or rather is directed to the centrosome through microtubule interactions.

III. Centrosome and Spindle Pole Body Dynamics through the Cell Cycle

To appreciate the need for tight control over centrosome function, it is worth giving a brief account of the changes that take place in centrosome structure and activity through a single cell cycle (more detailed descriptions of these events are given in other chapters within this volume). Cells entering G1 of a new cell cycle possess a single centrosome composed of two barrel-shaped centrioles surrounded by pericentriolar material (PCM). Early in G1, the two centrioles move a short distance apart and lose their characteristic orthogonal arrangement in a process of *disorientation.* Semiconservative *duplication* of centrosomes usually then begins around the time of S phase entry, as indicated by the appearance of two procentrioles. These elongate throughout S and G2, reaching full size by the next mitosis. In late G2, a centrosome *maturation* event takes place that is characterized by the recruitment of extra PCM, an increase in centrosome size, and a change in microtubule nucleation activity. As mitosis begins, *separation* of the two mature centrosomes toward either end of the cell occurs and the bipolar mitotic spindle is established. Finally, once chromosome segregation is complete, there is a decrease in centrosome size and cytokinesis ensures that each cell inherits one centrosome to begin the cycle afresh.

Fungal species, such as yeast and *Aspergillus,* use a structurally distinct organelle called the spindle pole body (SPB) as their major site of microtubule nucleation (Brinkley, 1985). This multilayered structure, which sits close to or within the nuclear envelope, is also subject to specific changes during the cell cycle. In budding yeast, the mechanism of SPB duplication seems, at first sight, to be quite different from centrosome duplication. It begins with the initial formation of a half-bridge structure that grows into a second SPB in time for bipolar spindle formation (Winey and Byers, 1993). However, proteins involved in microtubule nucleation have been found to be reasonably well conserved between organisms possessing either SPBs or centrosomes, and it seems probable that proteins regulating other aspects of centrosome function, including duplication, may also be conserved. The elegant molecular dissection of the budding yeast SPB by protein purification and MALDI mass spectrometry has revealed several proteins that are probably involved in regulating SPB structure and function (Wigge *et al.,* 1998). The complexity of phosphorylation at the SPB is revealed by the fact that a large number of SPB components appear to be phosphoproteins, at least as judged by their diffuse migration on SDS-polyacrylamide gels. Already defined as SPB phosphoproteins are the Spc42p protein of the central plaque (Donaldson and Kilmartin, 1996), Spc98p of the γ-tubulin complex (Pereira *et al.,* 1998), and the Spc110p coiled-coil protein that acts as a spacer on the nuclear side of the SPB (Friedman *et al.,* 1996). Phosphorylation of both Spc98p and Spc110p is cell-cycle-dependent and likely to contribute, albeit in an unknown way, to changes in SPB structure and activity through the cell cycle.

Unfortunately, genetic approaches have yielded comparatively little information so far on the regulation of SPB duplication in yeasts. The budding yeast Cdk, Cdc28p, is necessary for the completion of SPB duplication but not, apparently, for its initiation. Calmodulin, together with a related calcium binding protein, Cdc31p, is present within the yeast SPB and appears to be required for SPB duplication (Winey and Byers, 1993). One could speculate that these proteins work through activation of calmodulin-dependent protein kinases that localize to centrosomes of higher eukaryotes (Ohta *et al.,* 1990). The most closely related mammalian protein to Cdc31p is HsCen3 (human centrin gene 3), but there is no evidence yet that HsCen3 is involved in centrosome duplication (Middendorp *et al.,* 1997). Similarly elusive is the mechanism of action of the budding yeast protein kinase, Mps1p, which is necessary for a postmating factor arrest step is SPB duplication (Winey and Byers, 1993). Surprisingly, the same protein kinase also acts in the mitotic spindle checkpoint, raising the intriguing possibility that a link exists between these two important processes.

Later, we discuss at greater length some of the individual events of the centrosome cycle in higher eukaryotes and the protein kinases that have been implicated in their regulation (summarized in Fig. 1).

IV. In Control of Centrosome Duplication

To ensure equal segregation of chromosomes at mitosis, centrosome duplication must be coordinated with chromosome replication. In this way,

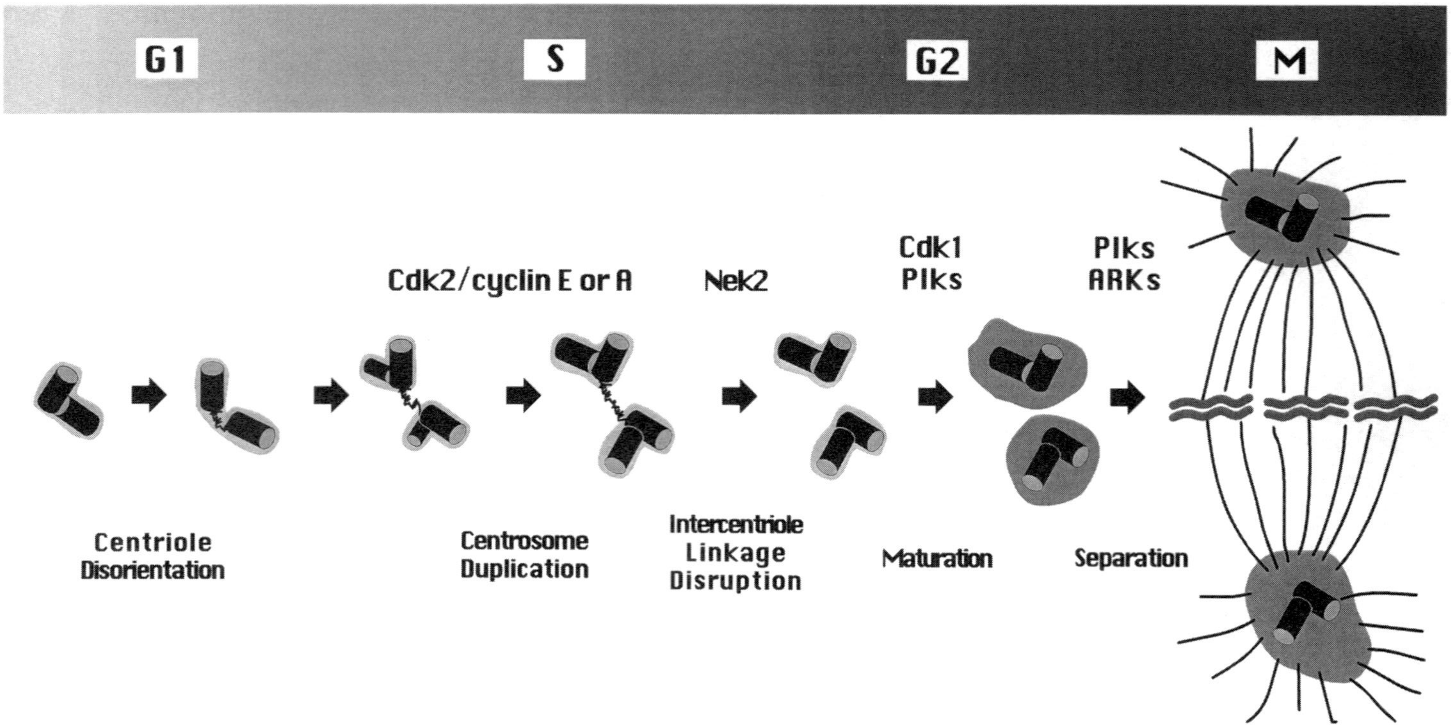

Figure 1 Control of the centrosome cycle by protein kinases. The major morphological changes in structure and organization of the animal cell centrosome are shown with respect to cell cycle position. Those kinases for which good evidence supports a role in a particular process within this cycle are indicated above the relevant point. Note that intercentriolar linkage disruption is drawn as occurring before centrosome maturation and separation. This is based on an apparent splitting of centrosomes being visible in some species even in early G2. However, it should be stressed that the precise timing of these events is likely to vary in different systems and may often take place simultaneously at the G2-to-M-phase transition. The intercentriolar linkage is indicated by the jagged black line between mother and daughter centrioles. Objects are not drawn to scale.

cells with replicated DNA enter mitosis with no more or less than two centrosomes and, hence, form a spindle that is bipolar. Clearly, one would predict that regulatory mechanisms should exist to ensure chromosomal and centrosomal synchronization under normal growth conditions. However, in some cell types, such as Chinese hamster ovary (CHO) cells, it has proved possible to uncouple DNA synthesis and centrosome replication using drugs, suggesting that to some extent the control of these pathways is independent (Balczon *et al.,* 1995). Similarly, mutations in a number of *Drosophila* genes, for instance Gnu (Freeman *et al.,* 1986), produce uncoupling of nuclear and centrosomal cycles. The demonstration that reduced PP2A levels cause the appearance of multiple centrosomes in cellularized embryos underscores the importance of phosphorylation in the coordination of these processes (Snaith *et al.,* 1996). Recent approaches in embryonic and somatic systems have provided a first glimpse into how protein kinases may coordinate DNA synthesis and centrosome duplication. Together, they point to Cdk2 as key player (Hinchcliffe *et al.,* 1999; Lacey *et al.,* 1999; Matsumoto *et al.,* 1999; Meraldi *et al.,* 1999).

In *Xenopus* cell-free egg extracts, it was found that Cdk2 is necessary to drive multiple rounds of centrosome duplication during an aphidicolin-induced S-phase arrest (Hinchliffe *et al.,* 1999). Addition of the Cdk inhibitor protein Xic1 to the extracts blocked centrosome duplication, but this could be restored by the addition of an excess of purified Cdk2/cyclin E. Data in line with this study were also obtained in *Xenopus* early embryos where microinjection of the Cdk inhibitors p21 or p27 blocked centrosome duplication (Lacey *et al.,* 1999). Recent results also support a requirement for Cdk2 in centrosome duplication in somatic mammalian cells. This was determined through overexpression of cell cycle regulators in CHO cells blocked in S-phase with hydroxyurea (Matsumoto *et al.,* 1999; Meraldi *et al.,* 1999). In somatic cells, though, cyclin A appears to be the major partner for Cdk2 in this process rather than cyclin E (Meraldi *et al.,* 1999). It is possible that this simply reflects variations between embryonic and somatic cell cycles as, for instance, Cdk2 associates with only cyclin E, and not cyclin A, prior to the mid-blastula stage of *Xenopus* embryo development. The somatic cell studies also indicate that centrosome duplication requires E2F transcriptional activity (Meraldi *et al.,* 1999). Unfortunately, little is known about the transcriptional regulation of those centrosomal proteins identified so far and, thus, it is not possible to tell whether any of these are targets of E2F-induced transcription.

A key role for Cdk2 in allowing repeated rounds of centrosome duplication fits with an idea, proposed by Sluder and colleagues, that centrosome duplication is "licensed" during S-phase (Hinchcliffe *et al.,* 1998). Maintaining cells in S-phase with artificially high Cdk2 activity might thus allow centrosome reduplication to occur while, presumably, in the normal cell cycle a comparatively short peak of Cdk2 activity in S-phase is sufficient to trigger only one round of duplication. It is worth remembering, though, that, unlike CHO cells, many other cell lines do not continue to duplicate their centrosomes in the presence of hydroxyurea, implying that they have retained some intrinsic mechanism to prevent the excessive accumulation of centrosomes during an S-phase arrest.

V. Regulating Microtubule Nucleation

The primary functions of the centrosome are to nucleate microtubule polymerization and to act as a focal point for the microtubule network. The same can be said of its fungal counterpart, the SPB. Considerable advances have been made into the mechanism of microtubule nucleation from the centrosome/SPB, revealing not least that many of the proteins involved show substantial similarity from yeast to humans. One of the most highly studied centrosomal proteins, γ-tubulin, is a distinct member of the tubulin superfamily and is essential for microtubule nucleation at the centrosome (for reviews, see Andersen, 1999; Zimmerman *et al.*, 1999; Murphy and Stearns, 1996; Pereira and Schiebel, 1997; and other chapters in this volume). Importantly, γ-tubulin has been shown to exist in progressively higher order complexes that include homologs of the yeast Spc97p/98p proteins as well as pericentrin (Dictenberg *et al.*, 1998; Martin *et al.*, 1998; Murphy *et al.*, 1998; Oegema *et al.*, 1999; Tassin *et al.*, 1998). These insights have led to the conclusion that centrosome organization and microtubule nucleation are interdependent processes and this is likely to be reflected in their regulation. There is a gradual increase in the amount of both pericentrin and γ-tubulin at the centrosome as cells progress toward mitosis before the components are abruptly lost as cells exit back into G1 (Dictenberg *et al.*, 1998). One mechanism for augmenting the extent of microtubule nucleation through the cell cycle, therefore, consists in stimulating the association of γ-tubulin complexes with the centrosome. Whether this is phosphorylation-dependent remains to be proven but is an attractive possibility.

The steady recruitment alone, though, cannot fully explain the sudden increase in microtubule nucleation that occurs at the onset of mitosis, and there is strong evidence to support an important role for protein phosphorylation. The phosphoepitope-specific antibody MPM-2 produces significant centrosome staining in mitosis, and either preincubation with MPM-2 antibodies or phosphatase treatment can block microtubule nucleation from mitotic centrosomes (Centonze and Borisy, 1990). Increased nucleation from isolated centrosomes can be directly stimulated in interphase *Xenopus* egg extracts by addition of Cdk1/cyclin complexes (Buendia *et al.*, 1992;

Ohta *et al.,* 1993; Verde *et al.,* 1990). These results indicate that Cdk1 is a prime candidate for modulating the microtubule nucleation capacity of the centrosome. Cdk1 is clearly associated with centrosomes/spindle poles from the onset of mitosis (Bailly *et al.,* 1989), although there is some debate over whether it may also be present earlier in the cell cycle (Pockwinse *et al.,* 1997). Precisely which Cdk1/cyclin complexes are involved is at present an unresolved issue. Using bacterially expressed cyclins, Buendia *et al.* (1992) found that cyclin A, but not cyclin B, could stimulate the nucleation capacity of centrosomes incubated in interphase *Xenopus* egg extracts. However, when Cdk1/cyclin B purified from starfish oocytes was used, one group reported no increase in nucleation activity (Buendia *et al.,* 1992), whereas another noted a marked increase (Ohta *et al.,* 1993). The method of interphase egg extract preparation might be critical to these results.

Other centrosomal kinases have been proposed to regulate microtubule nucleation. These include LK6, a *Drosophila* kinase, and PKA, both of which are present at the centrosome throughout the cell cycle and prominently detected on spindle poles (De Camilli *et al.,* 1986; Kidd and Raff, 1997; Nigg *et al.,* 1985). Constitutive high-level expression of active, but not inactive, LK6 leads to a variety of microtubule defects, including ectopic aster formation in *Drosophila* eggs and early embryos (Kidd and Raff, 1997). Protein phosphatase 1α (PP1α) and protein phosphatase 4 (PP4) associate with centrosomes and presumably act in competition with centrosomal kinases (Andreassen *et al.,* 1998; Brewis *et al.,* 1993). PP4 appears to directly influence either microtubule nucleation rates or microtubule attachment to centrosomes as PP4 gene disruption reduces the amount of γ-tubulin and number of microtubules at the centrosome (Helps *et al.,* 1998). The results with PP4 suggest that dephosphorylation of certain proteins stimulates microtubule growth or anchoring, emphasizing that not all phosphorylation events are necessarily of an activating nature.

It is apparent that a direct link between the core components of nucleation complexes and the enzymes that supposedly control nucleation activity is still absent. In fact, few candidate substrates for regulating microtubule nucleation rates have been identified. Phosphorylation of γ-tubulin itself has not been detected. On the other hand, yeast Spc98p, a conserved member of the γ-TuSC, is phosphorylated in a cell-cycle-dependent manner, specifically on the nuclear side of the SPB (Pereira *et al.,* 1998). Although it is attractive to speculate that this affects microtubule nucleation capacity, this notion remains to be tested. Besides Spc98p, other potential substrates include the *Drosophila* centrosomal phosphoprotein CP60 and the high molecular weight microtubule-associated protein *abnormal spindle* or Asp (Kellogg *et al.,* 1995; Saunders *et al.,* 1997). CP60 binds microtubules in a dephosphorylated form, and association with centrosomes is maximal during mitosis. However, it is also a good *in vitro* substrate of Cdk1. One interpretation of these results is that CP60 is involved in microtubule attachment to the mitotic spindle pole and that phosphorylation by Cdk1 stimulates microtubule release. Asp, on the other hand, was recently proposed to be involved in recruiting γ-TuRCs to the centrosome. It was seen that the ability of *Drosophila* embryo extracts to restore microtubule-nucleating activity to salt-stripped isolated centrosome preparations was lost by either Asp mutation or immunodepletion (Avides and Glover, 1999).

Phosphorylation is also believed to play a major role in regulating microtubule dynamics (McNally, 1996, 1999). Multiple kinases, including Cdk1, MAPK, and PKA, have all been implicated in regulating microtubule dynamics, as indeed have certain protein phosphatases. The most likely substrates are microtubule-associated proteins that are known to regulate the growth and stability of microtubule such as MAP4, XKCM1, and Opl8/stathmin. However, as phosphorylation of microtubule-associated proteins need not take place at the centrosome, a detailed discussion of this subject is beyond the scope of this review. Of course, it is likely that other mechanisms beyond phosphorylation are also important in the regulation of microtubule nucleation and dynamics. Recent elegant work has indicated that a major contribution to microtubule regulation comes through the action of the small GTPase Ran. Ran, in its GTP-bound form, can strongly stimulate aster formation from sperm centrosomes in *Xenopus* egg extracts (Ohba *et al.,* 1999; Wilde and Zheng, 1999). Moreover, a Ran-binding protein, RanBPM, accumulates on centrosomes, and incubation with RanBPM antibodies blocks aster formation (Nakamura *et al.,* 1998).

VI. Establishing a Bipolar Mitotic Spindle

Mitotic spindle formation is a fascinating and complex process that involves major contributions from centrosomes, microtubules, and chromosomes (Merdes and Cleveland, 1997). The G2/M transition is accompanied by two changes in centrosome organization that are discernible by light microscopy: first, centrosome maturation, characterized by an increase in centrosome volume as a result of protein recruitment; and second, centrosome separation, when the two centrosomes move apart to form the opposing spindle poles. For the sake of clarity, we have subdivided our discussion into the role of phosphorylation in these two events. However, it should be stressed that these processes are naturally interwoven, and hence, a protein kinase that regulates one process may significantly contribute to the other. In particular, the initiation of centrosome separation is likely to require prior recruitment of protein complexes to the centrosome; thus, kinases involved in the recruitment step may also appear necessary for separation.

A. Centrosome Maturation

Many proteins described as being centrosomal are only such during mitosis. Others, found at the centrosome throughout the cell cycle, are nevertheless present at the centrosome at much higher levels during mitosis, e.g., γ-tubulin. The regulators that stimulate this protein recruitment in late G2 remain poorly understood, although cyclin-dependent kinases are again among the prime candidates. Phosphorylation by Cdk1/cyclin B is necessary for centrosomal recruitment of Eg5, a member of the bimC kinesin-related motor protein family (Blangy *et al.,* 1995; Sawin and Mitchison, 1995). Following recruitment, Eg5 actively participates in centrosome separation and bipolar spindle assembly. Cdk1 itself also exhibits strong association with the centrosome at prophase and, in *Aspergillus nidulans,* it has been reported that another protein kinase, NIMA, may positively direct Cdk1/cyclin B to the centrosome/SPB, as well as to the nucleus (Wu *et al.,* 1998). A third kinase implicated in centrosome maturation is the polo-like kinase. First described in *Drosophila,* polo-like kinases (Plks) are found universally from fungi to vertebrates and studies in different organisms have indicated multiple roles for these kinases at different stages of mitosis (Glover *et al.,* 1998; Nigg, 1998). Importantly, in the context of this review, all known Plks display transient association with spindle poles in early mitosis and, in almost all species, inactivation leads to some form of defect in the establishment of the mitotic spindle. More specifically, Plk inactivation causes certain deficiencies in centrosome maturation suggesting that this might be the first function of these kinases at the onset of mitosis. In fly embryos, *polo* mutants fail to recruit the CP190 centrosomal protein (Sunkel and Glover, 1988) and, in a similar manner, microinjection of anti-Plk1 antibodies into immortalized human tissue culture cells leads to monopolar spindles with small immature centrosomes that fail to recruit γ-tubulin and MPM-2 phosphoepitopes (Lane and Nigg, 1996). As indicated earlier, the *Drosophila* protein Asp is required for γ-TuRC recruitment and, intriguingly, Asp might be a potential Plk substrate (Gonzalez *et al.,* 1998). Failure to phosphorylate a human homolog of Asp may, therefore, provide one explanation for the defect in γ-tubulin recruitment in the microinjection experiment. An early mitotic substrate of Plks is the dual-specificity phosphatase Cdc25, whose role is to activate Cdk1 through threonine and tyrosine dephosphorylation (Kumagai and Dunphy, 1996). This activation cascade is likely to occur in large part at the centrosome, as this is where Plks and Cdk1 are concentrated at this time. Thus, with both Plks and Cdk1 being implicated in maturation, it would seem that multiple protein kinases are required to bring about proper centrosome maturation (Fig. 2).

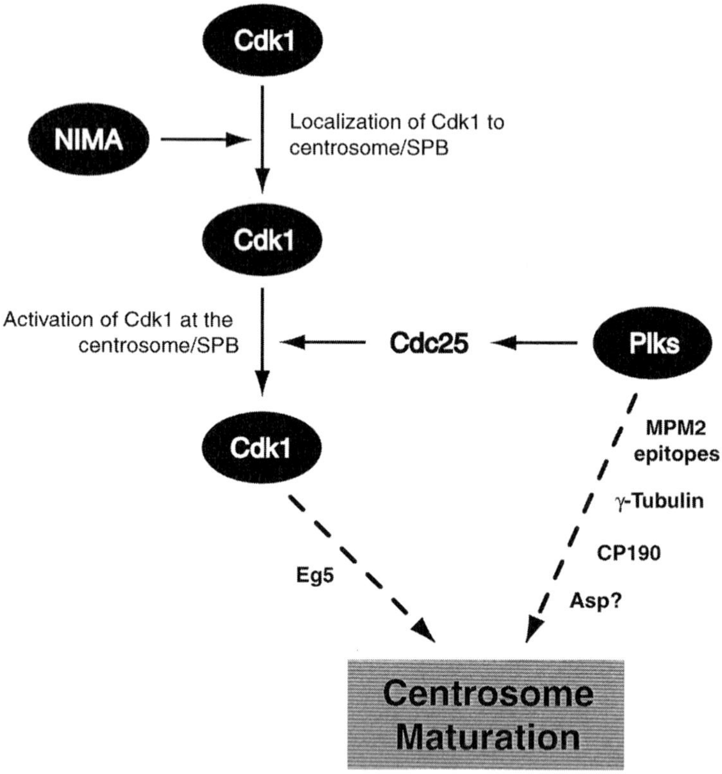

Figure 2 Protein kinases implicated in centrosome maturation. This model presents a speculative view of the interactions that may be required to achieve centrosome/SPB maturation and is based upon a number of studies in different experimental organisms. So far, members of at least three structurally distinct protein kinase families, Cdks, Plks, and NIMA-related kinases, have been implicated in centrosome/SPB maturation. In *Aspergillus nidulans,* NIMA is required for correct localization of Cdk1/cyclin B to the spindle pole body, as well as to the nucleus. Cdk1 itself has then been shown to stimulate centrosomal recruitment of the plus-end-directed kinesin-like motor protein Eg5 and is likely to be involved in recruiting other proteins as well. Human Plk1 is required for centrosome recruitment of γ-tubulin and the appearance of MPM-2 phosphoepitopes on the centrosome, whereas *Drosophila* polo is necessary for CP190 and, perhaps, Asp recruitment. In parallel, it has been demonstrated that Plks are required for activation of Cdk1 via the phosphatase Cdc25. Hence, protein phosphorylation is used to maintain a tightly coordinated control over the timing of centrosome maturation.

B. Centrosome Separation

Centrosome separation and spindle establishment occur simultaneously, and disturbance of one usually leads to disruption of the other. The reason for this is straightforward, as many of the proteins that actively move centrosomes apart, such as motor proteins, do so by binding and crosslinking microtubules in a manner that by consequence leads to spindle assembly. Disruption of the *Drosophila* gene *polo,* or its fission yeast counterpart *plo1,* induces the formation of monopolar spindles (Llamazares *et al.,* 1991; Ohkura *et al.,* 1995; Sunkel and Glover, 1988). This might suggest that Plks have a direct role in centrosome separation. Furthermore, identification of the kinesin-like motor protein *pavarotti* and some microtubule-associated proteins as candidate *polo* substrates in *Drosophila* falls in line with an active role for Plks in spindle assembly (Adams *et al.,* 1998; Tavares *et al.,* 1996). However, as noted in the previous section, Plks are also required for centrosome maturation, and so the failure to separate centrosomes could also be a consequence of a primary defect in maturation.

Aurora-related kinases (ARKs) have recently emerged as an important new family of mitotic regulators, some of which apparently play a specific role in centrosome separation. Named after the Aurora gene in *Drosophila,* ARKs have been identified in many species, including budding yeast, nematode, frog, mouse, rat, and human (as examples, see Chan and Botstein, 1993; Gopalan *et al.,* 1997; Kimura *et al.,* 1997; Roghi *et al.,* 1998; Schumacher *et al.,* 1998; Shindo *et al.,* 1998; Terada *et al.,* 1998). Based on sequence, localization, and function, ARKs appear to exist in at least two distinct subfamilies: the ARK1 subfamily (Ce-AIR1, Dm-aurora, X1-Eg2, Mm-ARK1/ayk1/IAK1, Hs-AIK/BTAK/aurora2/STK15), which is required for centrosome separation and mitotic spindle assembly; and the ARK2 subfamily (Ce-AIR2, Mm-ARK2/STK1, Rn-AIM1, Hs-STK12/AIK2/aurora1), which appears to be required for cytokinesis. Additional ARKs (Mm-AIE1, Hs-STK13/AIE2/AIK3) have been found in mammals that are expressed specifically in testis but that do not easily fit either the ARK1 or ARK2 criteria. Interestingly, AIK3 localizes to the spindle poles only during late mitosis, although the consequences of this remain obscure (Kimura *et al.,* 1999). Budding yeast possess a single gene homologous to Aurora, called Ipl1, mutation of which leads to chromosome segregation defects (Chan and Botstein, 1993).

Protein abundance, activity, and localization fit with a role for the ARK1 subfamily in centrosome separation. The protein level and kinase activity of ARK1 family members are low in G1/S, increase in G2/M, and peak in mitosis, while immunofluorescence microscopy reveals that these proteins generally associate with the centrosome from G2 until telophase. Mutation of the *Drosophila* Aurora gene leads to a failure in centrosome separation and the formation of monopolar spindles (Glover *et al.,* 1995). In *Xenopus* cell-free egg extracts, bipolar spindle formation is prevented when the *Xenopus* ARK1, Eg2, is inhibited by a dominant-negative mutant, a result that could well reflect failure of centrosome separation (Roghi *et al.,* 1998b). In *Caenorhabditis elegans,* results obtained by RNA interference suggest that AIR-1 is necessary for spindle assembly, but not for centrosome separation (Schumacher *et al.*, 1998). However, to confirm this result it will be important to analyze a bona fide AIR-1 mutant. The first ARK1 substrate has been reported and, perhaps not surprisingly, is a motor protein already mentioned in the context of centrosome separation, namely Eg5. Eg5, previously shown to be phosphorylated by Cdk1 in its tail domain to stimulate centrosome recruitment, is also phosphorylated by Eg2 in its stalk domain (Giet *et al.*, 1999). The key experiment now is to examine by mutation whether this phosphorylation event is specifically required for the centrosome separation activity of the Eg5 motor.

Ultrastructural studies have indicated that, during interphase, centrosomes are physically held together by an intercentriolar linkage that is composed of fibrous components of the PCM. This raises the question of how this linkage is severed to allow centrosome separation to proceed. Recent experiments have implicated a member of the NIMA protein kinase family in regulation of what would appear to be a preparatory step for centrosome separation. Nek2, for NIMA-related kinase 2, is a mammalian serine/threonine kinase that is structurally related to *Aspergillus* NIMA (Schultz *et al.*, 1994). Nek2 protein abundance and kinase activity are cell-cycle-regulated with peak levels in S/G2, although localization studies indicate that Nek2 is a core component of the centrosome throughout the cell cycle (Fry *et al.*, 1995, 1998b). Overexpression of active, but not inactive, Nek2 kinase in cultured cells induces a premature splitting of centrosomes that is not associated with spindle formation or with recruitment of motor proteins such as Eg5 to the centrosome. Hence, Nek2 activity may somehow stimulate a loss of cohesion between the duplicated centrosomes prior to organization of the mitotic spindle.

A candidate centrosomal substrate of Nek2, called C-Nap1, was identified through a yeast two-hybrid interaction screen with Nek2 as a bait (Fry *et al.*, 1998a). This novel 281-kDa coiled-coil protein was also isolated independently using human autoimmune sera reactive against centrosomal antigens (Mack *et al.*, 1998). C-Nap1 can be strongly phosphorylated by Nek2 on its carboxy-terminal domain, suggesting that it may be a bona fide substrate of this kinase. However, whereas Nek2 is detectable on centrosomes throughout the cell cycle, C-Nap1 staining disappears from centrosomes upon the onset of mitosis, only reappearing as cells enter G1. Most intriguingly, in interphase cells, both C-Nap1 and Nek2 specifically associate with the proximal ends of the two centrioles, where electron-dense fibrous material has been previously observed. Based on these results, a working model implicates Nek2 and C-Nap1 in regulating the intercentriolar linkage that binds the two centrosomes throughout interphase (discussed in more detail in Mayor *et al.*, 1999). C-Nap1 localization implies that it is in a strategic position to anchor the ends of this structure to the centrioles. At the G2/M transition, phosphorylation by Nek2 might alter C-Nap1 interactions or stability, leading to disappearance of C-Nap1 and dissolution of the intercentriolar linkage prior to the second step of motor-protein-driven centrosome separation.

Dissolution of the intercentriolar linkage might also take place at other times in the cell cycle in response to extracellular signals. EGF stimulation of quiescent 3T3 cells or neutrophil stimulation by chemoattractants both induce a transient splitting of the two centrioles, and neutrophil activation also leads to increased microtubule nucleation (Schliwa *et al.,* 1982; Sherline and Mascardo, 1982). These responses are more likely to be the result of receptor-mediated protein kinase signaling cascades than cell-cycle-regulated processes and may involve cytosolic kinases transiently associated with centrosomes during the brief time that the corresponding signals are on. Indeed, a number of kinases more usually associated with cell signaling than cell cycle have been described as centrosomal, including Ca^{2+}/calmodulin-dependent protein kinase, CK I, CK II, PKC-θ, PI-3-kinase, and the src family tyrosine kinase, fyn (Brockman *et al.,* 1992; Kapeller *et al.,* 1993; Krek *et al.,* 1992; Ley *et al.,* 1994; Ohta *et al.,* 1990; Passalacqua *et al.,* 1999). It is possible, though, that some of these localizations may reflect the apparent importance of the centrosome to act as a "meeting place," promoting, by means of increased local protein concentration, intracellular reactions that have little or no direct influence on centrosome structure or function. In support of such a model, two large scaffold proteins have recently been localized to the centrosome, AKAP450 and CG-NAP (Schmidt *et al.,* 1999; Takahashi *et al.,* 1999; Witczak *et al.,* 1999). Scaffold proteins exist to bring kinases into direct contact with their substrates. AKAP450 is an A-kinase anchoring protein, whereas CG-NAP binds several signaling molecules, including PKA, PKN, and PP2A. Future studies will be needed to distinguish between those pathways that lead directly to centrosome regulation, and those where the centrosome merely acts as a solid-phase platform on which catalytic reactions proceed.

VII. Creating the Zygotic Centrosome

The generation of the zygotic centrosome is an intriguing but poorly understood event (Schatten, 1994). In the vast majority of animal species, the centrosome is paternally inherited in the form of the sperm basal body. Fertilization leads to conversion of the basal body, which is capable of subtending only a single flagellum, into a structure fully competent to organize the bipolar spindle of the first embryonic division. Rare exceptions do exist, for instance during mouse development, where centrioles arise *de novo* after the first few cell divisions and, before which, spindle poles are organized from maternal stores of microtubule-nucleating material (Schatten *et al.*, 1986, 1991). In both cases, however, evidence is accumulating that centrosome assembly is not a haphazard event, but rather takes place in a highly organized and structured manner.

The sperm basal body, as well as containing two centrioles, contains some core centrosomal proteins, including centrin and pericentrin (Stearns and Kirschner, 1994). Other key proteins, such as γ-tubulin, are missing. Remodeling the sperm basal body into the zygotic centrosome, therefore, involves the ordered recruitment of both core and peripheral centrosomal components. That this process may in part depend upon phosphorylation comes from observations that when *Xenopus* sperm are incubated in egg extracts, as well as gaining the ability to nucleate microtubule asters, their centrosomes become positive for MPM-2 phosphoepitopes (Stearns and Kirschner, 1994). There is also a major recruitment of the γ-tubulin protein to the sperm centrosome, and this process depends upon ATP (Felix *et al.*, 1994; Stearns and Kirschner, 1994). Immunoblotting reveals that *Xenopus,* human, and bovine sperm do contain low levels of γ-tubulin prior to incubation in extract, but that the centrosome is not capable of nucleating microtubule asters in this condition (Navara *et al.*, 1997; Tassin *et al.*, 1998). Permeabilized human sperm incubated in *Xenopus* egg extracts also acquire phosphorylation on the centrosome, although to generate an aster they require an additional priming step of disulfide reduction (Schatten, 1994). Thus, to generate a competent MTOC from the sperm centrosome, it appears that both regulated recruitment and activation of preexisting proteins must occur, and that somewhere along the line protein phosphorylation is needed. A first step toward identifying this elusive requirement has come from recent studies on X-Nek2B, the *Xenopus* homolog of the human Nek2 kinase. X-Nek2B, absent from the basal body, is recruited extremely rapidly from egg extracts onto sperm centrosomes. Moreover, this recruitment may be a prerequisite for further recruitment of other proteins (Fry *et al.*, 2000). A complete understanding of this subject is particularly important, as failure to properly assemble the zygotic centrosome has been proposed as one major cause of infertility (Schatten, 1994).

VIII. Loss of Phosphorylation Control: Cancer Implications

Defects in processes such as centrosome duplication or separation will prevent the cell from making a bipolar spindle. Under these circumstances, cell cycle progression should ideally be arrested by the spindle checkpoint that detects faulty attachment of sister chromatid kinetochores to microtubules and perhaps other spindle defects (Gorbsky, 1997; Rudner and

Murray, 1996). However, following either adaptation to the checkpoint or acquisition of mutations in checkpoint genes, centrosome abnormalities take on a menacing potential for introducing gross chromosomal aberrations. Thus, following up on ideas originally voiced at the beginning of the century (Boveri, 1914), the cancer research community is showing renewed interest in the centrosome. Indeed, many primary tumors are characterized by the presence of supernumerary centrosomes (Lingle *et al.,* 1998; Pihan *et al.,* 1998). However, it is not yet clear whether a primary defect in centrosome regulation exists in these tumors or whether multiple centrosomes accumulate rather as an indirect result of defects in cell cycle progression. Molecular information is now emerging about pathways that could potentially lead to centrosome amplification in tumor cells. Abnormal duplication of centrosomes has been reported in mouse embryonic fibroblasts lacking the cyclin-dependent kinase inhibitor p21 (Mantel *et al.,* 1999) or either of the tumor suppressor genes p53 or BRCA1 (Fukasawa *et al.,* 1996; Xu *et al.,* 1999). BRCA1 has also been reported to associate with spindle poles, where its phosphorylation state might regulate an interaction with γ-tubulin (Hsu and White, 1998).

The human centrosomal kinase ARK1 (AIK/BTAK/aurora2/STK15) is overexpressed in both breast and colon tumors as well as a variety of different cancer cell lines, and this may result from the frequent amplification in cancers of the ARK1 locus on chromosome 20q13 (Bischoff *et al.,* 1998; Sen *et al.,* 1997; Zhou *et al.,* 1998). The critical question is whether ARK1 is the relevant "cancer" gene on this amplicon or merely an innocent bystander. Interestingly, ectopic expression of ARK1 in cultured cells maintained in low serum induces an increase in centrosome number and aneuploidy and can lead to transformation *in vitro* (Bischoff *et al.,* 1998; Zhou *et al.,* 1998). This raises the possibility that aberrant centrosome regulation induced by ARK1 overexpression could permit missegregation of chromosomes and thus contribute to genetic instability and tumor progression. Human Plk1 is another centrosomal kinase whose expression has been found to be elevated in a number of different tumor types (Holtrich *et al.,* 1994; Wolf *et al.,* 1997; Yuan *et al.,* 1997). Furthermore, its overexpression in mammalian cells is reported to lead to multinucleation and transformation (Mundt *et al.,* 1997; Smith *et al.,* 1997). However, as Plk1 has a wide variety of mitotic targets besides the centrosome, one cannot yet link these effects directly to centrosome deregulation. Meanwhile, inappropriate phosphorylation of centrosomal proteins has also been detected in human tumors. Notably, the small calcium-binding centrosomal protein centrin shows a number of abnormalities in breast tumor cells, including excessive hyperphosphorylation (Lingle *et al.,* 1998), and it is possible that the phosphorylation state of many centrosomal proteins will prove to be important in tumor cell progression.

IX. Perspectives: Into the Next Millennium

To conclude, the work discussed here clearly demonstrates that phosphorylation plays a vital role in regulating centrosome architecture and function during progression of the animal cell cycle. Moreover, many of the centrosome-associated protein kinases that catalyze these events have now been identified. What, then, are the major challenges for the next stage of research into centrosome regulation? First, it will be essential to make that all-important connection between centrosomal kinases and substrates so that we can understand how phosphorylation triggers the changes that we observe at the morphological level. For example, we know next to nothing about the targets of phosphorylation that allow the initiation of centriole duplication or lead to increased rates of microtubule nucleation. Second, we need to learn how centrosomal kinases themselves are regulated both in respect to cell cycle progression and in response to external signals. For instance, phosphorylation of centrosomal proteins is likely to account for many of the changes that take place at the centrosome during establishment of cell polarity, directional cell migration, and generation of a ciliated epithelium. Third, an important area for investigation is the role of the centrosome in determining mitotic spindle orientation and cleavage furrow positioning. Experiments in fission yeast suggest that asymmetric localization of a protein kinase, *S.p.* Cdc7p, to the SPB is necessary for correct septation (Sohrmann *et al.*, 1998). Plks, too, may have an additional role in signaling from parts of the spindle apparatus to the site of the future cleavage furrow (Glover *et al.*, 1998). It will be intriguing to learn how proteins concentrated at the centrosome/SPB can trigger events at some distance away within the cell. Finally, there is no doubt that many of these phosphorylation events are critical in preparing the centrosome for the establishment of the mitotic spindle and maintaining its bipolarity from one cell division to the next. The proposal that supernumerary centrosomes might provoke genetic instabilities and contribute to cancer cell development potentially implicates centrosome-associated protein kinases and phosphatases in tumorigenesis. Understanding how these enzymes regulate the centrosome cycle as well as elucidating the consequences of their deregulation is going to present a formidable task indeed for the start of the new millennium.

Acknowledgments

We thank colleagues in our laboratories and elsewhere in the centrosome community for stimulating discussions on the actions of protein kinases at the centrosome, and Patrick Meraldi for critical reading of the manuscript. Work in the authors' laboratories is supported by grants from The Wellcome Trust (056335) and the Cancer Research Campaign (SP2486/01/01) to A.M.F., and from the Swiss National Science Foundation (31-50576.97), the Swiss Cancer League (SKL 267-1-1996), and the Canton of Geneva to E.A.N.

References

Adams, R. R., Tavares, A. A., Salzberg, A., Bellen, H. J., and Glover, D. M. (1998). *Genes Dev.* **12,** 1483–1494.

Andersen, S. S. L. (1999). *Int. Rev. Cytol.* **187,** 51–109.

Andreassen, P. R., Lacroix, F. B., Villa-Moruzzi, E., and Margolis, R. L. (1998). *J. Cell Biol.* **141,** 1207–1215.

Avides, M. C., and Glover, D. M. (1999). *Science* **283,** 1733–1735.

Bailly, E., Doree, M., Nurse, P., and Bornens, M. (1989). *EMBO J.* **8,** 3985–3995.

Balczon, R., Bao, L., Zimmer, W. E., Brown, K., Zinkowski, R. P., and Brinkley, B. R. (1995). *J. Cell Biol.* **130,** 105–115.

Bischoff, J. R., Anderson, L., Zhu, Y., Mossie, K., Ng, L., Souza, B., Schryver, B., Flanagan, P., Clairvoyant, F., Ginther, C., Chan, C. S., Novotny, M., Slamon, D. J., and Plowman, G. D. (1998). *EMBO J.* **17,** 3052–3065.

Blangy, A., Lane, H. A., d'Herin, P., Harper, M., Kress, M., and Nigg, E. A. (1995). *Cell* **83,** 1159–1169.

Boveri, T. (1914). "The Origin of Malignant Tumors." Williams and Wilkins, Baltimore.

Brewis, N. D., Street, A. J., Prescott, A. R., and Cohen, P. T. (1993). *EMBO J.* **12,** 987–996.

Brinkley, B. R. (1985). *Ann. Rev. Cell Biol.* **1,** 145–172.

Brockman, J. L., Gross, S. D., Sussman, M. R., and Anderson, R. A. (1992). *Proc. Natl. Acad. Sci. USA* **89,** 9454–9458.

Buendia, B., Draetta, G., and Karsenti, E. (1992). *J. Cell Biol.* **116,** 1431–1442.

Centonze, V. E., and Borisy, G. G. (1990). *J. Cell Sci.* **95,** 405–411.

Chan, C. S., and Botstein, D. (1993). *Genetics* **135,** 677–691.

De Camilli, P., Moretti, M., Donini, S. D., Walter, U., and Lohmann, S. M. (1986). *J. Cell Biol.* **103,** 189–203.

Dictenberg, J. B., Zimmerman, W., Sparks, C. A., Young, A., Vidair, C., Zheng, Y., Carrington, W., Fay, F. S., and Doxsey, S. J. (1998). *J. Cell Biol.* **141,** 163–174.

Donaldson, A. D., and Kilmartin, J. V. (1996). *J. Cell Biol.* **132,** 887–901.

Felix, M.-A., Antony, C., Wright, M., and Maro, B. (1994). *J. Cell Biol.* **124,** 19–31.

Freeman, M., Nusslein-Volhard, C., and Glover, D. M. (1986). *Cell* **46,** 457–468.

Friedman, D. B., Sundberg, H. A., Huang, E. Y., and Davis, T. N. (1996). *J. Cell Biol.* **132,** 903–914.

Fry, A. M., Schultz, S. J., Bartek, J., and Nigg, E. A. (1995). *J. Biol. Chem.* **2705,** 12899–12905.

Fry, A. M., Mayor, T., Meraldi, P., Stierhof, Y. D., Tanaka, K., and Nigg, E. A. (1998a). *J. Cell Biol.* **141,** 1563–1574.

Fry, A. M., Meraldi, P., and Nigg, E. A. (1998b). *EMBO J.* **17,** 470–481.

Fry, A. M., Descombes, P., Twomey, C., Bacchieri, R., and Nigg, E. A. (2000). *J. Cell Sci.* **113,** 1973–1984.

Fukasawa, K., Choi, T., Kuriyama, R., Rulong, S., and Vande Woude, G. F. (1996). *Science* **271,** 1744–1747.

Giet, R., Uzbekov, R., Cubizolles, F., Le Guellec, K., and Prigent, C. (1999). *J. Biol. Chem.* **274,** 15005–15013.

Glover, D. M., Leibowitz, M. H., McLean, D. A., and Parry, H. (1995). *Cell* **81,** 95–105.

Glover, D. M., Hagan, I. M., and Tavares, A. A. (1998). *Genes Dev.* **12,** 3777–3787.

Gonzalez, C., Sunkel, C. E., and Glover, D. M. (1998). *Chromosoma* **107,** 452–460.

Gopalan, G., Chan, C. S., and Donovan, P. J. (1997). *J. Cell Biol.* **138,** 643–656.

Gorbsky, G. J. (1997). *BioEssays* **19,** 193–197.

Helps, N. R., Brewis, N. D., Lineruth, K., Davis, T., Kaiser, K., and Cohen, P. T. W. (1998). *J. Cell Sci.* **111,** 1331–1340.

Hinchcliffe, E. H., Cassels, G. O., Reider, C. L., and Sluder, G. (1998). *J. Cell Biol.* **140,** 1417–1426.

Hinchcliffe, E. H., Li, C., Thompson, E. A., Maller, J. L., and Sluder, G. (1999). *Science* **283,** 851–854.

Holtrich, U., Wolf, G., Brauninger, A., Karn, T., Bohme, H., Rubsamen-Waigmann, H., and Strebhardt, K. (1994). *Proc. Natl. Acad. Sci. USA* **91,** 1736–1740.

Hsu, L.-C., and White, R. L. (1998). *Proc. Natl. Acad. Sci. USA* **95,** 12983–12987.

Kapeller, R., Chakrabarti, R., Cantley, L., Fay, F., and Corvera, S. (1993). *Mol. Cell. Biol.* **13,** 6052–6063.

Kellogg, D. R., Moritz, M., and Alberts, B. M. (1994). *Ann. Rev. Biochem.* **63,** 639–674.

Kellogg, D. R., Oegema, K., Raff, J., Schneider, K., and Alberts, B. M. (1995). *Mol. Biol. Cell* **6,** 1673–1684.

Kidd, D., and Raff, J. W. (1997). *J. Cell Sci.* **110,** 209–219.

Kimura, M., Kotani, S., Hattori, T., Sumi, N., Yoshioka, T., Todokoro, K., and Okano, Y. (1997). *J. Biol. Chem.* **272,** 13766–13771.

Kimura, M., Matsuda, Y., Yoshioka, T., and Okano, Y. (1999). *J. Biol. Chem.* **274,** 7334–7340.

Krek, W., Maridor, G., and Nigg, E. A. (1992). *J. Cell Biol.* **116,** 43–55.

Kumagai, A., and Dunphy, W. G. (1996). *Science* **273,** 1377–1380.

Lacey, K. R., Jackson, P. K., and Stearns, T. (1999). *Proc. Natl. Acad. Sci. USA* **96,** 2817–2822.

Lane, H. A., and Nigg, E. A. (1996). *J. Cell Biol.* **135,** 1701–1713.

Ley, S. C., Marsh, M., Bebbington, C. R., Proudfoot, K., and Jordan, P. (1994). *J. Cell Biol.* **125,** 639–649.

Lingle, W. L., Lutz, W. H., Ingle, J. N., Maihle, N. J., and Salisbury, J. L. (1998). *Proc. Natl. Acad. Sci. USA* **95,** 2950–2955.

Llamazares, S., Moreira, A., Tavares, A., Girdham, C., Spruce, B. A., Gonzalez, C., Karess, R. E., Glover, D. M., and Sunkel, C. E. (1991). *Genes Dev.* **5,** 2153–2165.

Mack, G. J., Rees, J., Sandblom, O., Balczon, R., Fritzler, M. J., and Rattner, J. B. (1998). *Arthritis Rheumatol.* **41,** 551–558.

Mantel, C., Braun, S. E., Reid, S., Henegariu, O., Liu, L., Hangoc, G., and Broxmeyer, H. E. (1999). *Blood* **93,** 1390–1398.

Martin, O. C., Gunawardane, R. N., Iwamatsu, A., and Zheng, Y. (1998). *J. Cell Biol.* **141,** 675–687.

Matsumoto, Y., Hayashi, K., and Nishida, E. (1999). *Curr. Biol.* **9,** 429–432.

Mayor, T., Meraldi, P., Stierhof, Y.-D., Nigg, E. A., and Fry, A. M. (1999). *FEBS Lett.* **452,** 92–95.

McNally, F. J. (1996). *Curr. Opin. Cell Biol.* **8,** 23–29.

McNally, F. (1999). *Curr. Biol.* **9,** 274–276.

Meraldi, P., Lukas, J., Fry, A. M., Bartek, J., and Nigg, E. A. (1999). *Nature Cell Biol.* **1,** 88–93.

Merdes, A., and Cleveland, D. W. (1997). *J. Cell Biol.* **138,** 953–956.

Middendrop, S., Paoletti, A., Schiebel, E., and Bornens, M. (1997). *Proc. Natl. Acad. Sci. USA* **94,** 9141–9146.

Morgan, D. O. (1997). *Ann. Rev. Cell Dev. Biol.* **13,** 261–291.

Mundt, K. E., Golsteyn, R. M., Lane, H. A., and Nigg, E. A. (1997). *Biochem. Biophys. Res. Commun.* **239,** 377–385.
Murphy, S. M., and Stearns, T. (1996). *Curr. Biol.* **6,** 642–644.
Murphy, S. M., Urbani, L., and Stearns, T. (1998). *J. Cell Biol.* **141,** 663–674.
Nakamura, M., Masuda, H., Horii, J., Kuma, K., Yokoyama, N., Ohba, T., Nishitani, H., Miyata, T., Tanaka, M., and Nishimoto, T. (1998). *J. Cell Biol.* **143,** 1041–1052.
Navara, C. S., Hewitson, L. C., Simerly, C. R., Sutovsky, P., and Schatten, G. (1997). *Am. J. Reprod. Immunol.* **37,** 39–49.
Nigg, E. A. (1995). *BioEssays* **17,** 471–480.
Nigg, E. A. (1998). *Curr. Opini. Cell Biol.* **10,** 776–783.
Nigg, E. A., Schafer, G., Hilz, H., and Eppenberger, H. M. (1985). *Cell* **41,** 1039–1051.
Norbury, C., and Nurse, P. (1992). *Ann. Rev. Biochem.* **61,** 441–470.
Oegema, K., Wiese, C., Martin, O. C., Milligan, R. A., Iwamatsu, A., Mitchison, T. J., and Zheng, Y. (1999). *J. Cell Biol.* **144,** 721–733.
Ohba, T., Nakamura, M., Nishitani, H., and Nishimoto, T. (1999). *Science* **284,** 1356–1358.
Ohkura, H., Hagan, I. M., and Glover, D. M. (1995). *Genes Dev.* **9,** 1059–1073.
Ohta, Y., Ohba, T., and Miyamoto, E. (1990). *Proc. Natl. Acad. Sci. USA* **87,** 5341–5345.
Ohta, K., Shiina, N., Okumura, E., Hisanaga, S.-I., Kishimoto, T., Endo, S., Gotoh, Y., Nishida, E., and Sakai, H. (1993). *J. Cell Sci.* **104,** 125–137.
Passalacqua, M., Patrone, M., Sparatore, B., Melloni, E., and Pontremoli, S. (1999). *Biochem. J.* **337,** 113–118.
Pereira, G., and Schiebel, E. (1997). *J. Cell Sci.* **110,** 295–300.
Pereira, G., Knop, M., and Schiebel, E. (1998). *Mol. Biol. Cell* **9,** 775–793.
Pihan, G. A., Purohit, A., Wallace, J., Knecht, H., Woda, B., Quesenberry, P., and Doxsey, S. J. (1998). *Cancer Res.* **58,** 3974–3985.
Pockwinse, S. M., Krockmalnic, G., Doxsey, S. J., Nickerson, J., Lian, J. B., Van Wijnen, A. J., Stein, J. L., Stein, G. S., and Penman, S. (1997). *Proc. Natl. Acad. Sci. USA* **94,** 3022–3027.
Roghi, C., Giet, R., Uzbekov, R., Morin, N., Chartrain, I., Le Guellec, R., Couturier, A., Doree, M., Philippe, M., and Prigent, C. (1998). *J. Cell Sci.* **111,** 557–572.
Rudner, A. D., and Murray, A. W. (1996). *Curr. Opin. Cell Biol.* **8,** 773–780.
Saunders, R. D. C., M. C., A., Howard, T., Gonzalez, C., and Glover, D. M. (1997). *J. Cell Biol.* **137,** 881–890.
Sawin, K. E., and Mitchison, T. J. (1995). *Proc. Natl. Acad. Sci. USA* **92,** 4289–4293.
Schatten, G. (1994). *Dev. Biol.* **165,** 299–335.
Schatten, H., Schatten, G., Mazia, D., Balczon, R., and Simerly, C. (1986). *Proc. Natl. Acad. Sci. USA* **83,** 105–109.
Schatten, G., Simerly, C., and Schatten, H. (1991). *Proc. Natl. Acad. Sci. USA* **88,** 6785–6789.
Schliwa, M., Pryzwansky, K. B., and Euteneuer, U. (1982). *Cell* **31,** 705–717.
Schmidt, P. H., Dransfield, D. T., Claudio, J. O., Hawley, R. G., Trotter, K. W., Milgram, S. L., and Goldenring, J. R. (1999). *J. Biol. Chem.* **274,** 3055–3066.
Schultz, S. J., Fry, A. M., Sutterlin, C., Ried, T., and Nigg, E. A. (1994). *Cell Growth Differen.* **5,** 625–635.
Schumacher, J. M., Ashcroft, N., Donovan, P. J., and Golden, A. (1998). *Development* **125,** 4391–4402.
Sen, S., Zhou, H., and White, R. A. (1997). *Oncogene* **14,** 2195–2200.
Sherline, P., and Mascardo, R. N. (1982). *J. Cell Biol.* **93,** 507–511.
Shindo, M., Nakano, H., Kuroyanagi, H., Shirasawa, T., Mihara, M., Gilbert, D. J., Jenkins, N. A., Copeland, N. G., Yagita, H., and Okumura, K. (1998). *Biochem. Biophys. Res. Commun.* **244,** 285–292.

Smith, M., Wilson, M., Hamanaka, R., Chase, D., Kung, H., Longo, D., and Ferris, D. (1997). *Biochem. Biophys. Res. Commun.* **234,** 397–405.

Snaith, H. A., Armstrong, C. G., Guo, Y., Kaiser, K., and Cohen, P. T. W. (1996). *J. Cell Sci.* **109,** 3001–3012.

Sohrmann, M., Schmidt, S., Hagan, I., and Simanis, V. (1998). *Genes Dev.* **12,** 84–94.

Stearns, T., and Kirschner, M. (1994). *Cell* **76,** 623–637.

Sunkel, C. E., and Glover, D. M. (1988). *J. Cell Sci.* **89,** 25–38.

Takahashi, M., Shibata, H., Shimakawa, M., Miyamoto, M., Mukai, H., and Ono, Y. (1999). *J. Biol. Chem.* **274,** 17267–17274.

Tassin, A.-M., Celati, C., Moudjou, M., and Bornens, M. (1998). *J. Cell Biol.* **141,** 689–701.

Tavares, A. A., Glover, D. M., and Sunkel, C. E., (1996). *EMBO J.* **15,** 4873–4883.

Terada, Y., Tatsuka, M., Suzuki, F., Yasuda, Y., Fujita, S., and Otsu, M. (1998). *EMBO J.* **17,** 667–676.

Vandre, D. D., and Borisy, G. G. (1989). *J. Cell Sci.* **94,** 245–258.

Vandre, D. D., Davis, F. M., Rao, P. N., and Borisy, G. G. (1986). *Eur. J. Cell Biol.* **41,** 72–81.

Verde, F., Labbe, J.-C., Doree, M., and Karsenti, E. (1990). *Nature* **343,** 233–238.

Wigge, P. A., Jensen, O. N., Holmes, S., Soues, S., Mann, M., and Kilmartin, J. V. (1998). *J. Cell Biol.* **141,** 967–977.

Wilde, A., and Zheng, Y. (1999). *Science* **284,** 1359–1362.

Winey, M., and Byers, B. (1993). *Trends Genet.* **9,** 300–304.

Witczak, O., Skalhegg, B. S., Keryer, G., Bornens, M., Tasken, K., Jahnsen, T., and Orstavik, S. (1999). *EMBO J.* **18,** 1858–1868.

Wolf, G., Elez, R., Doermer, A., Holtrich, H., Ackerman, H., Stutte, H., Altmannsberger, H.-M., Rubsamne-Waigmann, H., and Strebhardt, K. (1997). *Oncogene* **14,** 543–549.

Wu, L., Osmani, S. A., and Mirabito, P. M. (1998). *J. Cell Biol.* **141,** 1575–1587.

Xu, X., Weaver, Z., Linke, S. P., Li, C., Gotay, J., Wang, X. W., Harris, C. C., Ried, T., and Deng, C. X. (1999). *Mol. Cell* **3,** 389–395.

Yuan, J., Horlin, A., Hock, B., Stutte, H., Rubsamen-Waigmann, H., and Strebhardt, K. (1997). *Am. J. Pathol.* **150,** 1165–1172.

Zhou, H., Kuang, J., Zhong, L., Kuo, W.-L., Gray, J., Sahin, A., Brinkley, B. R., and Sen, S. (1998). *Nature Genet.* **20,** 189–193.

Zimmerman, W., Sparks, C. A., and Doxsey, S. J. (1999). *Curr. Opin. Cell Biol.* **11,** 122–128.

14

The Role of the Centrosome in the Development of Malignant Tumors

Wilma L. Lingle[1,2] and *Jeffrey L. Salisbury*[1]
[1] Tumor Biology Program
[2] Division of Experimental Pathology
Mayo Clinic Foundation
Rochester, Minnesota 55905

I. Introduction

The fundamental characteristic of the centrosome is that it embodies the major microtubule organizing center (MTOC) of the cell. As such, the centrosome determines the number and polarity of cytoplasmic microtubules. Once in each cell cycle, the centrosome is duplicated to give rise to two centrosomes (i.e., the mitotic spindle poles) that organize the microtubule array of the mitotic spindle and thereby make possible equal segregation of sister chromatids into each of two daughter cells at the time of cell division. The centrosome also plays a role in organizing cytoplasmic structure in interphase cells through its influence on the spatial array of microtubules. Recent observations have implicated defects in centrosome behavior in the progression of malignant tumors. In particular, centrosome defects may be the causative basis of *aneuploidy* (inappropriate number and combination of chromosomes) and *anaplasia* (loss of tissue organization and architecture). New insights into centrosome composition, structure,

Current Topics in Developmental Biology, Vol. 49

0070-2153/00 $35.00

assembly, and regulation of centrosome duplication are bringing full circle early theories on the role of centrosome defects in the origin of malignant tumors (Andersen, 1999; Marshall and Rosenbaum, 1999; Karsenti, 1999; Brinkley and Goepfert, 1998; and elsewhere in this volume). Thus, a clearer understanding of centrosomes and cancer is beginning to emerge. This review concentrates specifically on centrosome defects seen in malignant tumors and the role that they may play in tumor progression.

A. Recognition of Cancer as a Disease of Cells and the Influence of Centrosomes in Tumor Progression

The earliest recorded historical mention of the problem of cancer is seen in the Edwin Smith Surgical Papyrus dating to the 17th century B.C. in Egypt (Breasted, 1930). In this ancient record a tumor (*ben.wet*) of the breast is described as a "swelling spread over the breast . . . (which is) without fever, with no granulation or fluid . . . having a touch or surface like that of a kind of ball, . . . likened to an unripe fruit which is hard and cool to the touch." Four millennia later in 1838, J. Müller found tumors to be "composed of cells, each containing nuclei and nucleoli" (as cited in Ewing, 1919). In his classic treatise on cellular pathology, Rudolph Virchow established the doctrine of *Omnis cellula e cellula* and recognized the cell as the basic element of pathological processes. Yet, Virchow, contrary to his own declaration on the origin of cells, believed that cancer originated from a fluid blastema of connective tissue (Virchow, 1860). It was not until early in this century that the cellular basis of cancer was generally accepted. Theodor Boveri (1914) wrote that "the problem of tumors is a cell problem" and that "every theory of malignant tumors is wrong which does not take into account its unicellular origins." Boveri further recognized that in order for normal development to proceed, cells must have a complete set of chromosomes. Based on his studies on the similarity between abnormal sea urchin embryo development following dispermic fertilization and cell anomalies seen in cancer, Boveri proposed the hypothesis that malignant tumors arise through centrosome defects that result in improper cell divisions and give rise to aneuploidy (1914). Galeotti also reported early in this century that asymmetric mitosis in tumors may result from "secondary subdivision of centrosomes, one of which may divide into as many as four parts, each forming attraction spheres. Under these conditions the migration of chromosomes is often delayed or unequal" (as cited in Ewing, 1919, page 39). Thus, early in the 20th century, the two defining properties of malignant tumors (alteration in tissue architecture and genetic instability) were established as hallmarks of cancer and their anatomic and causative roots were suspected to lie in inappropriate centrosome behavior. More recently, progress in understanding the molecular basis for tumor progression has focused on the role of dominant acting oncogenes, the loss of action of tumor suppressor genes, and their control of critical cell cycle events.

B. Tumor Biology Primer

A tumor is an autonomous new growth of tissue (Ewing, 1919). Beyond this seemingly simple definition, it is difficult to add characteristics of tumors that apply to all cancers. This is largely because human tumors vary greatly in their properties. *Neoplasms* (new growths) can arise in any tissue, and they differ from normal tissue regeneration and repair in that they are abnormal growths that persist and continue to grow after cessation of the stimuli that evoked their initial appearace (Lieberman and Lebvitz, 1996). Neoplasms may be *benign,* slow-growing, and relatively innocuous in that they remain at their site of origin. Benign tumors are generally encapsulated, well differentiated, and show normal nuclear structure. Nevertheless, if they secrete hormones or other factors affecting distant targets, hemorrhage, or otherwise compromise vital functions through location and sheer mass, "benign" tumors may have deleterious and even lethal consequences. *Malignant* tumors, on the other hand, are distinct from benign tumors in that malignant tumor cells migrate away from their original site of growth, initially through *invasion* of surrounding tissue, and then through *metastasis* to distant sites where they establish new tumors (Nowell, 1976; Loeb, 1991). The degree to which tumor cells retain differentiated characteristics of their tissue of origin is called *tumor grade.* Thus, tumor grade is a histopathological term and its designation is somewhat subjective depending on the training of the observer. A low-grade tumor is well differentiated, whereas a high-grade tumor tends to be *anaplastic.* Nuclear grade, another histopathological term, refers to the degree to which nuclear shape, staining, and location of heterochromatic DNA resembles that of normal tissue. *Tumor staging,* on the other hand, is a prognostic tool that is based on the size of the primary tumor, lymph node involvement, and the presence or absence of metastasis to distant sites.

Transformation is the term used to describe the conversion of normal cells to those with abnormalities in cellular appearance and growth regulation in tissue culture (*morphological transformation*) seen for cancer cells. This typically includes acquisition of unlimited growth potential, alteration in cell morphology, loss of contact inhibition of growth, growth in the soft agar colony assay, and loss of dependence on growth factors or serum, among others (Roberts, 1989). *Malignant transformation* has the further requirement that the cells can produce a tumor in an appropriate animal model. A confounding feature of tumors is their heterogeneity; tumors consist of different populations of cells with diverse characteristics (Fidler and Kripke, 1977). Although most tumors are monoclonal in origin, subpopulations of cells arise that differ in immunogenecity, growth rate, karyotype, receptor status, susceptibility to cytotoxic drugs, and metastatic potential. For the genesis of aggressive malignant tumors, continual selection operates to drive a cascade of (sequential) steps that involve multiple tumor–host interactions. These changes are often irreversible and typically require multiple genetic lesions in key oncogenes and tumor suppressor genes.

Because the nominal rate for genetic mutation in somatic cells appears to be insufficient to generate the number of genetic changes found in most cancers, a "mutator" phenotype has been proposed that acts to specifically increase the level of genomic instability during tumor progression (Loeb, 1991; Nowell, 1976). Recent experimental studies using chemically transformed Chinese hamster embryo cells and analysis of karyotypic instability in human colon cancer cell lines has demonstrated that the degree of genetic instability is proportional to the degree of aneuploidy (Duesberg *et al.*, 1998; Lengauer *et al.*, 1998; Li *et al.*, 1997). Aneuploidy alone may be sufficient to explain genetic instability and the resulting karyotypic and phenotypic heterogeneity seen in cancer cells. Genomic instability, that is, the alteration in chromosome number through loss or gain of whole chromosomes (*aneuploidy*), chromosome translocation, gene amplification, and mutation, is a characteristic of solid tumors (Cheng and Loeb, 1997). Although certain genetic alterations may be common for particular tumor types, multiple genetic alterations are required for the full development of the cancer phenotype, and in some tumors these changes may follow a progressive pattern of acquisition (Vogelstein *et al.*, 1988). While the most frequent outcome of aneuploidy is cell death, some aneuploid cells may gain selective growth advantage, and it is the descendants of these cells that go on to develop the tumor.

II. Centrosome Defects and Abnormal Mitoses in Cancer

Abnormal mitotic figures, including multipolar and monopolar mitoses and lagging chromosomes, are easily discerned in standard histological slides used for diagnosis, and their presence is often noted in tumors (Koss, 1992; Pritchard and Youngberg, 1993; Haferkamp *et al.*, 1999; Remstein *et al.*, 1999; Tomaszewski *et al.*, 1999; Zamecnik and Michal, 1999). Although not always recognized as such, multipolar and monopolar mitoses are a direct consequence of centrosomal defects. Respectively, multipolar and monopolar mitoses result from extra centrosome duplication and lack of centrosome duplication or separation prior to mitosis. Cells in which some components of cell cycle checkpoint control remain intact may arrest in mitosis and eventually undergo apoptosis or necrotic cell death in response to abnormal spindle formation. However, daughter cells that do result from such mitoses most often are aneuploid or polyploid, and it is likely that many are not viable. Among those aneuploid and polyploid cells that do survive exists the potential for enhanced growth that could lead to tumor progression. Thus, abnormal centrosome number and function may confer to cells a "mutator" phenotype as first described by Nowell (1976).

III. Amplified Centrosomes and Aneuploidy

The relationship between an abnormal number of centrosomes and development of aneuploidy has been demonstrated in a mouse model of pancreatic cancer induced by simian virus 40 tumor antigen (Levine *et al.,* 1991). This tumor model is characterized by the sequential appearance of tetraploid and then aneuploid cell populations. When examined by transmission electron microscopy, interphase tetraploid cells contained 5–11 centrioles, and 5 or more centrioles were observed in mitotic cells, at least 1 of which was tripolar. The authors speculated that extra centrioles predisposed cells to form multipolar mitotic spindles, yielding aneuploid daughter cells associated with subsequent tumor development. A similar phenomenon was observed in cultured cells transfected to express the *Vpr* gene of human immunodeficiency virus 1 (HIV-1) in the presence of tetracycline. Vpr expression induced multipolar mitoses and aneuploidy (Minemoto *et al.,* 1999). The effects of this viral gene on centrosome structure and function, leading to aneuploidy through the formation of multipolar mitoses, may be a mechanism for the cancer predisposition associated with HIV-1 infections.

The presence of extra centrosomes or MTOCs prior to mitosis does not necessarily commit the cell to multipolar mitosis. For example, cultured N115 mouse neuroblastoma cells contain many separate MTOCs during interphase (Sharp *et al.,* 1982; Ring *et al.,* 1982). During prometaphase, the MTOCs gather into clusters and chains, which by metaphase assemble into two groups, one group residing at each pole of the bipolar spindle (Ring *et al.,* 1982). Serial reconstruction of spindle poles in mitotic cells of human breast tumors has revealed a similar clustering of centrosomes (Lingle and Salisbury, 1999). However, because in one instance a cell retained three spindle poles in spite of centrosome clustering, it must be assumed that the clustering process is not 100% efficient in forming bipolar spindles when multiple centrosomes or MTOCs are present (Fig. 1). Never the less, clustering of extra centrosomes at the spindle poles during mitosis can minimize the possibility of aneuploidy in cells containing amplified centrosomes.

Figure 1 Reconstruction of multipolar mitosis. This reconstruction is from six transmission electron micrographs of nonadjacent serial sections. The condensed chromosomes are arranged in a triskelion with three spindle poles generating three sets of spindle microtubules. The spindle pole near the 12 o'clock position is actually a cluster of two groups of centrioles generating two slightly separate microtubule foci that function as one spindle pole. Even though some of the supernumerary centrioles were clustered at one spindle pole, this cell has three spindle poles. Daughter cells from mitoses such as these will be aneuploid (see also color insert). (From Lingle and Salisbury, *American Journal of Pathology,* (1999) **155,** 1941–1951.)

As shown by immunofluorescence and immunohistochemical studies, centrosome amplification is a feature common to many human tumors such as breast tumors (Lingle *et al.,* 1998; Pihan *et al.,* 1998; Carroll *et al.,* 1999), astrocytoma, lung tumors (Pihan *et al.,* 1998), neuroectodermal tumors (Weber *et al.,* 1998), squamous cell carcinomas of the head and neck (Carroll *et al.,* 1999), and pancreatic tumors (Sato *et al.,* 1999) (Fig. 2). These studies have utilized numerous marker antibodies to detect amplified centrosomes, including human autoimmune sera and antibodies against centrin, pericentrin, γ-tubulin, and MPM-2 epitope, indicating that numerous centrosomal proteins are concomitantly overexpressed in tumors. In human breast tumors, not only are tumor centrosomes amplified, they contain levels of phosphorylated centrosomal proteins that are inappropriately high for interphase cells (Lingle *et al.,* 1998). *In vitro* experiments have demonstrated that amplified tumor centrosomes are functional MTOCs (Lingle *et al.,*

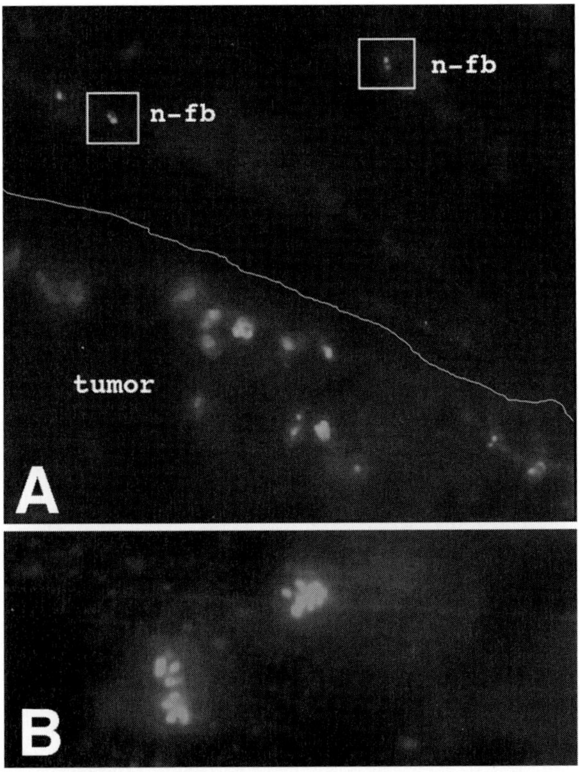

Figure 2 Immunofluorescence staining of centrosomes in human breast tumors. (A) A region of an adenocarcinoma of the breast stained for centrin showing excessive staining of centrosomes in the area of a tumor (below the marked line) and normal staining of pairs of centrioles in fibroblasts of the surrounding connective tissue. (B) Supernumerary centrioles in two adjacent cells of another tumor, stained for centrin. Each cell has eight or more centrioles.

1998; Pihan *et al.*, 1998), and breast tumor centrosomes exhibit a significantly greater than normal capacity to nucleate microtubules even at times when the tumor cells are not in mitosis (Lingle *et al.*, 1998). These facts indicate that in addition to the deregulation of centrosome duplication in tumor cells, the function and activity of centrosomes is not synchronized with the cell cycle.

IV. Excess Pericentriolar Material Is Associated with High Frequency of Abnormal Mitoses

In a recent electron microscopic comparison between centrosomes of normal human breast tissue and invasive breast tumor centrosomes, it was noted that tumors frequently displayed ultrastructurally abnormal centrosomes, whereas centrosome abnormalities were rarely observed in normal tissues (Lingle and Salisbury, 1999). Tumor-associated centrosome abnormalities included (1) supernumerary centrioles; (2) excess pericentriolar material; (3) disrupted centriole barrel structure; (4) unincorporated microtubule complexes; (5) centrioles of unusual length; (6) centrioles functioning as ciliary basal bodies; and (7) mispositioned centrosomes. These alterations are associated with changes in cell polarity, and changes in cell and tissue differentiation, in addition to chromosome missegregation through multipolar mitoses. Significantly, of these seven centrosome abnormalities, the presence of excess pericentriolar material had the strongest association with a high frequency of abnormal mitoses. Usually, but not always, supernumerary centrioles were present along with excess pericentriolar material (Fig. 3). However, the presence of supernumerary centrioles in the absence of excess pericentriolar material did not significantly correlate with higher frequencies of abnormal mitoses. This means that unregulated centriole duplication by itself is likely insufficient to create multipolar mitoses; pericentriolar proteins must accumulate in excess at the centrosome to support the formation of multipolar mitotic spindles.

V. Centrosome-Associated Kinases and Cancer

Centrosome-associated kinases, including members of the aurora kinase family and the Polo-like kinases (PLK), are likely candidates for increased activity and/or accumulation at tumor centrosomes. These kinases are located at the centrosome or spindle pole in cell-cycle-dependent manner. As discussed in more detail elsewhere in this volume, these kinases are involved in regulating centrosome function and duplication, and their overexpression is associated with the development of aneuploidy (Sen *et al.,* 1997; Bischoff *et al.,* 1998; Zhou *et al.,* 1998). It has been suggested that aurora2 (a.k.a. BTAK/STK15), along with aurora1 and PLK1, may form a centrosome-associated kinase cascade whose disruption leads to genomic instability and chromosome defects (Bischoff *et al.,* 1998). BTAK/STK15 is overexpressed in breast tumors (Tanaka *et al.,* 1999; Zhou *et al.,* 1998), colon tumors (Bischoff *et al.,* 1998; Katayama *et al.,* 1999), and numerous tumor-derived cell lines (Zhou *et al.,* 1998). Overexpression of BTAK/STK15 in NIH 3T3 cells induces centrosome amplification and cell transformation (Zhou *et al.,* 1998). Most importantly, overexpression of STK15/BTAK in MCF10A cells (which are near diploid and have a normal mitotic apparatus) results in centrosome amplification that leads to aneuploidy. This demonstrates that, in a near-normal background, STK15/BTAK overexpression leads to centrosome amplification, chromosomal instability, and transformation in mammalian cells (Zhou *et al.,* 1998). Centrosomes ampli-

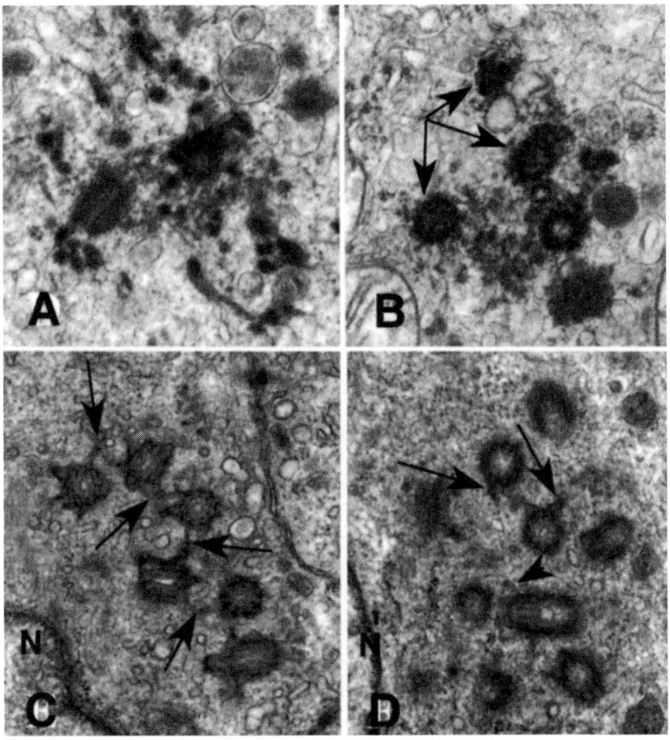

Figure 3 Transmission electron micrographs of amplified centrosomes in human breast tumors. (A) Excess pericentriolar material appears as darkly stained material appressed to the barrels of the centrioles and in small clumps next to the centrioles. Four centrioles are present in this thin section through an amplified centrosome. (B) One centriole encased in excess pericentriolar material and associated masses of pericentriolar material are present in the amplified centrosome. Centrosomes with excess pericentriolar material are associated with a high frequency of abnormal mitoses. (C, D) Clusters of six and nine centrioles linked together by fine fibers are present in these centrosomes. No excess pericentriolar material is present, and this type of centrosome is not associated with a high frequency of abnormal mitoses. (From Lingle and Salisbury, *American Journal of Pathology,* (1999) **155,** 1941–1951.)

fied by STK15/BTAK overexpression have not yet been ultrastructurally characterized; it will be of interest to learn if these centrosomes have excess pericentriolar material and/or other centrosome abnormalities seen in breast tumor centrosomes.

VI. Tumor Suppressor Proteins and the Centrosome

In the absence of tumor suppressor function through mutation, deletion, or disruption of the pathway in which they operate, cells are more likely to undergo malignant transformation. Recently, two known tumor suppressor proteins, BRCA1 and p53, have been localized at the centrosome of mammalian cells (Brown *et al.,* 1994; Hsu and White, 1998), and it has been speculated that some of their tumor suppressor functions take place at the centrosome.

Germline mutations of the BRCA1 tumor suppressor gene predispose women to breast and ovarian cancers (Irminger-Finger *et al.,* 1999). BRCA1 protein is a large protein with numerous functional domains, including binding sites for p53, Rad51, RNA polymerase II holoenzyme, RNA helicase A, CtBP-interacting protein, c-myc, BRCA1-associated RING domain protein, and BRCA2 (Chen *et al.,* 1998, 1999; Irminger-Finger *et al.,* 1999). Immunofluorescence microscopy and analysis of isolated centrosomes provide evidence that BRCA1 protein is associated with centrosomes during mitosis. BRCA1 localizes with the centrosome during mitosis and coimmunoprecipitates with γ-tubulin, indicating that it may be involved with regulation of microtubule nucleation (Hsu and White, 1998). Mouse embryonic fibroblasts expressing only mutant BRCA1 with a targeted deletion of exon 11 contain multiple, functional centrosomes and undergo unequal chromosome segregation, abnormal nuclear division, and aneuploidy (Xu *et al.,* 1999). Xu and co-workers (1999) speculate that BRCA1 has an essential role in maintaining genetic stability through the regulation of centrosome duplication, and that the action of BRCA1 at the centrosome provide a molecular basis for the role of BRCA1 in tumorigenesis.

p53, the most frequently mutated gene in human cancers, is involved in checkpoint functions at the G1/S and the G2/M cell cycle transitions (Cross *et al.,* 1995; Hollstein *et al.,* 1998; Prives and Hall, 1999; Yin *et al.,* 1999). p53 mutation and nullizygosity are associated with increased chromosomal instability (Fukasawa *et al.,* 1997; Boyle *et al.,* 1998; Gualberto *et al.,* 1998; Weber *et al.,* 1998; Carroll *et al.,* 1999). Although most p53 is nuclear, a portion of p53 is localized at the centrosome in established human cell lines (Brown *et al.,* 1994) and in primary cultures of normal mammary epithelial cells (Lingle, unpublished). Mouse embryo fibroblasts (MEF) null for p53 undergo centrosome amplification (Fukasawa *et al.,* 1996), as do cells in tissues of mice nullizygous for p53 (Fukasawa *et al.,* 1997). Immunofluorescence labeling of p53 null MEFs with antibodies against γ-tubulin shows that these cells contain numerous centriole-sized spots arranged in clusters (Fukasawa *et al.,* 1996), similar to that seen in some human breast tumors (Lingle and Salisbury, 1999). Tissues of p53 null mice frequently are aneuploid and contain multipolar mitotic figures. *In vivo,* p53-independent apoptosis eliminates many of these aneuploid cells that contain amplified centrosomes (Fukasawa *et al.,* 1997).

In order to test the hypothesis that p53 mutation is the cause of centrosome amplification in human tumors, Weber and co-workers (1998) examined three brain tumors for centrosome amplification and numerical chromosome aberration. Although the sample size was small—two neuroectodermal tumors with p53 mutations and one benign meningioma with wild-type p53 as a control tumor—the findings were consistent with the hypothesis. Centrosome amplification and evidence of multiple occurrences of chromosome segregation infidelity were found in the two tumors with mutant p53, whereas the control tumor had normal centrosomes and straightforward loss of three chromosomes. This is the first report correlating aneuploidy, centrosome amplification, and p53 mutation in human tumors.

Centrosome amplification also was associated with p53 mutation in breast carcinomas and squamous cell carcinomas of the head and neck (Carrol *et al.*, 1999). In this collection of tumors, however, not all tumors with amplified centrosomes had p53 mutations. Interestingly, those tumors with wild-type p53 and amplified centrosomes contained high levels of Mdm2. In normal cells, Mdm2 promotes rapid turnover of p53 by targeting it for ubiquitin-dependent degradation. Overexpression of Mdm2 causes an extremely short half-life of p53 and essentially makes the tumor cells behave as if they are p53 null, with attendant centrosome amplification and chromosome instability (Carroll *et al.*, 1999). Disruptions of other elements of the p53 pathway also have been shown to affect centrosome structure and function. p53-dependent induction of p21(cip-1/waf-1) is an important component of some cell cycle checkpoints. Reduced p21(cip-1/waf-1) expression results in gross nuclear abnormalities, centriole amplification, and polyploidy, most likely by uncoupling the centrosome cycle from the DNA cycle (Mantel *et al.*, 1999).

Although the most established function of p53 is as a tumor suppressor gene, certain mutations exhibit gain-of-function activities that increase oncogenic transformation through genomic instability (Gualberto *et al.*, 1998). Interestingly, the genomic instability occurs in the absence of transcriptional activation by p53; "thus p53 mutations can contribute to progression of a cancer cell not only by absence of p53 tumor suppressor activity but also by the presence of an activity that promotes genetic instability" (Gualberto *et al.*, 1998).

A similar gain-of-function was described in studies using a model of chemically induced papilloma in mice (Wang *et al.*, 1998). These studies used mice with wild-type p53 and knockout mice expressing no p53 or expressing mutant p53 under the control of a human keratin-1-based vector. The chemically induced tumors in mice with mutant p53 exhibited a less differentiated phenotype than those tumors elicited in p53 null mice or those in nontransgenic mice with wild-type p53. The p53 mutant tumors had a much higher frequency of centrosome anomalies than did p53 null or p53 wild-type tumors. The frequency of centrosome anomaly correlated positively with metastasis and anaplasia (Wang *et al.*, 1998). The finding that centrosome anomalies were associated with less differentiated phenotypes and metastasis indicates that the interphase functions of the centrosome may be compromised in cells expressing mutant p53.

Fidelity in chromosome segregation and preservation of diploidy requires proper structure and function of the duplicated centrosomes; maintenance of cell and tissue polarity requires proper structure and function of interphase centrosomes. The evidence suggests that the centrosome is a nexus of regulation of cell cycle and cell polarity imposed in part by the actions of the p53 and BRCA1 tumor suppressor proteins.

VII. Maintenance of Cell and Tissue Polarity Minimizes Tumor Aggression in Model Systems

In mammary cell culture systems, the establishment of cell–cell contact and epithelial polarity is sufficient to stimulate the expression of β-casein (Roskelley *et al.*, 1994), while whey acidic protein additionally requires the formation of alveolar structures (Chen and Bissell, 1989; Lin *et al.*, 1995). In these cultured mammary epithelial cells, cell and tissue polarity regulates the expression of proteins associated with cell differentiation. Normal cell and tissue behavior is determined in part by interactions between the intermediate filament cytoskeleton and the extracellular matrix, and alterations in tissue structure can lead to the progression of tumors (Schmeichel *et al.*, 1998). Indeed, the malignant phenotype of cultured cells can be reverted to normal without changing the genotype of the cells. This is accomplished simply by application of a β1-integrin inhibitory antibody to cells in the cell culture system (Weaver *et al.*, 1997). In response to application of the inhibitory antibody, cells essentially redifferentiate by forming three-dimensional acini with a basement membrane and reorganize cytoskeletons. Nude mice injected with antibody-treated tumor cells have tumors significantly reduced in number and size. Significantly, the observed phenotypes were reversible upon removal of the antibodies. These results show that in this model system the tissue phenotype (i.e., anaplastic vs. differentiated) is dominant over the cellular genotype (i.e., malignant vs. nontransformed) (Weaver *et al.*, 1997). Although not mentioned specifically by Bissel and co-workers, the centrosome, as the regulator of the microtubule cytoskeleton, is mechanically associated with the actin and intermediate cytoskeletons and the desmosomes that maintain tissue polarity. Perhaps disruptions of centrosome function that increase the anaplastic phenotype by adversely affecting tissue polarity are potentially correctable in the presence of malignant genetic lesions.

VIII. Conclusions

Centrosomes are involved with cancer in two possible ways. The first is through disruption of their function as the poles of the bipolar mitotic spindle apparatus. Centrosome defects that increase the chances of improper chromosome segregation during mitosis result in aneuploidy and lead to tumor progression. These defects include amplification of both the number and the size of centrosomes, hyperphosphorylation of centrosomal proteins, and an increase in their microtubule nucleating capacity (Lingle *et al.,* 1998; Pihan *et al.,* 1998; Weber *et al.,* 1998; Carroll *et al.,* 1999; Sato *et al.,* 1999). Centrosome-associated kinases and tumor suppressor proteins such as p53 and BRCA1 may be involved in regulating the transition of the centrosome from its interphase function to its mitotic function. Mutations in tumor suppressor genes and perturbations to the centrosome-associated kinase activities are likely causes of these centrosome defects that lead to aneuploidy. Recently, a century-old hypothesis first proposed by van Hansemann (1890, as cited in Rasnick and Duesberg, 1999), then by Boveri (1914), that aneuploidy is the cause of cancer has been resurrected (Brinkley & Goepfert, 1998; Rasnick and Duesberg, 1999). The hypothesis was refined to state that "cancer is the phenotype of cells above a certain threshold of aneuploidy" (Rasnick and Duesberg, 1999). It must be noted, however, that Duesberg suggests that centrosome abnormalities are a consequence of aneuploidy rather than aneuploidy being a consequence abnormal centrosomes (Duesberg, 1999).

Centrosomes also may be involved in cancer through their role in establishing and maintaining cell, and therefore tissue, polarity. Although the regulation of interphase functions of the centrosome is less well understood than are the mitotic functions, the centrosome and its interphase microtubule array are intimately involved with the actin and intermediate filament cytoskeletons. The three cytoskeletal systems act in concert to determine cell and tissue polarity. The fact that tissue phenotype (i.e., degree of polarity and differentiation) of cells cultured in a model system is dominant over the cellular genotype with regard to invasive potential (Weaver *et al.,* 1997) is a significant gain toward understanding malignant transformation in tumors.

Regardless of whether centrosome abnormalities are a cause or a consequence of cancer, the structure and function of the centrosome present promising opportunities for cancer therapy. A more thorough understanding of regulation of mitotic and interphase centrosome functions will facilitate exploitation of the centrosome as a target for effective therapeutic agents against cancer.

Acknowledgments

The authors laboratories are supported by the Mayo Clinic Foundation and by grants from NCI (CA72836) to J.L.S. and the Department of Defense (17-98-1-8122) to W.L.L.

References

Andersen, S. L. (1999). Molecular characteristics of the centrosome. *International Review of Cytology—a Survey of Cell Biology* **87,** 51–109.

Bischoff, J. R., Anderson, L., Zhu, Y. F., Mossie, K., Ng, L., Souza, B., Schryver, B., Flanagan, P., Clairvoyant, F., Ginther, C., Chan, C. M., Novotny, M., Slamon, D. J., and Plowman, G. D. (1998). A homologue of *Drosophila* aurora kinase is oncogenic and amplified in human colorectal cancers. *EMBO Journal* **17,** 3052–3065.

Boveri, T. (1914). "Zur Frage der Entstehung maligner Tumoren," Fischer Verlag, Jena. English translation by M. Boveri (1929). "The Origin of Malignant Tumors," Waverly Press, Baltimore, MD.

Boyle, J. M., Mitchell, E. D., Greaves, M. J., Roberts, S. A., Tricker, K., Burt, E., Varley, J. M., Birch, J. M., Scott, D., Li, F. (1998). Chromosome instability is a predominant trait of fibroblasts from Li-Fraumeni families. *British Journal of Cancer* **77,** 2181–2192.

Breasted, J. H. (1930). "The Edwin Smith Surgical Papyrus: Published in Facsimile and Hieroglyphic Transliteration with Translation and Commentary in Two Volumes." The University of Chicago, Chicago.

Brinkley, B. R., and Goepfert, T. M. (1998). Supernumerary centrosomes and cancer: Boveri's hypothesis resurrected. *Cell Motility and the Cytoskeleton* **41,** 281–288.

Brown, C. R., Doxsey, S. J., White, E., and Welch, W. J. (1994). Both viral (adenovirus E1B) and cellular (hsp 70, p53) components interact with centrosomes. *Journal of Cellular Physiology* **160,** 47–60.

Carroll, P. E., Okuda, M., Horn, H. F., Biddinger, P., Stambrook, P. J., Gleich, L. L., Li, Y. Q., Tarapore, P., and Fukasawa, K. (1999). Centrosome hyperamplification in human cancer: chromosome instability induced by p53 mutation and/or Mdm2 overexpression. *Oncogene* **18,** 1935–1944.

Chen, L. H., and Bissell, M. J. (1989). A novel regulatory mechanism for whey acidic protein gene expression. *Cell Regulation* **1,** 45–54.

Chen, J. J., Silver, D. P., Walpita, D., Cantor, S. B., Gazdar, A. F., Tomlinson, G., Couch, F. J., Weber, B. L., Ashley, T., Livingston, D. M., and Scully, R. (1998). Stable interaction between the products of the BRCA1 and BRCA2 tumor suppressor genes in mitotic and meiotic cells. *Molecular Cell* **2,** 317–328.

Chen, J. J., Silver, D., Cantor, S., Livingston, D. M., Scully, R. (1999). BRCA1, BRCA2, and Rad51 operate in a common DNA damage response pathway. *Cancer Research* **59,** 1752S–1756S.

Cheng, K. C., and Loeb, L. A. (1997). Genomic stability and instability: a working paradigm. *Current Topics in Microbiological Immunology* **221,** 5–18.

Cross, S. M., Sanchez, C. A., Morgan, C. A., Schimke, M. K., Ramel, S., Idzerda, R. L., Raskind, W. H., and Reid, B. J. (1995). A p53-dependent mouse spindle checkpoint. *Science* **267,** 1353–1356.

Duesberg, P. (1999). Are centrosomes or aneuploidy the key to cancer? *Science* **284,** 2091–2092.

Duesberg, P., Rausch, C., Rasnick, D., and Hehlmann, R. (1998). Genetic instability of cancer cells is proportional to their degree of aneuploidy. *Proceedings of the National Academy of Sciences of the United States of America* **95,** 13692–13697.

Ewing, J. (1919). "Neoplastic Diseases: A Text-book on Tumors," W. B. Saunders Co., Philadelphia.

Fidler, I. J., and Kripke, M. L. (1977). Metastasis results from preexisting variant cells within a malignant tumor. *Science* **197,** 893–895.

Fukasawa, K., Choi, T., Kuriyama, R., Rulong, S., and Vandevoude, G. F. (1996). Abnormal centrosome amplification in the absence of p53. *Science* **271,** 1744–1747.

Fukasawa, K., Wiener, F., Vandewoude, G. F., and Mai, S. B. (1997). Genomic instability and apoptosis are frequent in p53 deficient young mice. *Oncogene* **15,** 1295–1302.

Gualberto, A., Aldape, K., Kozakiewicz, K., and Tlsty, T. D. (1998). An oncogenic form of p53 confers a dominant, gain-of-function phenotype that disrupts spindle checkpoint control. *Proceedings of the National Academy of Sciences of the United States of America* **95,** 5166–5171.

Haferkamp, B., Bastian, B. C., Brocker, E. B., and Hamm, H. (1999). Pilomatrical carcinoma in an atypical location: case report and review of the literature. *Hautarzt* **50,** 355–359.

Hollstein, M., Moeckel, G., Hergenhahn, M., Spiegelhalder, B., Keil, M., Werleschneider, G., Bartsch, H., and Brickmann, J. (1998). On the origins of tumor mutations in cancer genes—insights from the p53 gene. *Mutation Research—Fundamental and Molecular Mechanisms of Mutagenesis* **405,** 145–154.

Hsu, L. C., and White, R. L. (1998). BRCA1 is associated with the centrosome during mitosis. *Proceedings of the National Academy of Sciences of the United States of America* **95,** 12983–12988.

Irminger-Finger, I., Siegel, B. D., and Leung, W. C. (1999). The functions of breast cancer susceptibility gene 1 (BRCA1) product and its associated proteins. *Biological Chemistry* **380,** 117–128.

Karsenti, E. (1999). Centrioles reveal their secrets. *Nature Cell Biology* **1,** E62–E64.

Katayama, H., Ota, T., Jisaki, F., Ueda, Y., Tanaka, T., Odashima, S., Suzuki, F., Terada, Y., and Tatsuka, M. (1999). Mitotic kinase expression and colorectal cancer progression. *Journal of the National Cancer Institute* **91,** 1160–1162.

Koss, L. G. (1992). "Diagnostic cytology and its histopathologic bases." Lippincott, Philadelphia.

Lengauer, C., Kinzler, K. W., and Vogelstein, B. (1998). Genetic instabilities in human cancers. *Nature* **396,** 643–649.

Levine, D. S., Sanchez, C. A., Rabinovitch, P. S., and Reid, B. J. (1991). Formation of the tetraploid intermediate is associated with the development of cells with more than four centrioles in the elastase-simian virus 40 tumor antigen transgenic mouse model of pancreatic cancer. *Proceedings of the National Academy of Sciences of the United States of America* **88,** 6427–6431.

Li, R., Yerganian, G., Duesberg, P., Kraemer, A., Willer, A., Rausch, C., and Hehlmann, R. (1997). Aneuploidy correlated 100% with chemical transformation of Chinese hamster cells. *Proceedings of the National Academy of Sciences of the United States of America* **94,** 14506–14511.

Lieberman, M. W., and Lebovitz, R. M. (1996). "Anderson's Pathology." Mosby, St. Louis.

Lin, C. Q., Dempsey, P. J., Coffey, R. J., and Bissell, M. J. (1995). Extracellular matrix regulates whey acidic protein gene expression by suppression of TGF-alpha in mouse mammary epithelial cells: studies on culture and in transgenic mice. *Journal of Cell Biology* **129,** 1115–1126.

Lingle, W. L., and Salisbury, J. L. (1999). Altered centrosome structure is associated with abnormal mitoses in human breast tumors. *American Journal of Pathology* **155,** 1941–1951.

Lingle, W. L., Lutz, W. H., Ingle, J. N., Maihle, N. J., and Salisbury, J. L. (1998). Centrosome hypertrophy in human breast tumors: implications for genomic stability and cell polarity. *Proceedings of the National Academy of Sciences of the United States of America* **95,** 2950–2955.

Loeb, L. A. (1991). Mutator phenotype may be required for multistage carcinogenesis. *Cancer Research* **51,** 3075–3079.

Mantel, C., Braun, S. E., Reid, S., Henegariu, O., Liu, L., Hangoc, G., and Broxmeyer, H. E. (1999). p21(cip-1/waf-1) deficiency causes deformed nuclear architecture, centriole overduplication, polyploidy, and relaxed microtubule damage checkpoints in human hematopoietic cells. *Blood* **93,** 1390–1398.

Marshall, W. F., and Rosenbaum, J. L. (1999). Cell division: The renaissance of the centriole. *Current Biology* **9,** R218–220.

Minemoto, Y., Shimura, M., Ishizaka, Y., Masamune, Y., and Yamashita, K. (1999). Multiple centrosome formation induced by the expression of Vpr gene of human immunodeficiency virus. *Biochemical and Biophysical Research Communications* **258,** 379–384.

Nowell, P. C. (1976). The clonal evolution of tumor cell populations. *Science* **194,** 23–28

Pihan, G. A., Purohit, A., Wallace, J., Knecht, H., Woda, B., Quesenberry, P., and Doxsey, S. J. (1998). Centrosome defects and genetic instability in malignant tumors. *Cancer Research* **58,** 3974–3985.

Pritchard, B. N., and Youngberg, G. A. (1993). A typical mitotic figures in basal cell carcinoma: A review of 208 cases. *American Journal of Dermatopathology* **15,** 549–552.

Prives, C., and Hall, P. A. (1999). The p53 pathway [Review]. *Journal of Pathology* **187,** 112–126.

Rasnick, D., and Duesberg, P. H. (1999). How aneuploidy affects metabolic control and causes cancer. *Biochemical Journal* **340,** 621–630.

Remstein, E. D., Arndt, C. S., and Nascimento, A. G. (1999). Plexiform fibrohistiocytic tumor: Clinicopathologic analysis of 22 cases. *American Journal of Surgical Pathology* **23,** 662–670.

Ring, D., Hubble, R., and Kirschner, M. (1982). Mitosis in a cell with multiple centrioles. *Journal of Cell Biology* **94,** 549–556.

Roberts, A. B. (1989). "Cancer Principles and Practice of Oncology," pp. 67–98. J. B. Lippincott Co., Philadelphia.

Roskelley, C. D., Desprez, P. Y., and Bissell, M. J. (1994). Extracellular matrix-dependent tissue specific gene expression in mammary epithelial cells requires both physical and biochemical signal transduction. *Proceedings of the National Academy of Sciences of the United States of America* **91,** 12378–12382.

Sato, N., Mizumoto, K., Nakamura, M., Nakamura, K., Kusumoto, M., Niiyama, H., Ogawa, T., and Tanaka, M. (1999). Centrosome abnormalities in pancreatic ductal carcinoma. *Clinical Cancer Research* **5,** 963–970.

Schmeichel, K. L., Weaver, V. M., and Bissell, M. J. (1998). Structural cues from the tissue microenvironment are essential determinants of the human mammary epithelial cell phenotype. *Journal of Mammary Gland Biology and Neoplasia* **3,** 201–213.

Sen, S., Zhou, H. Y., and White, R. A. (1997). A putative serine/threonine kinase encoding gene BTAK on chromosome 20q13 is amplified and overexpressed in human breast cancer cell lines. *Oncogene* **14,** 2195–2200.

Sharp, G. A., Osborn, M., and Weber, K. (1981). Ultrastructure of multiple microtubule initiation sites in mouse neuroblastoma cells. *Journal of Cell Science* **41,** 1–24.

Tanaka, T., Kimura, M., Matsunaga, K., Fukada, D., Mori, H., and Okano, Y. (1999). Centrosomal kinase AIK1 is overexpressed in invasive ductal carcinoma of the breast. *Cancer Research* **59,** 2041–2044.

Tomaszewski, M. M., Moad, J. C., and Lupton, G. P. (1999). Primary cutaneous Ki-1 (CD30) positive anaplastic large cell lymphoma in childhood. *Journal of the American Academy of Dermatology* **40,** 857–861.

Virchow, R. (1860). "Cellular Pathology as Based upon Physiological and Pathological Histology." London, John Churchill.

Vogelstein, B., Fearon, E. R., Hamilton, S. R., Kern, S. E., Presinger, A. C., Leppert, M., Nakamura, Y., White, R., Smits, A. M. M., and Bos, J. L. (1988). Genetic alterations during colorectal tumor development. *New England Journal of Medicine* **319,** 525–532.

Wang, X. J., Greenhalgh, D. A., Jiang, A., He, D. C., Zhong, L., Brinkley, B. R., and Roop, D. R. (1998). Analysis of centrosome abnormalities and angiogenesis in epidermal-targeted p53(172h) mutant and p53-knockout mice after chemical carcinogenesis—evidence for a gain of function. *Molecular Carcinogenesis* **23,** 185–192.

Weaver, V. M., Petersen, O. W., Wang, F., Larabell, C. A., Briand, P., Damsky, C., and Bissell, M. J. (1997). Reversion of the malignant phenotype of human breast cells in three-dimensional culture and in vivo by integrin blocking antibodies. *Journal of Cell Biology* **137,** 231–245.

Weber, R. G., Bridger, J. M., Benner, A., Weisenberger, D., Ehemann, V., Reifenberger, and Lichter, P. (1998). Centrosome amplification as a possible mechanism for numerical chromosome aberrations in cerebral primitive neuroectodermal tumors with TP53 mutations. *Cytogenetics and Cell Genetics* **83,** 266–269.

Xu, X. L., Weaver, Z., Linke, S. P., Li, C. L., Gotay, J., Wang, X. W., Harris, C. C., Ried, T., and Deng, C. X. (1999). Centrosome amplification and a defective G(2)-M cell cycle checkpoint induce genetic instability in BRCA1 exon 11 isoform-deficient cells. *Molecular Cell* **3,** 389–395.

Yin, X. Y., Grove, L., Datta, N. S., Long, M. W., and Prochownik, E. V. (1999). C-myc overexpression and p53 loss cooperate to promote genomic instability. *Oncogene* **18,** 1177–1184.

Zamecnik, M., and Michal, M. (1999). Malignant peripheral nerve sheath tumor with perineurial cell differentiation (malignant perineurioma). *Pathology International* **49,** 69–73.

Zhou, H. Y., Kuang, J., Zhong, L., Kuo, W. L., Gray, J. W., Sahin, A., Brinkley, B. R., and Sen, S. (1998). Tumour amplified kinase STK15/BTAK induces centrosome amplification, aneuploidy and transformation. *Nature Genetics* **20,** 189–193.

15

The Centrosome-Associated Aurora/Ipl-like Kinase Family

T. M. Goepfert and *B. R. Brinkley*
Department of Molecular and Cellular Biology
Baylor College of Medicine
Houston, Texas 77030

I. Introduction

The faithful and correct partitioning of the replicated genome to each daughter nucleus of dividing cells is the central theme of mitosis. Errors in the mitotic process can lead to catastrophic outcomes, including aneuploidy, tumorigenesis, or cell death. Although the duplication and partitioning of the genome is the central focus of the mitotic process, other cell components such as the centrosomes or pole determinants must also be duplicated and properly distributed, first and foremost, in order to assemble a functional bipolar mitotic spindle. In this chapter, we report on the status of an emerging family of serine/threonine (ser/thr) kinases generally identified as the aurora/Ipl-like kinase family that appears to affect centrosome activity, including duplication, maturation, and movement necessary for the proper segregation of the genome into daughter cells. The same kinases are often overexpressed in human tumors and function as oncogenes that ultimately lead to chromosome instability and aneuploidy. Many of the kinases are colocalized with the centrosome/spindle pole where they appear to regulate activities *in situ.*

As documented throughout this volume, centrosomes are discrete cytoplasmic domains characterized by a unique structural organization consisting of centrioles, a surrounding pericentriolar "matrix" (Schnackenberg *et al.,* 1998), and an assortment of electron-dense satellite bodies that collectively function as microtubule organizing centers (MTOCs). As illustrated

Current Topics in Developmental Biology, Vol. 49

0070-2153/00 $35.00

in Fig. 1, the centrosome is duplicated once in each cell cycle, culminating in the formation and separation of a pair of identical centrosomes that function as spindle poles in the mitosis. Following mitosis, each daughter cell inherits a single centrosome that will undergo the identical cycle in the next cell division unless the cells are programmed to enter G_0 and become differentiated, in which case most will retain only a single centrosome. Errors in the centrosome duplication process, an event that begins in S-phase of the cell cycle, are believed to be one of the principal pathways for the production of more than two centrosomes. Supernumerary centrosomes may also arise through the failure of cytokinesis, cell–cell fusions, or perhaps other pathways. Extra centrosomes may give rise to multipolar spindles that partition the genome asymmetrically, resulting in mitotic chaos and cell death or, in some cases, clonal selection of surviving cells and cancer (for review see Brinkley and Goepfert, 1998).

Defects that lead to errors in centrosome number are not mechanistically well defined, but the discovery that supernumerary centrosomes persist in

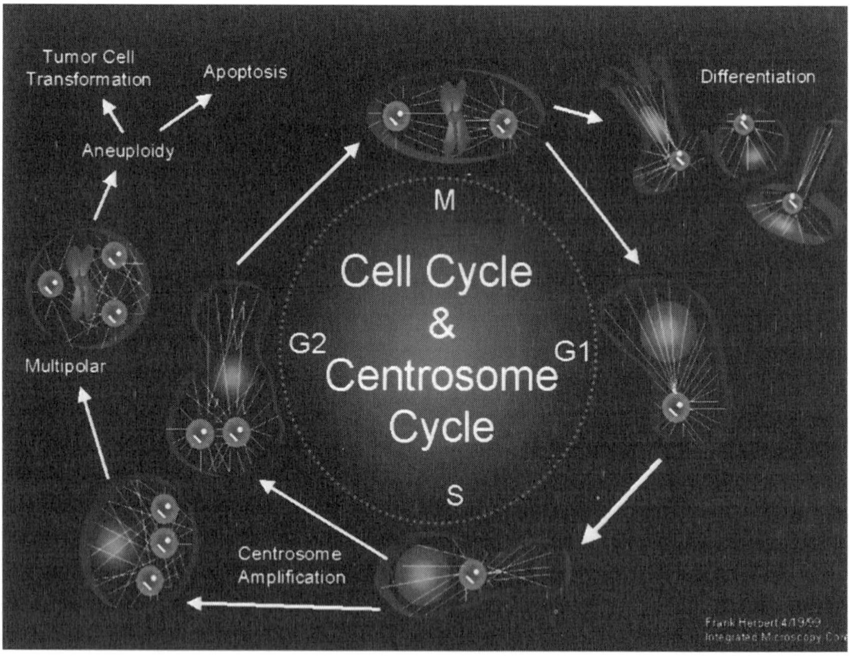

Figure 1 The centrosome cycle is superimposed on the cell cycle showing a bipolar spindle in M, a single centrosome in G_1, duplication of the centrosome in late S, and separation of centrosomes in G_2/M. When cells enter G_0 and become differentiated, the centrosome cycle is arrested. A scheme is proposed for the initiation of amplification due to errors in centrosome duplication in S-phase. Supernumerary centrosomes result in multipolar spindles that lead to aneuploidy, cell death, or transformation to tumor cells.

cells of a variety of solid tumors has led to the proposal that extra centrosomes may give rise to multipolar spindles that partition chromosomes incorrectly. In fact, the theory that malsegregation of chromosomes, caused by aberrant centrosome behavior, could lead to cancer was first proposed by Boveri (1914) almost a century ago.

More recently, Fukasawa and co-workers (1996) discovered that fibroblasts derived from mice lacking tumor suppressor gene p53 displayed extra centrosomes *in vitro.* In subsequent experiments involving transgenic mice, expression of mutation in the p53 gene was found to be linked to centrosome amplification in carcinogen-induced metastatic skin carcinomas (Wang *et al.,* 1998). The latter study was among the first to identify centrosome amplification in tumor cells *in vivo* and to link the phenomenon with genomic instability. Subsequently, several reports appeared that many human tumors *in vivo* display more than the expected one or two centrosomes (Lingle *et al.,* 1998; Pihan *et al.,* 1998) per cell, reviving the older hypothesis of Boveri (1914) that centrosome aberrations may be directly or indirectly involved in chromosome missegregation and aneuploidy.

Reports also appeared that some members of the aurora-like kinase gene family (aurora2/BTAK) were overexpressed in cells derived from colon (Bischoff *et al.,* 1998) and breast cancer (Zhou *et al.,* 1998). Further support for the hypothesis came with the discovery that many of the kinases colocalized with centrosomes and transfection of the respective cDNAs into normal diploid cells caused centrosome amplification, aneuploidy, and cell transformation. Thus, Aurora and Ipl-like kinases are now being intensely investigated in many species, as summarized in Table I, to learn more about them as possible regulators of mitosis and centrosome duplication.

This review summarizes the most recent developments in kinase functions of the Aurora/Ipl kinase family and related proteins. In addition, we review recent studies suggesting that various ser/thr kinase activity regulates key functions of centrosomes, and other components of the mitotic spindle.

II. The Aurora and Ipl-like Family: Structure, Regulation and Substrates

The founding members *Saccharomyces cerevisiae* Ipl-1 (increased in ploidy-1) and *Drosophila melanogaster* Aurora were both identified in mutant screens for mitotic defects. Severely aneuploid cells and, in the case of Aurora, monopolar spindles were found to arise from failure of centrosome separation at the onset of mitosis (Glover *et al.,* 1995). Homologous proteins have now been identified in *Xenopus,* mouse, rat, *Caenorhabditis elegans,* and human, suggesting a protein family that is generally similar

Table I The Centrosome-Associated Aurora/IPL-like Kinase Family[a]

Family	Target activity
Subfamily I	
AIE1 (mouse)	Centrosome localization
AIE2 (human)	
AIR-1 (*C. elegans*)	
ARK1 (mouse)	
IAK-1/STK6 (mouse)	
Nek2 (human)	Centrosome separation, phosphorylates C-Nap1
BTAK/STK15 (human)	Centrosome amplification
aur2 (human)	
AIK/AIK-1 (human)	
Eg2 (*Xenopus*)	Kinesin-related protein activity, phosphorylates XIEg5
Ipl-1 (*S. cerevisiae*)	Association with kinetochore, phosphorylates Ndc 10
aur2/ AIK (human)	Associates with Cdc 20
Subfamily II	
AIM-1 (rat)	Related to cytokinesis
AIK-2 (STK12) (human)	
AIK-3 (human)	
aur1 (human)	
STK-1 (mouse)	
ARK-2 (mouse)	
AIR-2 (*C. elegans*)	
Subfamily III	
AYK1 (mouse)	Related to meiosis
AIR-2 (*C. elegans*)	

[a] The aurora/Ipl kinase family associates with mitotic structures, including centrosomes, spindle microtubules and spindle midzone, kinetochores, and kinesin-related proteins (for references, see article).

and part of a group that has evolved to carry out diverse functions during mitosis.

As cells progress through the cell cycle, Aurora/Ipl-like kinases undergo major changes in abundance, activity, and localization. In somatic tissue culture cells, mRNA and/or protein levels are typically low in G_1–S-phase cells, rise and accumulate during G_2 and M, and fall rapidly following mitosis. Moreover, Aur/Ipl-like kinases undergo associations with one or more mitotic structures, including the spindle pole, spindle microtubules, the spindle midzone, and/or spindle microtubules that persist in the midbody during cytokinesis.

By sequence, the similarity of the N-terminal regulatory domains indicates that *Xenopus* Eg2 (Roghi *et al.,* 1998), human BTAK/aurora2/AIK1 (Sen *et al.,* 1997; Bischoff *et al.,* 1998; Kimura *et al.,* 1997) and ARK1 (Shindo *et al.,* 1998), and mouse IAK1 (Gopalan *et al.,* 1997) and AIE-1

(Tseng *et al.*, 1998) are more closely related and constitute a discrete subfamily. A second subfamily consists of human AIK2 (STK12) (Kimura *et al.*, 1998) and aur1 (Bischoff *et al.*, 1998), rat AIM-1 (Terada *et al.*, 1998), and mouse STK1 (Niwa *et al.*, 1996) and ARK2 (Shindo *et al.*, 1998). All members share 67–76% amino acid sequence identity in their catalytic domains.

More distantly related, in terms of both sequence and localization, are mouse Ayk1 (Yanai *et al.*, 1997) and AIR-2 (Schumacher *et al.*, 1998b). Ayk1 is specifically expressed in active meiotic cells and is also expressed in germ cells and mitotically active embryonic cells. Ayk1 transcripts accumulate just before the first meiotic cell division, suggesting a crucial role in mammalian meiotic divisions, probably in controlling some aspects of chromosome segregation. AIR-2 is associated with both meiotic and mitotic chromosomes. During both meiotic anaphase I and II, AIR-2 is present between the separating chromosomes. At each meiotic division, AIR-2 may coordinate the congression of metaphase chromosomes with the subsequent events of polar body extrusion and cytokinesis in *C. elegans.* Comparatively little is presently known about the physiological targets of Aurora/Ipl-like kinases, but a few candidate substrates have been identified (Table I).

A. Aur/Ipl-like Kinases Are Linked to Kinetochores

This possibility was originally suggested by studies in budding yeast. Recent data (Biggins *et al.*, 1999) show that Ipl1 regulates the interaction between microtubules and kinetochores. The protein Ndc10 is one of the four components of the multiprotein kinetochore complex CBF3. Regulation of kinetochore function is indicated by evidence that Ipl1 phosphorylates Ndc10 *in vitro.*

Aurora2/AIK associates with human Cdc20 (Farruggio *et al.*, 1999). Cdc20 is a protein involved in activation of the anaphase-promoting complex/cyclosome. Interestingly, both aur2/Aik and Cdc20 localize to the mitotic spindle poles. The demonstration that Cdc20 is associated with aurora2/Aik suggests that Cdc20 functions are in some way regulated through its association with this member of the aurora kinase subfamily. Further studies will no doubt illuminate this important signaling pathway, providing additional knowledge of a possible role in regulating kinetochore as well as centrosome activity.

B. Role in Cytokinesis

Evidence is building to indicate that Aur/Ipl-like kinases may also play a significant role in the regulation of cytokinesis. A group of kinases,

AIM-1 in rat (Terada *et al.*, 1998), aur1 (Bischoff *et al.*, 1998), AIK2 (Kimura *et al.*, 1998), AIK3 (Kimura *et al.*, 1999), and ARK2 (Shindo *et al.*, 1998) in human, STK1 (Niwa *et al.*, 1996) in mouse embryonic stem cells, and AIR-2 in *C. elegans* (Schumacher *et al.*, 1998b), seem to be required for late mitotic events and cytokinesis. It is also known that AIM-1, Aur1, AIK2, and AIK3 are related to cancer and their expression is altered in several cancer cell lines. AIM-1 overexpression produced multinuclearity in human cells, and increased ploidy and aneuploidy were induced as well. AIR-2, the related protein in *C. elegans,* is localized to mitotic centrosomes and metaphase chromosomes, but at anaphase it is associated with midbody microtubules where it persists at the cytokinesis remnant.

Among this subfamily, AIM-1 (Terada *et al.*, 1998) is the most widely studied. AIM-1 has been found to localize at the equator of central spindles during late anaphase, and at the midbody during telophase and cytokinesis. As a mitotic regulator, it appears to play a key role in the onset of cytokinesis during late anaphase and telophase. Transcriptionally controlled during G_2/M-phase, AIM-1 mRNA is induced in late S-phase and peaks near G_2/M. The protein is found in late anaphase as a distinct wide band extending across the midzone of the central spindle. The kinetics of protein expression coincides with this appearance, suggesting that AIM-1 might regulate the process from anaphase to telophase/cytokinesis (Tatsuka *et al.*, 1998). Overexpression of kinase-inactive AIM-1 (dominant-negative) disrupts cleavage furrow formation without affecting nuclear division. Cytokinesis frequently fails, resulting in cell polyploidy and subsequent cell death (Terada *et al.*, 1998).

It is noteworthy that all of the members of this subfamily appear to affect microtubule dynamics at different phases of mitosis. Those acting in earlier stages of mitosis appear to be involved in the disassembly of the cytoplasmic microtubule complex (CMTC) during the G_2–M phase transition, giving rise to arrays of shorter, more dynamic microtubules of the mitotic apparatus. Those acting later in mitosis appear to act in a pathway that stabilizes the midbody microtubules and their associated proteins, perhaps allowing a functional interaction between the midzone and the progressing cleavage furrow (Terada *et al.*, 1998). It will be interesting to explore whether phosphorylation by kinases regulates the function of motor proteins, and if so, how this relates to spindle dynamics, chromosome segregation, and cytokinesis.

C. Role in Centrosome Function

Centrosomes undergo dynamic changes and movements in a cell-cycle-dependent way. In cycling cells, the centrosome is replicated during

S phase, but its separation into a bipolar spindle is delayed until prophase/prometaphase of mitosis. Because of spindle elongation at anaphase B, the centrosomes move further apart during late anaphase. The mechanism that drives vertebrate centrosome separation is still not yet well defined. Genetic and biochemical studies point to a critical role of dynein and kinesin-related microtubule motors for generating the forces necessary for separating centrosomes (Endow *et al.*, 1999). However, centrosome movement is also known to be a cell-cycle-related process and therefore, reversible protein phosphorylation involving ser/thr kinases may also play a central role in controlling the separation, as implied in the following section.

1. Aur/Ipl-like Kinases Promote Centrosome Separation

The kinase Nek2, a mammalian ser/thr kinase structurally related to the mitotic regulator NIMA, induces a pronounced splitting of centrosomes. Localization studies and biochemical fractionation confirmed the importance of this result by demonstrating that Nek2 is a core component of the centrosome throughout the cell cycle (Fry *et al.*, 1998b). In the search for targets of the Nek2 kinase involved in centrosome–centrosome cohesion, a candidate, C-Nap1, was recently isolated. C-Napl (centrosomal Nek2-associated protein 1) is specifically associated with the proximal end of the two centrioles. The current working model (Mayor *et al.*, 1999) implicates C-Nap1 in a bridgelike structure that links the two centrosomes throughout interphase. At the G_2/M transition, phosphorylation by Nek2 might regulate C-Nap1 interaction or stability, leading to dissolution of the bridge prior to the step of motor-driven centrosome separation (Fry *et al.*, 1998a).

2. Aur/Ipl-like Kinases Are Involved in the Regulation of Kinesin-Related Protein Activities

Eg2, the *Xenopus* homolog, is associated with centrosomes and the mitotic spindle (Roghi *et al.*, 1998), and Eg2 is detected at the centrosome throughout mitosis. In prophase, Eg2 is localized to the PCM, but during metaphase, it appears to be repositioned from the centrosome to the microtubules of the bipolar spindle. During anaphase it seems to bind to the microtubules at the half-spindles. The relocation of Eg2 to the pole-directed microtubules occurs only during metaphase and anaphase, implying a role for Eg2 in the formation of a bipolar spindle formation and function (Roghi *et al.*, 1998). Recent data show a physical association of Eg2 with a microtubule-based motor protein X1Eg5 (Giet *et al.*, 1999), a kinesin-like protein required for mitotic spindle assembly in *Xenopus* egg extract. X1Eg5 and Eg2 were both detected around the centrosomes in prophase and on microtubules during metaphase and anaphase. Eg2 kinase associates with and phosphorylates the kinesin-related protein X1Eg5, which suggests that the aurora-related kinase family might also be involved in the regulation of kinesin-related motor protein activities. In addition, Eg2 appears to play a role in the progesterone-activated signaling pathway that triggers oocyte maturation (Andresson and Ruderman, 1998). These results argue that Eg2 is a component of the progesterone-activated signaling pathway that releases frog oocytes from cell cycle arrest.

3. Aur/Ipl-like Kinases Localize to Centrosomes

IAK-1 (STK6) (Gopalan *et al.,* 1997), AIE-1 (Tseng *et al.,* 1998), and ARK-1 (Shindo *et al.,* 1998) in mouse, and AIE2 (Tseng *et al.,* 1998) in human, are first detected at the centrosomes immediately after the duplicated centrosomes have separated. The protein remains with the centrosome and the centrosome/proximal region of the spindle throughout mitosis and is detected weakly on midbody microtubules at telophase and cytokinesis. Expression is tightly regulated temporally and spatially during the cell cycle. Transcripts first appear at the G_1/S boundary, they are elevated at M-phase, and disappear rapidly after completion of mitosis. The protein levels and kinase activities are also cell cycle regulated with a peak at M-Phase. IAK-1 protein has a distinct subcellular and temporal pattern of localization. AIR-1, the homolog in *C. elegans,* may also be a structural component of centrosomes; its kinase activity may function in the phosphorylation cascades that are required for the recruitment of centrosomal proteins, and for the formation of the mitotic spindle. Disruption of AIR-1 protein expression in *C. elegans* embryos results in severe aneuploidy and embryonic lethality. However, aneuploidy does not appear to arise as a result of failure in centrosome separation. Bipolar spindles are formed in the absence of AIR-1, but they appear to be disorganized and are nucleated by abnormal-looking centrosomes. In addition to its well-documented requirement during mitosis, AIR-1 may regulate microtubule-based developmental processes as well (Schumacher *et al.,* 1998a).

4. Aur/Ipl-like Kinases, Centrosome Amplification, and Cancer

Data compiled from human tumor tissue indicate that Aurora/Ipl-like kinases are involved in centrosome function and duplication, and that important substrates may be located at centrosomes and spindle pole. BTAK/aurora2/AIK1 (Zhou *et al.,* 1998; Bischoff *et al.,* 1998; Tanaka *et al.,* 1999), a centrosome-associated kinase, has been detected in breast tumor cell lines as well as in other tumors without evidence of gene amplification. BTAK/aurora2/AIK1 is implicated in the induction of centrosome duplication–distribution abnormalities and aneuploidy in mammalian cells. Immunofluorescence studies have shown that BTAK/aur2/AIK1 is localized to the spindle pole during mitosis, from prophase to anaphase, suggesting a role in centrosome function required for spindle formation and chromosome segregation. The kinase was shown capable of transforming rodent fibroblasts, indicating its oncogenic potential. Overexpression of BTAK in NIH3T3 cells led to the appearance of abnormal centrosome number, and transformation *in vitro.* A similar effect was also induced in near-diploid human breast epithelial cells where an increase in centrosome number correlated with the induction of aneuploidy (Zhou *et al.,* 1998).

Immunohistochemical analyses of breast invasive ductal adenocarcinomas revealed the occurrence of overexpression of AIK1 (hu STK6) (Tanaka *et al.,* 1999), whereas the protein was not detected in normal ductal and lobular cells. The molecular mechanism by which AIK1 protein is overexpressed in cancer cells has not yet been identified.

The importance of these kinases in tumorigenesis is also reflected in the fact that several kinases map to chromosome regions that are amplified, rearranged, or deleted in cancer tissue (i.e., BTAK on chromosome 20q13; AIK2 on chromosome 17p13.1; and AIK3 on chromosome 19q13.43).

Defects in centrosome number, organization, and behavior constitute a condition defined as "centrosome hypertrophy" or "centrosome amplification" (Lingle *et al.,* 1998; Brinkley and Goepfert, 1998). The amplification may result from an increase in centrosome protein expression or a decrease in turnover of centrosome components, and/or recruitment of centrosome precursors from cytoplasmic pools. Aberrant centrosome duplication may alter cell polarity and increase the incidence of multipolar mitosis, leading to chromosomal segregation abnormalities and aneuploidy. However, there are likely to be multiple pathways for achieving genomic instability in tumor cells. Aneuploidy and chromosome instability (CIN) are known to be associated with most human cancers, but some tumors, classified as those displaying DNA microsatellite instablity (MIN), are characterized as having diploid nuclei (Lengauer *et al.,* 1998). It will be important to determine if CIN-type tumors involve centrosome amplification, whereas those that remain diploid (MIN) have normal centrosome numbers and arise by an alternative mechanism.

III. Summary

Because of the well-known role of the centrosome and mitotic apparatus in genome partitioning in normal cells, defects in pathways essential for mitotic regulation are likely implicated in the cascade of events leading to aneuploidy and neoplasia. Exogenous overexpression of AIM-1, for example, produces multinuclearity in human cells and increased ploidy as well as aneuploidy (Tatsuka *et al.,* 1998). Overexpression in colorectal tumor cell lines is thought to have a causal relationship with multinuclearity and increased ploidy. Cytokinesis error caused by AIM-1 overexpression is a major factor in the predisposition to cancer. As previously mentioned, the involvement of BTAK/aur2/AIK in centrosome amplification and its oncogenic activity are compelling. Aur2 has also been implicated in oncogenesis, and defects in kinetochore function leading to chromosome instability in human tumors should not be minimized (Farruggio *et al.,* 1999). Further studies are needed to provide a clearer definition of how these kinetic proteins are linked and regulated in normal mitosis and cancer.

Thus, Boveri appears to have been correct in formulating his early hypothesis that a defective mitotic apparatus and centrosome number were central and causative in chromosome missegregation and cancer. One hundred years later, at the onset of a new millennium and with light-years of advanced technology in our favor, we are just now beginning to piece together the enzymes, substrates, and signaling pathways that support and explain his long-ignored but prophetic claim.

Acknowledgments

The authors are grateful for discussions and advice from Ilia Ouspenski, Subrata Sen, and Mike Mancini. We thank L. Zhong and Y. E. Adigun for technical assistance and F. Herbert for assistance with graphics and illustrations. The work was supported by NIH, NCI grants CA41424 and CA 64255.

References

Andresson, T., and Ruderman, J. V. (1998). The kinase Eg2 is a component of the *Xenopus* oocyte progesterone-activated signaling pathway. *EMBO J* **17**(19), 5627–5637.

Biggins, S., Severin, F. F., Bhalla, N., Sassoon, I., Hyman, A. A., and Murray, A. W. (1999). The conserved protein kinase Ipl1 regulates microtubule binding to kinetochores in budding yeast. *Genes Dev* **13**(5), 532–544.

Bischoff, J. R., Anderson, L., Zhu, Y., Mossie, K., Ng, L., Souza, B., Schryver, B., Flanagan, P., Clairvoyant, F., Ginther, C., Chan, C. S., Novotny, M., Slamon, D. J., and Plowman, G. D. (1998). A homologue of *Drosophila* aurora kinase is oncogenic and amplified in human colorectal cancers. *EMBO J* **17**(11), 3052–3065.

Boveri, T. (1914). *Zur Frage der Entstehung maligner Tumoren.* Jena:Fischer Verlag.

Brinkley, B. R., and Goepfert, T. M. (1998). Supernumerary centrosomes and cancer: Boveri's hypothesis resurrected. *Cell Motil Cytoskeleton* **41**(4), 281–288.

Endow, S. A. (1999). Microtubule motors in spindle and chromosome motility. *Eur J Biochem* **262**(1), 12–18.

Farruggio, D. C., Townsley, F. M., and Ruderman, J. V. (1999). Cdc20 associates with the kinase aurora2/Aik [In Process Citation]. *Proc Natl Acad Sci USA* **96**(13), 7306–7311.

Fry, A. M., Mayor, T., Meraldi, P., Stierhof, Y. D., Tanaka, K., and Nigg, E. A. (1998a). C-Nap1, a novel centrosomal coiled-coil protein and candidate substrate of the cell cycle-regulated protein kinase Nek2. *J Cell Biol* **141**(7), 1563–1574.

Fry, A. M., Meraldi, P., and Nigg, E. A. (1998b). A centrosomal function for the human Nek2 protein kinase, a member of the NIMA family of cell cycle regulators. *EMBO J* **17**(2), 470–481.

Fukasawa, K., Choi, T., Kuriyama, R., Rulong, S., and Vande Woude, G. F. (1996). Abnormal centrosome amplification in the absence of p53. *Science* **271**(5256), 1744–1747.

Giet, R., Uzbekov, R., Cubizolles, F., Le Guellec, K., and Prigent, C. (1999). The *Xenopus laevis* aurora-related protein kinase pEg2 associates with and phosphorylates the kinesin-related protein X1Eg5 [In Process Citation]. *J Biol Chem* **274**(21), 15005–15013.

Glover, D. M., Leibowitz, M. H., McLean, D. A., and Parry, H. (1995). Mutations in aurora prevent centrosome separation leading to the formation of monopolar spindles. *Cell* **81**(1), 95–105.

Gopalan, G., Chan, C. S., and Donovan, P. J. (1997). A novel mammalian, mitotic spindle-associated kinase is related to yeast and fly chromosome segregation regulators. *J Cell Biol* **138**(3), 643–656.

Kimura, M., Kotani, S., Hattori, T., Sumi, N., Yoshioka, T., Todokoro, K., and Okano, Y. (1997). Cell cycle-dependent expression and spindle pole localization of a novel human protein kinase, Aik, related to Aurora of *Drosophila* and yeast Ipl1. *J Biol Chem* **272**(21), 13766–13771.

Kimura, M., Matsuda, Y., Yoshioka, T., Sumi, N., and Okano, Y. (1998). Identification and characterization of STK12/Aik2: a human gene related to aurora of *Drosophila* and yeast IPL1. *Cytogenet Cell Genet* **82**(3–4), 147–152.

Kimura, M., Matsuda, Y., Yoshioka, T., and Okano, Y. (1999). Cell cycle-dependent expression and centrosome localization of a third human aurora/Ipl1-related protein kinase, AIK3. *J Biol Chem* **274**(11), 7334–7340.

Lengauer, C., Kinzler, K. W., and Vogelstein, B. (1998). Genetic instabilities in human cancers. *Nature* **396**(6712), 643–649.

Lingle, W. L. Lutz, W. H., Ingle, J. N., Maihle, N. J., and Salisbury, J. L. (1998). Centrosome hypertrophy in human breast tumors: implications for genomic stability and cell polarity. *Proc Natl Acad Sci USA* **95**(6), 2950–2955.

Mayor, T., Meraldi, P., Stierhof, Y. D., Nigg, E. A., and Fry, A. M. (1999). Protein kinases in control of the centrosome cycle. *FEBS Lett* **452**(1–2), 92–95.

Niwa, H., Abe, K., Kunisada, T., and Yamamura, K. (1996). Cell-cycle-dependent expression of the STK-1 gene encoding a novel murine putative protein kinase. *Gene* **169**(2), 197–201.

Pihan, G. A., Purohit, A., Wallace, J., Knecht, H., Woda, B., Quesenberry, P., and Doxsey, S. J. (1998). Centrosome defects and genetic instability in malignant tumors. *Cancer Res* **58**(17), 3974–3985.

Roghi, C., Giet, R., Uzbekov, R., Morin, N., Chartrain, I., Le Guellec, R., Couturier, A., Doree, M., Philippe, M., and Prigent, C. (1998). The *Xenopus* protein kinase pEg2 associates with the centrosome in a cell cycle-dependent manner, binds to the spindle microtubules and is involved in bipolar mitotic spindle assembly. *J Cell Sci* **111**(Pt 5), 557–572.

Schnackenberg, B. J., Khodjakov, A., Rieder, C. L., and Palazzo, R. E. (1998). The disassembly and reassembly of functional centrosomes *in vitro. Proc Natl Acad Sci USA* **95**(16), 9295–9300.

Schumacher, J. M., Ashcroft, N., Donovan, P. J., and Golden, A. (1998a). A highly conserved centrosomal kinase, AIR-1, is required for accurate cell cycle progression and segregation of developmental factors in *Caenorhabditis elegans* embryos. *Development* **125**(22), 4391–4402.

Schumacher, J. M., Golden, A., and Donovan, P. J. (1998b). AIR-2: An Aurora/Ipl1-related protein kinase associated with chromosomes and midbody microtubules is required for polar body extrusion and cytokinesis in *Caenorhabditis elegans* embryos. *J Cell Biol* **143**(6), 1635–1646.

Sen, S., Zhou, H., and White, R. A. (1997). A putative serine/threonine kinase encoding gene BTAK on chromosome 20q13 is amplified and overexpressed in human breast cancer cell lines. *Oncogene* **14**(18), 2195–2200.

Shindo, M., Nakano, H., Kuroyanagi, H., Shirasawa, T., Mihara, M., Gilbert, D. J., Jenkins, N. A., Copeland, N. G., Yagita, H., and Okumura, K. (1998). cDNA cloning, expression, subcellular localization, and chromosomal assignment of mammalian aurora homologues, aurora-related kinase (ARK) 1 and 2. *Biochem Biophys Res Commun* **244**(1), 285–292.

Tanaka, T., Kimura, M., Matsunaga, K., Fukada, D., Mori, H., and Okano, Y. (1999). Centrosomal kinase AIK1 is overexpressed in invasive ductal carcinoma of the breast. *Cancer Res* **59**(9), 2041–2044.

Tatsuka, M., Katayama, H., Ota, T., Tanaka, T., Odashima, S., Suzuki, F., and Terada, Y. (1998). Multinuclearity and increased ploidy caused by overexpression of the aurora- and Ipl1-like midbody-associated protein mitotic kinase in human cancer cells. *Cancer Res* **58**(21), 4811–4816.

Terada, Y., Tatsuka, M., Suzuki, F., Yasuda, Y., Fujita, S., and Otsu, M. (1998). AIM-1: a mammalian midbody-associated protein required for cytokinesis. *EMBO J.* **17**(3), 667–676.

Tseng, T. C., Chen, S. H., Hsu, Y. P., and Tang, T. K. (1998). Protein kinase profile of sperm and eggs: cloning and characterization of two novel testis-specific protein kinases (AIE1, AIE2) related to yeast and fly chromosome segregation regulators. *DNA Cell Biol.* **17**(10), 823–833.

Vaughn, J. P., Davis, P. L., Jarboe, M. D., Huper, G., Evans, A. C., Wiseman, R. W., Berchuck, A., Iglehart, J. D., Futreal, P. A., and Marks, J. R. (1996). BRCA1 expression is induced before DNA synthesis in both normal and tumor-derived breast cells. *Cell Growth Differ* **7**(6), 711–715.

Wang, X.-J., Greenbalgh, D. A. Jiang. A., He, D., Zhong, L., Medina, D., Brinkley, B. R. and Roop, D. R. (1998). Expression of a p53 mutant in the epidermis of transgenic mice accelerates chemical carcinogenesis chemical carcinogenesis. *Oncogene* **17,** 35–45.

Yanai, A., Arama, E., Kilfin, G., and Motro, B. (1997). Ayk1, a novel mammalian gene related to *Drosophila* aurora centrosome separation kinase, is specifically expressed during meiosis. *Oncogene* **14**(24), 2943–2950.

Zhou, H., Kuang, J., Zhong, L., Kuo, W. L., Gray, J. W., Sahin, A., Brinkley, B. R., and Sen, S. (1998). Tumour amplified kinase STK15/BTAK induces centrosome amplification, aneuploidy and transformation [see comments]. *Nat Genet* **20**(2), 189–193.

16

Centrosome Reduction during Mammalian Spermiogenesis

G. Manandhar,[2] C. Simerly,[1,2] and G. Schatten[1,2,*]
[1] Departments of Obstetrics–Gynecology and Cell and Developmental Biology
[2] Oregon Regional Primate Research Center
Oregon Health Sciences University
Beaverton, Oregon 97006

I. Introduction

Mammalian fertilization is essentially a union of male and female gametes, and the achievement of proper cytoplasmic conditions for successful cleavages. Sexual reproduction is endowed with a spectacular strategy that ensures inheritance of the cellular constituents from both parents, but avoids redundancy of cellular organelles that would interfere with the normal development of the progeny. Meiotic division of the germ cells reduces the genomic material by half so that the male and female gametes contribute the equivalent amount of genomic material to constitute the diploid genome of the zygote. Besides the genome, centrosomal inheritance is a vitally important event during sexual reproduction. The centrosomes play crucial roles by organizing the microtubular aster that brings the parental genomes into close apposition during fertilization (Schatten, 1982), and by forming bipolar spindles that ensure the equal distribution of the genome to the daughter cells during cleavages (reviewed by Schatten, 1994). Because of

* Author to whom the correspondence should be addressed.

Current Topics in Developmental Biology, Vol. 49

0070-2153/00 $35.00

these vital roles, zygotes must inherit a correct number of centrosomes and exert tight control over their replication. Improper centrosomal inheritance is one of the causes of fertilization failure and abnormal embryonic development in higher animals, including humans (Simerly *et al.,* 1995; Palermo *et al.,* 1997).

Whereas the male and female gametes contribute equal amounts of genomic material, their centrosomal contribution is not equivalent. The earliest hypothesis forwarded by Theodor Boveri (1901) suggested a paternal mode of centrosome inheritance. Thus, the oocytes were thought to lose functional centrosomes (Mazia, 1978, 1984). This viewpoint has been consolidated by the observation that centrioles disappear from oocytes during the extended period of meiotic arrest. Systematic electron microscopic studies of mouse oocytes have confirmed the absence of centrioles in the meiotic spindle poles (Szollosi *et al.,* 1972; Schatten *et al.,* 1989). The centriole degeneration of oocytes could be a universal feature, since they were not found in oocytes of phylogenetically diverse animals such as sea urchins (Paweletz *et al.,* 1984, 1987a), frogs (Huchon *et al.,* 1981), rabbits (Zamboni and Mastroianni, 1966), chickens (Greenfield, 1966), *Drosophila* (Mahowald and Strassheim, 1970), cows (Sathananthan *et al.,* 1997), and humans (Hertig and Adam, 1967).

Boveri's hypothesis that the sperm contribute the centrosomes has since been supported by subsequent electron microscopic studies of the sperm. Centrioles were observed in the sperm of the sea urchin (Longo and Anderson, 1968; Longo, 1973), *Xenopus* (Bernardini *et al.,* 1986), ostrich (Soley, 1994), rabbit (Zamboni and Stefanini, 1971), sheep (LeGuen and Crozet, 1989; Crozet, 1990), guinea pig (Fawcett, 1965), bull (Sathananthan *et al.,* 1997), monkey (Zamboni and Stefanini, 1971), and human (Zamboni and Stefanini, 1971; Sathananthan *et al.,* 1991, 1996). These centrioles are introduced into oocytes during fertilization, duplicated, and perpetuated to the daughter cells after each cleavage.

Recent studies have suggested that oocyte centrosomes do not degenerate completely. Though oocytes lack the centriole, they possess fibrous MTOCs that organize bipolar spindles during meiosis (Szollosi *et al.,* 1972; Schatten *et al.,* 1989). The latent MTOCs of the unfertilized oocytes reappear and function after parthenogenetic activation (Szollosi and Ozil, 1991). The oocytes contain an arsenal of centrosomal proteins (Holy and Schatten, 1991; Schatten *et al.,* 1992; Calarco-Gillam *et al.,* 1983; Maro *et al.,* 1985; Palacios *et al.,* 1993; Hiraoka *et al.,* 1989; Merdes *et al.,* 1996; Navara *et al.,* 1996), enough to assemble 1000–2000 active centrosomes (Gard *et al.,* 1990). Conversely, the sperm lose most of the centrosomal proteins (Sutovsky *et al.,* 1999) while they retain the centrioles. Thus, the centrosomes are reduced in male and female gametes in a reciprocal manner, which ensures that they contribute unequal yet complementary centrosomal components during fertilization.

It is obvious that centrosome reduction during gametogenesis plays a vital role to balance the number of centrosomes in the zygote. But this phenomenon has not yet been explained fully. Moreover, suppression/elimination of the activity of this enigmatic organelle in differentiated cells is still a mysterious phenomenon. Earlier works on centrosome reduction have provided important findings, such as the absence of centrioles in oocytes (Szollosi *et al.,* 1972; Schatten *et al.,* 1989) and rat sperm (Woolley and Fawcett, 1973), and the loss of the replicating capacity of the maternal centrosome during meiosis (Sluder *et al.,* 1989, 1993). Recently our laboratory has completed a series of studies to understand the cellular and molecular pathways of centrosome reduction during mouse, rhesus, and human spermiogenesis. This article synthesizes the fundamental events of this process.

II. Overview of Mammalian Spermiogenesis

Spermiogenesis is the terminal differentiation of postmeiotic germ cells into spermatozoa. A haploid cell formed after male meiosis is called the spermatid. The complicated morphogenetic process of spermiogenesis takes place inside the seminiferous tubules of the testis, in close association with the Sertoli cells. After completing the testicular stages of spermiogenesis, the spermatids move into the lumen of seminiferous tubules. Testicular spermiogenesis takes about 20 days in the rat (Oakberg, 1956; Clermont and Harvey, 1965), 15.1 days in mouse (Clermont and Trott, 1969), 17.5 days in the rhesus monkey (de Rooij *et al.,* 1986), and 18–19 days in man (Heller and Clermont, 1963). Further maturation takes place in the epididymis.

On the basis of morphological criteria, observable under a light microscope, the testicular phase of mouse spermiogenesis has been divided into 16 developmental steps (Oakberg, 1956). Mouse spermiogenesis steps have been further characterized by their ultrastructural features (Dooher and Bennett, 1973). Regardless of some minor variations, the developmental process as well as the relative duration of the various spermiogenesis steps are similar in different mammalian species. Mouse and human spermiogenesis has been described in detail in the original papers (Gardner and Holyoke, 1964; Gardner, 1966; Dooher and Bennett, 1973; Clermont, 1963) and in several review articles (Russell *et al.,* 1990; DeKretser and Kerr, 1994; Roosen-Runge, 1977). Earlier studies on spermiogenesis were mainly done using light microscopy on histological sections and by electron microscopy. Recent developments in immunocytochemistry and whole-mount single cell imaging techniques have provided a new possibility to investigate centrosomes during developmental processes at a subcellular resolution under a light microscope. Some important observations on centrosome reduction have been obtained from isolated spermatids using these techniques (Manandhar *et al.*, 1998, 1999). The developmental stages of isolated mouse spermatids that show remarkable events of centrosomal changes are described next.

A. Round Spermatid Stage

An early stage spermatid has isodiametric shape *in situ* and after isolation. It has a round nucleus and a distinct heterochromatic region in the center of the nucleus. Each spermatid has a long incipient tail comprising an extended microtubular axoneme. Since the tail is very fragile at this stage, it is easily broken during the isolation process. Other prominent markers of round spermatids are the chromatoid body and the proacrosomal granule that are readily visible under a phase-contrast microscope. In the latter stages of development, a vacuolelike acrosome is visible around the proacrosomal granule. The round spermatid stage is tentatively equivalent to steps 1 to 5 of mouse spermiogenesis.

B. Elongating Spermatid Stage

Nuclear elongation is a remarkable event during spermiogenesis. This stage begins with eccentric localization of the nucleus, disappearance of the heterochromatic region, flattening of the acrosomal region of the nucleus, and gradual morphogenesis of the nucleus into a pear shape, then a broad falciform shape. In later-stage elongating spermatids, the acrosome is visible under phase contrast on the anterior convex side of the nuclear curvature. The cytoplasm is displaced toward the posterior side. In isolated spermatids most of the cytoplasm is lost, leaving a thin layer around the nucleus and the neck region. Under phase contrast optics, the centriolar complex is visible in the neck region as a dark rodlike structure that extends a slender axoneme posteriorly. The elongating spermatid stage is approximately equivalent to steps 6 to 12 of mouse spermiogenesis.

C. Late Elongating Stage/Spermiation Stage

Late-stage testicular spermiogenesis is marked by nuclear compaction and DNA condensation. A dense region appears in the middle of the nucleus that gradually expands centripetally. This feature is visible under phase contrast and Hoechst fluorescence microscopy. The midpiece undergoes various stages of thickening because of the accumulation of mitochondria around the axoneme. Finally, the mature testicular sperm are released into the seminiferous tubule lumen, a process called spermiation. Before spermiation, the spermatids discard the excess cytoplasm as residual bodies. Frequently, small residual bodies can be seen attached to the neck region of some testicular sperm. The late elongating stage encompasses spermiogenesis steps 13 to 16.

III. Stages of Centrosome Reduction during Spermiogenesis

The process of centrosome reduction extends throughout the entire testicular and epididymal stages of spermiogenesis and comprises four stages: (a) loss of microtubule nucleating function, (b) formation of transitional MTOCs, (c) loss of centrosomal proteins, and (d) centriole degeneration.

A. Loss of Microtubule Nucleating Function

During meiotic divisions, the centrosomes localize at the spindle poles organizing bipolar spindles. The round spermatids possess two centrioles in orthogonal orientation, lying freely in the cytoplasm. The pericentriolar region does not possess a detectable amount of fibrous material. The distal centriole possesses inconspicuous basal feet that, however, do not nucleate microtubules (Manandhar *et al.*, 1998). The proximal region of the distal centriole is associated with a dense body of undefined function. It develops into a hollow spherical body during the late elongating stage (Yang *et al.*, 1997). Immunofluorescent cytochemistry and immunogold electron microscopy show the presence of important centrosomal proteins in the spermatid centrosomes. γ-Tubulin is associated with the pericentriolar region, and centrin is found directly on the microtubular triplets of the centrioles (Manandhar *et al.*, 1998, 1999). Thus, the spermatid centrosomes appear to be normal, structurally and compositionally, yet they seem to have lost the fundamental function of the centrosome, i.e., microtubule nucleating activity. The round spermatids display a random network of cortical microtubules that do not seem to originate from the centrosome (Figs. 1A, 1G; see color insert). At this stage, inner cytoplasmic microtubules are not detectable. It can be argued that the cortical microtubules are initially nucleated from the centrosome but detached and drifted away later (Keating *et al.*, 1997). But a meticulous analysis of serial ultrathin sections confirmed that the microtubules do not originate from the spermatid centrosome (Manandhar *et al.*, 1998).

The most dramatic function of round spermatid centrioles is the formation of the axoneme. The distal centrioles extend typical (9 + 2) axonemes that meander through the intercellular spaces. There is no remarkable transitional region between the centriole proper and the axoneme, as in the lower eukaryotic flagellar apparatus (Lang, 1963; Ringo, 1967; Sanders and Salisbury, 1994). The axis of the distal centriole, which produces the axoneme, crosses the proximal centriole. In this regard, the distal centriole should be equivalent to a daughter centriole of the cycling cells (Lange and Gull, 1996). However, it is the mother centriole of the cycling cells that produces procilium during G0/G1 stage.

The centrosomes do not nucleate microtubules throughout spermiogenesis. In the beginning of elongating stage, the centrioles migrate and associate with the posterior region of the nucleus. The nuclear membranes develop a specific thickening and a depression in this region, referred to as the implantation fossa (Fawcett and Phillips, 1969). During elongating stages, the spermatid centrosomes still have standard centrioles as well as detectable γ-tubulin and centrin (Figs. 1B, 1C, 1H, 1I).

Apparently normal centrosomes comprising centriolar duplexes, γ-tubulin, and centrin, but lacking microtubule nucleating activity, were also observed in rhesus spermatids (manuscript submitted). Likewise, the centrosomes of beetle spermatocytes and spermatids possess γ-tubulin, but do not nucleate microtubules (Wolf and Joshi, 1995). Thus, the loss of the microtubule nucleating function could be a conserved feature of centrosome reduction during spermiogenesis, and this process takes place before the centriole shows any sign of degeneration or loss of the centrosomal proteins from the pericentriolar lattice.

Experimentation utilizing nocodazole has provided further support that spermatid centrosomes do not nucleate microtubules. Spermatid microtubular systems depolymerized by nocodazole can be repolymerized by washing away the nocodazole. During repolymerization, the microtubule bundles and asters form randomly in the cytoplasm but do not originate from the γ-tubulin containing bodies, the centrosomes (Manandhar *et al.*, 1998).

During interphase, centrosomes of the culture cells display highly diminished microtubule nucleating activity (Kuriyama and Borisy, 1981); however, some microtubules originate from the basal feet of the mother centrioles (Vorobjev and Chentsov, 1982). Loss of microtubule nucleating function by G0 centrioles has been shown in cell fusion experiments (Peterson and Berns, 1979; Szollosi *et al.*, 1986; Manandhar and Onishchenko, 1995). Loss of microtubule nucleating activity by the G0 centrioles is reversible, since the G0-stage culture cells can again enter into the division cycle after appropriate stimulation. But spermiogenesis is a terminal differentiation process during which the centrosomes lose microtubule nucleating function permanently. Besides spermiogenesis, centrosome degeneration has also been reported during myotubule differentiation (Connolly *et al.*, 1985). In this process, the centrosomes cease to nucleate microtubules before they degenerate (Tassin *et al.,* 1985). However, the loss of microtubule nucleating function of centrosomes is not a common feature in the differentiation processes, particularly in those cases in which centrosomes do not degenerate completely. For example, in neural cells, the centrosomes organize microtubules toward the axons and dendrites (Bartlett and Banker, 1984; Stevens *et al.,* 1988). The basal bodies (centrosomes) of various terminally differentiated ciliated epithelial cells nucleate cytoplasmic microtubules toward the nucleus (reviewed by Joshi, 1993). In mouse cochlear epithelial cells, some microtubules focus toward the centrosome that contains centrioles, pericentrin, and γ-tubulin (Mogensen *et al.,* 1997). The microtubule nucleating activity of centrosomes is enhanced when the cochlear cells are cultured *in vitro* (Tucker *et al.,* 1992).

B. Transitional MTOCs

During the stage when the nucleus begins elongation, the random cortical microtubular network is replaced by highly polarized microtubular bundles called the manchette (Rather and Olson, 1973). It develops around the nucleus as a veil-like structure, emanating from a narrow membranous furrow (Fig. 2A). The manchette MTOC encircles the nucleus as a perinuclear ring below the lower edge of the acrosome. The manchette microtubules seem to emerge from a dense fibrous material in this region whose fine structure resembles the pericentriolar fibrous material of somatic culture cells (Gould and Borisy, 1977), sea urchin embryos (Paweletz *et al.,* 1984), and the cortical MTOC of higher plant cells (Gunning *et al.,* 1978; Panteris *et al.,* 1991). A whole-mount electron microscopy of the isolated manchette reveals that microtubule bundles associate with dense globular aggregates at their distal ends (Tres and Kierszenbaum, 1996). Normal-looking centrosomes with centrioles, γ-tubulin, and centrin are located in the neck region of the elongating spermatids. The manchette microtubule bundles extend close to that region, but none of them seems to converge toward the centrosome (Fig. 2B). γ-Tubulin and centrin are not detectable in the manchette MTOC by immunofluorescent microscopy or immunogold electron microscopy.

Despite its organized structure, the manchette MTOC is largely a mysterious organelle. The possibility cannot be ruled out that a diffuse, but undetectable, γ-tubulin may be present in this region. It has been recently shown that the majority of cellular γ-tubulin exists as a free cytosolic form (Moudjou *et al.,* 1996), which may play a role in microtubule nucleation and stabilization (Raff, 1996). Alternatively, the lack of recognizable γ-tubulin in the manchette region or at their distal ends might signify that those microtubules are nucleated through a mechanism independent of

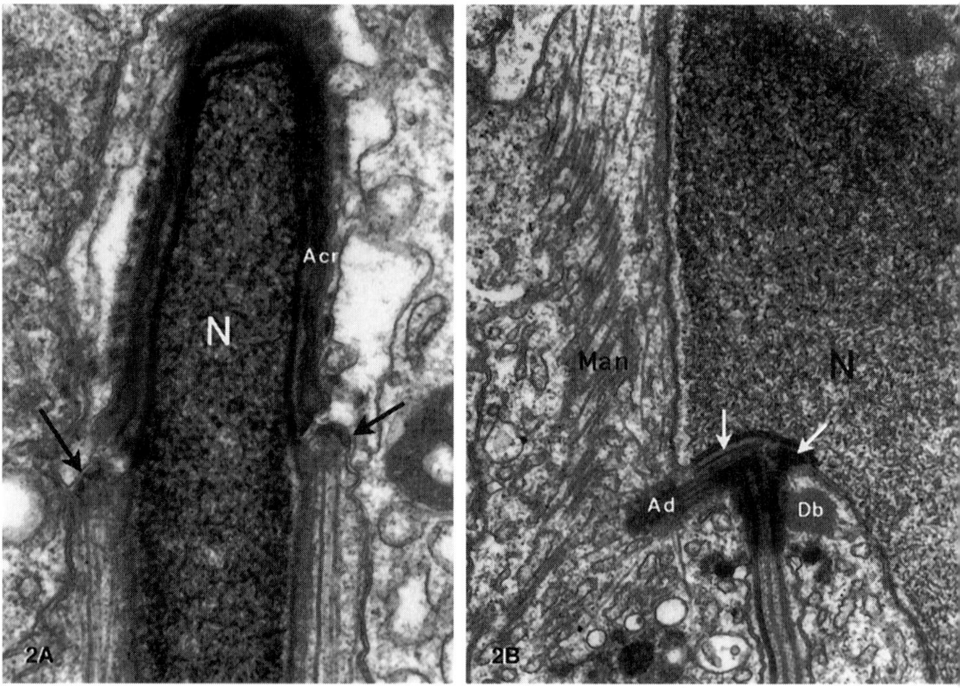

Figure 2 Transitional MTOCs of mouse elongating spermatids. (A) A section passing through the manchette microtubule nucleating region (arrows). The microtubules are deeply embedded in a fibrous material in the trough of a narrow membranous furrow around the nucleus. The manchette microtubule nucleating ring is located below the posterior edge of the acrosome (Acr). (B) A section of an elongating spermatid passing longitudinally through the distal and proximal centrioles and the adjunct (Ad). The latter is visibly decorated with fine fibrous material that emanates an aster of microtubules. The manchette microtubules (Man) extend close to the adjunct, but the adjunct MTOC does not nucleate them. The centrioles are associated with the implantation fossa with striated columns and a lamellar structure (arrows). The distal centriole has a dense body (Db) at one side. Bar, 0.5 γm. (B, from Manandhar *et al.* (1998). *Dev. Biol.* **203,** 424–434.)

γ-tubulin. Analogous microtubular nucleation outside the γ-tubulin spot has been observed in taxol-treated cells (Novakova *et al.,* 1996) and in brain cells from mutant *Drosophila* larvae (Wilson *et al.,* 1997). Similar to manchette, mouse cochlear cells exhibit longitudinal microtubule bundles nucleated from the subapical system of fibrillar aggregates (Tucker *et al.,* 1992) that does not contain detectable γ-tubulin or pericentrin (Mogensen *et al.,* 1997). As previously mentioned, the manchette originates from a narrow membranous furrow. Membranous folds of cytoplasm are known to play the role of MTOC in several types of cells (Dickinson and Sheldon,

1984; Bakhuizen *et al.,* 1985; Apostolakos and Galatis, 1985; Tucker *et al.,* 1992; Mogensen *et al.,* 1993), possibly creating a cellular compartment of favorable Ca^{2+} balance for microtubule polymerization (Hepler, 1989). Alternative mechanisms for noncentrosomal microtubule nucleation and stabilization in various cell types have been recently reviewed (Keating and Borisy, 1999).

The proximal centriole of elongating spermatids develops a relatively short outgrowth called the adjunct (Fig. 2B). Ultrastructurally an adjunct is similar to a centriole, formed by the extension of the centriolar microtubules; however, it is considered to be a new structure (Fawcett and Phillips, 1969). Their A- and B- tubules display complete rings in cross sections, but the C-tubule shows an open edge that joins the A-tubule of the preceding triplet. A dense material is deposited on the inner surface of the adjunct cylinder that displays periodic striations. Externally the adjunct is surrounded by a flocculent fibrillar material (Manandhar *et al.,* in press) resembling the pericentriolar centrosomal proteins of somatic cells (Kimble and Kuriyama, 1992). Light and electron microscopy have demonstrated that γ-tubulin associates with this region (Yang *et al.,* 1997; Manandhar *et al.,* 1998; Fouquet *et al.,* 1998). But centrin in not present in the adjunct (Manandhar *et al.,* 1999). The fibrous structure of the adjunct acts as a MTOC and nucleates the microtubule aster (Fig. 2B). In elongating spermatids, the adjunct aster is usually concealed by the manchette. It is evident in some late-stage elongating spermatids in which the manchette has been recently disassembled (Figs. 1E, 1I).

The manchette and adjunct are both transient structures that finally disorganize at the late elongating stage of spermiogenesis. An unusual event of spermiogenesis is the loss of the MTOC from the centriolar complex and the reappearance as the manchette and adjunct MTOC. Disjunction of MTOCs from the centriolar complex has been previously reported in culture cells exposed to X-rays (Sato *et al.,* 1983), laser irradiation (Berns *et al.,* 1977), treated with antimitotic drugs (Keryer *et al.,* 1984), and during myotubule biogenesis (Tassin *et al.,* 1985). These conditions lead to centrosome inactivation or degeneration. A direct correlation between centrosome degeneration and pericentriolar material scattering has been shown in an experiment in which the centrosomes of the cells were exposed to anti- glutamylated peptide antibody (Bobinnec *et al.,* 1998). The disjuncted pericentriolar material relocated randomly in the cytoplasm and around the nuclear area, but were still capable of nucleating microtubules. On these grounds it can be assumed that the disjunction of MTOC from the centrosome proper and redistribution to the perinuclear ring and the adjunct tip is a transitional event of centrosome reduction taking place during spermiogenesis.

C. Loss of Centrosomal Proteins: γ-Tubulin and Centrin Degeneration

Though apparently incapable of microtubule nucleation, γ-tubulin is localized in the neck region until the late elongating stage of mouse spermiogenesis. During this stage it dissociates from the pericentriolar region of the spermatids and is discarded in the residual bodies in the form of several dense spots and diffuse material (Fig. 1D). After this stage, the neck region no longer displays γ-tubulin (Figs. 1D–1F). The residual bodies often carry few bundles of discarded manchette microtubule bundles, but they do not associate with the γ-tubulin spots. Since centrioles also degenerate during mouse spermiogenesis, the γ-tubulin loss seems to correlate with centriole degeneration. But γ-tubulin loss takes place much earlier than centriole degeneration (described later). γ-Tubulin is discarded in a similar manner from rhesus spermatids, although centrioles degenerate partially in them. γ-Tubulin is not detectable in rhesus mature sperm (manuscript submitted) or in *Xenopus* sperm (Stearns and Kirschner, 1994; Felix *et al.,* 1994) that possess centrioles (Bernardini *et al.,* 1986; Felix *et al.,* 1994). These observations suggest that γ-tubulin loss is an independent step of the centrosome reduction process and it takes place regardless of whether centrioles degenerate.

Unlike γ-tubulin, centrin intimately associates with the centriole and its degradation also correlates with the centriolar disintegration. Mouse round and elongating spermatids display two punctate centrin spots in the neck region corresponding to two centrioles (Figs. 1H, 1I). Unlike γ-tubulin, they are not discarded with the residual bodies (Figs. 1J–1M). After the spermiation stage, centrin labeling gradually diminishes in mouse sperm. In immunofluorescent preparations, the fully formed testicular sperm display various intensities of centrin labeling, the majority displaying highly diminished fluorescence. The mouse sperm lose centrin completely by the time they reach the epididymis (Figs. 1L, 1M).

Nonrodent vertebrate sperm, which retain centrioles, also retain centrin. It is found in *Xenopus* (Stearns and Kirschner, 1994), human (Salisbury *et al.,* 1986), and rhesus sperm (our unpublished observation), all of which possess centrioles (DeKretser, 1969; Zamboni and Stefanini, 1971; Felix *et al.,* 1994; Sathananthan *et al.,* 1996; Manandhar *et al., in press*).

It is difficult to reconcile the fact that the typical nonrodent mammalian sperm possesses centrin but lacks γ-tubulin. One of the major functions of the sperm centrosome is to organize the sperm aster during fertilization, which leads to pronuclear apposition (Schatten, 1982; for review see Reinsch and Gonczy, 1998). Nonrodent mammalian sperm centrosomes possess centrin that may not be involved in microtubule nucleation (Stearns and Kirschner, 1994), whereas they lack γ-tubulin, which would be required to organize microtubule aster and spindle poles. γ-Tubulin directly nucleates microtubules by associating with some accessory proteins (Zheng *et al.,* 1995; Moritz *et al.,* 1995; Raff, 1996). The lack of γ-tubulin in the sperm centrosome is compensated by the oocyte cytoplasm, which contains a reserve of functional γ-tubulin (Wolf and Joshi, 1996; Vogel *et al.,* 1997; Felix *et al.,* 1994; Gard, 1994; Gueth-Hollonet *et al.,* 1993). The sperm centrosome recruits the maternal γ-tubulin (Stearns and Kirschner, 1994; Felix *et al.,* 1994) to form a blended biparental functional zygotic centrosome (Schatten, 1994). Moreover, it is likely that mammalian sperm carry some residual γ-tubulin (Fouquet *et al.,* 1998; Simerly *et al.,* 1999) that might act as a seed to attract more maternal γ-tubulin. Such residual γ-tubulin is revealed if sperm are "primed" before immunocytochemical detection (Zoran *et al.,* 1994). However, γ-tubulin of the mature sperm is a remnant of the functionally inactive γ-tubulin of spermatids (discussed above) and therefore, not likely to play a significant role during fertilization. *Xenopus* mature sperm are shown to possess some other centrosomal proteins such as pericentrin (Stearns and Kirschner, 1994). CTR2611 (Felix *et al.,* 1994), and Spc98p (Tassin *et al.,* 1998). Pericentrin may provide a structural scaffold that holds the γ-tubulin forming microtubule nucleating centers (reviewed in Dictenberg *et al.,* 1998). Therefore, although the mode of centrosome reduction during spermiogenesis comprises elimination of centrosomal proteins, it may not end up with complete loss of all of them. The functional role of such residual centrosomal proteins during fertilization is still unknown.

Besides higher animal spermiogenesis, investigations of centrosomal proteins in the mature and immature sperm cells of *Drosophila* have also provided important cues to the centrosome reduction process (Callaine *et al.,* 1999). Centrosomes of *Drosophila* meiotic germ cells and spermatids possess γ-tubulin, centrosomin, and centrin (Riparbelli *et al.,* 1997; Wilson *et al.,* 1997). During early spermiogenesis, γ-tubulin and centrin dissociate from the pericentriolar region and localize at a transitional region between the nucleus and the axoneme that might be involved in microtubule nucleation. Another *Drosophila* centrosomal protcin, CP190, also colocalizes in that region (Riparbelli *et al.,* 1997). These proteins are totally eliminated in the mature sperm. Centrosomin is also discarded from the mature sperm (Li and Kaufman, 1996).

D. Centriole Degeneration

Complete centriole degeneration has been previously reported in rat sperm (Woolley and Fawcett, 1973). Mouse epididymal sperm also lose centrioles completely (Manandhar *et al.,* 1998, 1999). The distal centrioles degenerate during the testicular stage, and by the time the sperm reach the epididymis, they are lost completely (Fig. 3A). The axonemal microtubules end bluntly in the distal centriolar vault. The center of the distal centriolar lumen, however, is occupied by a microtubule duplex that is a proximal extension of the axonemal central duplex. Ultrastructural changes of the distal centrioles during the degeneration process have been described by Fawcett and Phillips (1969). The proximal centrioles of mouse sperm degenerate in epididy-

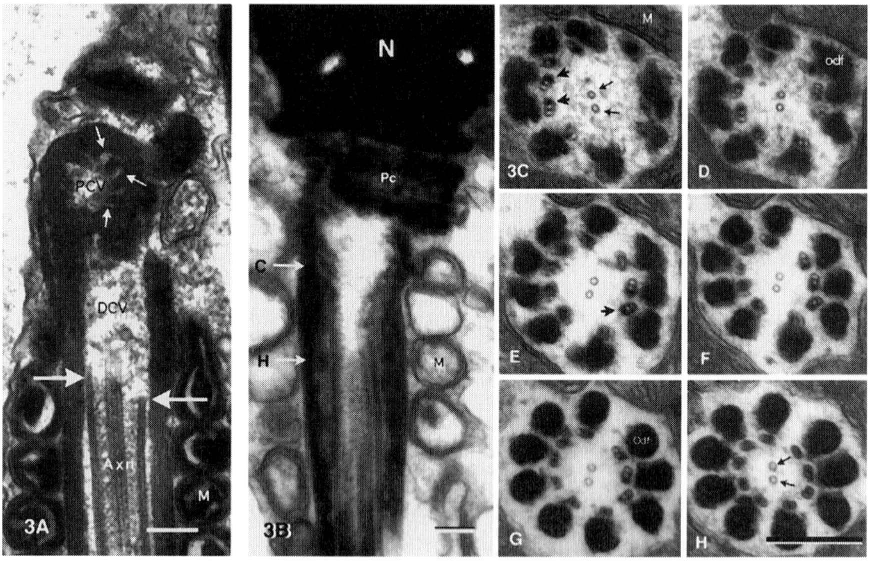

Figure 3 Centriole degeneration in mouse and rhesus sperm. (A) A section of mouse epididymal sperm passing longitudinally through the distal centriolar vault (DCV) and transversely through the proximal centriolar vault (PCV). The microtubules of both centrioles have been completely degenerated. The axoneme microtubules end bluntly (large arrows) in the distal centriolar vault. The proximal centriolar vault displays fingerlike projections (small arrows), which are the spaces previously occupied by the centriolar triplets. (From Manandhar *et al.* (1999). *Cell Motil. Cytoskel.* **43,** 137–144, with permission.) (B) A section of human ejaculated sperm passing longitudinally through the distal centriole (arrows) and the proximal centriole (Pc). The microtubules of the proximal centrioles are intact. The distal centriolar microtubules are degenerated and unrecognizable on the ventral side of the sperm (the side where the proximal centriole resides). (C–H) Consecutive serial transverse sections of the distal centriole of human sperm. The approximate planes of these sections are marked in B. (C, D). The rostral region of the distal centriole in which centriolar microtubules are missing or disorganized on the ventral side. The middle (E, F) and the caudal (G) regions of the distal centriole. Three centriolar triplets are recognizable on the ventral side, but the others have collapsed. (H) A section passing through the axonemal region. A microtubular duplex is visible in the center of all sections, extending from the axoneme to the base of the proximal centriolar vault (arrows in H and C). The A-tubule of the centriolar triplets appears dark because of deposition of dense material (arrowheads in C, E). N, nucleus; M, mitochondria; Axn, axoneme; PCV, proximal centriolar vault; DCV, distal centriolar vault; Pc, proximal centriole, Bar, 0.25 γm. (B–H, from Manandhar *et al., Hum. Reprod.,* **15,**256–263 (2000) with permission of Oxford University Press).

mis. Some epididymal sperm exhibit short microtubule stubs deeply embedded in the centriolar vault. Finally, in fully mature epididymal sperm, the proximal centrioles are lost completely, leaving empty proximal centriolar vaults (Fig. 3A).

The centriole loss in murine sperm signifies a complete elimination of the centrosome. Centriole degeneration is not complete in other nonrodent mammalian sperm. The proximal centriole remains intact in the ejaculated sperm of sheep (Crozet, 1990), rhesus (Zamboni and Stefanini, 1971), and human (Sathananthan *et al.,* 1991, 1996). But the distal centrioles undergo various degrees of degeneration in sperm of several species (Fawcett and Ito, 1965; Fawcett, 1965; Gordon, 1972). The distal centriole loss is more prominent in rodent species such as guinea pig, chinchilla, squirrel, and hamster (Fawcett and Phillips, 1969). A meticulous study of rhesus and human sperm using electron microscopy and immunogold labeling has shown that their distal centrioles exist in a highly degenerated form (Figs. 3B–3H; Manandhar *et al.,* in press). About 50% of their microtubular triplets are lost. Among those remaining, the majority are collapsed, and their A-tubules are filled with a dense material as in an axoneme. The central lumen is occupied by a microtubular duplex as in mouse sperm. The centriolar degeneration is more remarkable at the rostral region and on the ventral side of the sperm. Observation of highly degenerated and axonemal-like distal centrioles in rhesus and human sperm has raised a fundamental question of whether the oocytes would be able to convert them back into centrioles and reconstruct zygotic centrosomes during fertilization.

A partial or complete loss of the centriolar apparatus during spermiogenesis may be a conserved feature in phylogenetically diverse animals, but such studies are rare. Snail sperm (*Lumnease stagnalis*) lose both centrioles at maturity (Krioutchkova *et al.,* 1994; Krioutchkova and Onishchenko, 1999). The proximal centriole disappears during prometaphase II, so that the round spermatids possess only one centriole. During late spermiogenesis, the microtubular cylinder of the centriole is replaced by nine amorphous columns (Krioutchkova *et al.,* 1994). In insects, centrioles either disappear during spermiogenesis (Phillips, 1970) or become highly modified into various "acentriolar" structures (Callaine *et al.,* 1999).

The reason why distal centrioles are more susceptible to degeneration than the proximal centrioles in the mammalian sperm is not fully clear. As mentioned previously, the distal centrioles could be derived from the mother centrioles, which are one cell cycle generation older than the proximal centrioles (Vorobjev and Chentsov, 1982) and thus are prone to degenerate sooner. The centriolar microtubules are more stable than the cytoplasmic microtubules and may not exhibit a dynamic instability (Kochanski and Borisy, 1990). Moreover, the stability of the spermatid centrioles could be attributed to various additional posttranslational modifications (Fouquet *et al.*, 1997), or ensheathing with dense osmiophilic materials (Fawcett and Phillips, 1969). Because of such molecular and structural modifications, spermatid centrioles might endure the degeneration process over the long duration of spermiogenesis (15 days in mice). Centrioles of the culture cells can be experimentally induced to degenerate by exposing them to anti-glutamylated tubulin antibody (GT335). This antibody binds to the glutamylated sites of the α- and β-tubulins of centrioles, making them inaccessible to the centriolar organizing proteins (Bobinnec *et al.*, 1998). However, centriolar disintegration during spermiogenesis could be due to depletion of the cytoplasmic reserve of tubulins or centriolar organizing proteins, since the transcriptional activity of the nucleus is totally shut off during the late spermiogenesis period. Interestingly, the axonemal microtubules, including the centriolar duplex that extends into the disintegrated distal centriolar area, must have a different stabilizing factor, since they are not affected by the process that degenerates the centrioles.

As stated earlier, the prevailing hypothesis of centrosome inheritance assumes that the male gamete contributes two centrioles that replicate before each cleavage and propagate to the daughter cells. This hypothesis holds true during sea urchin fertilization (Paweletz *et al.*, 1987a, 1987b), but may not be valid in higher animal fertilization. In murine species, centrioles are not inherited from the sperm, but regenerate *de novo* in the trophectoderm cells of fully expanded blastocysts (Szollosi *et al.*, 1972; Abumuslimov *et al.*, 1994). Centrioles form spontaneously in rabbit oocytes during parthenogenetic development (Szollosi and Ozil, 1991), which shows that oocytes have all components for centriole assembly and are able to form centrioles spontaneously after an appropriate stimulation. In nonmurine mammalian species, the sperm introduces an intact proximal centriole that replicates to form one centriolar duplex and localizes at one of the spindle poles during the first zygotic division (Crozet, 1990; LeGuen and Crozet, 1989; Sathananthan *et al.*, 1996). However, distal centrioles are highly degenerated and aberrant in their sperm. The oocytes may replenish the lost microtubular triplets of this centriole and replicate it to create the second centriolar duplex. Alternatively, the second pair of centrioles might be formed spontaneously, without the participation of the distal centriole. This viewpoint has been supported by some electron microscopic studies. The spindle poles of the first zygotic cleavage are broad (Crozet, 1990; Wu *et al.*, 1996), very similar to the acentriolar spindle pole of the oocytes (Szollosi *et al.*, 1972). An extensive survey of human zygotes by electron microscopy has failed to find centriolar duplexes in all spindle poles during the first cleavage (Sathananthan *et al.*, 1996). Thus, it is probable that only the proximal centriole is replicated before the first cleavage and the second pair is regenerated during the later cleavage cycle. A systematic study of the spindle polar regions of the first zygotic cleavage with serial sectioning is necessary to make a definite conclusion.

IV. Conclusions

The observation of a systematic degeneration of distal centriole and γ-tubulin in nonrodent mammalian sperm implies that centrosome reduction could be a ubiquitous phenomenon taking place in all mammalian sperm to varying degrees, the mouse sperm representing the highest state and exhibiting complete loss of both centrioles and centrosomal proteins. The complete absence of centrosomes in mouse sperm further consolidates the hypothesis that murine oocytes can compensate for the lack of paternal centrosomal contribution. Likewise, the lack of a standard distal centriole and γ-tubulin in rhesus and human sperm does not interfere with the formation of the zygotic centrosome during fertilization and bipolar spindles during cleavages. Thus, it is one of the major challenges for future developmental cell biologists to investigate the marvelous centrosome regenerative capability of the oocytes.

Acknowledgments

We thank our collaborators Dr. H. C. Joshi (Emory University, Atlanta, GA), Dr. T. Stearns (Stanford University, CA) and Dr. J. L. Salisbury (Mayo Clinic, Rochester, MN) for providing γ-tubulin and anticentrin antibodies that were used in the studies. M. Webb of the EM lab of ORPRC provided technical help during TEM studies. We are highly grateful to D. Takahashi for critically reading the manuscript and making grammatical corrections.

References

Abumuslimov, S. S., Nadezhdina, E. S., and Chentsov, Yu. S. (1994). An electron microscopic study of centriole and centrosome morphogenesis in the early development of mouse. *Tsitologiia* **36,** 1054–1061.

Apostolakos, P., and Galatis, B. (1985). Studies on the development of the air pores and air chambers of *Marchantia paleacea.* III. Microtubule organization in preprophase–prophase initial apparatus cells—formation of incomplete preprophase microtubule bands. *Protoplasma* **128,** 120–135.

Bakhuizen, R., *et al.* (1985). Nuclear envelope radiating microtubules in plant cells during interphase mitotic transition. *Protoplasma* **128,** 43–51.

Bartlett, P. W., and Banker, G. A. (1984). An electron microscopic study of the development of axons and dendrites by hippocampal neurons in culture. I. Cells which develop without intercellular contacts. *J. Neurosci.* **4,** 1944–1953.

Bernardini, G., Stipani, R., and Melone, G. (1986). The ultrastructure of *Xenopus* spermatozoon. *J. Ultrast. Mol. Struct. Res.* **94,** 188–194.

Berns, M. W., Rattner, J. B., Brenner, S., and Meredith, S. (1977). The role of centriolar region in animal cell mitosis. *J. Cell Biol.* **72,** 351–365.

Bobinnec, Y., Khodjakov, A., Mir, L. M., Rieder, C. L., Edde, B., and Bornens, M. (1998). Centriole disassembly in vivo and its effect on centrosome structure and function in vertebrate cells. *J. Cell Biol.* **143,** 1575–1589.

Boveri, T. (1901). Zellen-Studien: Über die Natur der Centrosomen. IV. Fischer, Jena.

Calarco-Gillam, P. D., Siebert, M. C., Hubble, R., Mitchison, T., and Kirschner, M. (1983). Centrosome development in early mouse embryos as defined by autoantibody against pericentriolar material. *Cell* **35,** 621–629.

Callaine, G., Riparbelli, G. M., and Dallai, R. (1999). Centrosome inheritance in insects: Fertilization and parthenogenesis. *Biol. Cell* **91,** 355–366.

Clermont, Y. (1963). The cycle of the seminiferous epithelium in man. *Am. J. Anat.* **112,** 35–51.

Clermont, Y., and Harvey, S. C. (1965). Duration of the cycle of the seminiferous epithelium of normal, hypophysectomized-hormone treated albino rats. *Endocrinology* **76,** 80–89.

Clermont, Y., and Trott, M. (1969). Duration of the cycle of the seminiferous epithelium in the mouse and hamster determined by means of ^{3}H-thymidine and radioautography. *Fertl. Steril.* **20,** 805–817.

Connolly, J. A., Kiosses, B. W., and Kalnin, V. I. (1985). Centrioles are lost as embryonic myoblasts fuse into myotubules in vitro. *Eur. J. Cell Biol.* **39,** 341–345.

Crozet, N. (1990). Behavior of the sperm centriole during sheep oocyte fertilization. *Eur. J. Cell Biol.* **53,** 321–332.

DeKretser, D. M. (1969). Ultrastructural features of human spermiogenesis. *Z. Zellforsch.* **98,** 477–505.

DeKretser, D. M., and Kerr, J. B. (1994). The cytology of testis. In "The Physiology of Reproduction" (E. Khobil and J. D. Neill, Eds.), pp. 1177–1290. Raven Press, New York.

de Rooij, D. G., van Alphen, M. M. A., and van de Kant, H. J. G. (1986). Duration of the cycle of the seminiferous epithelium and its stages in the rhesus monkey (*Macaca mulatta*). *Biol. Reprod.* **35,** 587–591.

Dickinson, H. G., and Sheldon, J. M. (1984). A radial system of microtubules extending between the nuclear envelope and the plasma membrane during early male haplotene in flowering plants. *Planta* **161,** 86–90.

Dictenberg, J. B., Zimmerman, W., Sparks, C. A., Young, A., Vidair, C., Zheng, Y., Carrington, W., Fay, F. S., and Doxsey, S. J. (1998). Pericentrin and γ-tubulin form a protein complex and are organized into a novel lattice at the centrosome. *J. Cell Biol.* **141,** 163–174.

Dooher, G. B., and Bennett, D. (1973). Fine structural observations on the development of the sperm head in the mouse. *Am. J. Anat.* **136,** 339–362.

Fawcett, D. W. (1965). The anatomy of the mammalian spermatozoon with particular reference to the guinea pig. *Z. Zellforsch.* **67,** 279–296.

Fawcett, D. W., and Ito, S. (1965). The fine structure of bat spermatozoa. *Amer. J. Anat.* **116,** 567–609.

Fawcett, D. W., and Phillips, D. M. (1969). The fine structure and development of the neck region of the mammalian spermatozoon. *Anat. Rec.* **165,** 153–184.

Felix, M., Antony, C., Write, M., and Maro, B. (1994). Centrosome assembly in vitro: Role of γ-tubulin recruitment in *Xenopus* sperm aster formation. *J. Cell Biol.* **124,** 19–31.

Fouquet, J. P., Kann, M. L., Pechart, I., and Prigent, Y. (1997). Expression of tubulin isoforms during the differentiation of mammalian spermatozoa. *Tissue & Cell* **29,** 573–583.

Fouquet, J. P., Kann, M. L., Combeau, C., and Melki, R. (1998). γ-Tubulin during the differentiation of spermatozoa in various mammals and man. *Mol. Hum. Reprod.* **12,** 1122–1129.

Gard, D. L. (1994). γ-Tubulin is asymmetrically distributed in the cortex of *Xenopus* oocytes. *Dev. Biol.* **161,** 131–140.

Gard, D. L., Hafezi, S., Zhang, T., and Doxsey, S. J. (1990). Centrosome duplication in cycloheximide-treated *Xenopus* blastulae in the absence of detectable cell cycle. *J. Cell Biol.* **110,** 2033–2042.

Gardner, P. J. (1966). Fine structure of the seminiferous tubule of the Swiss mouse. The spermatid. *Anat. Rec.* **155,** 235–250.

Gardner, P. J., and Holyoke, E. A. (1964). Fine structure of the seminiferous tubule of the Swiss mouse. I. Limiting membrane, Sertoli cell, spermatogonia, and spermatocytes. *Anat. Rec.* **150,** 391–404.

Gordon, M. (1972). The distal centriole in Guinea pig spermiogenesis. *J. Ultrastruc. Res.* **39,** 364–388.

Gould, R. R., and Borisy, G. G. (1977). The pericentriolar material in chinese hamster ovary cells nucleates microtubule formation. *J. Cell Biol.* **73,** 601–615.

Greenfield, M. L. (1966). The oocyte of domestic chicken shortly after hatching: Studies by electron microscope. *J. Embryol. Exp. Morphol.* **15,** 297–316.

Gueth-Hallonet, C., Antony, C., Aghion, J., Santa-Maria, A., Lajoie-Mazenc, I., Wright, M., and Maro, B. (1993). γ-Tubulin is present in acentriolar MTOCs during mouse development. *J. Cell Sci.* **105,** 157–166.

Gunning, B. E. S., Hardham, A. R., and Hughes, J. E. (1978). Evidences for the initiation of microtubules in discrete regions of the cell cortex in *Azolla* root-tip cells, and an hypothesis on the development of cortical arrays of microtubules. *Planta* **143,** 161–179.

Heller, C. G., and Clermont, Y. (1963). Spermatogenesis in man: an estimate of its duration. *Science* **140,** 184–186.

Hepler, P. K. (1989). Membranes in the mitotic apparatus. In "Mitosis Molecules and Mechanisms" (J. S. Hyams and B. R. Brinkley, Eds.), pp. 241–271. Academic Press, New York.

Hertig, A. T., and Adams, E. (1967). Studies on the human oocyte and its follicle: ultrastructural and cytochemical observations on the primordial follicle. *J. Cell Biol.* **34,** 647–675.

Hiraoka, L., Golden, W., and Magnuson, T. (1989). Spindle-pole organization during early mouse development. *Dev. Biol.* **133,** 24–36.

Holy, J., and Schatten, G. (1991). Spindle poles of centrosomes of sea urchin embryos are partially composed of material recruited from maternal stores. *Dev. Biol.* **147,** 343–353.

Huchon, D., Crozet, N., Cantenot, N., and Ozon, R. (1981). Germinal vesicle breakdown in *Xenopus laevis* oocytes: Description of a transient microtubular structure. *Reprod. Nutr. Dev.* **21,** 135–148.

Joshi, H. C. (1993). γ-Tubulin: the hub of cellular microtubule assemblies. *BioEssays* **15,** 637–643.

Keating, T. J., and Borisy, G. G. (1999). Centrosomal and non-centrosomal microtubule. *Biol. Cell* **91,** 321–325.

Keating, T. J., Peloquin, J. G., Radionov, V. I., Momcilovic, D., and Borisy, G. G. (1997). Microtubule release from centrosome. *Proc. Natl. Acad. Sci. USA* **94,** 5078–5083.

Keryer, G., Ris, H., and Borisy, G. (1984). Centriole distribution during tripolar mitosis in chinese hamster ovary cells. *J. Cell Biol.* **98,** 2222–2229.

Kimble, M., and Kuriyama, R. (1992). Functional components of microtubule organizing centers. *Int. Rev. Cytol.* **136,** 1–50.

Kochanski, R. S., and Borisy, G. G. (1990). Mode of centriole duplication and distribution. *J. Cell Biol.* **110,** 1599–1605.

Krioutchkova, M. M., and Onishchenko, G. E. (1999). Structural and functional characteristics of centrosome in gametogenesis and early embryogenesis of animals. *Int. Rev. Cytol.* **185,** 107–156.

Krioutchkova, M. M., Onishchenko, G. E., and Chentsov (1994). An ultrastructural study of the centrosome and centrioles in gametogenesis and early embryogenesis of *Lumneae stagnalis* L. *J. Struct. Biol.* **112,** 49–58.

Kuriyama, R., and Borisy, G. (1981). Microtubule-nucleating activity of centrosome in Chinese hamster ovary cells is independent of the centriole cycle but coupled to the mitotic cycle. *J. Cell Biol.* **91,** 822–826.

Lang, N. (1963). An additional ultrastructural component of flagelia. *J. Cell Biol.* **19,** 631–634.

Lange, B. M. H., and Gull, K. (1996). Structure and function of the centriole in animal cells: progress and questions. *Trends Cell Biol.* **6,** 348–352.

Le Guen, P., and Crozet, N. (1989). Microtubule and centrosome distributions during sheep fertilization. *Eur. J. Cell Biol.* **48,** 238–249.

Li, K., and Kaufman, T. C. (1996). The homeotic target gene centrosomin encodes an essential centrosome component. *Cell* **85,** 585–596.

Longo, F. J., and Anderson, E. (1968). The fine structure of pronuclear development and fusion in sea urchin *Arbacia punctulata. J. Cell Biol.* **39,** 339–368.

Longo, F. J. (1973). Fertilization: A comparative ultrastructural review. *Biol. Reprod.* **9,** 149–215.

Mahowald, A. P., and Strassheim, J. M. (1970). Intracellular migration of centrioles in the germarium of *Drosophila melanogaster. J. Cell Biol.* **45,** 306–320.

Manandhar, G., and Onishchenko, G. E. (1995). Centriolar cycle of fused cells. *J. Cells Sci.* **108,** 667–674.

Manandhar, G., Sutovsky, P., Joshi, H. C., Stearns, T., and Schatten, G. (1998). Centrosome reduction during mouse spermiogenesis. *Dev. Biol.* **203,** 424–434.

Manandhar, G., Simerly, C., Salisbury, J. L., and Schatten, G. (1999). Centriole and centrin degeneration during mouse spermiogenesis. *Cell Motil. Cytoskelet,* **43,** 137–144.

Manandhar, G., Simerly, C., and Schatten, G. (2000). Distal centrioles are highly degenerated in rhesus and human sperm. *Hum. Reprod.* **15,** 256–263.

Maro, B., Howlett, S. K., and Webb, M. (1985). Non-spindle microtubule organizing centers in metaphase II arrested mouse oocyte. *J. Cell Biol.* **101,** 1665–1672.

Mazia, D. (1978). Origin of twoness in cell reproduction. In "Cell Reproduction: In Honor of Daniel Mazia" (E. R. Dirksen, D. M. Prescott and C. F. Fox, Eds.), pp. 1–14. Academic Press, New York.

Mazia, D. (1984). Centrosomes and mitotic poles. *Exp. Cell Res.* **153,** 1–15.

Merdes, A., Ramayar, K., Vechio, J. D., and Cleveland, D. W. (1996). A complex of NuMA and cytoplasmic dynein is essential for mitotic spindle assembly. *Cell* **87,** 447–458.

Mogensen, M. M., Tucker, J. B., Baggaley, T. B. (1993). Multiple plasma membrane associated MTOC systems in the acentrosomal cone cells of *Drosophila* ommatidia. *Eur. J. Cell Biol.* **60,** 67–75.

Mogensen, M. M., Mackie, J. B., Doxsey, S. J., Stearns, T., and Tucker, J. B. (1997). Centrosomal deployment of γ tubulin and pericentrin—Evidence for a microtubule-nucleating domain and a minus-end docking domain in mouse epithelial cells. *Cell Motil. Cytoskelet.* **36,** 276–290.

Moritz, M., Braunfeld, M. B., Fung, J. C., Sedat, J. W., Alberts, B. M., and Agard, D. A. (1995). Three-dimensional structural characterization of centrosomes from early *Drosophila* embryos. *J. Cell Biol.* **130,** 1149–1159.

Moudjou, M., Bordes, N., Paintrand, M., and Bornens, M. (1996). γ-Tubulin in mammalian cells: The centrosomal and the cytosolic forms. *J. Cell Sci.* **109,** 875–887.

Navara, C. S., Simerly, C., Compton, D. A., and Schatten, G. (1996). Maternal NuMA is required for nuclear formation and aster organization during fertilization: NuMA reorganization is microtubule independent. *Mol. Biol. Cell* **7** (Suppl.), 208a.

Novakova, M., Draberova, E., Schurman, W., Czihak, G., Viklicky, V., and Draber, P. (1996). γ-Tubulin redistribution in taxol-treated mitotic cells probed by monoclonal antibodies. *Cell Motil. Cytoskelet.* **33,** 38–51.

Oakberg, E. F. (1956). Duration of spermiogenesis in the mouse and timing of stages of the cycle of the seminiferous epithelium. *Am. J. Anat.* **99,** 507–516.

Palacios, M. J., Joshi, H. C., Simerly, C., and Schatten, G. (1993). γ-Tubulin reorganization during mouse fertilization and early development. *J. Cell Sci.* **104,** 383–389.

Palermo, G. D., Colombero, L. T., and Rozenwaks, Z. (1997). The human centrosome is responsible for syngamy and early embryonic development. *Rev. Reprod.* **2,** 19–27.

Panteris, E., Galatis, B., and Apostolakos, P. (1991). Patterns of cortical and perinuclear microtubule organization in meristematic root cells of *Adiantum capillus veneris. Protoplasma* **165,** 173–188.

Paweletz, N., Mazia, D., and Finze, E. (1984). The centrosome cycle in the mitotic cycle of sea urchin eggs. *Exp. Cell Res.* **152,** 47–65.

Paweletz, N., Mazia, D., and Finze, E. (1987a). Fine structural studies of the bipolarization of the mitotic apparatus in the fertilized sea urchin eggs. I. The structure and behavior of centrosomes before fusion of the pronuclei. *Eur. J. Cell Biol.* **44,** 195–204.

Paweletz, N., Mazia, D., and Finze, E. (1987b). Fine structural studies of the bipolarization of the mitotic apparatus in the fertilized sea urchin eggs. II. Bipolarization before the first mitosis. *Eur. J. Cell Biol.* **44,** 205–213.

Peterson, S. P., and Berns, M. W. (1979). Mitosis in flat PTK_2-human hybrid cells. *Exp. Cell Res.* **120,** 223–236.

Phillips, D. M. (1970). Insect sperm: Their structure and morphogenesis. *J. Cell Biol.* **44,** 243–277.

Raff, J. W. (1996). Centrosomes and microtubules: Welded with a ring. *Trends Cell Biol.* **6,** 248–251.

Rattner, J. B., and Olson, G. (1973). Observations of the fine structure of the nuclear ring of the mammalian spermatid. *J. Ultrastruc. Res.* **43,** 438–444.

Reinsch, S., and Gonczy, P. (1998). Mechanism of nuclear positioning. *J. Cell Sci.* **111,** 2283–2295.

Ringo, D. L. (1967). Flagellar motion and fine structure of the flagellar apparatus in *Chlamydomonas. J. Cell Biol.* **33,** 543–571.

Riparbelli, M. G., Whitfield, W. G. F., Dallai, R., and Callaini, G. (1997). Assembly of zygotic centrosome in fertilized *Drosophila* egg. *Mech. Dev.* **65,** 135–144.

Roosen-Runge, E. C. (1977). "The Process of Spermatogenesis in Mammals." Cambridge University Press, Cambridge.

Russell, L. D., Ettin, R. A., Hikim, A. P. S., and Clegg, E. D. (1990). In "Histological and Pathological Evaluation of the Testis." Cache River Press, Clearwater, FL.

Salisbury, J. L., Baron, A. T., Coling, D. E., Martindale, V. E., and Sanders, M. A. (1986). Calcium modulated contractile proteins associated with the eukaryotic centrosome. *Cell Motil. Cytoskelet.* **6,** 193–197.

Sanders, M. A., and Salisbury, J. L. (1994). Centrin plays an essential role in microtubule severing during flagellar excision in *Chlamydomonas reindardtii. J. Cell Biol.* **124,** 795–805.

Sathananthan, A. H., Kola, I., Trounson, A., Ng, S. C., Bongso, A., and Ratnam, S. S. (1991). Centrioles in the beginning of human development. *Proc. Natl. Acad. Sci. USA* **88,** 4806–4810.

Sathananthan, A. H., Ratnam, S. S., Ng, S. C., Tarim, J. J., Gianaroli, L., and Trounson, A. (1996). The sperm centriole: Its inheritance, replication and perpetuation in early human embryos. *Hum. Reprod.* **11,** 345–356.

Sathananthan, A. H., Tatham, B. Dharmawardena, V., Grills, B., Lewis, I., and Trounson, A. (1997). Inheritance of sperm centrioles and centrosomes in bovine embryos. *Arch. Androl.* **38,** 37–48.

Sato, C., Kuriyama, R., and Nishizawa, K. (1983). Microtubule-organizing centers abnormal in number, structure, and nucleating activity in X-ray irradiated mammalian cell. *J. Cell Biol.* **96,** 776–782.

Schatten, G. (1982). Motility during fertilization. *Int. Rev. Cytol.* **79,** 59–163.

Schatten, G. (1994). The centrosome and its mode inheritance: The reduction of the centrosome during gametogensis and its restoration during fertilization. *Dev. Biol.* **165,** 299–335.

Schatten, H., Thompson-Coffe, C., Coffe, G., Simerly, C., and Schatten, G. (1989). Centrosomes, centrioles and posttranslationally modified α-tubulin during fertilization. In "The Molecular Biology of Fertilization" (H. Schatten and G. Schatten, Eds.), pp. 189–210. Academic Press, San Diego.

Schatten, H., Walter, M., Biessmann, H., and Schatten, G. (1992). Activation of maternal centrosomes in unfertilized sea urchin eggs. *Cell Motil. Cytoskel.* **23,** 61–70.

Simerly, C., Wu, G. J., Zoran, S., Ord, T., Rawlins, R., Jones, J., Navara, C., Gerrity, M., Rinehart, J., and Binor, Z. (1995). The paternal inheritance of centrosome, the cell's microtubule organizing center, in humans and the implications for infertility. *Nat. Med.* **1,** 47–53.

Simerly, C., Zoran, S. S., Payne, C., Dominko, T., Sutovsky, P., Navara, C. S., Salisbury, J., and Schatten, G. (1999). Biparental inheritance of γ-tubulin during human fertilization: Molecular reconstitution of functional zygotic centrosomes in inseminated human oocytes and in cell free extracts nucleated by human sperm. *Mol. Biol. Cell* **10,** 2955–2969.

Sluder, G., Miller, F., Lewis, F. J., Davison, E. D., and Rieder, C. L. (1989). Centrosome inheritance in starfish zygotes: Selective loss of maternal centrosome after fertilization. *Dev. Biol.* **131.** 567–579.

Sluder, G., Miller, F., and Lewis, F. (1993). Centrosome inheritance in starfish zygotes II: Selective suppression of the maternal centrosome during meiosis. *Dev. Biol.* **155,** 58–67.

Soley, J. I. (1994). Centriole development and formation of the flagellum during spermiogenesis in the ostrich (*Struthio camelus*). *J. Anat.* **185,** 301–313.

Stearns, T., and Kirschner, M. (1994). In vitro reconstitution of centrosome assembly and function: The central role of γ-tubulin. *Cell* **76,** 623–637.

Stevens, J. K., Trogadis, J., and Jakobs, R. J. (1988). Development and control of axial neurite from: a serial electron microscopic analysis. In "Intrinsic Determinants of Neural Form and Function" (R. J. Lasek and M. M. Black, Eds.), pp. 115–146. Alan R. Liss, New York.

Sutovsky, P., Manandhar, G., and Schatten, G. (1999). Biogenesis of centrosome during mammalian gametogenesis and fertilization. *Protoplasma* **206,** 249–262.

Szollosi, D., and Ozil, J. (1991). De novo formation of centrioles in parthenogenetically activated, diploidized rabbit embryos. *Biol. Cell.* **72,** 61–66.

Szollosi, D., Calarco, P., and Donahue, R. P. (1972). Absence of centrioles in the first and second meiotic spindles of mouse oocytes. *J. Cell Sci.* **11,** 521–541.

Szollosi, D., Czolowska, R., Soltynska, M. S., and Tarkowski, A. K. (1986). Ultrastructure of cell fusion and premature chromosome condensation of thymocyte nuclei in metaphase II mouse oocyte. *Biol. Cell.* **56,** 239–250.

Tassin, A. M., Maro, B., and Bornens, M. (1985). Fate of microtubule-organizing centers during myogenesis in vitro. *J. Cell Biol.* **100,** 35–46.

Tassin, A. M., Celati, C., Moudjou, M., and Bornens, M. (1998). Characterization of human homologue of yeast Spc98p and its association with γ-tubulin. *J. Cell Biol.* **141,** 689–701.

Tres, L., and Kierszenbaun, A. L. (1996). *Sak57,* an acidic keratin initially present in the spermatid manchette before becoming a components of paraxonemal structures of the developing tail. *Mol. Reprod. Dev.* **44,** 395–407.

Tucker, J. B., Paton, C. C., Richardson, G. P., Mogensen, M. M., and Russel, I. J. (1992). A cell surface–associated centrosomal layer of microtubule-organizing material in the inner pillar cell of the mouse cochlea. *J. Cell Sci.* **102,** 215–226.

Vogel, J. M., Stearns, T., Rieder, C. L., and Palazzo, R. E. (1997). Centrosomes isolated from *Spisula solidissima* oocytes contain rings and an unusual stoichiometric ratio of α/β tubulin. *J. Cell Biol.* **137,** 193–202.

Vorobjev, I. A., and Chentsov, Yu. S. (1982). Centriole in the cell cycle. I. Epithelial cells. *J. Cell Biol.* **98,** 938–949.

Wilson, P. G., Fuller, M. T., and Borisy, G. (1997). Monastral bipolar spindles: implications for dynamic centrosome organization. *J. Cell Sci.* **110,** 451–464.

Wolf, K. W., and Joshi, H. C. (1995). Microtubule organization and the distribution of γ-tubulin in spermatogenesis of a beetle, *Tenebrio molitor* (Tenebrionidae, Coleoptera, Insecta). *J. Cell Sci.* **108,** 3855–3865.

Wolf, K. W., and Joshi, H. C. (1996). Microtubule organization and the distribution of gamma-tubulin in male meiosis of Lepidoptera. *Mol. Reprod. Dev.* **45,** 547–559.

Woolley, D. M., and Fawcett, D. W. (1973). The degeneration and disappearance of the centrioles during the development of the rat spermatozoon. *Anat. Rec.* **177,** 289–302.

Wu, G.-J., Simerly, C. Zoran, S., Funte, L. R., and Schatten, G. (1996). Microtubule and chromatin dynamics during fertilization and early development in rhesus monkey, and regulation by intracellular calcium ions. *Biol. Reprod.* **55,** 260–270.

Yang, Z., Gallicano, G. I., Yu, Q.-C., and Fuchs, E. (1997). An unexpected localization of basonuclein in the centrosome, mitochondria, and acrosome of developing spermatids. *J. Cell Biol.* **137,** 657–669.

Zamboni, L., and Mastroianni, L., Jr. (1966). Electron microscopic studies of rabbit ova I: The follicular oocyte. *J. Ultrastruc. Res.* **14,** 95–117.

Zamboni, L., and Stefanini, M. (1971). The fine structure of the neck of mammalian spermatozoa. *Anat. Rec.* **169,** 155–172.

Zheng, Y., Wong, M. L., Alberts, B., and Mitchison, T. (1995). Nucleation of microtubule assembly by a γ-tubulin-containing ring complex. *Nature* **378,** 578–583.

Zoran, S., Schoff, P., Stearns, T., Salisbury, J., and Schatten, G. (1994). Reconstitution of human sperm centrosome in vitro. *Mol. Biol. Cell* **5** (Suppl.), 38a.

17

The Centrosome of the Early *C. elegans* Embryo: Inheritance, Assembly, Replication, and Developmental Roles

Kevin F. O'Connell
Laboratory of Molecular Biology
University of Wisconsin
Madison, Wisconsin 53706-1596

I. Why *C. elegans?*

The main advantage of *C. elegans* as an experimental model lies in its simplicity. The full-grown adult is about 1 mm in length and composed of fewer than 1000 somatic nuclei. It has a short reproductive cycle of approximately 3 days and simple nutritional requirements, feeding primarily on bacteria. In the laboratory, worms can be grown in petri plates containing nutrient agar seeded with a lawn of *E. coli.* The species is sexually dimorphic, containing both hermaphrodites and males. Hermaphrodites produce both sperm and eggs and are capable of internal self-fertilization. Thus, lines homozygous for a mutation of interest may be maintained without the need to perform genetic crosses, an important advantage in cases where the mutation of interest affects the ability to copulate.

The *C. elegans* embryo is exceptionally well suited as an experimental system for centrosome research. The supply of embryos is essentially unlimited as each self-fertilizing hermaphrodite can produce up to 300 embryos (and more if mated). The embryos are enveloped in a chitinous eggshell and vitelline membrane that allows them to survive outside the uterus

Current Topics in Developmental Biology, Vol. 49

0070-2153/00 $35.00

where they may be studied in isolation. The early blastomeres are large (10–25 μm in diameter), and the embryos are unpigmented, facilitating cytological studies. Importantly, the centrosomes are visible by differential interference contrast (DIC) microscopy—both nuclei and centrosomes exclude the refractile granules that fill the egg and appear as cytoplasmic clearings—and thus can be viewed in living specimens (Fig. 1).

A further advantage of the nematode embryo is its rapid and stereotypical development: cell cycles are short, ranging from 10 to 40 minutes, and the dynamic behavior of nuclei, centrosomes, and spindles is essentially invariant from embryo to embryo (Fig. 1). Such stereotypic behavior allows easy detection of even subtle mutational effects on the centrosome. Lastly, mutational perturbation of centrosome function will not trigger cell cycle

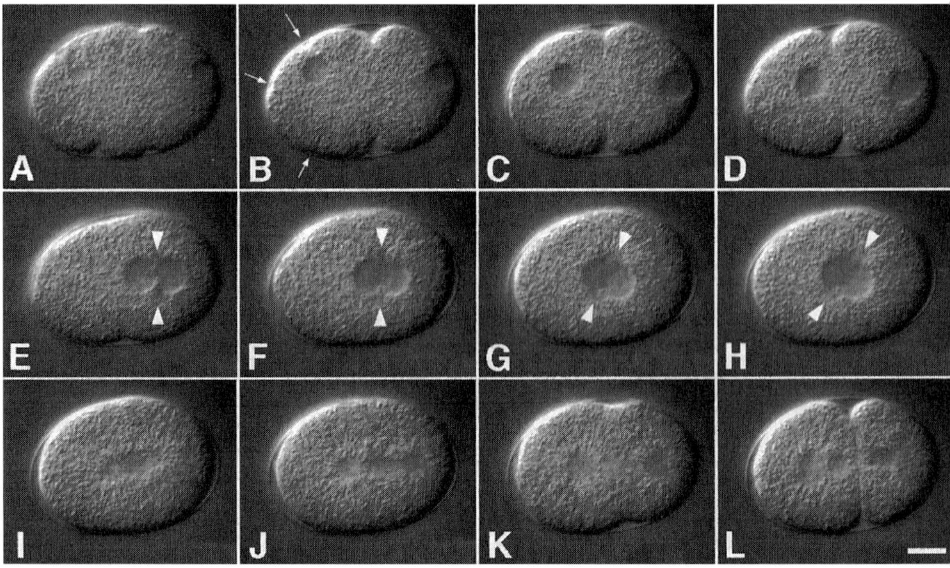

Figure 1 Time-lapse recording of a *C. elegans* zygote. Anterior to left. (A–E) PC stage. (A) As this period begins, cortical contractions ripple the cell surface. A prominent PC furrow forms at midegg length (B–E), and cell asymmetry becomes apparent as contractions become restricted to the anterior where they form shallow invaginations (B arrows). Typically the two pronuclei are situated at opposite ends of the zygote with the oocyte pronucleus at the anterior and the sperm pronucleus at the posterior. (B–E) During PC, the oocyte pronucleus migrates to the posterior. (C–E) Here it meets the sperm pronucleus, which has migrated a short distance anterior of its initial position. (F–L) Spindle alignment and first cleavage. As the two pronuclei meet, the centrosomes become visible next to the nuclei (arrowheads in E–H). Before spindle assembly, the centrosome–pronucleus complex moves to the center of the cell and rotates by 90° so that the metaphase spindle (I) forms at the center of the cell, parallel to the long axis. During anaphase (J, K) the spindle is displaced toward the posterior so that first division is asymmetric (L). Bar, 10 μm.

arrest, as the spindle assembly checkpoint is lax or altogether absent during the early divisions. Thus, one can investigate the effects of a centrosome mutation on downstream events.

Although these are highly desirable features, the recent boost in popularity of *C. elegans* may largely be attributed to two significant developments. First, essentially the complete sequence of the worm genome has been determined (The *C. elegans* Sequencing Consortium, 1998), allowing researchers to identify the worm homologs of important genes from other species. Second, powerful reverse genetic methods have been developed to transiently or permanently "knock out" a gene of interest. The most popular of these methods is RNA-mediated interference or RNAi (Fire *et al.,* 1998). RNAi involves the introduction of double-stranded RNA, derived from the gene of interest, into the germline of the worm. Although the mechanism is unknown, double-stranded RNA inhibits protein expression in a gene-specific manner, mimicking strong loss-of-function phenotypes. Thus, it is now possible to rapidly test the function of conserved centrosome associated proteins of heretofore unknown function.

II. Origin and Replication of the Zygotic Centrosomes

The centrosome or microtubule-organizing center (MTOC) is an unusual organelle in regard to the manner in which it is inherited. It is assembled after fertilization from an unequal contribution of maternal and paternal components (Archer and Solomon, 1994; Stearns and Winey, 1997). Typically, one parent contributes the centriole pair while both parents may contribute material to the pericentriolar matrix. Why such an elaborate lineage? One reason is to keep the number of centrosomes per cell constant. In dividing cells, a centrosome-per-cell ratio corresponding to 1 (prior to S phase) or 2 (after S phase) must be maintained. To achieve this, the centrosome is duplicated precisely once per cell cycle during S phase, and then each centrosome is parceled out to a daughter cell at mitosis. Fertilization, however, is problematic. The egg and the sperm cannot both donate a pair of centrioles. If they did, duplication during the ensuing S phase would produce four centrosomes and ultimately a tetrapolar spindle. Thus, mechanisms have evolved to ensure a uniparental mode of centriole inheritance. Another problem encountered at fertilization concerns the maturation divisions of the egg. These divisions occur after fertilization in the presence of the sperm centrosome, which conceivably could interfere with assembly of the female meiotic spindles. The following sections describe the strategies that *C. elegans* employs to deal with these problems, and how they compare to those of other species.

A. Inheritance of the Worm MTOC

In *C. elegans,* as in many species, including marine invertebrates (Sluder *et al.,* 1989; Sluder and Rieder, 1985) and mammals (Simerly *et al.,* 1995), the centrioles are paternally inherited. Each spermatozoon contains a single pair of centrioles embedded in an RNA-containing perinuclear material (Wolf *et al.,* 1978). Interestingly, ultrastructural analysis of *C. elegans* centrioles indicates that they are distinct from those of most other animals, being composed of nine singlet, rather than triplet, microtubules (Albertson and Thomson, 1993). It is not known if the sperm centrosome is competent to organize microtubules, but this point is moot; except for those composing the centrioles, nematode sperm lack microtubules—as well as microfilaments and myosin (Nelson *et al.,* 1982; Ward, 1986). Furthermore, nematode sperm lack a flagellum (Ward *et al.,* 1981; Wolf *et al.,* 1978), so the centrioles are not required to serve as basal bodies. Therefore, the centrioles are probably not essential for sperm function. Rather, they seem to be passengers who, upon fertilization, are liberated into the egg cytoplasm and reactivated.

In organisms such as *C. elegans* where the centrioles are strictly of paternal origin, mechanisms must exist to exclude the maternal MTOC from the zygotic lineage. This requires that the embryo be able to distinguish the maternal and paternal MTOCs. In starfish zygotes, the reproductive capacity of the maternal centrosome is selectively blocked; the MTOC does not duplicate between meiosis I and II, but instead splits to form two organizing centers, each with a single centriole (Sluder *et al.,* 1989). The residual egg MTOC is unable to replicate and fails to exist as an independent entity after meiosis (Sluder *et al.,* 1989). *C. elegans* zyotes must also limit the replication potential of maternal MTOCs. The mature egg completely lacks centrioles (Albertson and Thomson, 1993) and, as in *Drosophila,* the poles of both female meiotic spindles are acentriolar. The absence of centrioles may serve to limit replication of the maternal MTOC and perhaps its stability (Bobinnec *et al.,* 1998); following meiosis, the only MTOC evident in the zygote is the centriole pair–containing sperm centrosome. Centrioles are most assuredly present within the population of mitotic stem cells of the female germline but are lost before eggs reach maturity. Study of the centriole disposal mechanism should be amenable to genetic methods, but so far screens for female sterile and maternal-effect lethal mutants have not yielded any candidate genes.

Fertilization creates another potentially unstable situation; at the time the egg is undergoing meiosis, the sperm MTOC is present in the same cytoplasm. Somehow, the egg manages to keep the sperm MTOC from participating in female meiosis and forming multipolar spindles. Eggs might accomplish this by delaying assembly of the sperm centrosome until after meiosis is complete (Schatten, 1994; Sluder *et al.*, 1993). However, Wu and Palazzo (1999) have shown that in the surf clam *Spisula solidissima,* the sperm centrosome organizes an aster before meiosis. It is selectively deactivated during meiosis I, then reactivated before the mitotic divisions. Thus, the surf clam zygote must be able to distinguish between the maternal and paternal centrosomes to differentially regulate their activities. In contrast, the sperm aster in *C. elegans* zygotes does not form until after meiosis (Albertson, 1984), indicating that in some organisms, the sperm centrosome may somehow be masked to prevent its assembly while meiosis is in progress (Schatten, 1994; Sluder *et al.,* 1993).

B. Assembly of the Zygote Centrosome

At fertilization the paternal centriole pair is released into the maternal cytoplasm, an event that initiates a dramatic transformation of both the centrosome itself and the zygote (see below). The unfertilized *C. elegans* egg contains a dense cytoplasmic network of randomly oriented microtubules (Fig. 2A). Fertilization prompts the egg to begin remodeling this network (Albertson, 1984). As the egg resumes meiosis, microtubules organize around the maternal chromosomes to form the first, then the second, of the meiotic spindles (Albertson and Thomson, 1993). Two polar bodies are extruded at one pole of the oblong zygote, after which the female pronucleus forms nearby. The male pronucleus is typically located at the opposing pole in close association with the cortex (Fig. 1B). At a point between the male pronucleus and cortex, a small sperm aster forms. After the sperm centrosome duplicates, the two daughter centrosomes migrate apart and move onto the side of the sperm pronucleus facing the oocyte pronucleus (Fig. 2B). During this time, the asters enlarge presumably reflecting an increase in both the number and length of astral microtubules. Growth of the asters continues as the pronuclei migrate toward one another (Fig. 2C) and plateaus as the pronuclei meet (Fig. 2D) (Albertson, 1984). At this time, essentially all microtubules are associated with the centrosomes.

The gradual increase in microtubule nucleating capacity of the sperm centrosomes likely involves the recruitment of maternally provided components from the egg cytoplasm. One of these components is probably the worm equivalent of the γ-tubulin ring complexes (γ-TuRCs) that have been described in other systems (see Chapter 3). The γ-TuRCs (Oegema *et al.,* 1999; Zheng *et al.,* 1995) and their related subcomplexes (Oegema *et al.,* 1999) have the capacity to nucleate microtubules *in vitro* and are soluble forms of the microtubule nucleating sites present at the centrosome (Moritz *et al.,* 1998). *Xenopus* sperm lack γ-tubulin and must recruit γ-tubulin complexes from maternal pools in the egg cytoplasm before they are compe-

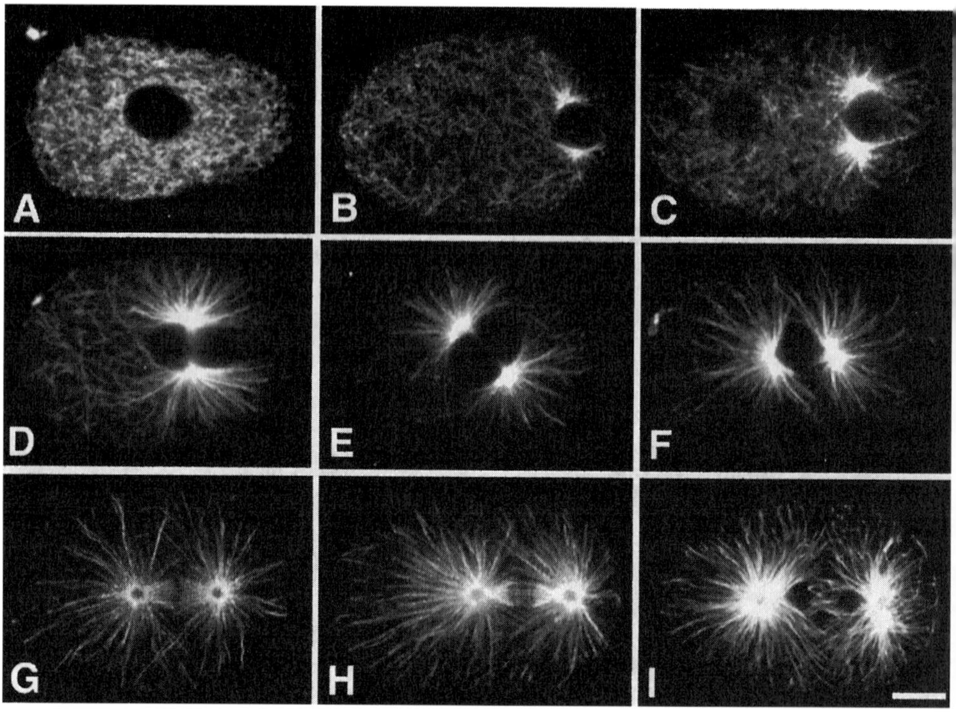

Figure 2 Formation of the sperm asters and the first mitotic spindle in the *C. elegans* zygote. Anterior to left. (A) Oocyte with cytoplasmic microtubule network. (B, C) The sperm asters associated with the male pronucleus steadily increase in size during pronuclear migration. (D) Pronuclear apposition. The asters have reached their maximum size. (E, F) Spindle alignment. The pronucleus–centrosome complex moves to the center of the cell and rotates onto the AP axis. (G) Metaphase. The spindle is positioned close to the center of the AP axis. (H) Anaphase. The spindle becomes displaced posteriorly. Note how much closer the posterior spindle pole is to the cell surface than is the anterior pole. (I) Telophase. The posterior centrosome assumes a flattened appearance, while the anterior centrosome remains spherical. Bar, 10 μm.

tent to organize microtubules (Stearns and Kirschner, 1994). A worm γ-TuRC has not been described, but the worm genome contains an ORF with homology to γ-tubulins from other species. A polyclonal serum directed against human γ-tubulin recognizes *C. elegans* centrosomes; the intensity of staining is weak immediately after fertilization and steadily increases in concert with sperm aster growth mutant (O'Connell *et al.*, 2000). This indicates that, in *C. elegans* as in other organisms, assembly of a functional centrosome around the sperm centrioles involves the recritment of maternally provided centrosome components including γ-tubulin.

Other proteins have been observed to associate with the sperm centrosomes as they mature. One of these is the product of the *zyg-9* (*zyg*ote-defective) gene (Matthews *et al.,* 1998). ZYG-9 localizes to meiotic and mitotic spindles where it is enriched at the poles. Mutations in *zyg-9* produce a maternal-effect embryonic lethal phenotype; meiotic spindles are disrupted and mitotic spindles are smaller than normal, being composed of unusually short microtubules (Albertson, 1984; Kemphues *et al.,* 1986; Matthews *et al.,* 1998). ZYG-9 shares sequence similarity with a human microtubule-binding protein ch-TOG and the product of the *Saccharomyces cerevisiae STU2* gene. ch-Tog localizes to mitotic spindles and centrosomes (Charrasse *et al.,* 1998) and is homologous to XMAP215 of *Xenopus,* a microtubule-associated protein that promotes the formation of long dynamic microtubules (Vasquez *et al.,* 1994). The shorter microtubules observed in *zyg-9* mutants suggest that ZYG-9 might perform a similar function in the *C. elegans* embryo. The Stu2p protein is a component of the yeast spindle pole body (SPB) (Wang and Huffaker, 1997). Based on its ability to bind both Spc72p, a γ-tubulin interacting SPB component, and microtubules, it has been suggested that Stu2p might be involved in regulating microtubule nucleation (Chen *et al.,* 1998; Zimmerman *et al.,* 1999). It is presently not known if ZYG-9 plays such a role in *C. elegans.*

Classical and reverse genetic approaches have been used to identify novel factors required for centrosome function and to probe the role of the worm homologs of centrosome-associated factors from other species. The *spd-2* (for *sp*indle-*d*efective) gene was isolated in a general screen for cell division mutants (O'Connell *et al.,* 1998). In *spd-2* mutant zygotes, formation of the sperm asters is inhibited, such that through most of the cell cycle, sperm asters are not present. Small astral arrays only appear shortly before nuclear envelope breakdown (NEB). As a result, *spd-2* mutant embryos exhibit defects in pronuclear migration, establishment of AP polarity, and mitotic spindle assembly (O'Connell *et al.,* 2000). Female meiotic spindles are not affected, and thus, *spd-2* appears to be specifically required for the formation of mitotic centrosomes. This defect may not be explained by the absence of sperm-derived centrioles, as centrioles have been detected in the mutant (K. F. O'Connell and J. G. White, unpublished results). Instead, *spd-2* mutations interfere with the ability of the sperm centrioles to recruit maternal factors; centrosomes are not detected with antibodies to γ-tubulin or ZYG-9 in the mutant (K. F. O'Connell and J. G. White, unpublished results).

With increasing frequency, RNAi has been used to assess potential centrosome related functions of molecularly defined genes. The worm genome sequence database contains matches to centrosome proteins identified in other species. The functions of a number of these putative worm homologs have already been tested. RNAi of *tbg-1,* the γ-tubulin gene, leads to defects in spindle structure and cell division during embryogenesis; however, astral microtubule arrays are still present (W. Saxton and S. Strome, personal communication). It is not known if this represents the null phenotype, as the effectiveness of RNAi in reducing γ-tubulin protein levels has not been assessed (i.e., by immunofluorescence staining). Nonetheless, these experiments demonstrate the involvement of worm γ-tubulin in organizing the embryonic microtubule cytoskeleton.

RNAi has also been used to analyze the functions of two *C. elegans* members of the Ipl1/Aurora family of serine/threonine kinases. The *Drosophila aurora* locus is required for bipolar spindle formation; *aurora* mutants exhibit a mitotic arrest phenotype in which centrosomes fail to separate, resulting in the formation of monopolar spindles (Glover *et al.,* 1995). In contrast, the function of the *S. cerevisiae* Ipl1p kinase is quite distinct from that of Aurora. Ipl1p is required for accurate chromosome segregation and regulates kinetochore–microtubule interactions by phosphorylating the kinetochore component Ncd10p (Biggins *et al.,* 1999; Chan and Botstein, 1993). The *C. elegans air-1* (*A*urora/*I*pl1 *r*elated) gene encodes a kinase with a high degree of similarity to Ipl1/Aurora family members (Schumacher *et al.,* 1998a). The product of the *air-1* gene is enriched at mitotic centrosomes, similar to the distribution of a mammalian Ipl1/Aurora related kinase (Gopalan *et al.,* 1997; Kimura *et al.,* 1997; Schumacher *et al.,* 1998a). Embryos in which *air-l* expression has been disrupted by RNAi fail to express AIR-1 protein, exhibit defects in spindle structure, and become aneuploid. Although bipolar, mitotic spindles in these embryos are disorganized and centrosomes appear incapable of nucleating normal sized astral arrays, suggesting that *air-1* kinase activity may function in the recruitment of centrosomal proteins (Schumacher *et al.,* 1998a). A second Ipl1/Aurora related kinase has been discovered, but it does not appear to have a centrosome related function; AIR-2 localizes to chromosomes and midbody microtubules and is required for the late stages of cytokinesis (Schumacher *et al.,* 1998b).

C. Replication of the MTOC

Shortly after fertilization, the single pair of sperm centrioles duplicates to produce two MTOCs (Albertson, 1984). These form the poles of the first mitotic spindle, and during cytokinesis, each is partitioned to a daughter cell, thus reestablishing the one centrosome per cell ratio. The subsequent divisions are rapid, and centrosome duplication must occur swiftly in order to keep pace. During these divisions, duplication appears to initiate late in the preceding cell cycle, about the time of anaphase (K. O'Connell and J. White, unpublished results). Antibodies raised to PLK-1, a *C. elegans* polo-like kinase (Chase *et al.,* 2000), have been found to label centrioles (K. F. O'Connell and J. White, unpublished results). Each pole of a metaphase spindle contains a single PLK-1-positive dot that presumably represents a single centriole pair. At telophase, each pole contains two dots. The doublets probably represent separated sister centrioles. During the rapid nuclear cycles of early *Drosophila* embryos, centrosome duplication exhibits similar timing; centriole separation occurs between late metaphase and telophase and procentriole formation during the ensuing interphase (Callaini and Riparbelli, 1990). Interestingly, in *C. elegans,* the two poles of a spindle do not always initiate duplication at the same time. In cells that divide asymmetrically (to produce nonidentical daughters), the centrioles at the posterior pole duplicate slightly before those at the anterior pole. This asynchrony may be promoted by a wave of cell cycle progression that has been detected in cells undergoing asymmetric cell division (Gonczy *et al.,* 1999).

Cell division processes are highly amenable to genetic analysis in *C. elegans* (Gonczy *et al.,* 1999; O'Connell *et al.,* 1998). Among the rich assortment of embryonic lethal mutants that now exists, *zyg-1* mutants are unique in that they show a defect in centrosome duplication (K. F. O'Connell, C. Caron, K. J. Kemphues, and J. G. White, unpublished results). Embryos born from *zyg-1* mothers produce monopolar spindles. Examination of these monastral arrays by immunofluorescence microscopy with anti-PLK-1 sera indicates the presence of a single pair of centrioles. The *zyg-1* gene encodes a novel protein with an N-terminal kinase domain. None of the six characterized *zyg-1* mutations map to the kinase portion, and thus it is presently not known if kinase activity is required for *zyg-1* function.

Interestingly, analysis of *zyg-1* mutants has revealed that centrosome duplication in the early embryo is regulated both paternally and maternally. In the absence of either maternal or paternal *zyg-1* activity, embryogenesis fails before the four-cell stage. Examination of these embryos by live DIC imaging or immunofluorescence microscopy reveals a block in centrosome duplication and bipolar spindle formation. These results indicate that a complex *zyg-1*-dependent regulatory scheme exists, and that in the early embryo, centrosome duplication, like centrosome assembly, is dependent upon maternal and paternal contributions.

Although *zyg-1* is presently the only gene known to be required for centrosome duplication in *C. elegans,* other candidates exist. Recently, a complex of the cyclin-dependent kinase cdk2 and cyclin E has been shown to be required for centrosome duplication in *Xenopus* embryos and extracts (Hinchcliffe *et al.,* 1999; Lacey *et al.,* 1999; Matsumoto *et al.,* 1999). The *C. elegans* genome encodes multiple cdk family members; the product of the *ncc-1* gene is required for entry into mitosis and appears to be the worm cdk1/cdc2 ortholog (Boxem *et al.,* 1999). The ORF K03E5.3 encodes a cdk that is equally related to human cdk1 and cdk2 (42% and 43% identity, respectively); RNAi silencing of this gene, however, has not revealed any obvious effecs on centrosome duplication (M. Boxem and S. van den Heuvel, personal communication). The worm genome also contains a single homolog of cyclin E encoded by ORF C37A2.4. RNAi of this gene leads to a late embryonic arrest phenotype, but an effect on centrosome duplication has not been observed (J. Liu and E. Kipreos, personal communication; M. Boxem and S. van den Heuvel, personal communication). Although it is possible that RNAi failed to sufficiently deplete the maternal pools of these proteins, RNAi is generally considered an effective and powerful gene silencing technique, particularly for maternally expressed genes (Boxem *et al.,* 1999; Fire *et al.,* 1998; Schumacher *et al.,* 1998a, 1998b; Skop and White, 1998). Thus, it is possible that other cdk–cyclin complexes function redundantly with the cdk2–cyclin E complex to regulate centrosome duplication.

III. The Central Role of the Centrosome–Pronucleus Complex in Specifying AP Polarity

Whereas the eggs of many species possess a highly polarized structure (Fernandez *et al.,* 1998; Goldstein and Freeman, 1997; Sardet *et al.,* 1994), the *C. elegans* egg does not exhibit any developmentally important asymmetries until after fertilization (Goldestein and Hird, 1996). Polarization of the anteroposterior (AP) axis occurs about three-quarters of the way through the first cell cycle when the zygote is reorganized (Hill and Strome, 1988, 1990). Developmental instructions are segregated along the AP axis as domains of developmentally important proteins are established in the cortex and cytoplasm (Kemphues and Strome, 1997). Goldstein and Hird (1996) have shown that the orientation of this axis is determined by the position of the sperm centrosome–pronucleus complex. During reorganization, this complex is positioned at the end of the egg destined to become the posterior. Recent work, outlined later, has pointed to the centrosomes or astral microtubules, rather than the sperm pronucleus, as playing an important role in this process.

The period of reorganization is referred to as pseudocleavage (PC), named after a transient and centrally located furrow that forms at this time (Figs. 1B–1D). PC is marked by the appearance of contractions on the cell surface; these briefly form surface contraction waves (SCWs) that sweep from the posterior toward the anterior. As the SCWs abate, the PC furrow forms at mid-egg length (Fig. 1B). At this time, one of the first asymmetries becomes apparent: In the anterior half of the zygote, the cortex continues to exhibit contractile activity and the surface appears ruffled, while in the posterior half, all contractile activity ceases and the cell surface appears smooth (Figs. 1B and 3A). These cortical activities are associated with an underlying fountain flow of cytoplasm in the posterior half of the embryo (Fig. 3A); near the surface of the zygote, cortical material flows anteriorly, while deep within, central cytoplasm streams in the opposite direction (Hird and White, 1993).

Hird and White (1993) found that the direction of the fountain flow is dictated by the position of the sperm centrosome–pronucleus complex. Flow is only observed when the complex is closely apposed to the cell cortex. Near the sperm pronucleus, cortical material is always directed away from the complex, while central cytoplasm is always directed toward the complex. If the sperm pronucleus moves along the cortex, the directions of the cytoplasmic and cortical movements will adjust to maintain this spatial relationship. To explain how the complex might direct the fountain flow, Hird and White (1993) proposed that astral microtubules create a tension gradient within the cortical actomyosin network by inducing localized relaxation of the cell cortex (Fig. 3A). This would result in an imbalance of forces. The cortical actomyosin network in the anterior would contract, producing ruffles in the cell surface and an area of high tension. The posterior cortical actomyosin network would relax, producing a smooth cell surface and an area of low tension. The cortical flow would be established as material in the posterior cortex is pulled up the tension gradient, and the cytoplasmic flow would be established to compensate volumetrically for the displacement of material within the posterior cortex. A similar model has been proposed to explain how a fountain flow in syncytial *Drosophila* embryos is generated (von Dassow and Schubiger, 1994), suggesting that all forms of cytoplasmic streaming may be mechanistically related (Bray and White, 1988).

The fountain flows have been thought to be involved in the asymmetric localization of cytoplasmic constituents. A set of germline (or P) granules is known to localize to the posterior during PC (Hill and Strome, 1988; Strome, 1983). P granules are ribonucleoprotein complexes and contain factors that are important in cell fate determination (Draper *et al.,* 1996; Gruidl *et al.,* 1996). They are initially distributed evenly in the zygote cytoplasm, then during PC, they circulate on the fountain flows and accumulate at the posterior where they are partitioned at first cleavage to the progenitor of the germline (Fig. 3A) (Hird *et al.,* 1996; Strome, 1983). Since both the fountain flows and the segregation of P granules can be inhibited by cytochalasin treatment, it has been suggested that the bulk movements of cytoplasm are involved in localizing P granules and perhaps other developmental determinants (Hill and Strome, 1988; Hird *et al.,* 1996; Strome, 1983). Recent results demonstrate that the fountain flow is not absolutely required for P granule localization. RNAi silencing of either a gene encoding a nonmuscle myosin heavy chain (*nmy-2*) or regulatory light (*mlc-4*)

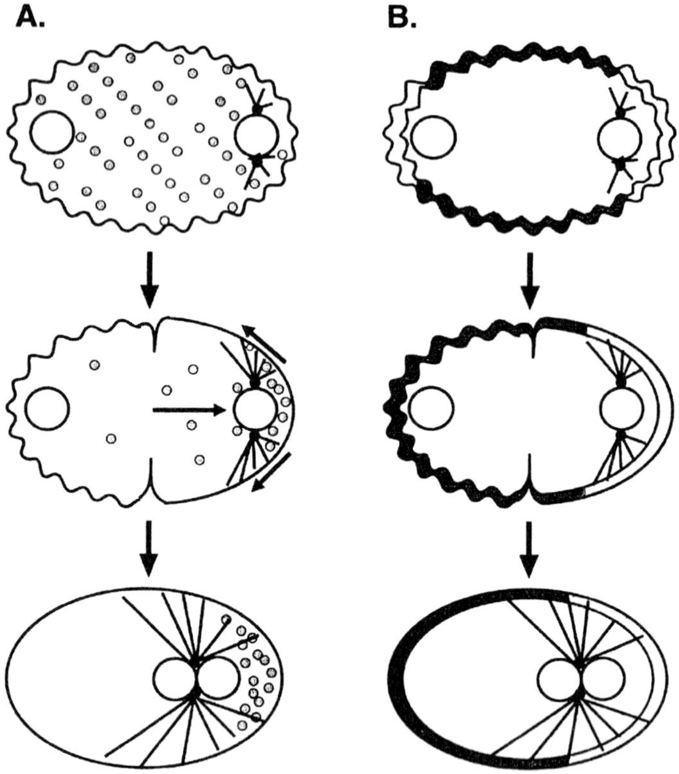

Figure 3 Establishment of AP polarity. (A) Cytoplasmic flow and segregation of P granules. (*Top*) Prior to PC, contractile activity is equally distributed in the cortex, directed cytoplasmic movements are absent, and P granules (small gray dots) are dispersed in the cytoplasm. (*Middle*) During PC, the cortical actomyosin network becomes polarized with tension highest in the anterior. A fountain flow is observed in the posterior as cytoplasm streams toward the sperm pronucleus–centrosome complex (internal arrow) and cortical material moves anteriorly away from the complex (external arrows). At the same time, P granules circulate on the fountain flow and begin accumulating at the posterior. The sperm centrosomes/asters may direct these rearrangements by signaling the posterior cortex to relax; this would create a tension gradient. The fountain flow would form in response to the gradient and promote the posterior localization of P granules. (*Bottom*) At the end of PC, the cortical contractile activity and flows cease, and P granule localization is complete. (B) Localization of PAR proteins. (*Top*) Prior to PC, PAR-2 (light shading) is localized at the poles of the zygote while PAR-3 (dark shading) is enriched along the lateral borders. (*Middle*) During PC, PAR-2 localizes to the pole nearest the sperm centrosome-pronucleus complex (the posterior pole), while PAR-3 localizes, in a reciprocal manner, to the anterior portion of the cortex. Localization of PAR-2 to the posterior may involve an interaction between the sperm centrosomes/asters and the adjacent cortex; this interaction may stabilize the position of PAR-2 and eventually lead to the exclusion of PAR-3. (*Bottom*) Once established, the anterior–posterior arrangement of PAR-2 and PAR-3 domains persists through first division.

chain blocks the cytoplasmic movements but does not completely block P granule localization (Shelton *et al.,* 1999). However, P granule localization is delayed relative to wild-type embryos, and *nmy-2* and *mlc-4* RNAi embryos fail to develop the full complement of AP asymmetries (Guo and Kemphues, 1996; Shelton *et al.,* 1999), indicating that the fountain flow is an essential aspect of the polarization process.

Analysis of PC in *spd-2* mutants indicates that the centrosomes/sperm asters are involved in the cytoplasmic fountain flow and the segregation of P granules (O'Connell *et al.,* 2000). The altered kinetics of centrosome assembly in *spd-2* mutants results in the absence of detectable sperm asters in PC-stage embryos. A sperm pronucleus, however, is present. In mutant *spd-2* zygotes, the cortical contractions of PC are present but are not localized to the anterior, indicating that a cortical tension gradient has not been established. A PC furrow does not form, and cortical and cytoplasmic flows are altogether absent. In addition, P granules fail to localize properly; initially they aggregate at the cell center, as occurs in cytochalasin-treated embryos (Strome, 1983), and later they localize randomly in the cell. Although these results support an essential role for the centrosomes and/or astral microtubules in defining AP polarity, other findings indicate that the sperm pronucleus is dispensable. Mutations in the *emb-27* and *emb-30* genes result in the formation of fertilization-competent anucleate sperm (Sadler and Shakes, 2000). Mutant zygotes, which inherit a centrosome but not a sperm pronucleus, are capable of AP polarization, bipolar spindle formation, and asymmetric cleavage.

Experimentally, in terms of where the polarity signal originates, it has been difficult to distinguish between the centrosomes proper or the sperm asters. Treatment of PC stage embryos with high concentrations of microtubule-disrupting agents affects neither the fountain flow (Hird and White, 1993) nor the segregation of P granules (Strome, 1983), suggesting that microtubules are not important. However, even at high concentrations of inhibitors some short stable microtubules may persist at the centrosome (Hyman and White, 1987), and thus it is possible these astral remnants are sufficient to drive AP remodeling. Alternatively, the polarity signal may originate at the centrosomes and be transmitted to the cell directly or via microtubules. Distinguishing between these possibilities will require molecular dissection of the signaling pathway.

Polarization of the AP axis also requires the *par* (*par*titioning-defective) genes (Kemphues and Strome, 1997). Similar to cytochalasin-treated embryos (Hill and Strome, 1988, 1990; Hird and White, 1993; Strome, 1983), *par* mutant embryos exhibit defects in the execution of PC, the segregation of P granules, and ultimately the distribution of developmental potential at the two-cell stage (Kirby *et al.,* 1990). Several of these genes encode proteins that localize to the cell cortex (Fig. 3B). PAR-1, a putative serine-threonine kinase, and PAR-2, a zinc-finger protein, localize to the posterior cortex (Guo and Kemphues, 1995), while two PDZ domain-containing proteins, PAR-3 and PAR-6, localize to the anterior, possibly as part of a complex (Etemad-Moghadam *et al.,* 1995; Hung and Kemphues, 1999). Given their localization patterns and mutant phenotypes, the PAR proteins likely interact with the actin cytoskeleton to reorganize the zygote during PC.

Analysis of *spd-2* mutants indicates that the centrosomes/sperm asters might also direct localization of the PAR proteins (O'Connell *et al.,* 2000). In *spd-2* mutant zygotes, PAR-3 is widely distributed throughout the cortex, its concentration dropping off toward the poles, while PAR-2 randomly localizes to one or both poles. Interstingly, prior to PC in wild-type zygotes, PAR-2 localizes to both polar regions, only later becoming restricted to the posterior pole (Fig. 3B) (Boyd *et al.,* 1996). It seems that before PC in wild-type zygotes, both poles are developmentally equivalent and are capable of becoming the posterior, as demonstrated by Goldestein and Hird (1996). The centrosome–pronucleus complex could break this symmetry by inducing the proximal pole to assume a posterior identity, perhaps by locally stabilizing cortical PAR-2. Cortical PAR-2 and PAR-3 spatially exclude each other, and thus, the entrenched PAR-2 at the posterior may restrict PAR-3 to the anterior, thereby creating developmentally distinct regions of the cortex.

The centrosome also acts as a shuttle for the asymmetric segregation of developmental factors. During embryogenesis, the germline is established by a series of asymmetric divisions, each producing a somatic founder cell and a germline precursor cell. Totipotency of the germline is maintained by PIE-1, a transcriptional repressor that inhibits somatic differentiation (Mello *et al.,* 1996). During the each of the asymmetric divisions, PIE-1 is segregated to the germline daughter. It first associates with both mitotic spindle poles. Subsequently, it is selectively lost from the pole destined for the somatic daughter cell, while it remains associated with the centrosome inherited by the germline daughter (Mello *et al.,* 1996).

IV. Nuclear Positioning

Between fertilization and first mitosis, several stereotypical nuclear movements occur (Figs. 1C–1H), and in all cases, the centrosomes play a prominent role. These movements fall into two categories. The first is pronuclear migration, which serves to bring together the two sets of parental chromosomes, and the second is spindle alignment, which serves to position the centrosomes on the AP axis so that the first mitotic spindle will be aligned with the zygote's axis of polarity.

A. Pronuclear Migration

Near the end of PC, the pronuclei, situated at opposing ends of the zygote, begin to migrate inward and converge upon one another (Figs. 1B–1E, 2B–2D). The velocity of the oocyte pronucleus is not constant (Albertson, 1984). Initially, it travels at a slow rate; then, as it approaches the sperm pronucleus, it accelerates approximately 10-fold. In contrast, the migrating sperm pronucleus maintains a constant velocity that is equal in magnitude to that of the slow phase of the oocyte pronucleus. Probably owing to its greater overall migration rate, the oocyte pronucleus travels further than the sperm pronucleus, and the two meet closer to the posterior pole (Figs. 1E, 2D).

Pronuclear migration follows a similar pattern in divergent species and involves two separate movements: migration of the oocyte pronucleus toward the center of the sperm aster(s), and migration of the sperm pronucleus toward the center of the cell (Fernandez *et al.,* 1998; Sardet *et al.,* 1994). It is thought that the movement of both pronuclei is directed by the sperm centrosome(s). Migration of the oocyte pronucleus might be driven by minus-end-directed microtubule motors attached to the pronuclear surface (Reinsch and Karsenti, 1997; Schatten and Schatten, 1981). By engaging astral microtubules, these motors would provide the necessary force and proper directionality for convergent migration. Centripedal migration of the sperm pronucleus is likely driven by forces acting on the astral microtubules and reflects the tendency of radial microtubule arrays to become positioned at the cell center (Rodionov and Borisy, 1997).

A recent *in vitro* study indicates that cytoplasmic dynein might provide the motive force for oocyte pronuclear migration. Reinsch and Karsenti (1997) have shown that nuclei assembled around chromatin beads translocate along microtubules toward the center of sperm asters and that such movements can be blocked with vanadate or antibodies to the intermediate chain of cytoplasmic dynein. Additional experimentation, perhaps using RNAi in *C. elegans,* will be needed to determine which motors are required *in vivo.*

B. Spindle Alignment

Shortly after polarity has been established and the pronuclei have met, the mitotic spindle forms. It is imperative that the spindle be aligned along the zygote's axis of polarity, so that at first division, P granules will be partitioned to the posterior (germline progenitor) cell. If the spindle fails to properly align, developmental instructions may not be correctly allocated to the daughter cells, ultimately causing development to fail. At pronuclear apposition, the centrosome–pronucleus complex is positioned in the posterior with the centrosomes on a transverse axis. Before the spindle forms, the complex moves to the center of the cell—a movement that appears continuous with the anteriorly directed migration of the sperm pronucleus—and rotates 90°, moving the centrosomes onto the AP axis (Figs. 1E–1H, 2D–2F).

Centrosome rotation occurs during each of the asymmetric divisions of the germline lineage, each time positioning the spindle on the AP axis. During rotation, one of the centrosomes—termed the leading centrosome and chosen at random—moves towards the anterior margin of the cell. The centrosomes and nucleus rotate as a unit, pivoting around the lagging centrosome (Hyman and White, 1987). The rotation mechanism is microtubule dependent and appears to involve a specialized region of the anterior cortex that captures and shortens astral microtubules. By preferentially engaging one set of astral microtubules, this site could provide the torque necessary for rotation. Consistent with this, transient laser-mediated disruption of the leading centrosome or microtubules running between the leading centrosome and the anterior cortex temporarily halts rotation (Hyman, 1989). Skop and White (1998) have shown that components of the dynactin complex, an activator of dynein-mediated motility, localize to the cortical site and are required for rotation. These results indicate that dynein might be tethered to the cortex where it would engage astral microtubules, and in concert with microtubule depolymerization, drive rotation.

V. Concluding Remarks

It is said that *C. elegans* has entered the postgenomics era, the complete genetic blueprint for the growth and development of this multicellular organism having been determined. The next great challenge will be to determine how the individual parts work together to form a fully functional organism. Encoded with in the blueprint is the full complement of proteins that are needed for formation, replication, and function of the centrosome. Ultimately, it should be possible to identify all the relevant factors, to understand the molecular mechanisms in which they function, and to appreciate how the centrosome might be integrated into the mechanisms of animal growth and development.

Acknowledgments

I thank M. Boxem, D. Chase, A. Golden, S. van den Heuvel, E. Kipreos, J. Liu, P. Sadler, W. Saxton, D. Shakes, and S. Strome for sharing unpublished data and J. White for critically reading the manuscript.

References

The *C. elegans* Sequencing Consortium (1998). Genome sequence of the nematode *C. elegans:* a platform for investigating biology. [published erratum appears in *Science* (1999). Jan 1;283(5398):35]. *Science* **282,** 2012–2018.

Albertson, D. G. (1984). Formation of the first cleavage spindle in nematode embryos. *Dev. Biol.* **101,** 61–72.

Albertson, D. G., and Thomson, J. N. (1993). Segregation of holocentric chromosomes at meiosis in the nematode *Caenorhabditis elegans. Chromosome Res.* **1,** 15–26.

Archer, J., and Solomon, F. (1994). Deconstructing the microtubule-organizing center [see comments]. *Cell* **76,** 589–591.

Biggins, S., Severin, F. F., Bhalla, N., Sassoon, I., Hyman, A. A., and Murray, A. W. (1999). The conserved protein kinase Ipl1 regulates microtubule binding to kinetochores in budding yeast. *Genes Dev.* **13,** 532–544.

Bobinnec, Y., Khodjakov, A., Mir, L. M., Rieder, C. L., Edde, B., and Bornens, M. (1998). Centriole disassembly in vivo and its effect on centrosome structure and function in vertebrate cells. *J. Cell Biol.* **143,** 1575–1589.

Boxem, M., Srinivasan, D. G., and van den Heuvel, S. (1999). *The Caenorhabiditis elegans gene ncc-1* encodes a cdc2-related kinase required for M phase in meiotic and mitotic cell divisions, but not for S phase. *Development* **126,** 2227–2239.

Boyd, L., Guo, S., Levitan, D., Stinchcomb, D. T., and Kemphues, K. J. (1996). PAR-2 is asymmetrically distributed and promotes association of P granules and PAR-1 with the cortex in *C. elegans* embryos. *Development* **122,** 3075–3084.

Bray, D., and White, J. G. (1988). Cortical flow in animal cells. *Science* **239,** 883–888.

Callaini, G., and Riparbelli, M. G. (1990). Centriole and centrosome cycle in the early *Drosophila* embryo. *J. Cell Sci.* **97,** 539–543.

Chan, C. S., and Botstein, D. (1993). Isolation and characterization of chromosome-gain and increase-in-ploidy mutants in yeast. *Genetics* **135,** 677–691.

Charrasse, S., Schroeder, M., Gauthier-Rouviere, C., Ango, F., Cassimeris, L., Gard, D. L., and Larroque, C. (1998). The TOGp protein is a new human microtubule-associated protein homologous to the *Xenopus* XMAP215. *J. Cell Sci.* **111,** 1371–1383.

Chase, D., Serafinas, C., Ashcroft, N., Kosinski, M., Longo, D., Ferris, D. K., and Golden, A. (2000). The polo-like kinase PLK-1 is required for nuclear envelope breakdown and the completion of meiosis in *Caenorhabditis elegans. Genesis* **1,** 26–41.

Chen, X. P., Yin, H., and Huffaker, T. C. (1998). The yeast spindle pole body component Spc72p interacts with Stu2p and is required for proper microtubule assembly. *J. Cell Biol.* **141,** 1169–1179.

Draper, B. W., Mello, C. C., Bowerman, B., Hardin, J., and Priess, J. R. (1996). MEX-3 is a KH domain protein that regulates blastomere identity in early *C. elegans* embryos. *Cell* **87,** 205–216.

Etemad-Moghadam, B., Guo, S., and Kemphues, K. J. (1995). Asymmetrically distributed PAR-3 protein contributes to cell polarity and spindle alignment in early *C. elegans* embryos. *Cell* **83,** 743–752.

Fernandez, J., Roegiers, F., Cantillana, V., and Sardet, C. (1998). Formation and localization of cytoplasmic domains in leech and ascidian zygotes. *Int. J. Dev. Biol.* **42,** 1075–1084.

Fire, A., Xu, S., Montgomery, M. K., Kostas, S. A., Driver, S. E., and Mello, C. C. (1998). Potent and specific genetic interference by double-stranded RNA in *Caenorhabditis elegans* [see comments]. *Nature* **391,** 806–811.

Glover, D. M., Leibowitz, M. H., McLean, D. A., and Parry, H. (1995). Mutations in aurora prevent centrosome separation leading to the formation of monopolar spindles. *Cell* **81,** 95–105.

Goldstein, B., and Freeman, G. (1997). Axis specification in animal development *BioEssays* **19,** 105–116.

Goldstein, B., and Hird, S. N. (1996). Specification of the anteroposterior axis in *Caenorhabditis elegans. Development* **122,** 1467–1474.

Gonczy, P., Schnabel, H., Kaletta, T., Amores, A. D., Hyman, T., and Schnabel, R. (1999). Dissection of cell division processes in the one cell stage *Caenorhabditis elegans* embryo by mutational analysis. *J. Cell Biol.* **144,** 927–946.

Gopalan, G., Chan, C. S., and Donovan, P. J. (1997). A novel mammalian, mitotic spindle-associated kinase is related to yeast and fly chromosome segregation regulators. *J. Cell Biol.* **138,** 643–656.

Gruidl, M. E., Smith, P. A., Kuznicki, K. A., McCrone, J. S., Kirchner, J., Roussell, D. L., Strome, S., and Bennett, K. L. (1996). Multiple potential germ-line helicases are components of the germ-line-specific P granules of *Caenorhabditis elegans. Proc. Natl. Acad. Sci. USA* **93,** 13837–13842.

Guo, S., and Kemphues, K. J. (1995). *par-1,* a gene required for establishing polarity in *C. elegans* embryos, encodes a putative Ser/Thr kinase that is asymmetrically distributed. *Cell* **81,** 611–620.

Guo, S., and Kemphues, K. J. (1996). A non-muscle myosin required for embryonic polarity in *Caenorhabditis elegans. Nature* **382,** 455–458.

Hill, D. P., and Strome, S. (1988). An analysis of the role of microfilaments in the establishment and maintenance of asymmetry in *Caenorhabditis elegans* zygotes. *Dev. Biol.* **125,** 75–84.

Hill, D. P., and Strome, S. (1990). Brief cytochalasin-induced disruption of microfilaments during a critical interval in 1-cell *C. elegans* embryos alters the partitioning of developmental instructions to the 2-cell embryo. *Development* **108,** 159–172.

Hinchcliffe, E. H., Li, C., Thompson, E. A., Maller, J. L., and Sluder, G. (1999). Requirement of Cdk2-cyclin E activity for repeated centrosome reproduction in *Xenopus* egg extracts [see comments]. *Science* **283,** 851–854.

Hird, S. N., and White, J. G. (1993). Cortical and cytoplasmic flow polarity in early embryonic cells of *Caenorhabditis elegans. J. Cell Biol.* **121,** 1343–1355.

Hird, S. N., Paulsen, J. E., and Strome, S. (1996). Segregation of germ granules in living *Caenorhabditis elegans* embryos: cell-type-specific mechanisms for cytoplasmic localization. *Development* **122,** 1303–1312.

Hung, T. J., and Kemphues, K. J. (1999). PAR-6 is a conserved PDZ domain-containing protein that colocalizes with PAR-3 in *Caenorhabditis elegans* embryos. *Development* **126,** 127–135.

Hyman, A. A. (1989). Centrosome movement in the early divisions of *Caenorhabditis elegans:* a cortical site determining centrosome position. *J. Cell Biol.* **109,** 1185–1193.

Hyman, A. A., and White, J. G. (1987). Determination of cell division axes in the early embryogenesis of *Caenorhabditis elegans. J. Cell Biol.* **105,** 2123–2135.

Kemphues, K. J., and Strome, S. (1997). Fertilization and establishment of polarity in the embryo. In *C. ELEGANS II* (D. L. Riddle, T. Blumenthal, B. J. Meyer, and J. R. Priess, Eds.), pp. 335–359. Cold Spring Harbor Laboratory Press, Plainview, NY.

Kemphues, K. J., Wolf, N., Wood, W. B., and Hirsh, D. (1986). Two loci required for cytoplasmic organization in early embryos of *Caenorhabditis elegans. Dev. Biol.* **113,** 449–460.

Kimura, M., Kotani, S., Hattori, T., Sumi, N., Yoshioka, T., Todokoro, K., and Okano, Y. (1997). Cell cycle-dependent expression and spindle pole localization of a novel human protein kinase, Aik, related to Aurora of *Drosophila* and yeast Ipl1. *J. Biol. Chem.* **272,** 13766–13771.

Kirby, C., Kusch, M., and Kemphues, K. (1990). Mutations in the *par* genes of *Caenorhabditis elegans* affect cytoplasmic reorganization during the first cell cycle. *Dev. Biol.* **142,** 203–215.

Lacey, K. R., Jackson, P. K., and Stearns, T. (1999). Cyclin-dependent kinase control of centrosome duplication. *Proc. Natl. Acad. Sci. USA* **96,** 2817–2822.

Matsumoto, Y., Hayashi, K., and Nishida, E. (1999). Cyclin-dependent kinase 2 (Cdk2) is required for centrosome duplication in mammalian cells. *Curr. Biol.* **9,** 429–432.

Matthews, L. R., Carter, P., Thierry-Mieg, D., and Kemphues, K. (1998). ZYG-9, a *Caenorhabditis elegans* protein required for microtubule organization and function, is a component of meiotic and mitotic spindle poles. *J. Cell Biol.* **141,** 1159–1168.

Mello, C. C., Schubert, C., Draper, B., Zhang, W., Lobel, R., and Priess, J. R. (1996). The PIE-1 protein and germline specification in *C. elegans* embryos [letter]. *Nature* **382,** 710–712.

Moritz, M., Zheng, Y., Alberts, B. M., and Oegema, K. (1998). Recruitment of the gamma-tubulin ring complex to *Drosophila* salt-stripped centrosome scaffolds. *J. Cell Biol.* **142,** 775–786.

Nelson, G. A., Roberts, T. M., and Ward, S. (1982). *Caenorhabditis elegans* spermatozoan locomotion: amoeboid movement with almost no actin. *J. Cell Biol.* **92,** 121–131.

O'Connell, K. F., Leys, C. M., and White, J. G. (1998). A genetic screen for temperature-sensitive cell-division mutants of *Caenorhabditis elegans. Genetics* **149,** 1303–1321.

O'Connell, K. F., Maxwell, K. N., and White, J. G. (2000). The *spd-2* gene is required for polarization of the anteroposterior axis and formation of the sperm asters in the *Caenorhabditis elegans* zygote. *Dev. Biol.* **222,** 55–70.

Oegema, K., Wiese, C., Martin, O. C., Milligan, R. A., Iwamatsu, A., Mitchison, T. J., and Zheng, Y. (1999). Characterization of two related *Drosophila* gamma-tubulin complexes that differ in their ability to nucleate microtubules. *J. Cell Biol.* **144,** 721–733.

Reinsch, S., and Karsenti, E. (1997). Movement of nuclei along microtubules in *Xenopus* egg extracts. *Curr. Biol.* **7,** 211–214.

Rodionov, V. I., and Borisy, G. G. (1997). Self-centering activity of cytoplasm. *Nature* **386,** 170–173.

Sadler, P. L., and Shakes, D. C. (2000). Anucleate *Caenorhabditis elegans* sperm can crawl, fertilize oocytes, and direct anterior-posterior polarization of the one-cell embryo. *Development* **127,** 355–366.

Sardet, C., McDougall, A., and Houliston, E. (1994). Cytoplasmic domains in eggs. *Trends Cell Biol.* **4,** 166–172.

Schatten, G. (1994). The centrosome and its mode of inheritance: the reduction of the centrosome during gametogenesis and its restoration during fertilization. *Dev. Biol.* **165,** 299–335.

Schatten, G., and Schatten, H. (1981). Effects of motility inhibitors during sea urchin fertilization: microfilament inhibitors prevent sperm incorporation and restructuring of fertilized egg cortex, whereas microtubule inhibitors prevent pronuclear migrations. *Exp. Cell Res.* **135,** 311–330.

Schumacher, J. M., Ashcroft, N., Donovan, P. J., and Golden, A. (1998a). A highly conserved centrosomal kinase, AIR-1, is required for accurate cell cycle progression and segregation of development factors in *Caenorhabditis elegans* embryos. *Development* **125,** 4391–4402.

Schumacher, J. M., Golden, A., and Donovan, P. J. (1998b). AIR-2: An Aurora/Ipl1-related protein kinase associated with chromosomes and midbody microtubules is required for polar body extrusion and cytokinesis in *Caenorhabditis elegans* embryos. *J. Cell Biol.* **143,** 1635–1646.

Shelton, C. A., Carter, J. C., Ellis, G. C., and Bowerman, B. (1999). The nonmuscle myosin regulatory light chain gene *mlc-4* is required for cytokinesis, anterior-posterior polarity, and body morphology during *Caenorhabditis elegans* embryogenesis *J. Cell Biol.* **146,** 439–452.

Simerly, C., Wu, G. J., Zoran, S., Ord, T., Rawlins, R., Jones, J., Navara, C., Gerrity, M., Rinehart, J., and Binor, Z. (1995). The paternal inheritance of the centrosome, the cell's microtubule-organizing center, in humans, and the implications for infertility [see comments] [published erratum appears in *Nat. Med.* (1995) Jun;1(6):599]. *Nat. Med.* **1,** 47–52.

Skop, A. R., and White, J. G. (1998). The dynactin complex is required for cleavage plane specification in early *Caenorhabditis elegans* embryos. *Curr. Biol.* **8,** 1110–1116.

Sluder, G., and Rieder, C. L. (1985). Experimental separation of pronuclei in fertilized sea urchin eggs: chromosomes do not organize a spindle in the absence of centrosomes. *J. Cell Biol.* **100,** 897–903.

Sluder, G., Miller, F. J., Lewis, K., Davison, E. D., and Rieder, C. L. (1989). Centrosome inheritance in starfish zygotes: selective loss of the maternal centrosome after fertilization. *Dev. Biol.* **131,** 567–579.

Sluder, G., Miller, F. J., and Lewis, K. (1993). Centrosome inheritance in starfish zygotes. II: Selective suppression of the maternal centrosome during meiosis. *Dev. Biol.* **155,** 58–67.

Stearns, T., and Kirschner, M. (1994). In vitro reconstitution of centrosome assembly and function: the central role of gamma-tubulin [see comments]. *Cell* **76,** 623–637.

Stearns, T., and Winey, M. (1997). The cell center at 100. *Cell* **91,** 303–309.

Strome, S., and Wood, W. (1983). Generation of asymmetry and segregation of germ-line granules in early *C. elegans* embryos. *Cell* **35,** 15–25.

Vasquez, R. J., Gard, D. L., and Cassimeris, L. (1994). XMAP from *Xenopus* eggs promotes rapid plus end assembly of microtubules and rapid microtubule polymer turnover. *J. Cell Biol.* **127,** 985–993.

von Dassow, G., and Schubiger, G. (1994). How an actin network might cause fountain streaming and nuclear migration in the syncytial *Drosophila* embryo [published erratum appears in *J. Cell Biol.* (1995). Sep; 130(5):1231–1234]. *J. Cell Biol.* **127,** 1637–1653.

Wang, P. J., and Huffaker, T. C. (1997). Stu2p: A microtubule-binding protein that is an essential component of the yeast spindle pole body. *J. Cell Biol.* **139,** 1271–1280.

Ward, S. (1986). The asymmetic localization of gene products during the development of *Caenorhabditis elegans.* In *Gametogenesis and the Early Embryo* (J. Gall, Ed.), pp. 55–75. A. R. Liss, New York.

Ward, S., Argon, Y., and Nelson, G. A. (1981). Sperm morphogenesis in wild-type and fertilization-defective mutants of *Caenorhabditis elegans. J. Cell Biol.* **91,** 26–44.

Wolf, N., Hirsh, D., and McIntosh, J. R. (1978). Spermatogenesis in males of the free-living nematode, *Caenorhabditis elegans. J. Ultrastruct. Res.* **63,** 155–169.

Wu, X., and Palazzo, R. E. (1999). Differential regulation of maternal vs. paternal centrosomes. *Proc. Natl. Acad. Sci. USA* **96,** 1397–1402.

Zheng, Y., Wong, M. L., Alberts, B., and Mitchison, T. (1995). Nucleation of microtubule assembly by a gamma-tubulin-containing ring complex. *Nature* **378,** 578–583.

Zimmerman, W., Sparks, C. A., and Doxsey, S. J. (1999). Amorphous no longer: the centrosome comes into focus. *Curr. Opin. Cell Biol.* **11,** 122–128.

18

The Centrosome in *Drosophila* Oocyte Development

Timothy L. Megraw[1] and *Thomas C. Kaufman*[2]
[1] Department of Biology
Indiana University
Bloomington, Indiana 47405

[2] Department of Biology
Howard Hughes Medical Institute
Indiana University
Bloomington, Indiana 47405

I. Introduction

In the fruit fly *Drosophila melanogaster,* microtubule organizing centers (MTOCs) coordinate a complex sequence of cellular changes during oocyte development. MTOCs play major roles at four different stages of oogenesis. These include the polarized divisions of germline stem cells and cystocytes, establishment of a single MTOC in the oocyte, organization of the oocyte's polar axes, and assembly of the meiotic spindles.

Asymmetric cell divisions establish a regularly patterned cyst. Early in oogenesis a cystoblast undergoes four synchronous divisions with arrested cytokinesis to produce a germ cell cyst of cytoplasmically bridged cells (for

Current Topics in Developmental Biology, Vol. 49

0070-2153/00 $35.00

reviews, see King, 1970; Mahowald and Kambysellis, 1980; Spradling, 1993; de Cuevas *et al.,* 1997). During these cystocyte divisions, asymmetrical anchoring of the spindles polarizes the cells into a precise and regular pattern (Storto and King, 1989). The oocyte develops from one of the two most branched, or interconnected, cells within the cyst. The remaining 15 cells develop into nurse cells. Critical to the determination of the oocyte, spindle orientation is controlled by the association of the spindles with the fusome; a cytoplasmic organelle composed of membrane cytoskeletal proteins. The precursor to the fusome is the spectrosome, which directs spindle orientation during the germline stem cell divisions.

An early event in oocyte differentiation is the establishment of a single MTOC in the cyst. At the end of the four cystocyte divisions the centrosomes are inactivated and the centrioles are transported into the pro-oocyte, where a single MTOC for the cyst is established. From this focus in the oocyte, microtubules emanate into all of the 15 nurse cells. This polarized microtubule array is essential for the transport of maternal factors required for oocyte differentiation and development (for reviews, see Theurkauf, 1994; Mahajan-Miklos and Cooley, 1994; Cooley and Theurkauf, 1994; Knowles and Cooley, 1994).

MTOCs play a pivotal role in the maternal specification of the polar axes that define the anterior/posterior and dorsal/ventral coordinates in the oocyte. These axes are set up during oogenesis in the oocyte by signaling cascades involving gurken, the *Drosophila* homolog of transforming growth factor-α (TGF-α). The signals are transduced between the oocyte and the epithelial sheath of follicle cells that surround the developing 16-cell cyst (for a review, see Nilson and Schupbach, 1999). These signals direct the organization of the microtubule cytoskeleton in the oocyte into polarized arrays along the anterior/posterior axis. The MTOC is localized at the posterior of the oocyte until midoogenesis, at which time gurken signals to the posterior follicle cells. The follicle cells then signal back to the oocyte, inducing a shift in the polarity of the microtubule array. Following this shift, the MTOC is localized at the anterior of the oocyte. The polar microtubule arrays enable maternal factors such as *bicoid* and *oskar* mRNAs to be localized to discrete regions of the oocyte. It is the pattern of localization of the maternal factors that specifies the dorsal/ventral and anterior/posterior axes. Disruption of the microtubule array abolishes the localization of determinants within the oocyte with a subsequent loss of axial specification.

Spindle assembly occurs in the absence of centrosomes during female meiosis. Meiosis is arrested in metaphase of meiosis I until ovulation. Ovulation activates the oocyte, and meiosis proceeds to generate four haploid nuclei. The nucleus most distal to the egg cortex becomes the oocyte pronucleus, while the remaining three condense their chromatin and become polar bodies that eventually degenerate. Meiosis I spindles are acentriolar.

Meiosis II spindles are assembled within the meiosis I spindle at anaphase. The meiosis II spindles are linked pole-to-pole with their axes perpendicular to the egg cortex. The shared meiosis II spindle pole is an acentriolar MTOC and contains several proteins that are also found in mitotic centrosomes. Genetic analysis has shown that at least one centrosomal protein, γ-tubulin, is required for meiosis.

In this chapter, we discuss the form and function of centrosomes and MTOCs beginning with the germline stem cells on through the development of the oocyte and end just before syngamy. Events after this point are covered in the following chapter, by Rothwell and Sullivan. For the purposes of this discussion we refer to the centrosome as that organelle made up of a pair of centrioles surrounded by pericentriolar material (PCM) in which the minus ends of microtubules are nucleated and embedded (for reviews on the centrosome, the reader is referred to Kalt and Schliwa, 1993; Kellogg *et al.,* 1994; Balczon, 1996; Gonzalez *et al.,* 1998; Zimmerman *et al.,* 1999; Andersen, 1999). There are several noncentrosomal microtubule organizing centers involved in oogenesis, and these will be referred to as simply MTOCs. The MTOCs in the stem cell and cystocyte divisions are centrosomes, whereas the MTOCs in the oocyte contain a cluster of centrioles early and become acentriolar later. In this chapter, we discuss the form and function of MTOCs in stem cell division, cystocyte cell divisions, oocyte determination, oocyte polarity, and meiosis. We focus primarily on the role of the microtubule cytoskeleton. For coverage of the roles played by the actin cytoskeleton in oocyte development we direct the reader to several reviews (Robinson and Cooley, 1997; Cooley and Theurkauf, 1994; Mahajan-Miklos and Cooley, 1994).

II. The Germarium: Centrosomes Are Anchored to Achieve Polarized Cell Divisions

The development of an oogenic stem cell into an egg chamber occurs in the germarium. The ovary is composed of an average of 16 ovarioles arranged in parallel and joined at their anterior ends by their terminal filaments (King, 1970; Mahowald and Kambysellis, 1980; Spradling, 1993). The ovariole is divided into the germarium and the vitellarium (Fig. 1A; see color insert). A schematic of the stages of oogenesis is shown in Figs. 1A and 1B (see color insert). The initial stages of oogenesis, the development of an oogenic stem cell into an egg chamber, occur in the germarium (Figure 1B; see color insert). The germarium is divided into four regions distinguished from one another by different oogenic stages. Region 1, at the distal end of the germarium, consists of germline stem cells, cystoblasts, and dividing cystocytes. Each germline stem cell divides to produce one stem cell and one cystoblast. The cystoblast divides four times with incomplete cytokinesis to produce a cyst of 16 cystocytes that are interconnected by cytoplasmic bridges. Mature 16-cell cysts are found in regions 2a and 2b. Somatic follicle cells migrate to positions between the 16-cell cysts in region 2a. In region 2a one of the cystocytes becomes determined to be the oocyte while the remaining 15 cystocytes become the nurse cells. The oocyte remains diploid and begins the meiotic program, whereas the nurse cell chromosomes endoreplicate and become highly polyploid. Flattened longitudinally, the cysts in region 2b span the width of the ovariole. The specified oocyte resides in the middle of each region 2b cyst. In region 3 the oocyte attains a posterior position in the cyst and remains there for the rest of oogenesis. The cyst in region 3 is referred to as a stage 1 egg chamber, containing 15 nurse cells and one oocyte surrounded by follicle cells. Egg chambers leave the germarium and mature as they progress posteriorly toward the oviduct. For an extensive review of oogenesis, see Spradling (1993).

The germ cells in region 1 divide asymmetrically. Polarized mitoses are achieved by anchoring spindles to the spectrosome in the stem cells and to the fusome in the cystocytes. The spectrosome is formed very early in the development of the germline. It first appears in the germ cells of gastrulating embryos (Lin and Spradling, 1997). The spectrosome is the precursor to the fusome (Lin *et al.*, 1994; Lin and Spradling, 1997; de Cuevas *et al.*, 1997). The fusome is a germline-specific cytoskeletal organelle that is associated with the synchronous, incomplete cell divisions that give rise to the clusters of interconnected germline cells, or cysts, in many insects (Telfer, 1975; Storto and King, 1989; Lin *et al.*, 1994; de Cuevas *et al.*, 1997; McKearin, 1997). The spectrosome and the fusome contain membrane cytoskeletal proteins such as α-spectrin, β-spectrin, the adducinlike product of the *hu-li tai shao* (*hts*) gene, and ankyrin (de Cuevas *et al.*, 1997; McKearin, 1997). Other components of the fusome include the minus-end-directed motor protein cytoplasmic dynein, the product of the *bag-of-marbles* (*bam*) gene, and cyclin A (McGrail and Hays, 1997; McKearin and Ohlstein, 1995; de Cuevas *et al.*, 1997).

A. The Spectrosome

The spectrosome orients the spindle during stem cell division. In region 1 of the germarium, an oogenic stem cell undergoes a polarized cell division that generates a new stem cell and a cystoblast. When a germline stem cell divides, the mitotic spindle is oriented with its axis parallel to the germarial axis. At metaphase, the spectrosome is situated next to the distal centrosome, the one located nearest the terminal filament cells. Here, the spectrosome sits opposite the site of cytokinesis (Deng and Lin, 1997). As the cell divides, the spectrosome extends asymmetrically across the dividing cells. Approximately one-third of the spectrosome material is inherited by the nascent cystoblast, while the rest remains in the daughter stem cell (Deng and Lin, 1997; de Cuevas *et al.,* 1997; de Cuevas and Spradling, 1998). Subsequent to the birth of the cystoblast, this structure is called the fusome.

In an *hts* null mutant, no spectrosome is formed in the germline stem cells (Lin and Spradling, 1997; Deng and Lin, 1997). Without the spectrosome, mitotic spindles become randomly oriented with respect to the germarial axis (Deng and Lin, 1997). Nevertheless, stem cells continue to divide and cystoblasts are produced. *hts* mutant cystocytes divide in the absence of fusomes, but form irregular cysts.

B. The Fusome and Cyst Development

The cystoblast undergoes four synchronous divisions to produce a regularly patterned 16-cell cyst (Fig. 1C; see color insert). With each division, incomplete cytokinesis results in cells connected to one another by a network of cytoplasmic bridges called ring canals. Ring canals are actin-based cytoskeletal structures. After four divisions, a syncytium, or cyst, of 16 interconnected cells is formed in which 2 cells contain four ring canals, 2 contain three ring canals, 4 contain two ring canals, and 8 contain a single ring canal junction (Figs. 1C and 1D; see color insert). The fusome grows with each division, remains as a plug in each ring canal, and extends through all the ring canals into every cyst cell (Fig. 1D; see color insert) (de Cuevas *et al.,* 1997; McKearin, 1997; de Cuevas and Spradling, 1998). The fusome breaks down following the completion of cyst formation. Although some fusome components are present in the ring canals, ring canals form properly in the absence of fusomes in *hts* mutants. Based on the observation that the fusome is segregated unequally at the first cystoblast division, it has been suggested that the oocyte may be determined and inherit oocyte determining factors as early as the first cystoblast division (Lin and Spradling, 1995; de Cuevas and Spradling, 1998). Indeed, the oocyte is always chosen from one of the first two cystocytes.

During each synchronous cystocyte division, the mitotic spindles are oriented with one pole toward the fusome (Storto and King, 1989; Lin and Spradling, 1995). The control of spindle orientation is an important means to achieve asymmetric cell divisions during development (see Strome, 1993, for a review). In *Caenorhabditis elegans,* as in *Drosophila,* spindle orientation plays an important role in asymmetric cell division and germline cell fate (White and Strome, 1996). In *C. elegans* the *pie-1* gene is required for germline specification and encodes a protein that is localized to the centrosome that will be inherited by the germline (Mello *et al.,* 1996), and this localization may require an aurora-type kinase (Schumacher *et al.*, 1998). Whether such factors are involved in oocyte differentiation in *Drosophila* is as yet unknown.

The fusome anchors and orients the spindles by attachment to the centrosome during cystoblast divisions. Although the fusome is not essential for cystocyte cell division (Lin *et al.*, 1994; de Cuevas *et al.*, 1996), mutations in genes that encode fusome components such as *hts, α-spectrin, bam,* and dynein heavy chain all affect cyst development and result in a failure of oocyte differentiation (Yue and Spradling, 1992; Lin *et al.*, 1994; de Cuevas *et al.*, 1996; Deng and Lin, 1997; McKearin and Ohlstein, 1995; McGrail and Hays, 1997). In *hts* mutants, where the fusome is absent, the spindles orient randomly, mitoses proceed asynchronously, the number of ring canals and cyst cells formed is irregular, and the oocyte fails to be specified (Deng and Lin, 1997). It appears that the primary role of the fusome may be to anchor spindles so that polarized divisions occur during cyst formation.

Mutations in the cytoplasmic dynein heavy chain gene (*Dhc64C*) demonstrate the importance of spindle attachment and asymmetric divisions in oocyte differentiation (McGrail and Hays, 1997; Theurkauf, 1997). Cytoplasmic dynein accumulates at the fusome at mitosis. McGrail and Hays examined germline mutant clones of *Dhc64C* and female sterile allelic combinations of the gene. Germline clones of strong *Dhc64C* mutants produced ovaries with almost no mature eggs, but instead contained egg chambers that held 2, 4, 8, or 16 nurse cells but no oocyte (McGrail and Hays, 1997). These data indicate that dynein is required for the cystocyte cell divisions. However, the female sterile dynein mutants regularly produced egg chambers with 16 nurse cells and no oocyte. This failure to block cell division is likely due to the leaky (hypomorphic) character of the sterile alleles. Nevertheless, in these sterile genotypes the usual abutment of one spindle pole against the fusome did not occur, resulting in apparently random spindle orientation in the mutant cysts. This observation indicates that cytoplasmic dynein is required for centrosome attachment to the fusome. Furthermore, these results suggest that control of spindle orientation, and hence polarized cell division, is required for oocyte specification. A role for dynein in spindle orientation has also been shown in the yeast *Saccharomyces cerevisiae* (Li *et al.*, 1993). One hypothesis proposes that the oocyte is chosen from among the two cells that contain the highest number of ring canals. Consistent with this idea, *encore* mutant ovaries undergo an extra round of mitosis to produce a 32-cell cyst with an oocyte that contains five ring canals (Hawkins *et al.*, 1996). The most posterior cell in *Dhc64C* mutant egg chambers did not contain four ring canals; however, it was not determined whether any cells contained four ring canals (McGrail and Hays, 1997).

C. Centrosomes

The role of centrosomes in stem cell and cystocyte divisions is not known. One of the best characterized components of centrosomes is γ-tubulin. Two γ-tubulin genes exist in *Drosophila: γTub23C* and *γTub37C* (Zheng *et al.,* 1991; Sunkel *et al.,* 1995; Tavosanis *et al.,* 1997; Wilson and Borisy, 1998). *γTub23C* is expressed in the germarium where its protein accumulates at the stem cell and cystocyte centrosomes (Wilson *et al.,* 1997). Mutations in *γTub23C* are lethal and animals die in the late larval stages (Sunkel *et al.,* 1995). Germline clone analysis of *γTub23C* will be required to address its role in early oogenesis. *γTub37C,* on the other hand, is expressed maternally after the cystocyte divisions, and mutations in that gene have consequences later in oogenesis (see Section IV,C).

III. The Oocyte MTOC

The development of the egg chamber has been divided into 14 stages (King, 1970) (Fig. 1A). (For a detailed description of the stages of oogenesis we refer the reader to King, 1970, and Mahowald and Kambysellis, 1980). During oogenesis the nurse cells synthesize most of the products required for oocyte development, which are transported vectorially into the oocyte through the ring canals. The oocyte and nurse cells grow until stage 10, when the oocyte occupies nearly half of the egg chamber (Fig. 1A). The oocyte grows more rapidly from stages 11 to 14, when the contents of the nurse cells are transferred or "dumped" into the oocyte and the nurse cells undergo programmed cell death.

The oocyte MTOC plays a central role in both the specification of the oocyte and the establishment of its polar coordinates. In this section we discuss the dynamics of the microtubule array from the early oocyte to the end of stage 10, when the polar coordinates are established. In Section A, we discuss the role of the MTOC as a key player in the development of the oocyte. In Section B, we describe the morphological characteristics of the MTOC as it changes through oocyte development. In Section C, we discuss the roles of two genes, *egalitarian* and *Bicaudal-D,* in MTOC function. We review the current knowledge regarding the molecular nature of the oocyte MTOC in Section D.

A. Microtubules Play a Central Role in Oocyte Development

Perhaps the first event in the determination of the oocyte is the formation of an MTOC (Theurkauf *et al.,* 1992, 1993). Another early event in oocyte development is the flow of the centrioles from the 15 sister cystocytes into the chosen oocyte (Mahowald and Strassheim, 1970). Here, the centrioles are clustered at the posterior of the oocyte. Three events—the unraveling of synaptonemal complexes in the cells destined to become nurse cells, the transport of centrioles and other macromolecules into the oocyte, and the formation of the single MTOC in the oocyte—are apparently concurrent in region 2a of the germarium (Theurkauf, 1994; Mahowald and Kambysellis, 1980).

The dynamics of the microtubule array and the temporal–spatial changes in the MTOC have been observed in the developing oocyte by staining microtubules with antibodies to α-tubulin (Theurkauf *et al.,* 1992). From stage 1 to stage 6, the microtubule array extends into the nurse cells from the MTOC, situated at the posterior of the oocyte (Theurkauf *et al.,* 1992, 1993) (see Fig. 2 in the color insert). At stage 6, gurken protein, the nucleus, and the MTOC are localized together at the posterior of the oocyte. Here, gurken signals to the posterior follicle cells through the EGF receptor (Schupbach, 1987; Neuman-Silberberg and Schupbach, 1993; Nilson and Schupbach, 1999). At stage 7, when the posterior follicle cells signal back to the oocyte, a shift in the polarity of the microtubule array in the oocyte occurs. At this time the oocyte nucleus and associated gurken migrate to the anterior of the oocyte. Nuclear migration, or positioning, is a microtubule-dependent process that in many systems is dependent upon MTOCs associated with the nucleus (Reinsch and Gonczy, 1998). After migration of the nucleus to a position that will mark the anterior-dorsal position of the developing oocyte, the MTOC becomes localized to the anterior cortex of the oocyte where it anchors the minus ends of the microtubule array (Theurkauf *et al.,* 1992). In *gurken* mutants the posterior follicle cells fail to signal properly to the oocyte and "bipolar" oocytes with MTOCs at both the posterior and anterior result (Gonzalez-Reyes *et al.,* 1995; Roth *et al.,* 1995). The reciprocal localization of β-galactosidase fusions of the plus-end-directed kinesin and the minus-end-directed NOD motor proteins in the oocyte is consistent with the assigned polarity of the microtubule array (Clark *et al.,* 1994, 1997). The cluster of centrioles present at the posterior of the oocyte degrade after stage 7 when the shift in microtubule polarity occurs (Mahowald and Strassheim, 1970; Theurkauf *et al.,* 1992). Thus, the MTOC at the anterior end of the oocyte is devoid of centrioles.

Disruption of the microtubule array with colchicine or other microtubule-destabilizing drugs inhibits oocyte development at different stages. Application of the drugs prevents the formation of the microtubule array, blocks the transport of mRNAs, and inhibits differentiation of the early oocyte (Theurkauf *et al.,* 1993; Koch and Spitzer, 1983). This results in an egg chamber with 16 nurse cells and no oocyte. In later stage oocytes, the drugs prevent nuclear migration and the localization of mRNAs and proteins important for patterning the oocyte (Theurkauf *et al.,* 1993; Pokrywka and Stephenson, 1991, 1995; Clark *et al.,* 1994; Mach and Lehmann, 1997). These data demonstrate the importance of microtubules and the MTOC in oocyte differentiation and patterning.

B. Morphology and Dynamics of the Oocyte MTOC

The location and morphology of the oocyte MTOC is revealed by staining ovaries for the centrosomal protein centrosomin. Here, we report the pattern of accumulation of the first known molecular component of the oocyte MTOC. Figure 2 shows the pattern of centrosomin accumulation at different stages of oogenesis. Centrosomin is a component of centrosomes and other MTOCs in many cell types (Heuer *et al.,* 1995; Li and Kaufman, 1996; Li *et al.,* 1998; Fitch and Wakimoto, 1998; Bonaccorsi *et al.,* 1998; Moritz *et al.,* 1998; Llamazares *et al.,* 1999; Megraw *et al.,* 1999). Centrosomin localizes to the centrosomes in a cell-cycle-dependent manner early in oogenesis. In the germarium centrosomin localizes to centrosomes in the stem cells and cystocytes at mitosis. At interphase centrosomin is dispersed into the cytoplasm (see Figs. 2A and 2B in the color insert). Centrosomin is abundant in regions 1 and 2a of the germarium, whereas the protein levels appear to drop off in region 2b (Figs. 2A and 2B; see color insert). Figure 2B shows an eight-cell-stage cyst in mitosis stained for centrosomin and the fusome marker HTS. Double staining for centrosomin and HTS reveals that centrosomes are in close proximity to the fusome. One molecule that occupies the intervening space is dynein, which binds to the fusome at mitosis (McGrail and Hays, 1997; Theurkauf, 1997). In later stages centrosomin is cytoplasmic in nurse cells and localizes to the periphery of the nurse cell nuclei. No obvious accumulation of centrosomin in the early oocyte following the cystocyte divisions is seen, but the protein clearly localizes at the posterior MTOC in later stage oocytes. Centrosomin accumulates at the posterior MTOC in stage 6 oocytes, and appears as one or more discrete bodies that reside between the nucleus and the posterior follicle cells (Fig. 2C; see color insert). At this stage, when gurken is signaling to the posterior follicle cells, centrosomin also accumulates in the follicle cells at the apical side, facing the oocyte. At stage 7, when the polarity of the oocyte shifts, centrosomin is localized to several discrete bodies on the periphery of the migrating nucleus (Fig. 2D; see color insert). After the nucleus has migrated to its new position, which will become the anterior dorsal corner of the oocyte, centrosomin becomes diffuse and punctate as it begins to accumulate at the MTOC on the anterior cortex (Fig. 2E; see color insert). By stage 10 of oogenesis centrosomin is localized at the anterior cortex of the oocyte (Fig. 2F; see color insert). Centrosomin localization at the anterior MTOC occurs independently of microtubules. Females fed 50 μg/ml colchicine for 24 hr retain localization of centrosomin at the MTOC (our unpublished data). This distinguishes centrosomin as a core component of the MTOC and not a microtubule-associated protein (MAP).

Centrosomin staining reveals some novel features of the oocyte MTOC. Overall, the localization of centrosomin at the posterior and its relocation to the anterior are consistent with the reported temporal and spatial flux of the MTOC in the oocyte determined by staining for α-tubulin (Theurkauf *et al.,* 1992, 1993). Centrosomin staining reveals the MTOC as one or more discrete bodies at the posterior of the stage 6 oocyte. Centrosomin was not detected at the MTOC before stage 5. The MTOC bodies remain on the nuclear periphery until nuclear migration is completed. Following nuclear migration, the discrete MTOC bodies disassemble, and the MTOC is reassembled at the anterior as a broad, cortex-associated structure.

Although centrosomin is localized to the centrosomes in dividing germline cells and at the oocyte MTOC, it is not required during oogenesis (Megraw *et al.,* 1999). A null mutation in *centrosomin* has maternal effects on embryonic centrosome assembly and function (Megraw *et al.,* 1999) but does not affect oogenic cell divisions or the patterning of the chorion. To date, centrosomin is the only centrosomal protein reported to localize to the oocyte MTOC, and none are known to function there.

C. Roles of *Bic-D* and *egl*

The importance of the MTOC in oocyte specification and development is supported by studies of the *egalitarian* (*egl*) and *Bicaudal-D* (*Bic-D*) genes. The BIC-D and EGL proteins accumulate in the early oocyte (Suter and Steward, 1991; Mach and Lehmann, 1997). Mutations in these genes disrupt the microtubule array and prevent oocyte differentiation (Theurkauf *et al.,* 1993). *Bic-D* and *egl* mutants are female sterile, producing egg chambers with 16 nurse cells and no oocyte (Suter *et al.,* 1989; Wharton and Struhl, 1989; Schupbach and Wieschaus, 1991; Ran *et al.,* 1994; Carpenter, 1994). In a *Bic-D* null mutant, mRNAs normally transported into the early oocyte, such as the *orb* and *osk* mRNAs, fail to arrive (Ran *et al.,* 1994). In addition, *Bic-D* null mutants do not position a cell with four ring canals at the posterior of the cyst (Ran *et al.,* 1994). In *Bic-D* loss-of-function mutants the microtubule array does not form in the early oocyte. In *egl* mutants, however, the microtubule array forms transiently and then breaks down (Theurkauf *et al.,* 1993). From these studies, it has been proposed that BIC-D may be required for the establishment of the MTOC and EGL for its maintenance in the early oocyte.

BIC-D and EGL colocalize during all stages of oogenesis and have been shown to immunoprecipitate together as a complex (Mach and Lehmann, 1997). However, BIC-D and EGL did not interact directly with each other in a yeast two-hybrid assay (our unpublished results), indicating that other components may be present in this complex. Although BIC-D and EGL are genetically interdependent upon one another for localization (Suter and Steward, 1991; Mach and Lehmann, 1997), at least one mutation in BIC-D permits the localization of both the EGL and the mutant BIC-D protein to the early oocyte without permitting oocyte determination (Mach and Lehmann, 1997).

BIC-D and EGL are clearly involved in the organization and maintenance of the oocyte microtubule array, but are they components of the MTOC? Under certain conditions, *Bic-D* mutants can produce defects in oocyte polarity without apparently disrupting the microtubule organization (Swan and Suter, 1996). A model was proposed in which BIC-D has a role in the transport or the anchoring of localized mRNAs and proteins in early and late oogenesis (Swan and Suter, 1996). The localization of centrosomin (Fig. 2) is different from the localization of EGL and BIC-D. Whereas EGL and BIC-D localize broadly across the posterior cortex (Mach and Lehmann, 1997; Suter and Steward, 1991), centrosomin accumulates in discrete bodies at the posterior of stage 6 oocytes. Furthermore, centrosomin was not detected in anti-EGL or anti-BIC-D immunoprecipitates, nor did it interact with EGL or BIC-D in two-hybrid assays (our unpublished results). Disrupting the microtubule cytoskeleton in the oocyte by feeding female flies colchicine blocks the localization of EGL to the posterior of the oocyte, indicating that EGL is not a core component of the MTOC (Mach and Lehmann, 1997).

D. What about Centrosomal Proteins?

No functional components of the oocyte MTOC are known. Despite the importance of microtubules and microtubule polarity in the development of the oocyte, no genetic studies at present address the function of the MTOC or the identity of its components. Centrosomin localizes to centrosomes in early oogenesis and to the MTOC in developing oocytes. However, as noted earlier, although centrosomin is required for the cell divisions in the embryo, it is not required at any stage of oogenesis (Megraw *et al.*, 1999). Other centrosomal proteins such as γ-tubulin, CP60, CP190, NUF, and ASP have not been reported to localize to the oocyte MTOC (Tavosanis *et al.*, 1997; Wilson *et al.*, 1997; Kellogg *et al.*, 1995; Whitfield *et al.*, 1995; Rothwell *et al.*, 1998; Saunders *et al.*, 1997). γ-Tubulin in *Drosophila* and other organisms is essential for MTOC function (Oakley and Oakley, 1989;

Horio *et al.,* 1991; Sobel and Snyder, 1995; Spang *et al.,* 1996; Marschall *et al.,* 1996; Sunkel *et al.,* 1995; Tavosanis *et al.,* 1997; Wilson and Borisy, 1998; Llamazares *et al.,* 1999).

As pointed out earlier, there are two γ-tubulin genes, γ-*Tub23C* and γ-*Tub37C,* in *Drosophila* (Zheng *et al.,* 1991; Sunkel *et al.,* 1995; Tavosanis *et al.,* 1997; Wilson and Borisy, 1998; Llamazares *et al.,* 1999). γ-TUB23C is expressed in the stem cells and cystocytes, where it accumulates at centrosomes (Tavosanis *et al.,* 1997; Wilson *et al.,* 1997). γ-TUB37C accumulates in the oocyte and is supplied maternally to the embryo where it is required for centrosome function (Tavosanis *et al.,* 1997; Wilson and Borisy, 1998; Llamazares *et al.,* 1999). γ-TUB37C, however, does not localize to the oocyte MTOC (Tavosanis *et al.,* 1997; Wilson *et al.,* 1997) and mutations in γ-*Tub37C* have no effect on oocyte development (Tavosanis *et al.,* 1997; Wilson and Borisy, 1998; Llamazares *et al.,* 1999). Lacking centrioles and γ-tubulin, the oocyte MTOC appears highly unorthodox.

Molecular motors play multiple roles in microtubule stability, dynamics, and spindle assembly (Walczak *et al.,* 1998; Walczak and Mitchison, 1996). Although the role of motors in oocyte microtubule organization and oocyte polarity is largely unexplored, the role of dynein in oocyte patterning has been approached. In female sterile allelic combinations of *Dhc64C* that produce egg chambers (hypomorphic alleles that bypass the need for dynein in the germarium; see Section II,B) mutant dynein does not localize to the posterior of the oocyte as it normally does. Also, the localization of Kinesin:ßgal to the posterior of the oocyte appeared reduced (McGrail and Hays, 1997). However, the morphogens staufen and *oskar* and *gurken* mRNAs are localized properly, and the nucleus migrates correctly to its anterior/dorsal position (McGrail and Hays, 1997).

IV. Meiosis

Meiosis is accomplished in the *Drosophila* female in the absence of centrosomes. The presumptive oocyte enters meiosis in the germarium before it is determined to be the oocyte (Spradling, 1993). All four of the cystocytes with 3 or 4 ring canals initiate meiosis and form synaptonemal complexes (SC) prior to the determination of one of the cystocytes with 4 ring canals to become an oocyte (Carpenter, 1975, 1994; Rasmussen, 1974). Once the oocyte has been determined in region 2a of the germarium, the other three cystocytes exit the meiotic program, their SCs are disassembled, and all 15 of the remaining cystocytes become polyploid nurse cells. By the time cysts reach region 2b, only the oocyte has SCs. Meiosis remains arrested in prophase of meiosis I until stage 13 of oogenesis.

At stage 13 the meiotic I spindle is assembled. At its poles, this spindle contains no centrioles, nor any proteins known to be components of mitotic centrosomes. Meiosis is arrested in stage 14 oocytes at metaphase I until a process called activation occurs. Oocyte activation can be achieved *in vivo* by ovulation, or *in vitro* by hydration (Doane, 1960; Mahowald *et al.*, 1983; Page and Orr-Weaver, 1997). Once activated, the oocyte proceeds through meiosis I and meiosis II. The meiotic end products are a haploid female pronucleus and three polar bodies. None of these are associated with centrosomes or known centrosomal proteins, but the polar bodies are associated with a nest of microtubules surrounding their chromosomes. The polar bodies remain at the cortex, while the pronucleus migrates toward the center of the oocyte where it will fuse with the sperm pronucleus at fertilization.

In this section we discuss the assembly and dynamics of the meiosis I spindle. We then describe the assembly of the meiosis II spindles, and the characteristics of their unique astral MTOC. Finally, we discuss the role of γ-tubulin in meiosis.

A. The Meiosis I Spindle Is Acentriolar

The metaphase meiosis I spindle lies near the egg cortex and is aligned parallel to it (Theurkauf and Hawley, 1992). Upon activation, the meiosis I spindle undergoes a series of rotating and elongating movements before it pivots 90° to an alignment perpendicular to the cell cortex (Endow and Komma, 1997; Riparbelli and Callaini, 1996).

The details of these meiotic spindle dynamics have been revealed *in vivo* using NCD-GFP (nonclaret disjunctional–green fluorescent protein) as a marker (Endow and Komma, 1997, 1998). In nonactivated oocytes, the spindle oscillates but does not pivot at the cell cortex (Endow and Komma, 1997). Figure 3 (see color insert) illustrates the steps in meiotic spindle progress from metaphase of meiosis I to metaphase of meiosis II.

The spindle that forms in meiosis I is acentriolar. Proteins that are components of mitotic centrosomes, such as γ-tubulin, centrosomin, CP60, and CP190, are not localized to the meiosis I spindle poles (Matthies *et al.*, 1996; Tavosanis *et al.*, 1997; Theurkauf and Hawley, 1992; McKim and Hawley, 1995; Riparbelli and Callaini, 1996). Meiotic spindles form during oogenesis without centrosomes in many, but not all, animal species, including *Drosophila*, mouse, and *Xenopus* (Sonnenblick, 1950; Szollosi *et al.*, 1972; Gerhart, 1980; Krioutchkova and Onishchenko, 1999).

In acentriolar spindles, it is thought that focusing of microtubules at the spindle poles is accomplished through the activity of molecular motors. Whereas the primary role of centrosomes is to organize spindle microtubules into a bipolar array, motor proteins play multiple roles in spindle assembly (see Walczak *et al.,* 1998; Walczak and Mitchison, 1996). In *Drosophila,* the minus-end-directed KinC family kinesin motor protein NCD functions in the assembly of the bipolar spindle at meiosis I (Kimble and Church, 1983; Hatsumi and Endow, 1992; Matthies *et al.,* 1996; Endow and Komma, 1997). However, in *Xenopus* the minus-end-directed motor protein cytoplasmic dynein has been shown *in vitro* to be essential for the focusing of spindle poles in the presence or absence of centrosomes (Heald *et al.,* 1996, 1997; Walczak *et al.,* 1998). In *Saccharomyces cerevisiae,* dynein is required for proper spindle orientation (Li *et al.,* 1993). This is analogous to dynein's role in *Drosophila* cystocyte divisions (McGrail and Hays, 1997) described earlier (Section II,B). The role of cytoplasmic dynein in meiotic spindle assembly in *Drosophila* has not been directly addressed. Perhaps the central role of motor proteins in acentrosomal spindle assembly is an adaptive result of the elimination of centrioles from the egg prior to fertilization. Centriole degradation in the oocyte ensures that the first mitotic spindle in the embryo uses a centriole contributed through a sexual union from the sperm. In animals where maternal centrosomes are present, other mechanisms such as centrosome inactivation must be employed (Wu and Palazzo, 1999).

Wilson and Borisy have reported that asters were observed at the cortex in wild-type stage 14 oocytes activated *in vitro,* but not in laid eggs (Wilson and Borisy, 1998). They observed multiple asters of microtubules that were nonuniform in size and distribution at the surface of oocytes fixed within 10 min or less after activation. In addition, the poles of meiosis I spindles were associated with these cortical asters. Thus, meiosis I spindles may be astral, at least transiently. Although γTUB37C was not detected at the cortical asters, no asters were seen in a γ*Tub37C* null mutant. Perhaps the cortical asters represent novel, transient MTOCs in the early activated egg. The process of *in vivo* activation may be too rapid to allow these asters to be seen in laid oocytes, while *in vitro* activation might retard meiosis sufficiently for observation of these novel MTOCs.

B. The Meiosis II Astral/Anastral Spindles

Two meiosis II spindles are assembled from the meiosis I spindle. As the meiosis I spindle elongates at anaphase, a "pucker" appears in the middle of the spindle that develops into a spindle pole (Riparbelli and Callaini, 1996, 1998; Endow and Komma, 1997, 1998). γ-Tubulin is found at this MTOC early in its development (Endow and Komma, 1997, 1998; Riparbelli and Callaini, 1998). This central MTOC becomes the shared astral pole of the tandemly arranged meiosis II spindles (see Fig. 3 in the color insert).

Whereas the distal poles of the meiosis II spindles are anastral, the central pole forms a ring structure surrounded by astral microtubules (Puro, 1991). When revealed by α-tubulin or γ-tubulin staining, or when observed live using Ncd-GFP, the morphology of the shared meiosis II MTOC changes through the cell cycle (Riparbelli and Callaini, 1998, Endow and Komma, 1998). The distal poles of the meiosis II spindles are acentriolar and lack detectable levels of any known centrosomal proteins. In contrast, the central pole contains CP190, centrosomin, and γ-tubulin (Riparbelli and Callaini, 1996, 1998; Fitch and Wakimoto, 1998; Llamazares *et al.*, 1999). The association of centrosomal proteins with the central pole is fundamentally different from their association with centrosomes. Although CP190 and γ-tubulin do not depend on microtubules for their association with the mitotic centrosome, they disappear from the central pole in the meiosis II spindle upon treatment with the microtubule-destabilizing drug colchicine (Riparbelli and Callaini, 1998). Thus, the meiosis II central pole is an acentriolar MTOC utilizing several proteins found in mitotic centrosomes.

C. γ-Tubulin Function in Meiosis

γ-Tubulin, a conserved tubulin, is a core component of centrosomes and other MTOCs (Oakley and Oakley, 1989; Burns, 1995) (for reviews, see Zimmerman *et al.*, 1999; Balczon, 1996). In the centrosome, γ-tubulin is assembled into ring complexes in which the minus ends of microtubules are embedded and from which microtubule growth is nucleated (Zheng *et al.*, 1995; Moritz *et al.*, 1995). As noted earlier, in *Drosophila* there are two isoforms of γ-tubulin encoded by the γ*Tub23C* and the γ*Tub37C* genes (Zheng *et al.*, 1991; Sunkel *et al.*, 1995; Tavosanis *et al.*, 1997; Wilson and Borisy, 1998; Llamazares *et al.*, 1999). γTUB23C is expressed in most tissues (Sunkel *et al.*, 1995; Wilson *et al.*, 1997). In region 1 of the germarium, where it is localized to the centrosomes, γTUB23C is expressed in stem cells and cystocytes. γTUB23C is not detected *in situ* in later stages of oogenesis in either nurse cells or the oocyte, or in centrosomes of early embryos, but is detected on Western blots of early embryos (Tavosanis *et al.*, 1997; Wilson *et al.*, 1997).

γ*Tub37C* encodes the maternal form of γ-tubulin. γ*Tub37C* is expressed in the nurse cells and oocyte during oogenesis and is supplied maternally to the egg (Wilson *et al.*, 1997; Tavosanis *et al.*, 1997; Llamazares *et al.*, 1999; Wilson and Borisy, 1998). Although it is expressed in oocytes, γTUB37C was not detected at the oocyte MTOC at any stage (Tavosanis *et al.*, 1997; Wilson *et al.*, 1997). Mutations in γ*Tub37C* affect the configuration of the meiosis I spindle, but the aberrant spindle can resolve and advance to meiosis II (Tavosanis *et al.*, 1997; Wilson and Borisy, 1998;

Llamazares *et al.,* 1999). Two groups recently analyzed the effects of *γTub37C* null and hypomorphic alleles in meiosis and early mitosis (Wilson and Borisy, 1998; Llamazares *et al.,* 1999). The examined null mutants were *fs(2)TW1*[1] (Llamazares *et al.,* 1999) and *fs(2)TW1*[3] (Wilson and Borisy, 1998). Both alleles appear to have the same lesion: a stop codon that truncates the carboxyl end of γTUB37C by 106 amino acids, producing an unstable protein. In these mutants, defective meiosis II spindles and sets of irregular bipolar microtubule arrays were observed. *γTub37C* null mutants arrest meiosis, preventing the formation of polar bodies in laid oocytes. Curiously, *in vitro* activated oocytes from null mutants seem to complete meiosis normally and produce polar bodies (Wilson and Borisy, 1998). In the hypomorphic mutants, meiosis proceeded normally, but the first mitotic division was arrested. Aberrant bipolar spindles with abnormally distributed centrosomes form in embryos of *γTub37C* hypomorphs (Wilson and Borisy, 1998; Llamazares *et al.,* 1999). In addition, DNA replication advances in the absence of nuclear proliferation in the hypomorphs (Llamazares *et al.,* 1999). The cortical asters found in *in vitro* activated oocytes appeared in hypomorphic, but not null, *γTub37C* mutants (Wilson and Borisy, 1998).

V. Summary

The *Drosophila* oocyte is a highly specialized cell type whose development utilizes MTOCs in various contexts. Figure 4 (see color insert) summarizes the characteristics of the MTOCs at different stages of oogenesis.

Polarized mitoses are required to achieve oocyte determination. In the asymmetric germ-cell divisions that culminate in the egg chamber, the mitotic centrosomes are anchored to the spectrosome or fusome in order to produce the regular branching pattern of the cyst cells. It appears that the primary role of the fusome is to orchestrate the polarity and synchrony of oogenic mitoses. In the absence of fusomes or anchored spindles, the regular interconnected cyst network is lost and the oocyte does not differentiate. It is not known if the spindle itself is asymmetric, or whether either centrosome has equal potential to interact with the fusome. Several models can explain the need for polarized mitoses for oocyte differentiation. In one, an unequal distribution of unknown oocyte differentiation factors occurs from as early as the first cystoblast division. Here, the fusome may be required for the distribution of the factors. In another model, there is a mechanism that measures the number of ring canals in the cell, limiting the choice of oocyte to two potential pro-oocytes. In this model, polarized, synchronous divisions must occur to produce only two cells with the highest number of ring canals. In both of these models the centrosome plays an indirect role.

A critical event in the determination of the oocyte is the formation of the MTOC. The oocyte MTOC forms shortly after completion of the germ cell mitoses and establishes a microtubule array along which factors required for oocyte determination are transported. It is unclear how this single MTOC forms in the 16-cell cyst, how the centrosomes become inactivated in the adjoining 15 nurse cells, or why the inactivated centrioles are transported into the oocyte. No molecular components of the MTOC are known except for centrosomin, which accumulates at the MTOC relatively late, at approximately stage 5 or 6 of oogenesis.

The MTOC plays a central role in establishing the oocyte's polar coordinates. The oocyte microtubule array is required for the polar localization of axis-determining factors. At midoogenesis the MTOC appears to mediate the reversal of the microtubule array and the migration of the nucleus in the oocyte. The posterior follicle cells signal this reversal after receiving the gurken signal. What changes occur at the MTOC to trigger this cytoskeletal rearrangement? A better understanding of the MTOC's molecular components is necessary before we can begin to unravel the mechanisms underlying these events.

The morphology of the MTOC changes after it shifts to the oocyte anterior. Staining with anti-centrosomin antibodies shows that the MTOC changes from discrete nucleus-associated bodies into a broad structure associated with the anterior cortex. The molecular mechanisms underlying this structural rearrangement of the MTOC at midoogenesis are presently unknown.

Meiosis I occurs in the absence of centrosomes, but meiosis II spindles are linked by a shared, acentriolar, astral MTOC. The organization of the meiosis I spindle poles requires the NCD motor protein; however, the meiosis I spindle poles are acentriolar and contain no known centrosomal core proteins. The meiosis II astral spindle pole has a unique ring-shaped morphology and contains centrosomal proteins, such as γ-tubulin. Strong mutations in the maternal γ*Tub37C* gene do not block meiosis I, but prevent the progression of meiosis II.

Acknowledgments

We thank Ann Scherer for the analysis of Bic-D and Egl interactions with each other and with centrosomin in the yeast two-hybrid system. We are grateful to Drs. Lynn Cooley, Tony Mahowald, Salud Llamazares, and Cayetano Gonzalez for providing figures. We extend special thanks to Sandhya Kilaru for providing her expert skills in the editing of this work, and to Drs. Ling-Rong Kao, Kathy Matthews, and Susan Glueck for their comments on the manuscript. T.

Megraw is supported by a grant from the National Science Foundation T. C. Kaufman is an investigator of the Howard Hughes Medical Institute.

References

Andersen, S. S. (1999). Molecular characteristics of the centrosome. *Int. Rev. Cytol.* **187,** 51–109.

Balczon, R. (1996). The centrosome in animal cells and its functional homologs in plant and yeast cells. *Int. Rev. Cytol.* **169,** 25–82.

Bonaccorsi, S., M. G. Giansanti, and M. Gatti (1998). Spindle self-organization and cytokinesis during male meiosis in asterless mutants of *Drosophila melanogaster. J. Cell Biol.* **142**(3), 751–761.

Burns, R. G. (1995). Analysis of the gamma-tubulin sequences: implications for the functional properties of gamma-tubulin. *J. Cell Sci.* **108**(Pt 6), 2123–2130.

Carpenter, A. T. (1975). Electron microscopy of meiosis in *Drosophila melanogaster* females. I. Structure, arrangement, and temporal change of the synaptonemal complex in wild-type. *Chromosoma* **51**(2), 157–182.

Carpenter, A. T. (1994). Egalitarian and the choice of cell fates in *Drosophila melanogaster* oogenesis. Ciba Found. Symp. **182,** 223–246.

Clark, I., E. Giniger, H. Ruohola-Baker, L. Y. Jan, and Y. N. Jan (1994). Transient posterior localization of a kinesin fusion protein reflects anteroposterior polarity of the *Drosophila* oocyte. *Curr. Biol.* **4**(4), 289–300.

Clark, I. E., L. Y. Jan, and Y. N. Jan (1997). Reciprocal localization of Nod and kinesin fusion proteins indicates microtubule polarity in the *Drosophila* oocyte, epithelium, neuron and muscle. *Development* **124**(2), 461–470.

Cooley, L., and W. E. Theurkauf (1994). Cytoskeletal functions during *Drosophila* oogenesis. *Science* **266**(5185), 590–596.

de Cuevas, M., and A. C. Spradling (1998). Morphogenesis of the *Drosophila* fusome and its implications for oocyte specification. *Development* **125**(15), 2781–2789.

de Cuevas, M., J. K. Lee, and A. C. Spradling (1996). alpha-spectrin is required for germline cell division and differentiation in the *Drosophila* ovary. *Development* **122**(12), 3959–3968.

de Cuevas, M., M. A. Lilly, and A. C. Spradling (1997). Germline cyst formation in *Drosophila. Annu. Rev. Genet.* **31,** 405–428.

Deng, W., and H. Lin (1997). Spectrosomes and fusomes anchor mitotic spindles during asymmetric germ cell divisions and facilitate the formation of a polarized microtubule array for oocyte specification in *Drosophila. Dev. Biol.* **189**(1), 79–94.

Doane, W. (1960). Completion of meiosis in uninseminated eggs of *Drosophila* melanogaster. *Science* **132,** 677–678.

Endow, S. A., and D. J. Komma (1997). Spindle dynamics during meiosis in *Drosophila* oocytes. *J. Cell Biol.* **137**(6), 1321–1336.

Endow, S. A., and D. J. Komma (1998). Assembly and dynamics of an anastral:astral spindle: the meiosis II spindle of *Drosophila* oocytes. *J. Cell Sci.* **111**(Pt 17), 2487–2495.

Fitch, K. R., and B. T. Wakimoto (1998). The paternal effect gene ms(3)sneaky is required for sperm activation and the initiation of embryogenesis in *Drosophila* melanogaster. *Dev. Biol.* **197**(2), 270–282.

Gerhart, J. C. (1980). Mechanisms regulating pattern formation in the amphibian egg and early embryo. *Biological Regulation and Development, vol. 2.* (Robert F. Goldberger, Ed.), Plenum Press, New York.

Gonzalez, C., G. Tavosanis, and C. Mollinari (1998). Centrosomes and microtubule organisation during *Drosophila* development. *J. Cell Sci.* **111**(Pt 18), 2697–2706.

Gonzalez-Reyes, A., H. Elliott, and D. St Johnston (1995). Polarization of both major body axes in *Drosophila* by gurken-torpedo signalling. *Nature* **375**(6533), 654–658.

Hatsumi, M., and S. A. Endow (1992). Mutants of the microtubule motor protein, nonclaret disjunctional, affect spindle structure and chromosome movement in meiosis and mitosis. *J. Cell Sci.* **101**(Pt 3), 547–559.

Hawkins, N. C., J. Thorpe, and T. Schupbach (1996). Encore, a gene required for the regulation of germ line mitosis and oocyte differentiation during *Drosophila* oogenesis. *Development* **122**(1), 281–290.

Heald, R., R. Tournebize, T. Blank, R. Sandaltzopoulos, P. Becker, A. Hyman, and E. Karsenti (1996). Self-organization of microtubules into bipolar spindles around artificial chromosomes in *Xenopus* egg extracts [see comments]. *Nature* **382**(6590), 420–425.

Heald, R., R. Tournebize, A. Habermann, E. Karsenti, and A. Hyman (1997). Spindle assembly in *Xenopus* egg extracts: respective roles of centrosomes and microtubule self-organization. *J. Cell Biol.* **138**(3), 615–628.

Heuer, J. G., K. Li, and T. C. Kaufman (1995). The *Drosophila* homeotic target gene centrosomin (cnn) encodes a novel centrosomal protein with leucine zippers and maps to a genomic region required for midgut morphogenesis. *Development* **121**(11), 3861–3876.

Horio, T., S. Uzawa, M. K. Jung, B. R. Oakley, K. Tanaka, and M. Yanagida (1991). The fission yeast gamma-tubulin is essential for mitosis and is localized at microtubule organizing centers. *J. Cell Sci.* **99**(Pt 4), 693–700.

Kalt, A., and M. Schliwa (1993). Molecular components of the centrosome. *Trends Cell Biol.* **3,** 118–128.

Kellogg, D. R., M. Moritz, and B. M. Alberts (1994). The centrosome and cellular organization. *Annu. Rev. Biochem.* **63,** 639–674.

Kellogg, D. R., K. Oegema, J. Raff, K. Schneider, and B. M. Alberts (1995). CP60: a microtubule-associated protein that is localized to the centrosome in a cell cycle-specific manner. *Mol. Biol. Cell.* **6**(12), 1673–1684.

Kimble, M., and K. Church (1983). Meiosis and early cleavage in *Drosophila melanogaster* eggs: effects of the claret-non-disjunctional mutation. *J. Cell Sci.* **62,** 301–318.

King, R. C. (1970). *Ovarian Development in Drosophila melanogaster.* Academic Press, New York.

Knowles, B. A., and L. Cooley (1994). The specialized cytoskeleton of the *Drosophila* egg chamber. *Trends Genet.* **10**(7), 235–241.

Koch, E. A., and R. H. Spitzer (1983). Multiple effects of colchicine on oogenesis in *Drosophila:* Induced sterility and switch of potential oocyte to nurse-cell developmental pathway. *Cell Tissue Res.* **228,** 21–32.

Krioutchkova, M. M., and G. E. Onishchenko (1999). Structural and functional characteristics of the centrosome in gametogenesis and early embryogenesis of animals. *Int. Rev. Cytol.* **185,** 107–156.

Li, K., and T. C. Kaufman (1996). The homeotic target gene centrosomin encodes an essential centrosomal component. *Cell* **85**(4), 585–596.

Li, Y. Y., E. Yeh, T. Hays, and K. Bloom. (1993). Disruption of mitotic spindle orientation in a yeast dynein mutant. *Proc. Natl. Acad. Sci. USA* **90**(21), 10096–10100.

Li, K., E. Y. Xu, J. K. Cecil, F. R. Turner, T. L. Megraw, and T. C. Kaufman (1998). *Drosophila* centrosomin protein is required for male meiosis and assembly of the flagellar axoneme. *J. Cell Biol.* **141**(2), 455–467.

Lin, H., and A. C. Spradling (1995). Fusome asymmetry and oocyte determination in *Drosophila. Dev Genet.* **16**(1), 6–12.

Lin, H., and A. C. Spradling (1997). A novel group of pumilio mutations affects the asymmetric division of germline stem cells in the *Drosophila* ovary. *Development* **124**(12), 2463–2476.

Lin, H., L. Yue, and A. C. Spradling (1994). The *Drosophila* fusome, a germline-specific organelle, contains membrane skeletal proteins and functions in cyst formation. *Development* **120**(4), 947–956.

Llamazares, S., G. Tavosanis, and C. Gonzalez (1999). Cytological characterisation of the mutant phenotypes produced during early embryogenesis by null and loss-of-function alleles of the gammaTub37C gene in *Drosophila. J. Cell Sci.* **112**(Pt 5), 659–667.

Mach, J. M., and R. Lehmann (1997). An Egalitarian–BicaudalD complex is essential for oocyte specification and axis determination in *Drosophila. Genes Dev.* **11**(4), 423–435.

Mahajan-Miklos, S., and L. Cooley (1994). Intercellular cytoplasm transport during *Drosophila* oogenesis. *Dev. Biol.* **165**(2), 336–351.

Mahowald, A. P., and M. P. Kambysellis (1980). Oogenesis. In *The Genetics and Biology of Drosophila.* (M. Ashburner and T. R. F. Wright, Eds.), Academic Press, New York.

Mahowald, A. P., T. J. Goralski, and J. H. Caulton (1983). In vitro activation of *Drosophila* eggs. Dev. Biol. **98**(2), 437–445.

Mahowald, A. P., and J. M. Strassheim (1970). Intercellular migration of centrioles in the germarium of *Drosophila melanogaster.* An electron microscopic study. *J. Cell Biol.* **45**(2), 306–320.

Marschall, L. G., R. L. Jeng, J. Mulholland, and T. Stearns (1996). Analysis of Tub4p, a yeast gamma-tubulin-like protein: implications for microtubule-organizing center function. *J. Cell Biol.* **134**(2), 443–454.

Matthies, H. J., H. B. McDonald, L. S. Goldstein, and W. E. Theurkauf (1996). Anastral meiotic spindle morphogenesis: role of the non-claret disjunctional kinesin-like protein. *J. Cell Biol.* **134**(2), 455–464.

McGrail, M., and T. S. Hays (1997). The microtubule motor cytoplasmic dynein is required for spindle orientation during germline cell divisions and oocyte differentiation in *Drosophila. Development* **124**(12), 2409–2419.

McKearin, D. (1997). The *Drosophila* fusome, organelle biogenesis and germ cell differentiation: if you build it. *Bioessays* **19**(2), 147–152.

McKearin, D., and B. Ohlstein (1995). A role for the *Drosophila* bag-of-marbles protein in the differentiation of cystoblasts from germline stem cells. *Development* **121**(9), 2937–2947.

McKim, K. S., and R. S. Hawley (1995). Chromosomal control of meiotic cell division. *Science* **270**(5242), 1595–1601.

Megraw, T. L., K. Li, L. R. Kao, and T. C. Kaufman (1999). The Centrosomin protein is required for centrosome assembly and function during cleavage in *Drosophila. Development* **126**(13), 2829–2839.

Mello, C. C., C. Schubert, B. Draper, W. Zhang, R. Lobel, and J. R. Priess (1996). The PIE-1 protein and germline specification in *C. elegans* embryos [letter]. *Nature* **382**(6593), 710–712.

Moritz, M., M. B. Braunfeld, J. W. Sedat, B. Alberts, and D. A. Agard (1995). Microtubule nucleation by gamma-tubulin-containing rings in the centrosome. *Nature* **378**(6557), 638–640.

Moritz, M., Y. Zheng, B. M. Alberts, and K. Oegema (1998). Recruitment of the gamma-tubulin ring complex to *Drosophila* salt-stripped centrosome scaffolds. *J. Cell Biol.* **142**(3), 775–786.

Neuman-Silberberg, F. S., and T. Schupbach (1993). The *Drosophila* dorsoventral patterning gene gurken produces a dorsally localized RNA and encodes a TGF alpha-like protein. *Cell* **75**(1), 165–174.

Nilson, L. A., and T. Schupbach (1999). EGF receptor signaling in *Drosophila* oogenesis. *Curr. Top. Dev. Biol.* **44,** 203–243.

Oakley, C. E. and B. R. Oakley (1989). Identification of gamma-tubulin, a new member of the tubulin superfamily encoded by mipA gene of *Aspergillus nidulans. Nature* **338**(6217), 662–664.

Page, A. W., and T. L. Orr-Weaver (1997). Activation of the meiotic divisions in *Drosophila* oocytes. *Dev. Biol.* **183**(2), 195–207.

Peifer, M., S. Orsulic, D. Sweeton, and E. Wieschaus (1993). A role for the *Drosophila* segment polarity gene armadillo in cell adhesion and cytoskeletal integrity during oogenesis. *Development* **118**(4), 1191–1207.

Pokrywka, N. J., and E. C. Stephenson (1991). Microtubules mediate the localization of bicoid RNA during *Drosophila* oogenesis. *Development* **113**(1), 55–66.

Pokrywka, N. J., and E. C. Stephenson (1995). Microtubules are a general component of mRNA localization systems in *Drosophila* oocytes. *Dev. Biol.* **167**(1), 363–370.

Puro, J. (1991). Differential mechanisms governing segregation of a univalent in oocytes and spermatocytes of *Drosophila melanogaster. Chromosoma* **100**(5), 305–314.

Ran, B., R. Bopp, and B. Suter (1994). Null alleles reveal novel requirements for Bic-D during *Drosophila* oogenesis and zygotic development. *Development* **120**(5), 1233–1242.

Rasmussen, S. W. (1974). Studies of the development of the synaptonemal complex in *Drosophila melanogaster. C. R. Trav. Lab. Carlsberg.* **39,** 443–468.

Reinsch, S., and P. Gonczy (1998). Mechanisms of nuclear positioning. *J. Cell Sci.* **111**(Pt 16), 2283–2295.

Riparbelli, M. G., and G. Callaini (1996). Meiotic spindle organization in fertilized *Drosophila* oocyte: presence of centrosomal components in the meiotic apparatus. *J. Cell Sci.* **109**(Pt 5), 911–918.

Riparbelli, M. G., and G. Callaini (1998). "gamma-Tubulin is transiently associated with the *Drosophila* oocyte meiotic apparatus. *Eur. J. Cell Biol.* **75**(1), 21–28.

Robinson, D. N., and L. Cooley (1997). Genetic analysis of the actin cytoskeleton in the *Drosophila* ovary. *Annu. Rev. Cell. Dev. Biol.* **13,** 147–170.

Roth, S., F. S. Neuman-Silberberg, G. Barcelo, and T. Schupbach (1995). cornichon and the EGF receptor signaling process are necessary for both anterior–posterior and dorsal–ventral pattern formation in *Drosophila. Cell* **81**(6), 967–978.

Rothwell, W. F., P. Fogarty, C. M. Field, and W. Sullivan (1998). Nuclear-fallout, a *Drosophila* protein that cycles from the cytoplasm to the centrosomes, regulates cortical microfilament organization. *Development* **125**(7), 1295–1303.

Saunders, R. D., M. C. Avides, T. Howard, C. Gonzalez, and D. M. Glover (1997). The *Drosophila* gene abnormal spindle encodes a novel microtubule-associated protein that associates with the polar regions of the mitotic spindle. *J. Cell Biol.* **137**(4), 881–890.

Schumacher, J. M., N. Ashcroft, P. J. Donovan, and A. Golden (1998). A highly conserved centrosomal kinase, AIR-1, is required for accurate cell cycle progression and segregation of developmental factors in *Caenorhabditis elegans* embryos. *Development* **125**(22), 4391–4402.

Schupbach, T. (1987). Germ line and soma cooperate during oogenesis to establish the dorsoventral pattern of egg shell and embryo in *Drosophila melanogaster. Cell* **49**(5), 699–707.

Schupbach, T., and E. Wieschaus (1991). Female sterile mutations on the second chromosome of *Drosophila melanogaster.* II. Mutations blocking oogenesis or altering egg morphology. *Genetics* **129**(4), 1119–1136.

Sobel, S. G., and M. Snyder (1995). A highly divergent gamma-tubulin gene is essential for cell growth and proper microtubule organization in *Saccharomyces cerevisiae. J. Cell Biol.* **131**(6 Pt 2), 1775–1788.

Sonnenblick, B. P. (1950). The early embryology of *Drosophila melanogaster.* In *Biology of Drosophila.* (M. Demerec, Ed.), Hafner Pub. Co., New York.

Spang, A., S. Geissler, K. Grein, and E. Schiebel (1996). gamma-Tubulin-like Tub4p of *Saccharomyces cerevisiae* is associated with the spindle pole body substructures that organize microtubules and is required for mitotic spindle formation. *J. Cell Biol.* **134**(2), 429–441.

Spradling, A. C. (1993). Developmental genetics of oogenesis. In *The Development of Drosophila melanogaster.* Cold Spring Harbor Laboratory Press, Plainview, NY.

Storto, P. D., and R. C. King (1989). The role of polyfusomes in generating branched chains of cystocytes during *Drosophila* oogenesis. *Dev. Genet.* **10**(2), 70–86.

Strome, S. (1993). Determination of cleavage planes. *Cell* **72**(1), 3–6.

Sunkel, C. E., R. Gomes, P. Sampaio, J. Perdigao, and C. Gonzalez (1995). Gamma-tubulin is required for the structure and function of the microtubule organizing centre in *Drosophila* neuroblasts. EMBO J. **14**(1), 28–36.

Suter, B., and R. Steward (1991). Requirement for phosphorylation and localization of the Bicaudal-D protein in *Drosophila* oocyte differentiation. *Cell* **67**(5), 917–926.

Suter, B., L. M. Romberg, and R. Steward (1989). Bicaudal-D, a *Drosophila* gene involved in developmental asymmetry: localized transcript accumulation in ovaries and sequence similarity to myosin heavy chain tail domains. *Genes Dev.* **3**(12A), 1957–1968.

Swan, A., and B. Suter (1996). Role of Bicaudal-D in patterning the *Drosophila* egg chamber in mid-oogenesis. *Development* **122**(11), 3577–3586.

Szollosi, D., P. Calarco, and R. P. Donahue (1972). Absence of centrioles in the first and second meiotic spindles of mouse oocytes. *J. Cell Sci.* **11**(2), 521–541.

Tavosanis, G., S. Llamazares, G. Goulielmos, and C. Gonzalez (1997). Essential role for gamma-tubulin in the acentriolar female meiotic spindle of *Drosophila.* EMBO J. **16**(8), 1809–1819.

Telfer, W. H. (1975). Development and physiology of the oocyte–nurse cell syncytium. *Adv. Insect Physiol.* **11,** 223–319.

Theurkauf, W. E. (1994). Microtubules and cytoplasm organization during *Drosophila oogenesis. Dev. Biol.* **165**(2), 352–360.

Theurkauf, W. E. (1997). Oocyte differentiation: a motor makes a difference. *Curr. Biol.* **7**(9), R548–R551.

Theurkauf, W. E., and R. S. Hawley (1992). Meiotic spindle assembly in *Drosophila* females: behavior of nonexchange chromosomes and the effects of mutations in the nod kinesin-like protein. *J. Cell Biol.* **116**(5), 1167–1180.

Theurkauf, W. E., S. Smiley, M. L. Wong, and B. M. Alberts (1992). Reorganization of the cytoskeleton during *Drosophila* oogenesis: implications for axis specification and intercellular transport. *Development* **115**(4), 923–936.

Theurkauf, W. E., B. M. Alberts, Y. N. Jan, and T. A. Jongens (1993). A central role for microtubules in the differentiation of *Drosophila* oocytes. *Development* **118**(4), 1169–1180.

Walczak, C. E., and T. J. Mitchison (1996). Kinesin-related proteins at mitotic spindle poles: function and regulation. *Cell* **85**(7), 943–946.

Walczak, C. E., I. Vernos, T. J. Mitchison, E. Karsenti, and R. Heald (1998). A model for the proposed roles of different microtubule-based motor proteins in establishing spindle bipolarity. *Curr. Biol.* **8**(16), 903–913.

Wharton, R. P., and G. Struhl (1989). Structure of the *Drosophila* BicaudalD protein and its role in localizing the the posterior determinant nanos. *Cell* **59**(5), 881–892.

White, J., and S. Strome (1996). Cleavage plane specification in *C. elegans:* how to divide the spoils. *Cell* **84**(2), 195–198.

Whitefield, W. G., M. A. Chaplin, K. Oegema, H. Parry, and D. M. Glover (1995). The 190 kDa centrosome-associated protein of *Drosophila melanogaster* contains four zinc finger motifs and binds to specific sites on polytene chromosomes. *J. Cell Sci.* **108**(Pt 11), 3377–3387.

Wilson, P. G., and G. G. Borisy (1998). Maternally expressed gamma Tub37CD in *Drosophila* is differentially required for female meiosis and embryonic mitosis. *Dev. Biol.* **199**(2), 273–290.

Wilson, P. G., Y. Zheng, C. E. Oakley, B. R. Oakley, G. G. Borisy, and M. T. Fuller (1997). Differential expression of two gamma-tubulin isoforms during gametogenesis and development in *Drosophila. Dev. Biol.* **184**(2), 207–221.

Wu, X., and R. E. Palazzo (1999). Differential regulation of maternal vs. paternal centrosomes. *Proc. Natl. Acad. Sci. USA* **96**(4), 1397–1402.

Yue, L., and A. C. Spradling (1992). hu-li tai shao, a gene required for ring canal formation during *Drosophila* oogenesis, encodes a homolog of adducin. *Genes Dev.* **6**(12B), 2443–2454.

Zheng, Y., M. K. Jung, and B. R. Oakley (1991). Gamma-tubulin is present in *Drosophila melanogaster* and *Homo sapiens* and is associated with the centrosome. *Cell* **65**(5), 817–823.

Zheng, Y., M. L. Wong, B. Alberts, and T. Mitchison (1995). Nucleation of microtubule assembly by a gamma-tubulin-containing ring complex. *Nature* **378**(6557), 578–583.

Zimmerman, W., C. A. Sparks, and S. J. Doxsey (1999). Amorphous no longer: the centrosome comes into focus. *Curr. Opin. Cell Biol.* **11**(1), 122–128.

19

The Centrosome in Early *Drosophila* Embryogenesis

W. F. Rothwell and *W. Sullivan*
Sinsheimer Laboratories
Department of Biology
University of California
Santa Cruz, California 95064

Current Topics in Developmental Biology, Vol. 49

0070-2153/00 $35.00

I. Introduction

The centrosome was first described over a century ago. Terms such as the "central body" and "division center of the cell" indicated the functional importance attributed to this organelle even during these initial observations (Wilson, 1925). Over the past decade much progress has been made in elucidating centrosome structure and composition, and components are rapidly being identified. In spite of these advances, function has been determined for only a handful of centrosome proteins. Consequently, systems amenable to genetic analysis or other functional approaches will play a major role in the next phase of centrosome research. The early *Drosophila* embryo stands out as a particularly useful system for functional studies of the centrosome. This organelle plays essential and diverse roles in guiding nuclear movement and cytoskeletal reorganization during the initial 3 hours of development.

Drosophila embryogenesis begins with 13 synchronous nuclear divisions that occur without accompanying cytokinesis (for review see Foe *et al.*, 1993; Fig. 1). The divisions immediately following fertilization proceed as an expanding ball in the interior of the embryo. During nuclear cycles 5, 6, and 7 the nuclei undergo an axial expansion so that they are evenly distributed along the length of the embryo. Migration of the nuclei to the cortex occurs during cycles 8 through 10. The first nuclei arrive at the posterior end of the embryo and form pole cells, the precursors to the germline. The majority of nuclei reach the cortex during interphase of nuclear cycle 10 and undergo four rounds of division as a monolayer just beneath the plasma membrane. These nuclei and their associated centrosomes induce dramatic reorganizations of the cortical cytoskeleton that are essential for normal nuclear divisions (Karr and Alberts, 1986; Warn, 1986; Sullivan and Theurkauf, 1995). During interphase of nuclear cycle 14, membranes invaginate around each nucleus to produce a cellularized embryo (reviewed in Schejter and Wieschaus, 1993).

This process is extremely rapid, with many of the nuclear cycles occurring in less than 10 minutes (Foe and Alberts, 1983). In addition, during the late cortical divisions, thousands of nuclei are dividing in a confined monolayer and space is clearly limited. In spite of the speed and crowding, the fidelity of these divisions is not significantly compromised. Much work, especially in the past decade, indicates that this is in large part due to the centrosome-based activities that coordinate nuclear and cytoskeletal dynamics during these divisions. Because *Drosophila* is amenable to molec-

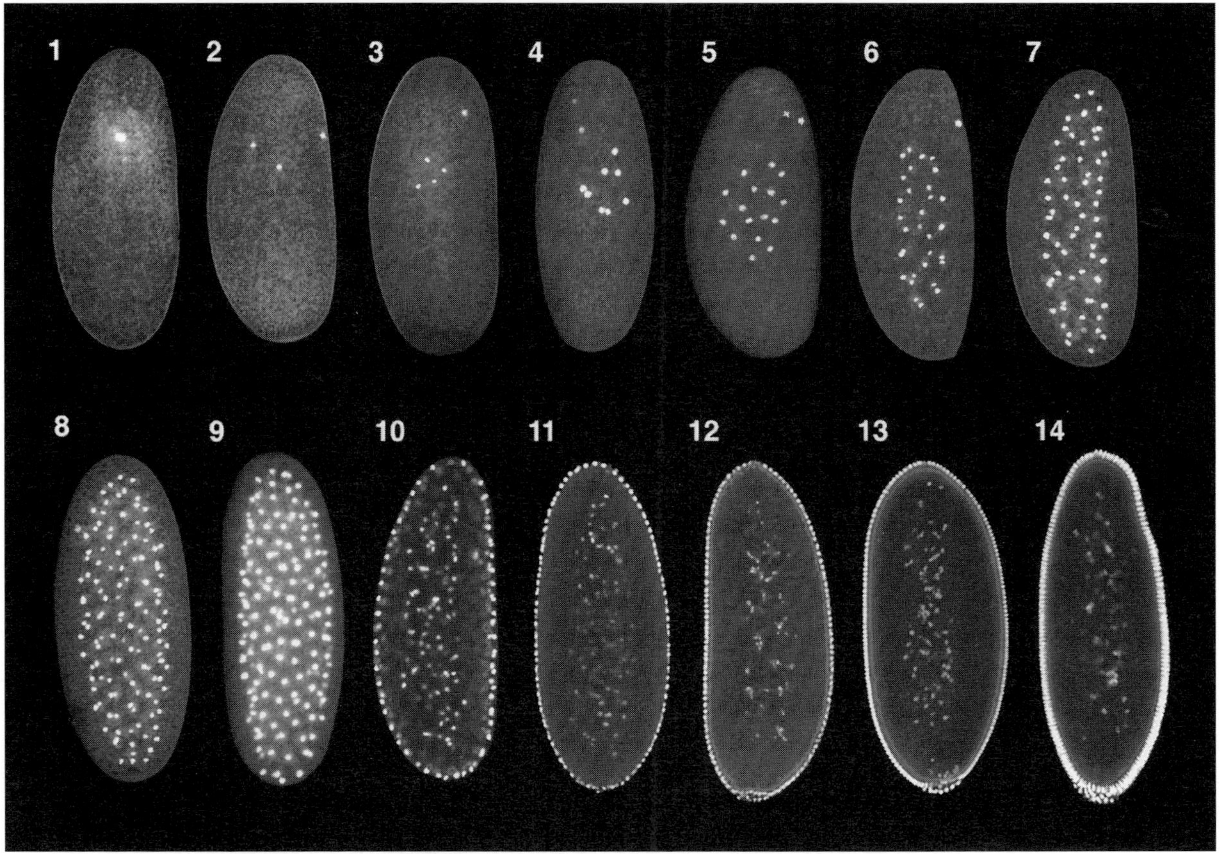

Figure 1 The first 13 divisions are rapid, synchronous, and occur without accompanying cytokinesis. Cellularization occurs during interphase of nuclear cycle 14. Centrosomes play a key role in the differentiation of three types of nuclei during these divisions: the posterior pole cells, the cortical nuclei, and the internal polyploid yolk nuclei.

ular genetic, biochemical, and a variety of cellular approaches, the mechanisms underlying many activities of the centrosome in the early embryo are now being addressed. The following issues are of particular interest: What are the mechanisms that control centrosome movement and migration? What is the role of the centrosome in coordinating nuclear and cytoskeletal events of the cell cycle? How does the centrosome direct the dramatic microfilament and plasma membrane rearrangements at the cortex? What is the mechanism of centrosome duplication and how is it regulated? Although we do not have definitive answers, recent studies are providing important insights into many of these questions. This chapter is not intended to be comprehensive, but instead highlights work that addresses these issues and provides insight into centrosome function.

II. Ultrastructure

A. Unique Centriole Structure in the *Drosophila* Embryo

Centrosomes in animal cells typically consist of a pair of perpendicularly oriented centrioles surrounded by an electron-dense pericentriolar material (PCM) (for reviews see Kalnins, 1992; Glover *et al.,* 1993; Kellogg *et al.,* 1994; Balczon, 1996). In most vertebrates, centrioles are characterized as short cylinders with clearly distinguishable proximal and distal ends and nine triplet microtubules arranged along the cylinder wall (for review see Lange and Gull, 1996). The structure of centrioles in the syncytial *Drosophila* embryo differs from their mammalian counterparts in that they are shorter (0.08–0.18 vs 0.18–0.7 μm long) and have morphologically similar proximal and distal ends (McDonald and Morphew, 1993; Moritz *et al.,* 1995a). In addition, embryonic *Drosophila* centrioles contain fewer peripheral microtubules than the triplets observed in mammalian cells (Moritz *et al.,* 1995a; Callaini *et al.,* 1997; Debec *et al.,* 1999). Mature centrioles in the early *Drosophila* embryo are morphologically similar to immature, newly forming centrioles in mammalian cells (Wheatley, 1982; Vorobjev and Nadezhdina, 1987; Moritz *et al.,* 1995a). This suggests that the rapid nuclear division cycles of the early *Drosophila* embryo do not allow for more than a minimal centriole to be built. This idea is supported by the finding that *Drosophila* tissue culture cells (Kc23) with complete cell cycles contain more complex centrioles with well defined proximal and distal ends (Debec and Marcaillou, 1997). However, it has been shown that centrioles maintain their "immature" structure during the longer cell division cycles in cellularized embryos (Callaini *et al.,* 1997). Therefore, a consistent correlation between cell cycle length and centriole architecture has not been clearly demonstrated.

B. γ-Tubulin Ring Complexes in the PCM

The PCM represents a complex collection of proteins and has been described as constituting a structural matrix with functional and regulatory proteins embedded within it (Kalt and Schliwa, 1993). Although much remains to be learned about the PCM, its role in the nucleation of microtubules has been known for some time (Gould and Borisy, 1977). Insight into the mechanism of microtubule nucleation has come from ultrastructural analysis of the PCM in isolated *Drosophila* centrosomes (Moritz *et al.*, 1995a, 1995b). Three-dimensional electron tomographic reconstructions demonstrate that the PCM in the centrosomes without microtubules (naked centrosomes) is more dense and prominent than PCM from centrosomes with regrown microtubules (Moritz *et al.*, 1995a). The authors suggest that this difference may reflect the redistribution of PCM along the microtubules as they extend. By identifying the origin of specific microtubules, nucleation of microtubules was observed to occur throughout the PCM but was present to the greatest degree in areas closest to the centrioles. Ringlike structures with a similar diameter to that of a microtubule were identified through reconstitutions of naked centrosomes (Moritz *et al.*, 1995a). Immuno–electron tomography reveals that these rings contain γ-tubulin (Moritz *et al.*, 1995b), a highly conserved protein involved in microtubule nucleation (for reviews see Joshi, 1994; Murphy and Stearns, 1996; Pereira and Schiebel, 1997). This structure, now known as the "γ-tubulin-containing ring complex" (γ-TuRC), has since been found to localize to microtubule minus ends and to be required for microtubule nucleation in *Drosophila* (Moritz *et al.*, 1995b, 1998).

C. Centrosomes Maintain Close Nuclear Envelope Association

During the syncytial divisions, the centrosomes maintain an extremely close association with the nuclear envelope (Fig. 2). Regions of the nuclear envelope directly adjacent to the centrosomes are deformed (Stafstrom and Staehelin, 1984; Paddy *et al.*, 1996). This may be the result of centrosome-based astral microtubles impinging on the envelope. At prometaphase, once sister centrosomes are fully separated and spindle formation has begun, the regions directly adjacent to the nuclear envelope break down while the rest of the envelope remains intact throughout metaphase (although nuclear pore complexes are lost from the envelope during metaphase) (Stafstrom and Staehelin, 1984; Fig. 2). It may be that centrosome-generated microtubules or other activities associated with the centrosome are responsible for localized nuclear envelope breakdown.

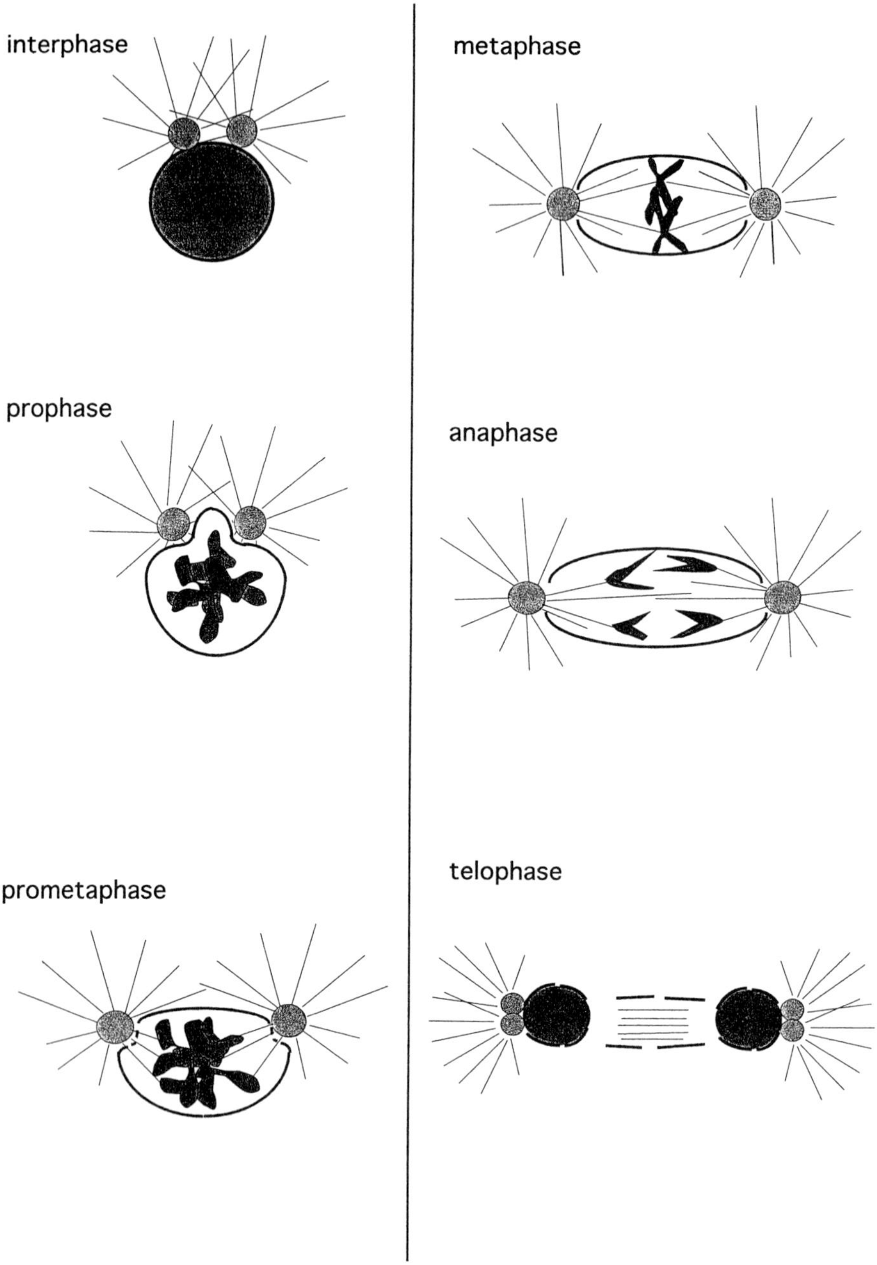

Figure 2 Centrosomes are closely associated with and deform the nuclear envelope. At prometaphase, the nuclear envelope breaks down only in regions adjacent to the centrosomes. This indicates that centrosome-associated activities may be responsible for nuclear envelope breakdown. Large regions of the nuclear envelope remain intact as the nuclei progress through mitosis.

The nuclear lamina is also deeply invaginated in regions directly adjacent to the centrosomes (Paddy *et al.,* 1996). These invaginations correspond to the position of the separating centrosomes but become more shallow as separation proceeds. During prometaphase, the lamina remains intact and localized along the entire nuclear periphery even at the polar regions where the nuclear envelope has broken down (as assayed by its permeability to macromolecules) (Paddy *et al.,* 1996). At this time, microtubules penetrate through the lamina into the nuclear interior. As the nuclei progress into metaphase, the lamina breaks down completely. The effects of the centrosomes on lamina structure suggest that the two may interact. One possibility suggested by Paddy *et al.* (1996) is that the lamina could provide a structural framework to direct centrosome migration.

III. Composition

A. Centrosome Proteins Enriched in Microtubule Affinity Columns

The centrosome is a complex organelle estimated to contain more than 100 different proteins (for review see Kalt and Schliwa, 1993; Kellogg *et al.,* 1994). Those proteins that retain their centrosomal localization in the absence of polymerized microtubules are defined as core centrosomal components (Whitfield *et al.,* 1995). A number of core components have been identified through microtubule affinity chromatography (Kellogg *et al.,* 1989) (Table I). This procedure yielded more than 50 microtubule-associated proteins (MAPs) from extracts prepared from early *Drosophila* embryos. Immunofluorescent analysis demonstrated that the majority of these proteins localized to the centrosome (17/24). Most of these (12/17) showed a cell-cycle-dependent localization, residing at the centrosome specifically during mitosis.

B. CP60 and CP190 Form a Complex That Exhibits Cell-Cycle-Dependent Centrosome and Nuclear Localization

Two MAPS that localize to the centrosomes during mitosis, CP60 and CP190 have been characterized molecularly and shown to represent core components (Whitfield *et al.,* 1988; Kellogg *et al.,* 1991; Kellogg and Alberts, 1992; Raff *et al.,* 1993; Oegema *et al.,* 1995; Table I). CP190 concentrates at the centrosomes from prophase through telophase and localizes to the nuclei during interphase in early *Drosphila* embryos (Kellogg and Alberts, 1992; Whitfield *et al.,* 1995; Oegema *et al.,* 1997). Three domains directing the localization of CP190 have been identified: nuclear localization, centro-

Table I Centrosomal Components in the Early *Drosophila* Embryo[a]

Protein	Size (kDa)	Domain(s)	Method of isolation	Proposed function	References[b]
Core components at the centrosome throughout the division cycle:					
γ-Tubulin37C	50.7		PCR, library screens/rescue analysis	mt nucleation	1, 2
LK6	185 220[c]	Ser-Thr protein kinase	mt affinity column, expression library screen	mt organization	3, 4
CNN	150	coiled-coils/leucine zippers	Immunopurification of targets of ANTP	Assembly and function of centrosomes	5–7
Core components at the centrosome during mitosis, nuclear at interphase:					
CP190	120	NLS, zfs	Expression library screen, mt affinity column	?	8–11
CP60	47.6	cdk phos., des. box	Eluted from CP190 immunoaffinity column	?	12, 13
Noncore components at the centrosomes only during mitosis					
Ncd	77.5	kinesin-like motor	Mutagenesis	Maintenance of spindle poles	14–17
Nuf	57	coiled-coils	Screen of fs lines/cDNA library screen	Organization of the cortex	18–20
Not known if core, at centrosome mainly during mitosis					
Asp	220	cdk phos., actin and calmodulin binding	Mutagenesis, chromosome walking, cDNA library screen	Organization of γ-tubulin/ spindle structure	21–23
Cyclin B	59.2	Cyclins signature	PCR, cDNA library screen	Regulation of cdc2 kinase	24, 25

[a] mt = microtubule; ANTP = Antennapedia protein; NLS = nuclear localization sequence; zfs = zinc fingers; cdk phos. = consensus sites for phosphorylation by cyclin dependent kinases; des. box = destruction box (targets cyclins for proteolysis); fs = female sterile.

[b] 1, Tavosanis *et al.* (1997); 2, Wilson *et al.* (1997); 3, Kellogg *et al.* (1989); 4, Kidd and Raff (1997); 5, Heuer *et al.* (1995); 6, Li and Kaufman (1996); 7, Megraw *et al.* (1999); 8, Whitfield *et al.* (1988); 9, Kellogg *et al.* (1989); 10, Whitfield *et al.* (1995); 11, Oegema *et al.* (1995); 12, Kellogg and Alberts (1992); 13, Kellogg *et al.* (1995); 14, Sturtevant (1929); 15, Endow *et al.* (1990); 16, Endow *et al.* (1994); 17, Endow and Komma (1996); 18, Sullivan *et al.* (1993b); 19, Rothwell *et al.* (1998); 20, Rothwell *et al.* (1999); 21, Gonzalez *et al.* (1990); 22, Saunders *et al.* (1997); 23, do Carmo Avides and Glover (1999); 24, Lehner and O'Farrell (1990); 25, Maldonado-Codina and Glover (1992).

[c] Rare.

some binding, and microtubule binding (Oegema *et al.,* 1995). Although the latter two domains overlap, CP190 does not require microtubules for its movement to or maintenance at the centrosome. Furthermore, immunofluorescence reveals no obvious colocalization of CP190 along the lengths of microtubules (Oegema *et al.,* 1995). Therefore, the role of microtubule binding for CP190 remains unknown. Fusion proteins containing only the centrosomal localization domain remain at the centrosome throughout the division cycle (Oegema *et al.,* 1995). This suggests that recruitment away from the centrosome to the nucleus is an active process. A single NLS present in CP190 is capable of directing its nuclear localization (Oegema *et al.,* 1995). CP190 also contains zinc finger motifs and associates with many sites on the polytene chromosomes (Whitfield *et al.,* 1995). This raises the possibility that CP190 may act as a transcription factor or may play a role in regulating chromatin structure. Alternatively, CP190 may be a component of the nuclear matrix. Injection of fluorescently labeled CP190 into living *Drosophila* embryos reveals that it forms a fibrous intranuclear network that does not colocalize with the DNA (Oegema *et al.,* 1997).

CP60 is part of a large protein complex that includes CP190 and other components (Kellogg *et al.,* 1991; Kellogg and Alberts, 1992). CP60 tightly associates with CP190 and localizes in a pattern similar to, yet temporally distinct from, that of CP190 (Kellogg *et al.,* 1995; Oegema *et al.,* 1997). CP60 is phosphorylated by cyclinB/$p34^{cdc2}$ kinase *in vitro* and contains a destruction box sequence suggesting that it is targeted for proteolysis at the end of mitosis (Glotzer *et al.,* 1991; Kellogg *et al.,* 1995). This, combined with fixed and live analysis indicating that cyclin B localizes to centrosomal regions in syncytial embryos (Maldonado-Codina and Glover, 1992; Huang and Raff, 1999), suggests that CP60 may be an *in vivo* substrate for cyclinB/ $p34^{cdc2}$ kinase.

C. Enzymes Associated with the Centrosome

LK6, another MAP identified by Kellogg *et al.* (1989), is a core centrosomal protein that encodes a serine/threonine kinase (Kidd and Raff, 1997; Table I). Unlike CP60 and CP190, LK6 localizes to the centrosomes at all stages of the nuclear cycle in early *Drosophila* embryos (Kellogg *et al.,* 1989; Kidd and Raff, 1997). The LK6 protein has a short half-life (approx. 10–20 min), suggesting that its production is regulated with each cell cycle (Kidd and Raff, 1997). Therefore, constitutive overexpression of LK6 would be expected to cause defects. Indeed, 20–50% of the eggs laid by females overexpressing LK6 remain unfertilized and a significant proportion of the developing embryos show mitotic defects, including centrosomes that detach from spindle poles during metaphase (Kidd and Raff, 1997). Therefore, LK6 may play a role in regulating microtubule function.

The potential importance of phosphorylation at the centrosome is further underscored by studies on the protein phosphatase 4 (PP4). In early *Drosophila* embryos, PP4 localizes to centrosomes, and analysis of *pp4*-derived embryos reveals that it is also required for proper microtubule organization (Helps *et al.,* 1998).

D. Centrosomin, a Target of Antennapedia, Is a Core Centrosomal Component

Centrosomin (CNN) is another core component that localizes to centrosomes throughout the division cycle in a variety of cell types (Heuer *et al.,* 1995; Li and Kaufman, 1996; Megraw *et al.,* 1999; Table I). CNN was initially identified as a target of the homeotic selector gene Antennapedia (ANTP) (Heuer *et al.,* 1995). This finding provides an important clue into the cellular mechanisms through which transcription factors such as Antennapedia may direct morphogenesis. For example, by targeting centrosomal components, these factors could direct morphogenesis by influencing spindle orientation.

CNN contains extensive regions of predicted alpha helical coiled-coil structure (Heuer *et al.,* 1995). These regions contain potential leucine zipper motifs that may mediate protein–protein interactions. Unlike CP60 and CP190, centrosomin does not bind microtubules (Li and Kaufman, 1996). Analysis of embryos derived from females hemizygous for a female sterile *cnn* mutation reveals that it plays a role in both the assembly and function of the centrosome (Megraw *et al.,* 1999). The centrosomes of these embryos contained little or no detectable level of the core centrosomal components CP60, CP190, and γ-tubulin. However, CP60 and CP190 are present in these embryos and continue to cycle normally into the nuclei during interphase. Embryos from these females exhibit a progressive increase in nuclear division errors as syncytial development proceeds. It is likely that these division errors are a direct result of defective centrosomes. In embryos lacking CNN, the astral microtubules are dramatically reduced and spindle microtubules fail to converge at the poles (Megraw *et al.,* 1999). Studies in unfertilized *Sciara* embryos have demonstrated that the astral microtubules are essential for maintaining the separation between dividing nuclei during the initial syncytial divisions (de Saint Phalle and Sullivan, 1998). It may be that reduced levels of astral microtubules in *cnn* embryos are responsible for some of the early nuclear division errors. In spite of these errors, many nuclei successfully migrate to the cortex. However, once at the cortex, shared spindle poles and nuclear fusions are common (Megraw *et al.,* 1999). This suggests that the improperly assembled centrosomes are unable to organize the metaphase furrows responsible for maintaining separation between adjacently dividing nuclei at the embryo cortex (Karr and Alberts, 1986; Sullivan *et al.,* 1990).

E. The γ-Tubulin Ring Complex Initiates Microtubule Formation

γ-Tubulin is a highly conserved centrosomal protein involved in microtubule nucleation (for reviews see Joshi, 1994; Murphy and Stearns, 1996; Pereira and Schiebel, 1997). Two γ-tubulin genes (γ-tubulin37C and γ-tubulin23C) exist in *Drosophila* (Zheng *et al.,* 1991; Tavosanis *et al.,* 1997; Wilson *et al.,* 1997). γ-Tubulin23C is expressed zygotically, in ovaries and in the male germline, whereas γ-tubulin37C expression is restricted to ovaries and early embryos (Tavosanis *et al.,* 1997; Wilson *et al.,* 1997). In syncytial embryos, no specific pattern of γ-tubulin23C staining is apparent. γ-Tubulin37C, however, concentrates at the centrosomes throughout the syncytial nuclear division cycles and, to a lesser degree, is observed as punctate particles scattered throughout the cytoplasm (Tavosanis *et al.,* 1997; Wilson *et al.,* 1997; Table I). Analysis of embryos derived from females mutant for γ-tubulin37C indicates that it is required early in syncytial development; embryos arrest prior to the blastoderm stage (Tavosanis *et al.,* 1997). Mitotic spindles present in embryos lacking γ-tubulin37C contain no asters (Wilson and Borisy, 1998; Llamazares *et al.,* 1999) and do not accumulate the core centrosomal component, CNN (Heuer *et al.,* 1995; Li and Kaufman, 1996; Llamazares *et al.,* 1999). This result is consistent with a role for γ-tubulin37C in building the mitotic centrosome. That spindles form without centrosomes most likely is due to the ability of bipolar microtubule arrays to form around chromatin, a phenomenon that has been observed in *Xenopus* eggs and in unfertilized *Sciara* embryos (Karsenti *et al.,* 1984; Heald *et al.,* 1996; de Saint Phalle and Sullivan, 1998).

γ-Tubulin exists as a member of the γ-tubulin ring complex (γ-TuRC), which is conserved in all animals examined (reviewed in Pereira and Schiebel, 1997). To study the molecular makeup required for microtubule nucleation and the role of γ-TuRC, Moritz *et al.* (1998) developed an *in vitro* assay for nucleation of microtubules from salt-stripped centrosomes. Treatment with high salt (2 M KI) removes the proteins CP60, CP190, CNN, and γ-tubulin from centrosomes and destroys their ability to nucleate microtubules. Nucleation capability is restored upon addition of extract from 0–2 hr old embryos. By immunodepleting specific proteins from the extract, the authors showed that γ-tubulin, but not CP60 or CP190, is required for microtubule nucleation. Because it was not possible to sufficiently immunodeplete CNN from the extract, its role in nucleation remains unknown.

Using gel filtration chromatography and sucrose gradient sedimentation, Moritz *et al.* (1998) determined that γ-tubulin exists both in a large (about 20 nm, 36.9S, about 3000 kDa) and a small (6.9 nm, 8.5S, 240 kDa) complex in the early embryo. CP60 and CP190 are not present in either complex. The fraction containing the large γ-TuRC complex, but not the small complex, is able to restore microtubule nucleating ability to salt-stripped centrosomes (Moritz *et al.*, 1998). Purified γ-TuRC alone will not restore this activity, indicating that the extract contains an additional factor(s) required for microtubule nucleation. Initial characterization indicates that this factor has an estimated size of 220 kDa (Moritz *et al.*, 1998). One possible candidate is Asp, a microtubule-associated protein that localizes to spindle poles during metaphase/anaphase and the midbody at telophase in early embryos (Saunders *et al.*, 1997; Table I). Extracts prepared from *asp*-derived mutant embryos are unable to restore microtubule nucleating activity to salt-stripped centrosomes (do Carmo Avides and Glover, 1999). Addition of purified Asp protein to these extracts restored microtubule nucleating activity. Because the Asp protein does not directly associate with the γ-TuRC, it may not play a direct role in microtubule nucleation (do Carmo Avides and Glover, 1999). Instead, Asp may be required for activation of γ-TuRC's microtubule nucleating activity. Other factors in addition to the γ-TuRC and Asp protein, such as pericentrin, may be required for microtubule nucleation.

F. Other Centrosome-Associated Proteins

Ncd is a kinesin-like motor protein that localizes to both the centrosomes and spindles during prometaphase–anaphase and to the midbody at telophase (Endow *et al.*, 1990, 1994; McDonald and Goldstein, 1990; Endow and Komma, 1996). Microtubules are required for its centrosomal localization; thus, it is not a core component. In mutant *ncd* embryos, centrosomes often split prematurely in metaphase and detach from the spindle poles (Endow *et al.*, 1994; Endow and Komma, 1996). A role for Ncd in the maintenance of centrosome positioning during mitosis has also been suggested by recent observations in living embryos (Sharp *et al.*, 1999; see Section IV).

Nuf, another noncore centrosome protein, contains extensive coiled-coil domains and concentrates at the centrosomes specifically during prophase; it is generally cytoplasmic throughout the remainder of the nuclear cycle (Rothwell *et al.*, 1998; Rothwell, unpublished observation). Genetic and cellular analyses indicate that it is not required during the mitotic divisions prior to nuclear migration. However, once the nuclei reach the cortex, Nuf acts at the centrosome to recruit actin and membrane to form the furrows that encompass each spindle during the cortical divisions (Sullivan *et al.*, 1993b; Rothwell *et al.*, 1998, 1999; see Section VII).

IV. Duplication and Separation

A. As in Other Cells, Centriole Duplication Coincides with the Initiation of S Phase

Analysis of the centrosome cycle in a human lymphoblastic cell line leads to the following generalized view of the centrosome cycle (Chretien *et al.*, 1997; Fig. 3). At metaphase, the poles contain a pair of orthogonally oriented centrosomes. As division proceeds, each interphase daughter nucleus inherits a single centriole pair in which this orthogonal orientation is maintained. During G1, this orientation is lost. In late G1 and S phase, centriole splitting and duplication occur. The newly formed procentrioles are observed at the base of each centriole. Depending on the cell type, maturation of the procentriole is completed anywhere from G2 to telophase (Robbins *et al.*, 1968; Rattner and Phillips, 1973; Rieder and Borisy, 1982). Centrosome separation and polar migration occurs during the onset of mitosis (Chretien *et al.*, 1997).

In the rapid nuclear cycles of the *Drosophila* embryo, the centriole cycle appears to be modified to accommodate the lack of gap phases (Stafstrom and Staehelin, 1984; Callaini and Riparbelli, 1990; Callaini *et al.*, 1997; Fig. 3). The following description applies to the early cortical cycles (10–12) where the centrosome cycle has most extensively been studied. As in other cell types, the metaphase spindle poles maintain an orthogonally oriented centriolar pair. In late metaphase/early anaphase, the orthogonal configuration is lost (Callaini and Riparbelli, 1990). At telophase, sister centrioles split, producing two distinct astral arrays. Separation of sister centrioles toward opposing poles of the nuclear envelope occurs during late telophase/early interphase of the early cortical syncytial divisions. Centriole duplication occurs during interphase (Callaini and Riparbelli, 1990). Thus, as in conventional cycles, centriole duplication roughly corresponds with S phase during the syncytial nuclear divisions. However, because the syncytial cycles lack G1, centriolar behavior typically associated with G1 in preparation for duplication occurs during anaphase and telophase in the syncytial embryo (Fig. 3).

Studies in sea urchins demonstrate that centriole splitting and separation do not depend on duplication. Tetrapolar spindles are formed in sea urchin embryos treated with mercaptoethanol (Mazia *et al.*, 1960; Sluder and Rieder, 1985). Each pole contains a single rather than a normal pair of centrioles, indicating that splitting occurred without duplication. Following the

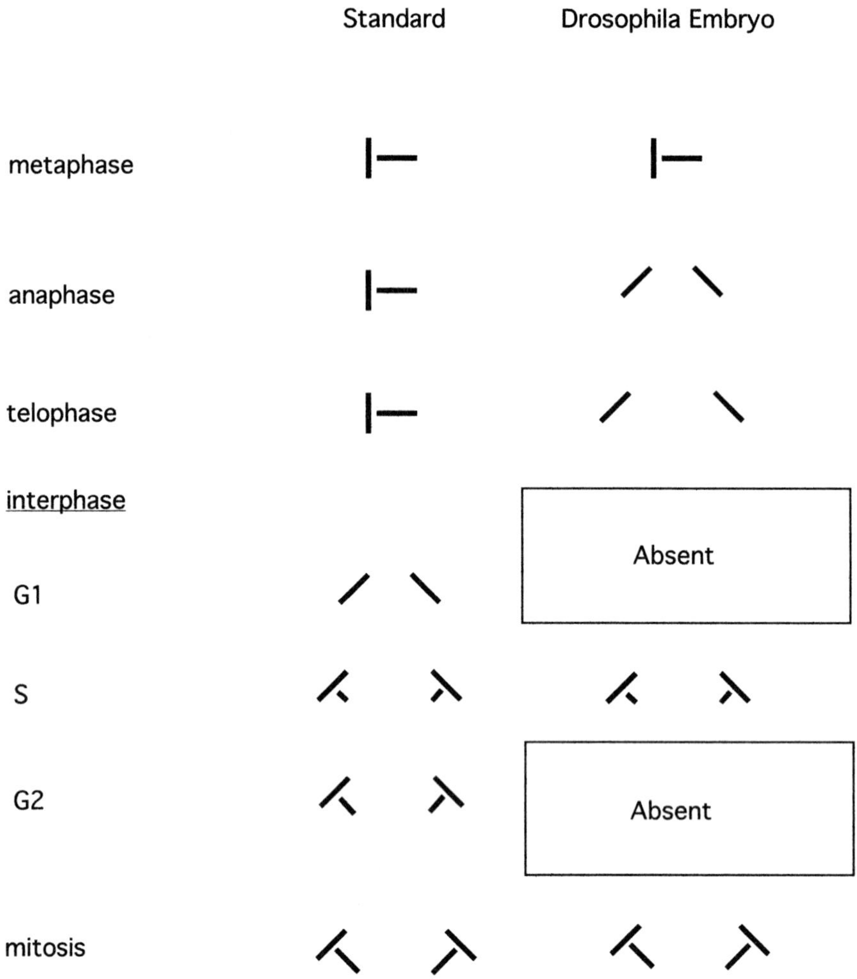

Figure 3 In the typical cell cycle, loss of centriole orientation and splitting of sister centrioles occurs during G1. Because the syncytial nuclear cycles lack gap phases, these events occur during anaphase and telophase. In both division cycles, centriole duplication is correlated with S phase.

progress of this cell into the next mitosis demonstrated that centrosomes with a single centriole are not competent to replicate. These conclusions may also apply in the syncytial *Drosophila* embryo. Embryos treated with the protein synthesis inhibitor cycloheximide arrest in interphase with four centrosomes evenly spaced along the nuclear envelope (Sullivan, unpublished observation). Although it has not been demonstrated directly, it is likely that each contains a single centriole, indicating that separation occurs without duplication.

B. Centrosome Duplication Is Dependent on DNA Synthesis but Not Nuclear Division

The eukaryotic cell cycle contains dependency relationships ensuring that one event in the cell cycle is properly completed before the next is initiated (reviewed in Hartwell and Weinert, 1989). Many of these dependency relationships are enforced by checkpoint mechanisms extrinsic to the events themselves. A key function of checkpoints is to maintain the fidelity of the cell cycle. If a cell cycle event is not properly executed, checkpoints delay progression of the cell cycle to provide time for repair or completion of the event. For instance, if DNA replication is interrupted, checkpoints delay entry into mitosis until it has been properly completed (Schlegal and Pardee, 1986). Eukaryotic cells possess a large number of checkpoints that monitor a variety of events throughout the cell cycle (Murray, 1993).

At first glance, the speed and synchrony of the syncytial divisions may seem incompatible with delays induced by cell cycle checkpoints. However, it is clear the these nuclear cycles possess a number of dependency relationships, many of which are enforced by cell cycle checkpoints (Yu *et al.*, 1998). These embryonic checkpoints rely on cell cycle delays to identify and discard the products of division errors, thus maintaining the fidelity of the entire nuclear population (Sullivan *et al.*, 1993a).

Whether these dependency relationships and checkpoints regulate centrosome duplication in the syncytial cycles has been an issue of much debate. This is in part due to studies in *Xenopus* embryos demonstrating that multiple cycles of centrosome duplication occur in the absence of protein synthesis (Gard *et al.*, 1990). Whether centrosome duplication in syncytial embryos is dependent on proper DNA replication has been an issue of particular interest. In mononucleate cells, this is addressed by determining whether centrosome number increases after injection with an S-phase inhibitor such as aphidicolin. Because it is large and multinucleate, the equivalent analysis in the *Drosophila* embryo requires determining centrosome/nuclear ratios. It is difficult to obtain accurate values, however, because in aphidicolin-treated embryos nuclei but not centrosomes recede into the interior of the embryo (Debec *et al.*, 1996). Therefore, live analysis in which the behavior of individual centrosomes and nuclei are followed is the best means of examining this issue.

A key insight into the relationship between the centrosome and nuclear cycles came from studies by Raff and Glover (1988) demonstrating that aphidicolin injection resulted in a dramatic slowing of the centrosome cycle. This analysis indicated that the rate of centrosome duplication depended on proper DNA replication. Subsequent live anlysis extended these results by demonstrating that not only is centrosome duplication delayed, but the majority of individual centrosomes examined are incapable of undergoing multiple rounds of duplication in aphidicolin-treated embryos (Debec *et al.*, 1996). Taken together, these studies demonstrate that centrosome duplication depends on proper DNA replication, and this may be enforced by a cell cycle checkpoint. This dependency relationship may explain why centrosome duplication requires protein synthesis in *Drosophila* embryos but not *Xenopus* embryos (Gard *et al.*, 1990; Sullivan, unpublished observation). Disrupting protein synthesis prevents DNA replication and activates a *Drosophila* checkpoint that inhibits centrosome duplication. In early *Xenopus* embryos, the checkpoint enforcing a dependency relationship between DNA replication and centrosome duplication is present but not activated (Dasso and Newport, 1990).

Although centrosome duplication depends on DNA replication, it does not depend on proper nuclear division. In the maternal-effect mutation *gnu,* continued cycles of DNA replication occur in the absence of nuclear division, resulting in embryos with a few giant polyploid nuclei (Freeman *et al.*, 1986). The cytoplasm immediately surrounding each nucleus contains large numbers of centrosomes. Therefore, in spite of failed nuclear division, the centrosomes continue to duplicate. Centrosomes associated with yolk nuclei in normal embryos provide a similar example. A population of yolk nuclei remain in the embryo interior as the somatic nuclei migrate to the cell periphery during nuclear cycles 8 and 9 (Rabinowitz, 1944; Zalokar and Erk, 1976). After nuclear cycle 10, these nuclei cease dividing, but DNA replication continues and they become polyploid (Foe and Alberts, 1983). As in *gnu* embryos, centrosomes become detached from their associated nuclei and undergo repeated rounds of duplication (Callaini and Dallai, 1991).

These studies also indicate that centrosome duplication does not require association with the nuclear envelope. This is supported by live analysis of normal embryos in which a centrosome has lost its close association with the nuclear envelope (Debec *et al.*, 1996). These free centrosomes duplicate, separate, and modulate microtubule dynamics in synchrony with neighboring centrosomes properly associated with the nuclear envelope. However, those free centrosomes at the greatest distance from normally dividing nuclei remain quiescent and do not exhibit any of the foregoing behaviors. It is possible that diffusable signals generated from the nuclei drive centrosome dynamics and duplication. Centrosomes at large distances from the nuclei may not receive these signals.

C. Mechanism of Separation

Separation of sister centrosomes has been most closely studied during the cortical divisions in which centrosome movements can be followed in living embryos (Kellogg *et al.,* 1988). During the telophase-to-interphase transition, it appears that the split sister centriole pair becomes apically positioned on the newly formed nuclear envelope (Debec *et al.,* 1996). Continued separation of the centrosomes describes an arc (the centrosomes migrate directly away from each other) resulting in each being positioned near opposite poles of the nucleus. Thus, not only is a force required to separate the centrosomes, but a mechanism is also required to maintain the angle of separation at 180° such that they describe an arc (Fig. 4). Analysis of the maternal effect mutation *daughterless-abo-like* (*dal*) indicates that these mechanisms may be distinct (Sullivan *et al.,* 1990). During the cortical divisions in *dal*-derived embryos, it appears that the angle of separation is often much less than 180° (Fig. 4). This results in separated sister centrosomes residing too close to one another but still on the same plane. Consequently, abnormally short or monopolar spindles are produced in the following metaphase. Interestingly, increasing the dosage of heterochromatin partially rescues the *dal* maternally induced lethality. Although it has not been examined directly, this raises the possibility that heterochromatin may influence the angle of centrosome separation in *dal*-derived embryos.

The force driving sister centrosome separation is probably microtubule based. Opposing arrays of astral microtubules are readily observed between separating sister centrosomes that have lost their nuclear envelope association (Debec *et al.,* 1996). In addition, cytoplasmic dynein, a minus-end-directed microtubule motor protein, is required for proper centrosome separation during the cortical divisions (Robinson *et al.,* 1999). In embryos lacking the dynein heavy chain, the initial stages of centrosome separation are disrupted. This produces severe spindle defects as the nuclei progress into metaphase. Dynein is also required to maintain attachment of the centrosomes to mitotic spindle poles (Robinson *et al.,* 1999). Once centrosome separation is complete, a balance of plus- and minus-end-directed motors is required to maintain centrosome separation in the mature spindle (Sharp *et al.,* 1999).

Centrosomes in syncytial embryos lacking the serine–threonine kinase aurora often form monopolar spindles, indicating that centrosomes did not separate appropriately (Glover *et al.,* 1995). Proteins similar to aurora have been identified in mammals, *Xenopus, Caenorhabditis elegans,* and yeast (Gopalan *et al.,* 1997; Kimura *et al.,* 1997; Yanai *et al.,* 1997; Bernard *et al.,* 1998; Roghi *et al.,* 1998; Schumacher *et al.,* 1998; Shindo *et al.,* 1998; Kimura *et al.,* 1999). Many of these proteins localize at the centrosome specifically

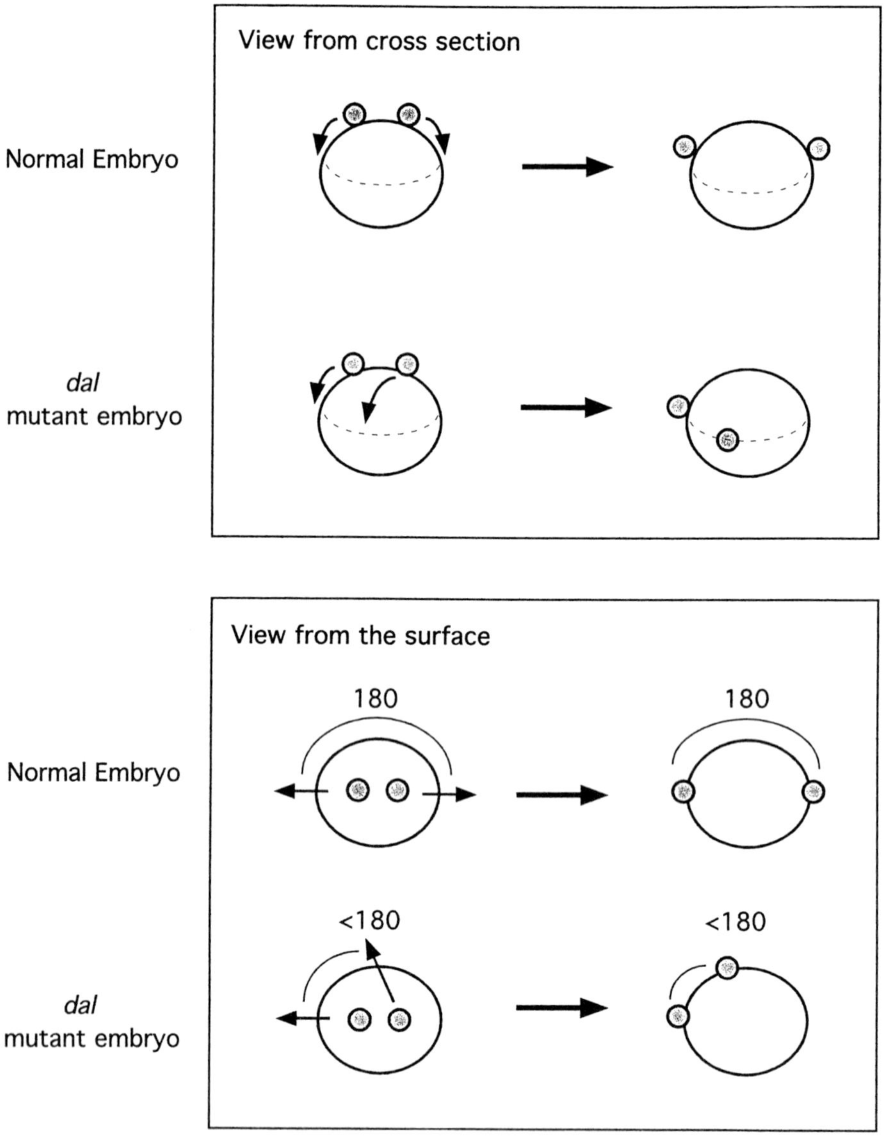

Figure 4 During the cortical divisions centrosome separation occurs along an arc: That is, the angle of separation is 180°. In embryos derived from *daughterless-abo-like* (*dal*) females, the angle is often much reduced. This results in centrosomes residing too close to one another and the formation of abnormally short or monopolar spindles.

during mitosis. Whether aurora localizes to centrosomes in the early *Drosophila* embryo remains to be determined.

V. Fertilization and the First Mitotic Division

A. Centrosomes Are Formed from Maternal and Paternal Factors and Provide Asters Necessary for Pronuclei Fusion

In *Drosophila,* the centrosome is supplied from the sperm basal body (Huettner, 1924; Sonnenblick, 1950). Immediately after entry, the basal body and the axoneme dissociate and the sperm head undergoes a dramatic metamorphosis to form the male pronucleus (for review see Fitch *et al.,* 1998). The basal body rapidly associates with the maternally provided centrosomal proteins CNN and CP190 and sports a small aster (Heuer *et al.,* 1995; Callaini and Riparbelli, 1996). These, and probably other maternal components play key roles in the biochemical transformation of the basal body into a centrosome. At this point, the newly formed centrosome is competent to duplicate, and two centrosomes, each producing robust astral arrays, are associated with the male pronucleus (Callaini and Riparbelli, 1996).

At approximately the same time of fertilization, the egg becomes fully hydrated and is released from its block in metaphase of meiosis I (Doane, 1960; Mahowald *et al.,* 1983). Completion of meiosis produces four haploid nuclei arranged perpendicularly with respect to the plasma membrane. The innermost nucleus migrates along the astral arrays emanating from the centrosomes associated with the male pronucleus (Rabinowitz, 1944). Pronuclear fusion may be driven by microtubule-based motor proteins. Mutant forms of Ncd, a minus-end-directed motor protein, enhance interactions between haploid meiotic products and may also be involved in pronuclear fusion (Komma and Endow, 1995). Mutations in the kinesin-like protein KLP3A also disrupt pronuclear fusion (Williams *et al.,* 1997). However, KLP3A is structurally related to a plus-end-directed motor, and migration of the female pronucleus toward the male pronucleus requires a minus-end-directed motor. If KLP3A proves to be a plus-end-directed motor, this suggests that it may be required in other capacities such as stabilization of the sperm aster. Alternatively, pronuclear migration may also require plus-end-directed migration of the male pronucleus along the astral microtubules (Williams *et al.,* 1997).

Pronuclear fusion is a unique developmental event, as the male pronucleus must integrate its genome, as well as its centrosomes, with the female pronucleus. Once the male and female pronuclei reside next to one another, the centrosomes separate to form the poles of the future spindle (Huettner,

1924; Sonnenblick, 1950). This is followed by partial nuclear envelope breakdown and entry into metaphase. Large sections of intact nuclear envelope maintain the maternal and paternal chromosome complements in distinct regions of the metaphase plate (Huettner, 1924; Sonnenblick, 1950). This is known as a gonomeric division, and integration of the maternal and paternal genomes does not occur until after the first anaphase.

B. Functional Centrosomes Are Not Required for the Formation of the First Metaphase Spindle

The first meiotic division occurs in the absence of a functional centrosome (Theurkauf and Hawley, 1992). The poles of these spindles are not well focused and centrosomal components are difficult to detect (Riparbelli and Callaini, 1996). Live analysis indicates that the spindles form by organizing around the chromosomes (Theurkauf and Hawley, 1992). In contrast, the first mitotic division is centrosome-based with focused spindles, core centrosomal components, and large astral arrays (Callaini and Riparbelli, 1996). In spite of its conspicuous presence, the role of the centrosome during the initial mitotic divisions remains unclear as embryos lacking core centrosomal components such as CNN still produce spindles and proceed through mitosis (Megraw *et al.,* 1999).

This issue has been addressed using another Diptera, *Sciara coprophila.* Although *Sciara* embryos acquire their centrosome from the basal body of the sperm, mitotic cycling is activated in unfertilized eggs (Ruder *et al.,* 1987; de Saint Phalle and Sullivan, 1998). Robust spindles are formed during metaphase in these unfertilized embryos in spite of the fact that they lack centrosomes. The spindles contain pole-to-pole and pole-to-kinetochore microtubules but lack astral microtubules. They are functional and proceed normally through anaphase with properly segregated sister chromosomes (de Saint Phalle and Sullivan, 1998). Anaphase B also occurs normally, demonstrating that at least in these embryos this process is independent of centrosomes and astral microtubules. These results are in accord with studies in *Xenopus* extracts demonstrating that chromosomes alone are competent to form bipolar spindles (Heald *et al.,* 1996, 1997). As described in a later section, a primary role of the centrosome during the initial divisions is to provide astral microtubules to maintain separation between syncytial nuclei.

In some cell types such as grasshopper spermatocytes, centrosomes are required for spindle formation (Zhang and Nicklas, 1995a, 1995b). To explain why spindle formation is dependent on the presence of a centrosome only in some cell types, it has been suggested that centrosomes lower the critical concentration required to promote microtubule assembly. The tubulin concentration in cells, such as the *Sciara* embryo, may be great enough that centrosomes are not required to promote microtubule assembly. Alternatively, cells may differ in whether they possess checkpoints that make spindle assembly dependent on the presence of centrosomes. Identifying a condition (mutation or drug) that eliminates this dependency relationship would provide strong support for this second alternative (Hartwell and Weinert, 1989).

C. Centrosomes and Parthenogenesis in *Drosophila*

Successful parthenogenesis is extremely rare in *D. melanogaster.* Of 500,000 unfertilized eggs examined, two inviable larva and no adults were produced (Stalker, 1954). Parthenogenesis is also rare across the entire family Drosophilidae. Of 29 species examined, only four (*D. parthenogenetica, D. polymorpha, D. affinis,* and *D. mercatorium*) produce adult progeny from unfertilized eggs (Stalker, 1954; Carson, 1967). Even in these species, less than 1% of unfertilized eggs from natural populations successfully develop to adults. The relative lack of parthenogenesis in *Drosophila* is surprising because many blocks to parthenogenesis are absent. Unfertilized *Drosophila melanogaster* embryos successfully complete meiosis and are translationally activated (Doane, 1960; Mahowald *et al.,* 1983). In contrast to mammals, the paternal genome is not required for successful development in *Drosophila* (Fuyama, 1984, 1986a, 1986b; Komma and Endow, 1995). Achieving diploidy through the fusion of haploid nuclei is a common event in *Drosophila.* Meiotic products in both unfertilized and fertilized embryos often fuse to produce diploid nuclei (Fuyama, 1984, 1986a, 1986b; Komma and Endow, 1995). The maternal-effect mutation *maternal-haploid* and the paternal-effect mutation *msk81* initiate mitotic cycles in the absence of pronuclear fusion (Edgar *et al.,* 1986; Yasuda *et al.,* 1995). However, even in these mutations, initiation of the mitotic cycles requires fertilization.

These observations indicate that there is an absolute requirement of the sperm to initiate progression of successful mitotic cycles. It is likely that one of the essential sperm-supplied factors is the basal body. In contrast to other species, it does not appear that centrosomes can be generated maternally in *D. melanogaster.* Although there have been some reported cases, these are rare and have not been reproduced. In fact, the only case of successful mitotic cycling without fertilization has been through the transplantation of blastoderm nuclei into unfertilized eggs. These will occasionally develop into adults (Illmensee, 1968; Schubiger and Schneiderman, 1971). It may be that centrosomes associated with these transplanted nuclei account for the successful development. In starfish embryos, injected centrosomes alone are able to induce mitotic cycling (Picard *et al.,* 1987). This experiment has not been performed in *Drosophila.*

An interesting idea concerning the role of the centrosome in the initiation of mitotic cycling comes from studies on a set of maternal-effect mutations that regulate DNA replication in the early embryo (Freeman *et al.,* 1986; Freeman and Glover, 1987; Shamanski and Orr-Weaver, 1991). For example, mutations in the gene *plutonium* (*plu*) undergo multiple rounds of S phase without an intervening mitosis, and embryos typically contain a few large polyploid nuclei (Shamanski and Orr-Weaver, 1991). In wild-type unfertilized embryos, all four meiotic products arrest with condensed chromosomes that have not undergone multiple rounds of S phase. In contrast, all four meiotic products continue to replicate their DNA in unfertilized *plu*-derived embryos. This indicates that Plu normally acts on all the meiotic products to prevent multiple rounds of S phase. This replication block is specifically released for the innermost meiotic product when it fuses with the male pronucleus. Axton and colleagues (1994) speculate that, as found in the later syncytial cycles, the centrosome promotes a localized breakdown of the nuclear envelope and thus specifically releases the female pronucleus from the repressive effects of Plu.

VI. Precortical Divisions

A. Centrosomes Provide Asters for Proper Nuclear Spacing and Orientation during the Initial Syncytial Divisions

Pronuclei fusion occurs in the anterior third of the embryo, and consequently the first mitotic divisions form an anterior-positioned ball of nuclei (Hatanaka and Okada, 1991; Foe *et al.,* 1993). A primary function of the centrosome during these initial divisions may be to generate astral microtubules that orient the nuclear divisions and prevent collisions between neighboring syncytial nuclei (de Saint Phalle and Sullivan, 1998; Megraw *et al.,* 1999). Embryos lacking core centrosomal components such as CNN develop spindles and undergo several rounds of mitosis, but large numbers of fused nuclei are often observed (Megraw *et al.,* 1999). This phenotype is similar to unfertilized *Sciara* embryos (de Saint Phalle and Sullivan, 1998). Lacking centrosomes, the dividing nuclei form kinetochore microtubules and pole-to-pole microtubules, but not astral microtubules. The initial divisions occur normally, but by the third and fourth nuclear cycles, neighboring telophase nuclei often collide and fuse. This suggests that as nuclear density increases, astral microtubules serve as fenders to prevent collisions. Supporting this interpretation is the fact that these collisions occur primarily during telophase when the astral microtubules are most prominent. The astral microtubules also function to establish overall nuclear orientation and distribution. Even in the initial divisions prior to extensive nuclear fusions, the unfertilized embryos exhibit abnormal nuclear configurations.

B. Centrosome-Generated Antiparallel Microtubule Arrays Are Required for Cortical Nuclear Migration

During the precortical divisions, the nuclei undergo two distinct patterns of migration: axial expansion followed by cortical migration. At nuclear cycle 4, eight nuclei form a sphere toward the anterior end of the embryo (as described above). During the next two cycles, the nuclei move axially to produce an even distribution in the interior of the embryo (Hatanaka and Okada, 1991). This process is sensitive to cytochalasin but not colcemid, indicating that the underlying mechanism requires microfilaments and not microtubules (Hatanaka and Okada, 1991). Supporting this conclusion is the fact that expansion occurs during prophase and metaphase when astral and interpolar microtubules are minimal.

Migration of the nuclei to the cortex begins during cycle 7, with the nuclei distributed in an evenly spaced ellipsoid in the interior of the embryo (Foe *et al.,* 1993). During the next two nuclear cycles, cortical migration proceeds through expansion of this ellipsoid of nuclei. The expansion force may be generated through plus-end-directed motors acting on neighboring arrays of overlapping astral microtubules (Baker *et al.,* 1993). This model is based on the observation that migration occurs only during telophase when each nucleus contains a centrosome pair with extensive astral microtubule arrays (Foe and Alberts, 1983). Also, in contrast to axial expansion, microtubule inhibitors, but not microfilament inhibitors, disrupt cortical nuclear migration (Zalokar and Erk, 1976). In this model, interactions between the centrosome-based astral arrays provide the force for migration. The nuclei do not play an active role in cortical migration; they are simply "passengers" on the migrating centrosomes. This is in accord with observations in which centrosomes unassociated with nuclei migrate to the cortex (Raff and Glover, 1989). Currently, no microtubule-based motors have been implicated in the process of cortical migration, but the maternally supplied bipolar kinesin KLP61F is a possible candidate (Sharp *et al.,* 1999).

C. Centrosomes and Yolk Nuclei

Establishment of different cell types during *Drosophila* embryogenesis first occurs with the formation of the yolk nuclei during cycles 8 and 9 (Rabinowitz, 1944; Zalokar and Erk, 1976; Foe and Alberts, 1983). These nuclei remain in the interior and divide in synchrony with the migrating nuclei. Once the somatic nuclei reach the cortex, this division synchrony is lost; the yolk nuclei are delayed entering telophase of nuclear cycle 10. After this point, mitotic divisions are no longer observed and the number of yolk nuclei remains constant from nuclear cycles 11 through 14 at about 200, although this number is variable because they often fuse (Foe and Alberts, 1983). Why these nuclei fail to migrate is unclear, but observations by Foe *et al.* (1993) indicate that they were initially part of the ellipsoid of expanding nuclei that had receded into the interior. The function of these nuclei is unknown, although they are a common feature of syncytial insect development.

The abrupt halt of mitosis in the yolk nuclei is probably a direct consequence of abnormal centrosomal behavior that begins after nuclear cycle 10 (Callaini and Dallai, 1991). The centrosomes duplicate but do not move to opposing poles. This results in the formation of nonfunctional monopolar spindles. Following this, large numbers of centrosomes are observed in the cytoplasm, indicating that they have lost their association with the nuclear envelope. Although the yolk nuclei no longer undergo mitosis, DNA replication continues and they become polyploid (Zalokar and Erk, 1976; Foe and Alberts, 1983). Whether this unusual centrosome behavior is determined by differentially distributed cytoplasmic factors or by factors intrinsic to each nucleus is not known. We favor the former, as differentially localized cytoplasmic factors influence nuclear behavior at the posterior pole (see next section) (Mahowald, 1962; Okada *et al.,* 1974).

VII. Cortical Divisions

A. Centrosomes and Pole Cell Formation

Through the mechanisms described earlier, the majority of nuclei reach the cortex during interphase of nuclear cycle 10. However, a few nuclei reach the posterior cortex during interphase of nuclear cycle 9. These nuclei and their closely associated centrosomes impinge on the plasma membrane, producing localized protrusions called pole buds (Foe and Alberts, 1983). The buds disappear as the nuclei enter mitosis and new buds form during interphase of nuclear cycle 10. During telophase of this cycle, the buds pinch off and form a cluster of posteriorly localized pole cells, the germline precursors. Special properties of the cytoplasm limit these events to the posterior pole. The cytoplasm at the posterior pole contains morphologically distinct polar granules consisting of RNA and protein (Counce, 1963; Mahowald, 1962, 1968, 1971). Transplantation of this cytoplasm is sufficient to generate ectopic pole cell formation (Okada *et al.,* 1974; Illmensee and

Mahowald, 1974). The posterior pole also contains extra plasma membrane in the form of extensive microvillar projections and high levels of contractile components such as myosin. This suggests that the posterior region of the cortex is primed to form cells, requiring only the proper stimulus.

An intriguing experiment indicates that arrival of the centrosomes to the cortex is sufficient to promote pole cell formation (Raff and Glover, 1989). Embryos were injected with the DNA synthesis inhibitor aphidicolin during cortical nuclear migration. This often resulted in centrosomes losing their association with the nuclear envelope and migrating freely to the cortex. Once at the cortex, many of these centrosomes maintained their ability to form astral microtubules and induce reorganization of the cortical actin and plasma membrane. Remarkably, centrosomes unassociated with a nucleus induced polar bud formation and pole cell formation. This centrosome-based activity required an intact actin cytoskeleton (Raff and Glover, 1989). Based on mechanisms by which the centrosome coordinates other rearrangements of the cortical cytoskeleton (see next section), centrosome-induced pole cell formation may be achieved through its ability to organize astral microtubule arrays.

As described, embryos deficient for the centrosomal protein CNN lack other core components, including γ-tubulin, and have severely reduced astral microtubules (Megraw *et al.,* 1999). Although the syncytial divisions are severely disrupted in these mutant embryos, pole cells (with nuclei) are often formed (Megraw and Kaufman, unpublished observations). This raises the possibility that the centrosome-based pole cell formation just described may occur through a mechanism that does not require astral microtubules. Alternatively, nuclei without functional centrosomes may also be sufficient to induce pole cell formation.

B. Mutual Interactions between the Centrosomes and Cortical Cytoskeleton

The majority of nuclei arrive at the cortex during interphase of nuclear cycle 10. Each nucleus maintains a pair of apically positioned sister centrosomes that induce rearrangements in the cortical cytoskeleton (Warn *et al.,* 1984; Karr and Alberts, 1986; Kellogg *et al.,* 1988; Figs. 5 and 6). Prior to nuclear migration, actin is concentrated in a homogeneous 3-μm layer just beneath the plasma membrane. Upon arrival of the nuclei, the actin becomes concentrated in caps centered above each centrosome pair. As sister centrosomes migrate toward opposite poles, the actin redistributes to form two smaller caps centered over each centrosome and its astral arrays. As prophase proceeds, actin-rich membrane invaginates to form furrows. At metaphase, these furrows encompass each newly formed spindle

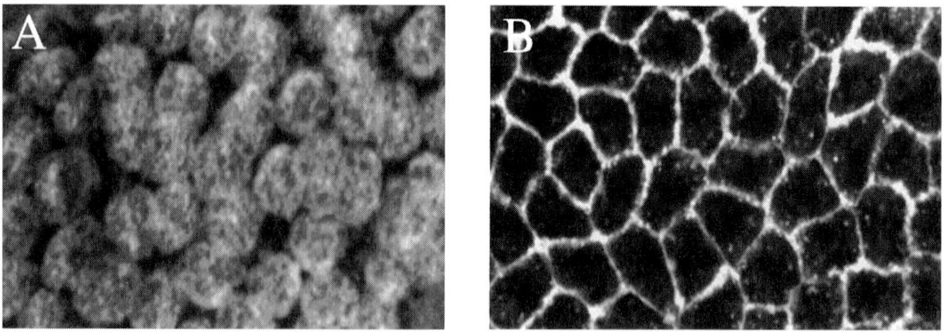

Figure 5 Centrosomes induce a dramatic reorganization of the cortical actin in syncytial *Drosophila* embryos. Actin is organized into caps above each interphase nucleus and its apical centrosome pair. As the nuclei enter metaphase, the actin reorganizes to form furrows encompassing the metaphase spindles. These furrows are required to prevent interactions between neighboring spindles.

in a half-shell. These structures, termed metaphase furrows, block inappropriate interactions between neighboring spindles (Sullivan *et al.,* 1990; Schejter and Wieschaus, 1993; Sullivan and Theurkauf, 1995). As the nuclei move through anaphase/telophase, the furrows regress and actin caps once again form in the following interphase. This cycling from caps to furrows continues until interphase of nuclear cycle 14, when the embryos undergo cellularization (Karr and Alberts, 1986; Foe *et al.,* 1993).

Centrosomes play a key role in directing these cortical cytoskeletal rearrangements. Free centrosomes unassociated with a nucleus maintain an ability to organize the actin in a cell-cycle-dependent manner (Raff and Glover, 1989; Yasuda *et al.,* 1991). During interphase, free centrosomes organize actin into caps. As the embryo progresses into metaphase, these caps enlarge to form a ring of actin around each centrosome (Raff and Glover, 1988; Yasuda *et al.,* 1991). Free centrosomes also direct cycles of cytoplasmic budding and pole cell formation, indicating that these cortical cytoskeletal reorganizations are functional and capable of contraction (Raff and Glover, 1988, 1989).

The mechanism concentrating actin at the centrosomes is unknown. Fluorescein phalloidin, a molecule that specifically binds polymerized actin, labels the actin caps, indicating that these consist primarily of microfilaments (Warn *et al.,* 1984). Therefore, a polymerizing activity closely associated with the centrosome is a possible mechanism of concentrating actin. Currently, no such acitivity has been identified, but there are examples of other types of actin organizing activity at the centrosome (see next section).

Injection of antibodies against tubulin into early embryos causes a disruption of microtubules and results in disorganization of the actin caps (Warn

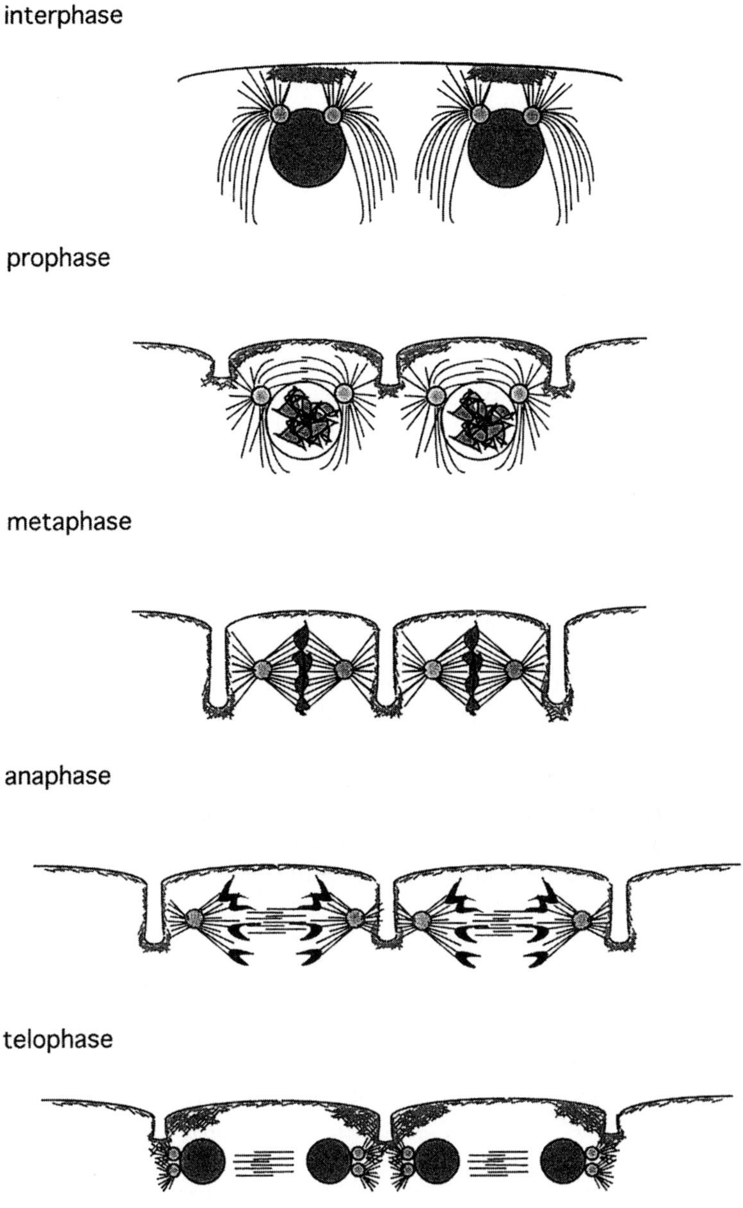

Figure 6 During interphase, actin caps are centered above each centrosome and its astral microtubules. As the centrosomes separate, the actin caps undergo a corresponding redistribution. The actin based furrows invaginate during late prophase and metaphase. During anaphase and telophase these furrows retract, and actin caps begin to reform around the newly duplicated centrosomes.

et al., 1987). This suggests an alternative microtubule-based mechanism of concentrating the actin into caps. For example, minus-end-directed motor proteins may associate with microfilaments and concentrate them into centrosome-based caps.

In the maternal-effect mutation *sponge,* the cortical cytoskeleton does not respond to the arrival of the centrosomes and actin caps do not form (Postner *et al.,* 1992). This mutation therefore provides an entry point for dissecting the signal and response elements of this centrosome-based cytoskeletal reorganization. Another cortical component, the actin binding protein 13D2 (Miller *et al.*, 1989), forms relatively normal caps in *sponge*-derived embryos. This suggests that 13D2 forms a cap in response to centrosome-based signals and then mediates actin cap formation (Postner *et al.*, 1992; Schejter and Wieschaus, 1993). Alternatively, 13D2 and actin may form caps through different mechanisms, with the *sponge* product being specially required for the latter. In view of the observation that intact microtubules are required for cap formation, it will be interesting to examine the distribution of microtubules in *sponge*-derived embryos.

Analysis of the maternal-effect mutation *daughterless-abo-like* (*dal*) provides further support for the role of centrosomes in orchestrating cortical cytoskeletal events (Sullivan *et al.,* 1990). Centrosome migration is disrupted in *dal* mutant embryos. This leads to closely positioned centrosomes that are unable to form a functional spindle. There are also large gaps in the metaphase furrows in *dal*-derived embryos (Sullivan *et al.,* 1993b). These gaps tend to occur in regions of the embryo in which the distance between centrosomes is abnormally large, indicating that intercentrosomal distance is critical for furrow establishment.

A number of observations indicate that the reciprocal relationship also holds: The cytoskeleton influences centrosome behavior. The centrosomes are intimately associated with the actin caps during the cortical divisions (Karr and Alberts, 1983; Foe and Odell, unpublished observations). In fact, the products of abnormal cortical nuclei recede into the interior of the embryo, but their centrosomes remain tightly associated with the cortical cytoskeleton (Sullivan *et al.,* 1993a). If these free centrosomes are near normally dividing nuclei, separation of sister centrosomes occurs. Remarkably, this separation occurs on a plane parallel to the plasma membrane (Debec *et al.,* 1996). This indicates that the cortical cytoskeleton guides centrosome separation. In fact, disruption of the cortical actin with cytochalasin prevents prophase separation of sister centrosomes (Callaini and Riparbelli, 1992). In other systems, it is evident that the cytoskeleton guides centrosome position and movement. Spindle pole body position and migration rely on interactions between the actin cytoskeleton and the astral microtubules (Palmer *et al.,* 1992). In *Drosophila* spermatocytes mutant for profilin (*chickadee*) and cofilin (*twinstar*), centrosome migration and separation are disrupted (Gunsalus *et al.,* 1995; Giansanti *et al.,* 1998), indicating that microfilaments are important for this process.

C. The Centrosome-Associated Protein Nuclear-fallout Is Required for Long-Range Transport of Furrow Components

As described earlier, the cortical nuclei with their apically localized centrosome pair induce cytoskeletal reorganizations essential for normal nuclear division. For example, treatment of embryos with cytochalasin B, a microfilament inhibitor, disrupts actin cap and metaphase furrow formation and results in collisions and fusions between neighboring nuclei (Zalokar and Erk, 1976). This observation provided the rationale for a genetic screen to isolate centrosomal and cytoskeletal components specifically required for these cytoskeletal rearrangements (Sullivan *et al.,* 1993b). Maternal-effect mutations were identified in which the embryos develop normally until completion of nuclear migration to the cortex. Once at the cortex, however, extensive division abnormalities occur. This screen should enrich for mutations in components specifically required for the cortical divisions. To date, more than 10 mutations have been identified that specifically disrupt the cortical divisions, and the cortical cytoskeletal defects have been analyzed in detail for a number of these (Sullivan *et al.,* 1990; Postner *et al.,* 1992; Sullivan *et al.,* 1993b; Fogarty *et al.,* 1994; Zhang *et al.,* 1996; Rothwell *et al.,* 1998, 1999). Analysis of two of these mutations, *scrambled* (*sced*) and *nuclear-fallout* (*nuf*), indicates that they both encode proteins closely associated with the centrosome that are required for proper metaphase furrow formation (Sullivan *et al.,* 1993b; Kramer *et al.,* 1997; Rothwell *et al.,* 1998, 1999).

Nuclear fallout (Nuf) is a highly phosphorylated protein with extensive coiled-coils predicted to form near its C-terminal end (Rothwell *et al.,* 1998). During prophase of the syncytial divisions, Nuf concentrates at the centrosomes and is generally cytoplasmic throughout the rest of the nuclear cycle (Rothwell *et al.,* 1998). Nuf is not a core centrosomal component, as microtubule depolymerization eliminates its centrosomal localization (Rothwell, unpublished observation). In embryos lacking Nuf, although microtubule structure and dynamics are normal (Sullivan *et al.,* 1993b), actin is not properly recruited from the interphase caps to the furrows (Rothwell *et al.,* 1998). During prophase and metaphase in *nuf*-derived embryos, actin remains abnormally concentrated around centrosomes, and there are large gaps in the metaphase furrows (Sullivan *et al.,* 1993b; Rothwell *et al.,* 1998). This suggests that Nuf acts at the centrosome to recruit actin from caps to furrows.

Nuf is also required for recruiting membrane to the furrows (Rothwell *et al.*, 1999). In wild-type embryos, membrane-bearing particles are recruited to the site of furrow formation during prophase. In equivalently staged *nuf*-derived embryos, many of these membrane particles are not properly recruited and remain in a punctate distribution around the nuclei and their associated centrosomes. This leads to a model in which Nuf acts at the centrosomes to initiate transport of membrane and associated components such as actin along the microtubules to the furrows (Fig. 7). This model accounts for the observation that furrow formation fails primarily in regions most distant from the centrosomes (Rothwell *et al.*, 1998), where the greatest demand would be placed on the transport process. Actin and the membrane-bearing particles are often closely associated, suggesting that they are transported to the furrows as a unit (Rothwell *et al.*, 1999). Nuf remains concentrated at the centrosomes throughout the recruitment process and may be involved in the recruitment of these particles onto microtubules for subsequent transport to the furrows.

D. Centrosomes Are Required for Cellularization

During interphase of nuclear cycle 14, each of the cortical nuclei becomes enclosed in a separate cell in a process called cellularization (for reviews

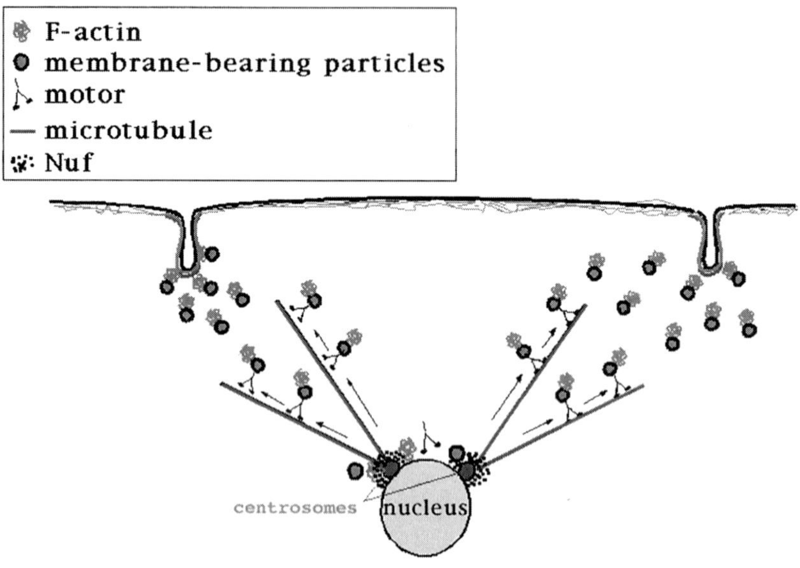

Figure 7 A model proposing that Nuf acts at the centrosome to load vesicles and associated proteins such as actin for transport along the microtubules to the cortex.

see Karr and Alberts, 1986; Foe *et al.*, 1993; Schejter and Wieschaus, 1993). At cellularization, the centrosomes separate slightly and nucleate a basket of microtubules that forms around each of the cortical nuclei. Actin-rich cellularization furrows then form and the nuclei are partitioned into separate cells. A role for the centrosome in initiating cellularization is indicated by the finding that disruption of their associated astral microtubule arrays disrupts the process (Zalokar and Erk, 1976; Foe and Alberts, 1983; Edgar *et al.*, 1987; Warn *et al.*, 1987). Free centrosomes are not able to induce cellularization (Raff and Glover, 1989). This indicates that cellularization requires nuclei in addition to functional centrosomes.

Nuf is stably associated with the centrosomes throughout cellularization (Rothwell *et al.*, 1998). Embryos lacking Nuf produce a cellularization phenotype in which there are extensive gaps where furrows did not initiate formation (Sullivan *et al.*, 1993b; Rothwell *et al.*, 1998). This phenotype is similar to that observed for the zygotic mutations *nullo* and *serendipity* (Schweisguth *et al.*, 1990; Simpson and Wieschaus, 1990; Rose and Wieschaus, 1992). These genes encode novel proteins that localize to the invaginating furrows and appear to be required for the stabilization of the growing furrow. In contrast, the maternally supplied Nuf centrosomal protein acts upstream of these zygotic genes in initiating furrow formation.

E. Centrosomes Play a Central Role in the Elimination of Abnormal Cortical Nuclei

A number of studies demonstrate that cell cycle checkpoints are present and operational during the syncytial cortical divisions (for review see Yu *et al.*, 1998). This is unexpected because checkpoint-induced cell cycle delays would disrupt division synchrony necessary for normal syncytial development. However, in contrast to checkpoints that rely on delays to correct the division errors, it appears that many checkpoints operating in the syncytial embryo rely on cell cycle delays to eliminate abnormal division products (Sullivan *et al.*, 1993a). For example, in embryos bearing an abnormally long chromosome, individual nuclei often delay anaphase initiation. The telophase products of these delayed nuclei recede into the interior and are removed from the dividing population of cortical nuclei. As described earlier, when these nuclei recede into the interior, their centrosomes remain on the cortex.

These observations suggest that nuclei are maintained at the cortex via their centrosomes. If the nuclear/centrosome association is disrupted, the nuclei will recede from the cortex. Telophase is a critical point in the cortical nuclear cycles because the centrosomes must reestablish their link to the newly reformed nuclear envelope. Telophase is also the stage in the division cycle at which abnormal nuclei tend to recede from the cortex (Sullivan *et al.,* 1993a). This suggests that checkpoints may operate in metaphase to delay the chromosome cycle but not the centrosome/spindle cycle of damaged nuclei. When these nuclei enter telophase, their chromosome and centrosome cycles are no longer in synchrony and they are eliminated. The net result is an efficient mechanism of maintaining the integrity of the dividing population of cortical nuclei.

References

Axton, J. M., Shamanski, F. L., Young, L. M., Henderson, D. S., Boyd, J. B., and Orr-Weaver, T. L. (1994). The inhibitor of DNA replication encoded by the *Drosophila* gene *plutonium* is a small, ankyrin repeat protein. *EMBO J.* **13,** 462–470.

Baker, J., Theurkauf, W. E., and Schubiger, G. (1993). Dynamic changes in microtubule configuration correlate with nuclear migration in the preblastoderm *Drosophila* embryo. *J. Cell Biol.* **122,** 113–121.

Balczon, R. (1996). The centrosome in animal cells and its functional homologs in plant and yeast cells. *Int. Rev. Cytol.* **169,** 25–82.

Bernard, M., Sanseau, P., Henry, C., Couturier, A., and Prigent, C. (1998). Cloning of STK13, a third human protein kinase related to *Drosophila* aurora and budding yeast Ip11 that maps on chromosome 19q 13.3-ter. *Genomics* **53,** 406–409.

Callaini, G., and Dallai, R. (1991). Abnormal behavior of the yolk centrosomes during early embryogenesis of *Drosophila melanogaster. Exp. Cell Res.* **192,** 16–21.

Callaini, G., and Riparbelli, M. G. (1990). Centriole and centrosome cycle in the early *Drosophila* embryo. *J. Cell Sci.* **97,** 539–543.

Callaini, G., and Riparbelli, M. G. (1992). Involvement of microtubules and microfilaments in centrosome dynamics during the syncytial mitoses of the early *Drosophila* embryo. *Exp. Cell Res.* **201,** 241–244.

Callaini, G., and Riparbelli, M. G. (1996). Fertilization in *Drosophila melanogaster:* centrosome inheritance and organization of the first mitotic spindle. *Dev. Biol.* **176,** 199–208.

Callaini, G., Whitfield, W. G., and Riparbelli, M. G. (1997). Centriole and centrosome dynamics during the embryonic cell cycles that follow the formation of the cellular blastoderm in *Drosophila. Exp. Cell Res.* **234,** 183–190.

Carson, H. L. (1967). Selection for parthenogenesis in *Drosophila mercatorum. Genetics* **55,** 157–171.

Chretien, D., Buendia, B., Fuller, S. D., and Karsenti, E. (1997). Reconstruction of the centrosome cycle from cryoelectron micrographs. *J. Struct. Biol.* **120,** 117–133.

Counce, S. J. (1963). Developmental morphology of polar granules in *Drosophila* including observation on pole cell behavior and distribution during embryogenesis. *J. Morphol.* **112,** 129–145.

Dasso, M., and Newport, J. W. (1990). Completion of DNA replication is monitored by a feedback system that controls the initiation of mitosis in vitro: studies in *Xenopus. Cell* **61,** 811–823.

Debec, A., and Marcaillou, C. (1997). Structural alterations of the mitotic apparatus induced by the heat shock response in *Drosophila* cells. *Biol. Cell* **89,** 67–78.

Debec, A., Kalpin, R. F., Daily, D. R., McCallum, P. D., Rothwell, W. F., and Sullivan, W. (1996). Live analysis of free centrosomes in normal and aphidicolin-treated *Drosophila* embryos. *J. Cell Biol.* **134,** 103–115.

Debec, A., Marcaillou, C., Bobinnec, Y., and Borot, C. (1999). The centrosome cycle in syncytial *Drosophila* embryos analyzed by energy filtering transmission electron microscopy. *Biol. Cell.* **91**(4–5), 379–391.

Doane, W. (1960). Completion of meiosis in uninseminated eggs of *Drosophila melanogaster. Science* **132,** 677–678.

do Carmo Avides, M., and Glover, D. M. (1999). Abnormal spindle protein, Asp, and the integrity of mitotic centrosomal microtubule organizing centers. *Science* **283,** 1733–1735.

Edgar, B. A., Kiehle, C. P., and Schubiger, G. (1986). Cell cycle control by the nucleo-cytoplasmic ratio in early *Drosophila* development. *Cell* **44,** 365–372.

Edgar, B. A., Odell, G. M., and Schubiger, G. (1987). Cytoarchitecture and the patterning of fushi tarazu expression in the *Drosophila* blastoderm. *Genes Dev.* **1,** 1226–1237.

Endow, S. A., and Komma, D. J. (1996). Centrosome and spindle function of the *Drosophila* Ncd microtubule motor visualized in live embryos using Ncd-GFP fusion proteins. *J. Cell Sci.* **109,** 2429–2442.

Endow, S. A., Henikoff, S., and Soler-Niedziela, L. (1990). Mediation of meiotic and early mitotic chromosome segregation in *Drosophila* by a protein related to kinesin. *Nature* **345,** 81–83.

Endow, S. A., Chandra, R., Komma, D. J., Yamamoto, A. H., and Salmon, E. D. (1994). Mutants of the *Drosophila* Ncd microtubule motor protein cause centrosomal and spindle pole defects in mitosis. *J. Cell Sci.* **107,** 859–867.

Fitch, K. R., Yasuda, G. K., Owens, K. N., and Wakimoto, B. T. (1998). Paternal effects in *Drosophila:* implications for mechanisms of early development. *Curr. Top. Dev. Biol.* **38,** 1–34.

Foe, V. E., and Alberts, B. M. (1983). Studies of nuclear and cytoplasmic behavior during the five mitotic cycles that precede gastrulation in *Drosophila* embryogenesis. *J. Cell Sci.* **61,** 31–70.

Foe, V. E., Odell, G., and Edgar, B. A. (1993). Mitosis and morphogenesis in the *Drosophila* embryo: point and counterpoint. In *The Development of Drosophila melanogaster* (E. M. Bate and A. Martinez Arias, Eds.). Cold Spring Harbor Laboratory Press, Plainview, NY.

Fogarty, P., Kalpin, R. F., and Sullivan, W. (1994). The *Drosophila* maternal-effect mutation *grapes* causes a metaphase arrest at nuclear cycle 13. *Development* **120,** 2131–2142.

Freeman, M., Nusslein-Volhard, C., and Glover, D. M. (1986). The dissociation of nuclear and centrosomal division in *gnu,* a mutation causing giant nuclei in *Drosophila. Cell* **46,** 457–468.

Freeman, M., and Glover, D. M. (1987). The *gnu* mutation of *Drosophila* causes inappropriate DNA synthesis in unfertilized and fertilized eggs. *Genes Dev.* **1,** 924–930.

Fuyama, Y. (1984). Gynogenesis in *Drosophila melanogaster. Jap. J. Genet.* **59,** 91–96.

Fuyama, Y. (1986a). Genetics of parthenogenesis in *Drosophila melanogaster.* II. Characterization of a gynogenetically reproducing strain. *Genetics* **114,** 4995–5009.

Fuyama, Y. (1986b). Genetics of parthenogenesis in *Drosophila melanogaster.* I. The modes of diploidization in the gynogenesis induced by male-sterile mutant, *ms(3)K81. Genetics* **112,** 237–248.

Gard, D. L., Hafezi, S., Zhang, T., and Doxsey, S. J. (1990). Centrosome duplication continues in cycloheximide-treated *Xenopus* blastulae in the absence of a detectable cell cycle. *J. Cell Biol.* **110,** 2033–2042.

Giansanti, M. G., Bonaccorsi, S., Williams, B., Williams, E. V., Santolamazza, C., Goldberg, M. L., and Gatti, M. (1998). Cooperative interactions between the central spindle and the contractile ring during *Drosophila* cytokinesis. *Genes Dev.* **12,** 396–410.

Glotzer, M., Murray, A. W., and Kirschner, M. W. (1991). Cyclin is degraded by the ubiquitin pathway. *Nature* **349,** 132–138.

Glover, D. M., Gonzalez, C., and Raff, J. W. (1993). The centrosome. *Sci. Am.* **268,** 62–68.

Glover, D. M., Leibowitz, M. H., McLean, D. A., and Parry, H. (1995). Mutations in aurora prevent centrosome separation leading to the formation of monopolar spindles. *Cell* **81,** 95–105.

Gonzalez, C., Saunders, R. D., Casal, J., Molina, I., Carmena, M., Ripoll, P., and Glover, D. M. (1990). Mutations at the *asp* locus of *Drosophila* lead to multiple free centrosomes in syncytial embryos, but restrict centrosome duplication in larval neuroblasts. *J. Cell Sci.* **96,** 605–616.

Gopalan, G., Chan, C. S., and Donovan, P. J. (1997). A novel mammalian, mitotic spindle-associated kinase is related to yeast and fly chromosome segregation regulators. *J. Cell Biol.* **138,** 643–656.

Gould, R. R., and Borisy, G. G. (1977). The pericentriolar material in Chinese hamster ovary cells nucleates microtubule formation. *J. Cell Biol.* **73,** 601–615.

Gunsalus, K. C., Bonaccorsi, S., Williams, E., Verni, F., Gatti, M., and Goldberg, M. L. (1995). Mutations in *twinstar,* a *Drosophila* gene encoding a cofilin/ADF homologue, result in defects in centrosome migration and cytokinesis. *J. Cell Biol.* **131,** 1243–1259.

Hartwell, L. H., and Weinert, T. A. (1989). Checkpoints: controls that ensure the order of cell cycle events. *Science* **246,** 629–634.

Hatanaka, K., and Okada, M. (1991). Retarded nuclear migration in *Drosophila* embryos with aberrant F-actin reorganization caused by maternal mutations and by cytochalasin treatment. *Development* **111,** 909–920.

Heald, R., Tournebize, R., Blank, T., Sandaltzopoulos, R., Becker, P., Hyman, A., and Karsenti, E. (1996). Self-organization of microtubules into bipolar spindles around artificial chromosomes in *Xenopus* egg extracts. *Nature* **382,** 420–425.

Heald, R., Tournebize, R., Habermann, A., Karsenti, E., and Hyman, A. (1997). Spindle assembly in *Xenopus* egg extracts: respective roles of centrosomes and microtubule self-organization. *J. Cell Biol.* **138,** 615–628.

Helps, N. R., Brewis, N. D., Lineruth, K., Davis, T., Kaiser, K., and Cohen, P. T. (1998). Protein phosphatase 4 is an essential enzyme required for organisation of microtubules at centrosomes in *Drosophila* embryos. *J. Cell Sci.* **111,** 1331–1340.

Heuer, J. G., Li, K., and Kaufman, T. C. (1995). The *Drosophila* homeotic target gene *centrosomin* (*cnn*) encodes a novel centrosomal protein with leucine zippers and maps to a genomic region required for midgut morphogenesis. *Development* **121,** 3861–3876.

Huang, J., and Raff, J. W. (1999). The disappearance of cyclin B at the end of mitosis is regulated spatially in *Drosophila* cells. *EMBO J.* **18,** 2184–2195.

Huettner, A. F. (1924). Maturation and fertilization in *Drosophila melanogaster. J. Morphol.* **39,** 249–265.

Illmensee, K. (1968). Transplantation of embryonic nuclei into unfertilized eggs of *Drosophila melanogaster. Nature* **219,** 1268–1269.

Illmensee, K., and Mahowald, A. P. (1974). Transplantation of posterior polar plasm in *Drosophila.* Induction of germ cells at the anterior pole of the egg. *Proc. Natl. Acad. Sci. USA* **71,** 1016–1020.

Joshi, H. C. (1994). Microtubule organizing centers and gamma-tubulin. *Curr. Opin. Cell Biol.* **6,** 54–62.

Kalnins, V. I. (1992). *The Centrosome.* Academic Press, San Diego.

Kalt, A., and Schliwa, M. (1993). Molecular components of the centrosome. *Trends Cell Biol.* **3,** 118–128.

Karr, T. L., and Alberts, B. M. (1986). Organization of the cytoskeleton in early *Drosophila* embryos. *J. Cell Biol.* **102,** 1494–1509.

Karsenti, E., Newport, J., Hubble, R., and Kirschner, M. (1984). Interconversion of metaphase and interphase microtubule arrays, as studied by the injection of centrosomes and nuclei into *Xenopus* eggs. *J. Cell Biol.* **98,** 1730–1745.

Kellogg, D. R., and Alberts, B. M. (1992). Purification of a multiprotein complex containing centrosomal proteins from the *Drosophila* embryo by chromatography with low-affinity polyclonal antibodies. *Mol. Biol. Cell* **3,** 1–11.

Kellogg, D. R., Mitchison, T. J., and Alberts, B. M. (1988). Behavior of microtubules and actin filaments in living *Drosophila* embryos. *Development* **103,** 675–686.

Kellogg, D. R., Field, C. M., and Alberts, B. M. (1989). Identification of microtubule-associated proteins in the centrosome, spindle, and kinetochore of the early *Drosophila* embryo. *J. Cell Biol.* **109,** 2977–2991.

Kellogg, D. R., Sullivan, W., Theurkauf, W., Oegema, K., Raff, J. W., and Alberts, B. M. (1991). Studies on the centrosome and cytoplasmic organization in the early *Drosophila* embryo. *Cold Spring Harbor Symposia on Quantitative Biology* **56,** 649–662.

Kellogg, D. R., Moritz, M., and Alberts, B. M. (1994). The centrosome and cellular organization. *Annu. Rev. Biochem.* **63,** 639–674.

Kellogg, D. R., Oegema, K., Raff, J., Schneider, K., and Alberts, B. M. (1995). CP60: a microtubule-associated protein that is localized to the centrosome in a cell cycle-specific manner. *Mol. Biol. Cell* **6,** 1673–1684.

Kidd, D., and Raff, J. W. (1997). LK6, a short lived protein kinase in *Drosophila* that can associate with microtubules and centrosomes. *J. Cell Sci.* **110,** 209–219.

Kimura, M., Kotani, S., Hattori, T., Sumi, N., Yoshioka, T., Todokoro, K., and Okano, Y. (1997). Cell cycle-dependent expression and spindle pole localization of a novel human protein kinase, Aik, related to Aurora of *Drosophila* and yeast Ip11. *J. Biol. Chem.* **272,** 13766–13771.

Kimura, M., Matsuda, Y., Yoshioka, T., and Okano, Y. (1999). Cell cycle-dependent expression and centrosome localization of a third human aurora/Ip11-related protein kinase, AIK3. *J. Biol. Chem.* **274,** 7334–7340.

Komma, D. J., and Endow, S. A. (1995). Haploidy and androgenesis in *Drosophila. Proc. Natl. Acad. Sci. USA.* **92,** 11884–11888.

Kramer, J., Kuhn, J., and Theurkauf, W. E. (1997). The *Drosophila scrambled* gene product may provide a link between the microtubule and actin cytoskeleton: *Mol. Biol. Cell* **8,** 367a.

Lange, B. M. H., and Gull, K. (1996). A structural study of isolated mammalian centrioles using negative staining electron microscopy. *J. Struct. Biol.* **117,** 222–226.

Lehner, C. F., and O'Farrell, P. H. (1990). The roles of *Drosophila* cyclins A and B in mitotic control. *Cell* **61,** 535–547.

Li, K., and Kaufman, T. C. (1996). The homeotic target gene *centrosomin* encodes an essential centrosomal component. *Cell* **85,** 585–596.

Llamazares, S., Tavosanis, G., and Gonzalez, C. (1999). Cytological characterisation of the mutant phenotypes produced during early embryogenesis by null and loss-of-function alleles of the gammaTub37C gene in *Drosophila. J. Cell Sci.* **112,** 659–667.

Mahowald, A. P. (1962). Fine structure of pole cells and polar granules in *Drosophila melanogaster. J. Exp. Zool.* **151,** 201–205.

Mahowald, A. P. (1968). Polar granules of *Drosophila.* II. Ultrastructural changes during early embryogenesis. *J. Exp. Zool.* **167,** 237–262.

Mahowald, A. P. (1971). Polar granules of *Drosophila.* IV. Cytochemical studies showing loss of RNA from polar granules during early stages of embryogenesis. *J. Exp. Zool.* **176,** 345–352.

Mahowald, A. P., Goralski, T. J., and Caulton, J. H. (1983). In vitro activation of *Drosophila* eggs. *Dev. Biol.* **98,** 437–445.

Maldonado-Codina, G., and Glover, D. M. (1992). Cyclins A and B associate with chromatin and the polar regions of spindles, respectively, and do not undergo complete degradation at anaphase in syncytial *Drosophila* embryos. *J. Cell Biol.* **116,** 967–976.

Mazia, D., Harris, P. J., and Bibring, T. (1960). The multiplicity of the mitotic centers and the time course of their duplication and separation. *J. Biophys. Biochem. Cytol.* **7,** 1–20.

McDonald, H. B., and Goldstein, L. S. (1990). Identification and characterization of a gene encoding a kinesin-like protein in *Drosophila. Cell* **61,** 991–1000.

McDonald, K., and Morphew, M. K. (1993). Improved preservation of ultrastructure in difficult-to-fix organisms by high pressure freezing and freeze substitution: I. *Drosophila melanogaster* and *Strongylocentrotus purpuratus* embryos. *Microsc. Res. Techn.* **24,** 465–473.

Megraw, T. L., Li, K., Kao, L. R., and Kaufman, T. C. (1999). The centrosomin protein is required for centrosome assembly and function during cleavage in *Drosophila. Development* **126,** 2829–2839.

Miller, K. G., Field, C. M., and Alberts, B. M. (1989). Actin-binding proteins from *Drosophila* embryos: a complex network of interacting proteins detected by F-actin affinity chromatography. *J. Cell Biol.* **109,** 2963–2975.

Moritz, M., Braunfeld, M. B., Fung, J. C., Sedat, J. W., Alberts, B. M., and Agard, D. A. (1995a). Three-dimensional structural characterization of centrosomes from early *Drosophila* embryos. *J. Cell Biol.* **130,** 1149–1159.

Moritz, M., Braunfeld, M. B., Sedat, J. W., Alberts, B., and Agard, D. A. (1995b). Microtubule nucleation by gamma tubulin–containing rings in the centrosome. *Nature* **378,** 638–640.

Moritz, M., Zheng, Y., Alberts, B. M., and Oegema, K. (1998). Recruitment of the gamma-tubulin ring complex to *Drosophila* salt-stripped centrosome scaffolds. *J. Cell Biol.* **142,** 775–786.

Murphy, S. M., and Stearns, T. (1996). Cytoskeleton: microtubule nucleation takes shape. *Curr. Biol.* **6,** 642–644.

Murray, A., Ed. (1993). *The Cell Cycle: An Introduction.* Freeman, New York.

Oegema, K., Whitfield, W. G. F., and Alberts, B. (1995). The cell-cycle dependent localization of the CP190 centrosomal protein is determined by the coordinate action of two separable domains. *J. Cell Biol.* **131,** 1261–1273.

Oegema, K., Marshall, W. F., Sedat, J. W., and Alberts, B. M. (1997). Two proteins that cycle asynchronously between centrosomes and nuclear structures: *Drosophila* CP60 and CP190. *J. Cell Sci.* **110,** 1573–1583.

Okada, M., Kleinman, I. A., and Schneiderman, H. A. (1974). Restoration of fertility in sterilized *Drosophila* eggs by transplantation of polar cytoplasm. *Dev. Biol.* **37,** 43–54.

Paddy, M. R., Saumweber, H., Agard, D. A., and Sedat, J. W. (1996). Time-resolved, in vivo studies of mitotic spindle formation and nuclear lamina breakdown in *Drosophila* early embryos. *J. Cell Sci.* **109,** 591–607.

Palmer, R. E., Sullivan, D. S., Huffaker, T., and Koshland, D. (1992). Role of astral microtubules and actin in spindle orientation and migration in the budding yeast, *Saccharomyces cerevisiae. J. Cell Biol.* **119,** 583–593.

Pereira, G., and Schiebel, E. (1997). Centrosome-microtubule nucleation. *J. Cell Sci.* **110,** 295–300.

Picard, A., Karsenti, E., Dabauvalle, M. C., and Doree, M. (1987). Release of mature starfish oocytes from interphase arrest by microinjection of human centrosomes. *Nature* **327,** 170–172.

Postner, M. A., Miller, K. G., and Wieschaus, E. (1992). Maternal-effect mutations of the *sponge* locus affect cytoskeletal rearrangements in *Drosophila melanogaster* embryos. *J. Cell Biol.* **119,** 1205–1218.

Rabinowitz, M. (1944). Studies on the cytology and early embryology of the egg of *Drosophila melanogaster. J. Morph.* **69,** 1–49.

Raff, J. W., and Glover, D. M. (1988). Nuclear and cytoplasmic mitotic cycles continue in *Drosophila* embryos in which DNA synthesis is inhibited with aphidicolin. *J. Cell Biol.* **107,** 2009–2019.

Raff, J. W., and Glover, D. M. (1989). Centrosomes, and not nuclei, initiate pole cell formation in *Drosophila* embryos. *Cell* **57,** 611–619.

Raff, J. W., Kellogg, D. R., and Alberts, B. M. (1993). *Drosophila* gamma-tubulin is part of a complex containing two previously identified centrosomal MAPs. *J. Cell Biol.* **121,** 823–835.

Rattner, J. B., and Phillips, S. G. (1973). Independence of centriole formation and DNA synthesis. *J. Cell Biol.* **57,** 359–372.

Rieder, C. R., and Borisy, G. G. (1982). The centrosome cycle in PtK2 cells: Asymetric distribution and structural changes in the pericentriolar material. *Biol. Cell.* **44,** 117–132.

Riparbelli, M. G., and Callaini, G. (1996). Meiotic spindle organization in fertilized *Drosophila* oocyte: presence of centrosomal components in the meiotic apparatus. *J. Cell Sci.* **109,** 911–918.

Robbins, E., Jentzsch, G., and Micali, A. (1968). The centriole cycle in synchronized HeLa cells. *J. Cell Biol.* **36,** 329–339.

Robinson, J. T., Wojcik, E. J., Sanders, M. A., McGrail, M., and Hays, T. S. (1999). Cytoplasmic dynein is required for the nuclear attachment and migration of centrosomes during mitosis in *Drosophila. J. Cell Biol.* **146,** 597–608.

Roghi, C., Giet, R., Uzbekov, R., Morin, N., Chartrain, I., Le Guellec, R., Couturier, A., Doree, M., Philippe, M., and Prigent, C. (1998). The *Xenopus* protein kinase pEg2 associates with the centrosome in a cell cycle-dependent manner, binds to the spindle microtubules and is involved in bipolar mitotic spindle assembly. *J. Cell Sci.* **111,** 557–572.

Rose, L. S., and Wieschaus, E. (1992). The *Drosophila* cellularization gene *nullo* produces a blastoderm-specific transcript whose levels respond to the nucleocytoplasmic ratio. *Genes Dev.* **6,** 1255–1268.

Rothwell, W. F., Fogarty, P., Field, C. M., and Sullivan, W. (1998). Nuclear-fallout, a *Drosophila* protein that cycles from the cytoplasm to the centrosomes, regulates cortical microfilament organization. *Development* **125,** 1295–1303.

Rothwell, W. F., Zhang, C. X., Zelano, C., Hsieh, T. S., and Sullivan, W. (1999). The *Drosophila* centrosomal protein Nuf is required for recruiting Dah, a membrane associated protein, to furrows in the early embryo. *J. Cell Sci.* **112,** 2885–2893.

Ruder, F. J., Frasch, M., Mettrenleiter, T. C., and Buesen, W. (1987). Appearance of two maternally directed histone H2A variants precedes zygotic ubiquitination of H2A in early embryogenesis of *Sciara coprophila* (Diptera). *Dev. Biol.* **122,** 568–576.

de Saint Phalle, B., and Sullivan, W. (1998). Spindle assembly and mitosis without centrosomes in parthenogenetic *Sciara* embryos. *J. Cell Biol.* **141,** 1383–1391.

Saunders, R. D. C., do Carmo Avides, M., Howard, T., Gonzalez, C., and Glover, D. M. (1997). The *Drosophila* gene *abnormal spindle* encodes a novel microtubule-associated protein that associates with the polar regions of the mitotic spindle. *J. Cell Biol.* **137,** 881–890.

Schejter, E. D., and Wieschaus, E. (1993). Functional elements of the cytoskeleton in the early *Drosophila* embryo. *Annu. Rev. Cell Biol.* **9,** 67–99.

Schlegel, R., and Pardee, A. B. (1986). Caffeine-induced uncoupling of mitosis from the completion of DNA replication in mammalian cells. *Science* **232,** 1264–1266.

Schubiger, M., and Schneiderman, H. A. (1971). Nuclear transplantation in *Drosophila melanogaster. Nature* **230,** 185–186.

Schumacher, J. M., Golden, A., and Donovan, P. J. (1998b). AIR-2: An Aurora/Ip11-related protein kinase associated with chromosomes and midbody microtubules is required for polar body extrusion and cytokinesis in *Caenorhabditis elegans* embryos. *J. Cell Biol.* **143,** 1635–1646.

Schweisguth, F., Lepesant, J., and Vincent, A. (1990). The *serendipity alpha* gene encodes a membrane-associated protein required for the cellularization of the *Drosophila* embryo. *Genes Dev.* **4,** 922–931.

Shamanski, F. L., and Orr-Weaver, T. L. (1991). The *Drosophila plutonium* and *pan gu* genes regulate entry into S phase at fertilization. *Cell* **66,** 1289–1300.

Sharp, D. J., McDonald, K. L., Brown, H. M., Matthies, H. J., Walczak, C., Vale, R. D., Mitchison, T. J., and Scholey, J. M. (1999). The bipolar kinesin, KLP61F, cross-links microtubules within interpolar microtubule bundles of *Drosophila* embryonic mitotic spindles. *J. Cell Biol.* **144,** 125–138.

Shindo, M., Nakano, H., Kuroyanagi, H., Shirasawa, T., Mihara, M., Gilbert, D. J., Jenkins, N. A., Copeland, N. G., Yagita, H., and Okumura, K. (1998). cDNA cloning, expression, subcellular localization, and chromosomal assignment of mammalian aurora homologues, aurora-related kinase (ARK) 1 and 2. *Biochem. Biophys. Res. Commun.* **244,** 285–292.

Simpson, L., and Wieschaus, E. (1990). Zygotic activity of the *nullo* locus is required to stabilize the actin–myosin network during cellularization in *Drosophila. Development* **110,** 851–863.

Sluder, G., and Rieder, C. L. (1985). Centriole number and the reproductive capacity of spindle poles. *J. Cell Biol.* **100,** 887–896.

Sonnenblick, B. P. (1950). The early embryology of *Drosophila melanogaster.* In *Biology of Drosophila* (M. Demerec, Ed.), pp. 62–167. Wiley, New York.

Stafstrom, J. P., and Staehelin, L. A. (1984). Dynamics of the nuclear envelope and of nuclear pore complexes during mitosis in the *Drosophila* embryo. *Eur. J. Cell Biol.* **34,** 179–189.

Stalker, H. D. (1954). Parthenogenesis in *Drosophila. Genetics* **39,** 4–33.

Sturtevant, A. H. (1929). The claret mutant type of *Drosophila simulans:* a study of chromosome elimination and of cell-lineage. *Z. Wiss. Zool.* **135,** 323–356.

Sullivan, W., and Theurkauf, W. E. (1995). The cytoskeleton and morphogenesis of the early *Drosophila* embryo. *Curr. Opin. Cell Biol.* **7,** 18–22.

Sullivan, W., Minden, J. S., and Alberts, B. M. (1990). *daughterless-abo-like,* a *Drosophila* maternal-effect mutation that exhibits abnormal centrosome separation during the late blastoderm divisions. *Development* **110,** 311–323.

Sullivan, W., Daily, D. R., Fogarty, P., Yook, K., and Pinpinelli, S. (1993a). Delays in anaphase initiation occur in individual nuclei of the syncytial *Drosophila* embryo. *Mol. Biol. Cell* **4,** 885–896.

Sullivan, W., Fogarty, P., and Theurkauf, W. (1993b). Mutations affecting the cytoskeletal organization of syncytial *Drosophila* embryos. *Development* **118,** 1245–1254.

Tavosanis, G., Llamazares, S., Goulielmos, G., and Gonzalez, C. (1997). Essential role for gamma-tubulin in the acentriolar female meiotic spindle of *Drosophila. EMBO J.* **16,** 1809–1819.

Theurkauf, W. E., and Hawley, R. S. (1992). Meiotic spindle assembly in *Drosophila* females: behavior of nonexchange chromosomes and the effects of mutations in the nod kinesin-like protein. *J. Cell Biol.* **116,** 1167–1180.

Vorobjev, I. A., and Nadezhdina, E. S. (1987). The centrosome and its role in the organization of microtubules. *Int. Rev. Cytol.* **106,** 227–293.

Warn, R. M. (1986). The cytoskeleton of the early *Drosophila* embryo. *J. Cell Sci.* (Suppl.) **5,** 311–328.

Warn, R. M., Magrath, R., and Webb, S. (1984). Distribution of F-actin during cleavage of the *Drosophila* syncytial blastoderm. *J. Cell Biol.* **98,** 156–162.

Warn, R. M., Flegg, L., and Warn, A. (1987). An investigation of microtubule organization and function in living *Drosophila* embryos by injection of a fluorescently labeled antibody against tyrosinated alpha-tubulin. *J. Cell Biol.* **105,** 1721–1730.

Wheatley, D. N. (1982). *Cell Growth and Division.* E. Arnold, London.

Whitfield, W. G. F., Millar, S. E., Saumweber, H., Frasch, M., and Glover, D. M. (1988). Cloning of a gene encoding an antigen associated with the centrosome in *Drosophila. J. Cell Sci.* **89,** 467–480.

Whitfield, W. G. F., Chaplin, M. A., Oegema, K., Parry, H., and Glover, D. M. (1995). The 190 kDa centrosome-associated protein of *Drosophila melanogaster* contains four zinc finger motifs and binds to specific sites on polytene chromosomes. *J. Cell Sci.* **108,** 3377–3387.

Williams, B. C., Dernburg, A. F., Puro, J., Nokkala, S., and Goldberg, M. L. (1997). The *Drosophila* kinesin-like protein KLP3A is required for proper behavior of male and female pronuclei at fertilization. *Development* **124,** 2365–2376.

Wilson, E. B. (1925). In *The Cell in Development and Heredity* (J. A. Moore, Ed.). Garland Publishing, Inc., New York.

Wilson, P. G., and Borisy, G. G. (1998). Maternally expressed gamma Tub37CD in *Drosophila* is differentially required for female meiosis and embryonic mitosis. *Dev. Biol.* **199,** 273–290.

Wilson, P. G., Zheng, Y., Oakley, C. E., Oakley, B. R., Borisy, G. G., and Fuller, M. T. (1997). Differential expression of two gamma-tubulin isoforms during gametogenesis and development in *Drosophila. Dev. Biol.* **184,** 207–221.

Yanai, A., Arama, E., Kilfin, G., and Motro, B. (1997). ayk1, a novel mammalian gene related to *Drosophila* aurora centrosome separation kinase, is specifically expressed during meiosis. *Oncogene* **14,** 2943–2950.

Yasuda, G. K., Baker, J., and Schubiger, G. (1991). Independent roles of centrosomes and DNA in organizing the *Drosophila* cytoskeleton. *Development* **111,** 379–391.

Yasuda, G. K., Schubiger, G., and Wakimoto, B. T. (1995). Genetic characterization of *ms (3) K81,* a paternal effect gene of *Drosophila melanogaster. Genetics* **140,** 219–229.

Yu, K. R., Duronio, R. J., and Sullivan, W. (1998). *Cell Cycle Checkpoints: Safe Passage through Mitosis.* Oxford University Press.

Zalokar, M., and Erk, I. (1976). Division and migration of nuclei during early embryogenesis of *Drosophila melanogaster. J. Micro. Cell* **25,** 97–106.

Zhang, D., and Nicklas, R. B. (1995a). Chromosomes initiate spindle assembly upon experimental dissolution of the nuclear envelope in grasshopper spermatocytes. *J. Cell Biol.* **131,** 1125–1131.

Zhang, D., and Nicklas, R. B. (1995b). The impact of chromosomes and centrosomes on spindle assembly as observed in living cells. *J. Cell Biol.* **129,** 1287–1300.

Zhang, C. X., Lee, M. P., Chen, A. D., Brown, S. D., and Hsieh, T. (1996). Isolation and characterization of a *Drosophila* gene essential for early embryonic development and formation of cortical cleavage furrows. *J. Cell Biol.* **134,** 923–934.

Zheng, Y., Jung, M. K., and Oakley, B. R. (1991). gamma-Tubulin is present in *Drosophila melanogaster* and *Homo sapiens* and is associated with the centrosome. *Cell* **65,** 817–823.

20

Centrosome Maturation

Robert E. Palazzo,[1,2] *Jacalyn M. Vogel,*[1] *Bradley J. Schnackenberg,*[1] *Dawn R. Hull,*[1] and *Xingyong Wu*[1]
[1] Department of Molecular Biosciences
University of Kansas
Lawrence, Kansas 66045

[2] The Marine Biological Laboratory
Woods Hole, Massachusetts 02543

I. Introduction

The centrosome is one member of a family of structures that organize microtubules [microtubule organizing centers (MTOCs)] in eukaryotic cells (Pickett-Heaps, 1969; Wheatley, 1982; McIntosh, 1983; Brinkley, 1985; Kalnins, 1992; Kellogg *et al.,* 1994; Balczon, 1996). Considerable research has shown that cells can replicate centrosomes, assemble them *de novo,* dissociate them and/or destroy their ability to replicate, and even revive them after destruction. The cellular control of centrosome number and the centrosome's ability to nucleate microtubules, as well as the control of a centrosome's ability to pass from existence to nonexistence and back again, make for ponderous challenges in understanding centrosome regulation in the future.

The significance of centrosomes has been challenged by a renewed interest in anastral spindle assembly, a result of demonstrations that chromatin can direct the organization of microtubules to assemble bipolar spindles (Bajer and Mole-Bajer, 1982; Bastmeyer *et al.,* 1986; Steffen *et al.,* 1986;

Current Topics in Developmental Biology, Vol. 49

0070-2153/00 $35.00

Theurkauf and Hawley, 1992; Albertson and Thomson, 1993; McKim and Hawley, 1995; Heald *et al.,* 1996, 1997; Bonaccorsi *et al.,* 1998; de Saint Phalle and Sullivan, 1998), and that cell fragments can organize astral microtubule arrays in the absence of bona fide centrosomes (McNiven *et al.,* 1984; Rodionov and Borisy, 1997). However, concerns regarding a cellular dependence on centrosome function can be satisfied by the consideration that these organelles have been implicated in numerous fundamental cell processes, including the maintenance of cell shape and polarity (McIntosh, 1983), control of the distribution of organelles (Lee *et al.,* 1989; Soltys and Gupta, 1992; Cole *et al.,* 1996; Takahashi *et al.,* 1999), directional vesicle transport (Schroer and Sheetz, 1991), and cell migration and chemotaxis (Albrecht-Buehler, 1977; Gotlieb *et al.,* 1981; Nemere *et al.,* 1985), in addition to spindle assembly during meiosis and mitosis in most animal cell types (McIntosh, 1983, 1987; Mazia, 1984, 1987; Brinkley, 1985; Sluder and Rieder, 1985; Kirschner and Mitchison, 1986; Mitchison, 1989; Rieder and Alexander, 1990; Rieder, 1991; Zhang and Nicklas, 1995; Palazzo *et al.,* 1999). Importantly, when present, centrosomes control the position and polarity of microtubules, and thus indirectly influence the organization of other cellular components that interact with microtubules, including other cytoskeletal filaments (McIntosh, 1983; Brinkley, 1985). Thus, the early cytologist's view of the centrosome (central body) as the "dynamic center of the cell" (Wilson, 1925) still holds today.

Within the same cell type, the number of centrosomes, their morphology, the distribution of their components, and their ability to nucleate microtubules can vary (Conklin, 1902; Kuriyama and Borisy, 1981b; Sharp *et al.,* 1981; Mazia, 1984, 1987; Schatten *et al.,* 1985; Maro *et al.,* 1988; Ferreira *et al.,* 1993; Balczon *et al.,* 1995; also see Rothwell and Sullivan, and Megraw and Kaufman, this volume). Typical animal cells have only one centrosome that duplicates after the cells commit to replication, so that during mitosis they normally contain two centrosomes that direct the assembly of the bipolar spindle required for proper chromosome segregation. The amplification of centrosome number and/or the overexpression of centrosome components can have disastrous consequences, since the assembly of multiple or ectopic MTOCs during mitosis could interfere with the fidelity of chromosome segregation (Fukasawa *et al.,* 1996; Brinkley and Goepfert, 1998; Lingle *et al.,* 1998; Pihan *et al.,* 1998; Carroll *et al.,* 1999; Duesberg, 1999), contributing to genetic instability, aneuploidy, and possibly tumor progression (see chapters by Goepfert and Brinkley and by Lingle and Salisbury, this volume).

In spite of their significance and the long history of centrosome research, centrosome composition and the regulators that govern centrosome function remain poorly understood. However, significant progress has been made in the identification of protein components of the *S. cerevisiae* MTOC, the spindle pole body (Kilmartin *et al.,* 1993; Kilmartin and Goh, 1996; Wigge *et al.,* 1998; Adams and Kilmartin, 1999; also see chapters by Vogel and Snyder and by Francis and Davis, this volume). Further, functional centrosomes and related structures can now be isolated from mammalian cells (Mitchison and Kirschner, 1984, 1986; Bornens *et al.,* 1987; Komesli *et al.,* 1989), *Drosophila* (Moritz *et al.,* 1995a), *Spisula* (Vogel *et al.,* 1997; Schnackenberg *et al.,* 1998; Palazzo and Vogel, 1999), sea urchin (Thompson-Coffe *et al.,* 1996), and *Dictyostelium* (Gräf *et al.,* 1998; also see chapter by Gräf et al., this volume), suggesting that future research will lead to a comprehensive chemical analysis of these MTOCs.

In this chapter, the concept of centrosome maturation is partially reviewed. In the past, centrosome maturation has been thought of simply as the morphological and biochemical transformations that result in an increase in centrosome-dependent microtubule nucleation potential. However, recent progress has indicated that a number of other changes take place in centrosomes during the cell cycle that do not appear to be related to microtubule nucleation, and MTOC markers that are either transient or only acquired as a function of various phases of the cell have been identified (Alfa *et al.,* 1990; Lange and Gull, 1995; Bridge *et al.,* 1998; Rothwell *et al.,* 1998, 1999; Cerrutti and Simanis, 1999; also see chapters by Lange *et al.* and by Hagan and Peterson, this volume). Indeed, the early cytologist's description of the centrosome as a "central organizer" implies a potential that extends beyond that of microtubule organization. This view is supported by current information. Thus, the centrosome may be better viewed as a specialized site for the coordination of complex molecular interactions, in addition to those required for microtubule nucleation. For example, the centrosome could serve as a site for the concentration and coordination of the binding of subunits for the assembly of higher-order structures (Fig. 1). Once assembled, these could be dispatched to distant sites of the cell for participation in other complex and seemingly independent events, such as the formation of the cleavage furrow (Rothwell *et al.,* 1999). In addition, centrosome associated proteins could act to mediate enzyme–substrate interactions and coordinate posttranslational modifications (Nigg *et al.,* 1985; Keryer *et al.,* 1993; Takahashi *et al.,* 1999) to integrate centrosome control with other cellular functions required for proper cell cycle progression (see Alfa *et al.,* 1990; Hinchcliffe *et al.,* 1999; Lacey *et al.,* 1999). Thus, an exclusive view of the centrosome as a mere MTOC may be limiting, and perhaps the concept of centrosome maturation needs to be expanded.

II. Centrosome Maturation

The term centrosome maturation has been used to describe the change in microtubule nucleation potential that accompanies the G_2/M transition of the somatic cell cycle. It has long been known that during mitosis centrosomes nucleate at least five times more microtubules than do interphase centrosomes (Kuriyama and Borisy, 1981b; Brinkley, 1985; Kellogg *et al.,* 1994; Balczon, 1996; Paoletti and Bornens, 1997). This increase in microtubule nucleation potential is believed to play an important role in assuring contacts between at least a subset of centrosomal microtubules and the chromosome kinetochores. Such contacts are achieved through a stochastic process that is dependent on the dynamics of microtubule polymers (Kirschner and Mitchison, 1986; Mitchison, 1989). Thus, by increasing the number of microtubules that a centrosome nucleates, cells can ensure the proper alignment and segregation of chromosomes.

In somatic cells, centrosome maturation is preceded by centrosome replication. Extensive discussion of the regulation of somatic centrosome reproduction is found within chapters elsewhere in this volume, so brevity is in order here. Most animal centrosomes contain two orthogonally arranged centrioles that are associated with a variety of structures collectively known as the pericentriolar material (PCM) (Wheatley, 1982). Centrioles themselves are composed of short microtubules arranged in a unique nine-triplet array (Wheatley, 1982; Kalnins, 1992; also see chapters by Preble *et al.* and Lange *et al.,* this volume). Centrioles can nucleate microtubules; however, most cytoplasmic microtubules are nucleated by the PCM (Tilney and Goddard, 1970; Berns and Richardson, 1977; Gould and Borisy, 1977; Telzer and Rosenbaum, 1979; Kuriyama and Borisy, 1981b; Rieder and Borisy, 1982). Near the onset of S-phase, the two centrioles separate and each spawns a "procentriole" from the lateral-basal region of a preexisting centriole. Centriole duplication has been broken down into three discrete phases (a) initiation, (b) elongation, and (c) termination (Kuriyama and Borisy, 1981a). Thus, the number of MTOCs doubles as the cell passes through S-phase of the cell cycle (Brinkley, 1985; Kellogg *et al.,* 1994; Balczon, 1996; Hinchcliffe *et al.,* 1999; Lacey *et al.,* 1999). If mammalian cells (Balczon, 1996) or cycling *Xenopus* oocyte lysates (Hinchcliffe *et al.,* 1999; Lacey *et al.,* 1999) are held in S-phase, centrosome replication continues even though the rest of the cell cycle is halted, indicating that all the machinery needed to execute centrosome reproduction is active during this phase (see chapter by Sluder and Hinchcliffe, this volume). Although centrosome assembly occurs during S-phase, which includes the recruitment of PCM components and the acquisition of the ability to nucleate some microtubules, centrosomes mature as they cross the G_2/M border, resulting in a dramatic increase in their microtubule nucleation potential.

Two models have been proposed to explain centrosome maturation. One model suggests that centrosomes contain a linear "ribbon" structure to which microtubule nucleation sites are attached. In this model, the unfolding of such a complex exposes cryptic sites within the centrosome during M-phase (Mazia, 1987). An interpretation of this model is that interphase centrosome microtubule nucleation potential is actually higher than expressed, and the full microtubule nucleation potential is only revealed as a result of exposure of cryptic nucleation sites during M-phase. Another model is the recruitment model, which predicts that the increase in microtubule nucleation potential is a result of incorporation of new microtubule nucleation sites and assembly of additional PCM (reviewed in Schnackenberg and Palazzo, 1999).

There is some confusion regarding whether centrosome maturation is a gradual or relatively sudden event in somatic cells. Studies tracing the dynamics of proteins by fluorescent microscopy suggest that recruitment of PCM antigens is gradual, occurring steadily during S-, G_2, and M-phases (Dictenberg *et al.*, 1998). In contrast, another study that traced the content and turnover of a GFP-labeled PCM protein (γ-tubulin) indicates that recruitment of this antigen to the centrosome is a relatively sudden event that occurs during the G_2/M transition (Khodjakov and Rieder, 1999). Importantly, this study also provides the first evidence for the dynamic exchange of a centrosome component with a soluble cytoplasmic pool.

The control of centrosome maturation is also important for gametogenesis and early embryogenesis. For example, maternal centrosome assembly and maturation occurs during the completion of meiosis in many oocytes (Kuriyama *et al.*, 1986; Sluder *et al.*, 1989, 1993; Palazzo *et al.*, 1992; Schatten, 1994). In addition, following fertilization, paternal centrosome maturation is necessary for the microtubule-dependent guidance of the maternal pronucleus toward its paternal partner for syngamy, and also for the normal mitotic cleavage divisions during embryogenesis (Sluder *et al.*, 1989, 1993; Schatten, 1994; Wu and Palazzo, 1999). Finally, differential control of maternal and paternal centrosomes is required to assure that embryos inherit the correct number of replication-competent centrosomes for normal development, "one and only one" (Sluder *et al.*, 1989, 1993; Schatten, 1994).

To understand the mechanisms that underly centrosome maturation, we must understand centrosome composition and centrosome structure. Our understanding of centrosome structure has advanced significantly (Wheatley, 1982; Moritz *et al.*, 1995a; Vogel *et al.*, 1997; Schnackenberg *et al.*, 1998); however, our ability to identify the molecules responsible for centrosome function has lagged behind (Kellogg *et al.*, 1994; Balczon, 1996).

III. Centrosome Structure and Microtubule Nucleation Potential

A. Microtubule Nucleation

Considerable progress has been made in defining the molecular and structural requirements for centrosome-dependent microtubule nucleation, and models to explain how substructures within the centrosome's PCM organize α/β tubulin subunits to initiate microtubule polymerization have been proposed (for reviews see chapters by Oakley and by Gunawardane *et al.*, this volume; Schnackenberg and Palazzo, 1999; Wiese and Zheng, 1999). These models are based on the arrangement of an important PCM protein, γ-tubulin, a member of the tubulin gene family that localizes to and copurifies with centrosomes and is essential for centrosome-dependent microtubule nucleation (Oakley *et al.*, 1990; Stearns *et al.*, 1991; Zheng *et al.*, 1991; Joshi *et al.*, 1992; Horio and Oakley, 1994; Moritz *et al.*, 1995b, 1998; Sunkel *et al.*, 1995; Moudjou *et al.*, 1996; Vogel *et al.*, 1997; Schnackenberg *et al.*, 1998). However, a significant pool of γ-tubulin is also found in the cell cytoplasm, where it is a component of a variety of oligomeric protein complexes (Zheng *et al.*, 1995; Moudjou *et al.*, 1996; Akashi *et al.*, 1997; Detraves *et al.*, 1997; Knop *et al.*, 1997; Knop and Schiebel, 1997; Martin *et al.*, 1998; Moritz *et al.*, 1998; Murphy *et al.*, 1998; Oegema *et al.*, 1999). *In vitro* reconstitution studies have revealed that at least some of these γ-tubulin complexes can interact with microtubules in the absence of their association with centrosomes (Zheng *et al.*, 1995; Oegema *et al.*, 1999; Schnackenberg *et al.*, 2000). One form in particular has captured the attention of centrosome researchers since under the electron microscope it appears as a 25 nm diameter ring (γ-tubulin ring complex, γ-TuRC), which is the exact diameter of a microtubule (Zheng *et al.*, 1995).

The PCM of isolated *Drosophila* and *Spisula* centrosomes is studded with similar 25 nm ring structures (Moritz et al., 1995a,b; Vogel *et al.*, 1997; Schnackenberg *et al.*, 1998), and immunogold localization and electron microscopy of isolated *Drosophila* centrosomes proved that these rings contain γ-tubulin (Moritz *et al.*, 1995b). Further, biochemical analysis revealed that γ-tubulin is the major tubulin type to copurify with γ-TuRCs (Zheng *et al.*, 1995) and centrosomes (Vogel *et al.*, 1997), indicating that these rings are not likely to be short microtubule fragments, but rather bona fide templates for microtubule nucleation. In addition, the fact that isolated γ-TuRCs increase the rate of microtubule polymerization in a defined spectroscopic turbidity assay indicates their ability to nucleate microtubules independently (Zheng *et al.*, 1995). Thus, γ-TuRCs could induce the assembly of microtubules in cytoplasm as well as from centrosomes (reviewed in Wiese and Zheng, 1999). Therefore, the mechanism(s) by which γ-TuRCs are concentrated to a specific site(s) within the cell during the formation of an MTOC becomes an important issue.

B. The Centromatrix

A series of studies has led to the identification of a filamentous lattice within the centrosome PCM that appears to anchor γ-TuRCs. Deconvolution fluorescence microscopy revealed that γ-tubulin and another important centrosome protein, pericentrin, are distributed throughout the PCM in the form of a lattice (Dictenberg *et al.,* 1998). Further, analysis of salt-stripped *Spisula* centrosomes by intermediate voltage electron microscopy tomography revealed the presence of a matrix structure (the centromatrix) that is composed of 12- to 15-nm filaments (Schnackenberg *et al.,* 1998). Salt extraction of isolated centrosomes removes γ-tubulin, revealing the centromatrix, which retains the ability to bind γ-tubulin and other elements that are required for microtubule nucleation after such treatments.

Treatment of isolated *Spisula* (Schnackenberg *et al.,* 1998), *Drosophila* (Moritz *et al.,* 1998), or mammalian (Klotz *et al.,* 1990; Ohta *et al.,* 1993) centrosomes with chaotropic agents such as KI (among others) removes their ability to nucleate microtubules. In the case of *Spisula* centrosomes, KI removes γ-tubulin and 25-nm rings from the PCM (Schnackenberg *et al.,* 1998). Importantly, these salt-stripped centrosomes bind γ-tubulin and recover their ability to nucleate microtubules when incubated in cytoplasmic extracts prepared from a variety of organisms, including *Drosophila, Xenopus,* and mammalian sources (Schnackenberg *et al.,* 2000). In addition, salt-stripped *Drosophila* centrosomes can bind isolated γ-TuRCs and recover microtubule nucleation potential, but reconstitution of microtubule nucleation potential requires at least one additional factor (Moritz *et al.,* 1998; discussed later). Finally, treatment of salt-stripped mammalian centrosomes with extracts prepared from *Xenopus* oocytes (Ohta *et al.,* 1993) or tobacco cell cultures (Stoppin-Mellet *et al.,* 1999) also results in the recovery of microtubule nucleation potential. Taken together, these results suggest that animal centrosomes contain a substructure, the centromatrix, whose function is highly conserved, possibly across kingdoms. Thus, identification of centromatrix components and deciphering the mechanism(s) that govern centromatrix assembly will be an important objective for the future.

Corresponding to the recovery of microtubule nucleation potential by the *Spisula* centromatrix is the binding of γ-tubulin and 25-nm diameter rings (Schnackenberg *et al.,* 1998; also see Moritz *et al.,* 1998). Surprisingly, the recovery of microtubule nucleation potential does not require small organic molecules such as nucleotides, divalent cations, or protein kinase activity. However, studies using *Drosophila* salt-stripped centrosomes revealed that binding of affinity-purified γ-TuRCs and recovery of microtubule nucleation potential requires an additional soluble cytoplasmic factor of 220,000 MW (Moritz *et al.,* 1998), which has been referred to as an adapter (Schnackenberg and Palazzo, 1999). This adapter does not copurify with γ-TuRCs, but can bind to salt-insoluble centrosome remnants independently (Moritz *et al.,* 1998; Schnackenberg *et al.,* 2000).

Recent studies in *Drosophila* suggest that a mitotic spindle protein, Asp, might serve an adapter function. Asp was first identified as a 220-kDa microtubule binding protein required for spindle assembly in *Drosophila* embryos (Saunders *et al.,* 1997; do Carmo Avides and Glover, 1999). Salt-stripped *Drosophila* centrosomes, which recover microtubule nucleation potential when treated with cytoplasmic extracts prepared from wild-type embryos, fail to recover microtubule nucleation potential when treated with extracts prepared from asp mutants (do Carmo Avides and Glover, 1999). Importantly, addition of wild-type Asp isolated by microtubule-affinity methods to mutant embryo extracts restores the mutant extract's ability to support recovery of microtubule nucleation potential by KI-stripped centrosomes. This work suggests that Asp may be the adapter or one component of an adapter complex. However, earlier studies revealed that Asp does not localize to interphase centrosomes (Saunders *et al.,* 1997), which presumably nucleate microtubules and contain γ-tubulin. Further, Asp localizes to both the spindle poles and the central spindle during mitosis, and appears to bind microtubules directly with high affinity (Saunders *et al.,* 1997). Thus, although Asp may be the adapter or one component of an adapter, important questions remain regarding the role and the mechanism of Asp function in linking γ-TuRCs to the centromatrix.

C. A Model for Centrosome Assembly and Maturation

The results of these functional reconstitution studies allow the conceptual breakdown of the centrosome PCM into at least three independent entities: γ-TuRCs that nucleate microtubules, the centromatrix that binds and concentrates γ-TuRCs to a discrete site(s) within the cell, and an adapter(s) that links γ-TuRCs to the centromatrix (reviewed in Schnackenberg and Palazzo, 1999; see Fig. 2). Since cytoplasm contains an abundance of γ-TuRCs capable of nucleating microtubules independently of the centrosome (Zheng *et al.,* 1995; Schnackenberg *et al.,* 1998, 2000; Wiese and Zheng, 1999), it was proposed that centromatrix assembly is required for centrosome maturation (Schnackenberg and Palazzo, 1999). In this model, as centromatrix assembly ensues, an increasing number of binding sites become available for the docking of γ-TuRCs (Fig. 2). Thus, as the mass of centromatrix increases, γ-TuRCs are sequestered from the cytoplasm to generate a localized concentration of elements capable of nucleating microtubules, the MTOC.

Based on this model, wherever centromatrix components assemble, an MTOC will appear. Further, given the apparent stability of the centromatrix, evidenced by a resistance to treatment with strong chaotropes, it could serve as a relatively stable scaffold for the binding and turnover of centrosome components such as γ-tubulin (see Khodjakov and Rieder, 1999), or for the recruitment and transient association of regulatory proteins, kinases, substrates (Keryer *et al.,* 1993; Takahashi *et al.,* 1999), and precursor subunits for the assembly of higher-order structures that, once assembled, could dissociate and travel to distant parts of the cell to serve a final function (Fig. 1; Rothwell *et al.,* 1998, 1999). Recruitment of precursors and dispatch of processed ensembles could involve microtubule-dependent motor proteins, dynein and kinesin, respectively. Already, anchoring proteins that bind enzymes and substrates and apparently enhance catalytic interactions have been localized to the centrosome (Keryer *et al.,* 1993; Takahashi *et al.,* 1999). Some of these share at least some homology to the known centrosome protein pericentrin (Takahashi *et al.,* 1999).

IV. The Role of the Centriole

It is important to understand how centromatrix assembly is initiated and controlled. A recent study suggests that centrioles are required for maintaining the integrity of the PCM (Bobinnec *et al.,* 1998). This study demonstrated that antibodies to glutamylated tubulin isoforms stain centrioles and disrupt centriole structure when introduced into living cells. As these centrioles disappear, the PCM becomes disorganized and PCM components disperse throughout the cell. However, with time after antibody delivery, as antibody is presumably degraded by the cell cytoplasm, the centrioles reappear. As centrioles reassemble, the integrity of the PCM returns, and thus the centrosome is reassembled.

These experiments provide strong evidence that maintenance of PCM organization at a focal point within the cell is dependent on glutamylated tubulin isoforms within the centriole and/or the integrity of the centriole triplet structure (Bobinnec *et al.,* 1998; also see chapter by Küntziger and Bornens, this volume). This study also implies a close relationship between the centriole and the centromatrix. An intriguing possibility is that centriole components or molecules capable of binding to centriolar glutamaylated tubulin serve to initiate centromatrix assembly, thus providing a physical linkage between centrioles and the PCM (Fig. 2, D–F; see Küntziger and Bornens, this volume).

The role of the centriole in centrosome assembly and maturation has long been debated. Some oocytes, particularly vertebrate oocytes, are able to complete meiosis in the absence of centrioles and centrosomes (reviewed by Schatten, 1994), a result of their ability to assemble anastral spindles that are capable of segregating chromosomes (see Sawin and Mitchison, 1991; Verde *et al.,* 1991; Heald *et al.,* 1996, 1997). However, during the earliest stages of gametogenesis, the oogonia that ultimately give rise to these oocytes are products of mitotic divisions. Therefore, these cells presumably inherit centrosomes, complete with centrioles, before the completion of oocyte differentiation, but during differentiation they somehow lose these structures (Schatten, 1994). Whether these cells have active mechanisms for maternal centrosome disassembly reflecting a "centriole intolerance," or whether the loss of centrosome structure is a passive result of shutting down centrosome support mechanisms is not clear. Regardless, upon fertilization, the sperm provides the centrioles needed for the development of the organism. Thus, paternal centrioles serve as sites for the recruitment of the maternal components that are required for assembly of a fully functional centrosome (Schatten, 1994).

V. The Procentrosome

In contrast to the vertebrate systems, many invertebrate oocytes either retain or resurrect their maternal centrioles and centrosomes to guide astral spindle assembly during meiosis (Sluder and Rieder, 1985; Sluder *et al.*, 1989, 1993; Palazzo *et al.*, 1992, 1999; Schatten, 1994; and references therein). One of the elegant descriptions of centrosome assembly and maturation in invertebrates is the classic study of karyokinesis in *Crepidula fornicata* by E. G. Conklin (1902). This study indicates that *Crepidula* oocytes assemble maternal centrosomes during meiosis I. These maternal centrosomes do not assemble *de novo,* but rather emerge from the maturation of two precursor structures that will be referred to as procentrosomes. These procentrosomes might best be viewed as vestigial maternal centrosomes that have been partially, but not completely, disassembled during oogenesis.

During late prophase of meiosis I, *Crepidula* oocytes were shown to contain two small maternal asters, each containing a relatively few astral rays (microtubules). As meiosis I ensued, these maternal asters grew, increased their microtubule number, and displayed corresponding changes in centrosome morphology characterized by an increase in the volume of the PCM. Importantly, as these maternal centrosomes matured, a dense body appeared within the center, which Conklin described as the "centriole of Boveri" (Conklin, 1902). Thus, the two maternal procentrosomes apparently retained integrity even in the absence of centrioles, and yet were capable of nucleating microtubules and forming small asters. Further, as described, these procentrosomes served to guide centriole assembly within their core structure during maternal centrosome maturation.

Conklin's *Crepidula* study represents one of the earliest suggestions of *de novo* centriole assembly and suggests that some PCM components can retain integrity even in the absence of centrioles. The idea of a structural centrosome precursor that can direct the assembly of centrioles is also found in the descriptions of spermatogenesis in ferns and differentiation of respiratory and reproductive epithelia, where an amorphous "fibrogranular substance," which has no resemblance to centrioles, apparently guides centriole assembly (Dirksen, 1978, and references therein; also see Küntziger and Bornens and Lange *et al.*, this volume). Given the increasing availability of antibodies that bind PCM proteins, it would be interesting to return to such systems in an attempt to define the PCM antigens that remain as components of these presumably minimal centrosome structures. The proteins identified could play a key role in centrosome assembly and function.

Similar to *Crepidula, Spisula* oocytes are arrested in late prophase of meiosis I, at the G_2/M border, until fertilized or activated artificially with KC1 (Allen, 1953). Unlike *Crepidula,* centrosomes, asters, procentrosomes, or centrioles have yet to be conclusively identified in *Spisula* oocytes before activation, since transmitted light microscopy (Allen, 1953), electron microscopy (Longo and Anderson, 1970; Palazzo *et al.*, 1992), and fluorescence microscopy using tubulin antibodies (Kuriyama *et al.*, 1986) have failed to detect them. In addition, earlier studies indicated that lysates prepared from unactivated oocytes were incapable of forming asters, suggesting that they did not contain centrosomes (Weisenberg and Rosenfeld, 1975; also see Palazzo *et al.*, 1992). In contrast, a recent report showed that unactivated *Spisula* oocyte lysates are capable of forming small asters (Palazzo *et al.*, 1999). Preliminary evidence suggests that these are not cytasters since they contain certain centrosome-specific antigens at their core that are not found in cytasters (work in progress). Thus, *Spisula* oocytes may contain procentrosomes, consistent with Conklin's findings in *Crepidula.*

Given the methods already in place for the isolation of *Spisula* centrosomes from activated M-phase oocytes (Palazzo and Vogel, 1999), this system could also prove to be a rich source of procentrosomes. Regardless, like *Crepidula,* activation of *Spisula* oocytes induces rapid maternal centrosome maturation, which includes centriole duplication and a time-dependent increase in microtubule nucleation potential (Fig. 2, A–C). Importantly, all the major events of centriole duplication and maternal centrosome maturation can be induced in a cell-free system using *Spisula* oocyte lysates, offering opportunities for a biochemical approach to understand the mechanisms involved (Palazzo *et al.*, 1992). Finally, it has been proposed that the time-dependent increase in microtubule nucleation potential of *Spisula* centrosomes is directly related to the rate of centromatrix assembly (Schnackenberg and Palazzo, 1999).

In summary, there is evidence to support the idea that the centriole is required for maintenance of PCM and centrosome integrity in somatic cells. In contrast, there is ample evidence from invertebrate and plant systems in support of the idea that centrosomes can retain structure in the absence of centrioles. Clearly, there is much to be learned about centriole–PCM interactions.

VI. Paternal vs Maternal Centrosomes

Boveri recognized that eggs typically lose their centrosomes during oogenesis, and that the sperm replaces the centrosome during fertilization (1901). This theory of fertilization is now taken for granted, since it has been verified in many organisms (Wilson, 1925; Sluder and Rieder, 1985; Le Guen and Crozet, 1989; Sluder *et al.,* 1989; Crozet, 1990; Luykx, 1991; Schatten *et al.,* 1991; Sluder *et al.,* 1993; Schatten, 1994). Clearly, if oocytes were to retain a fully functional maternal centrosome capable of replication, introduction of a paternal centrosome would result in the presence of two replication-competent centrosomes and lead to the formation of tetrapolar spindles and disruption of chromosome segregation. Thus, the maternal centrosome's ability to replicate is lost either during oocyte differentiation or during the completion of meiosis, leaving the paternal centrosome to replicate as needed for coordinated development (see Sluder *et al.,* 1989, 1993; reviewed in Schatten, 1994).

In starfish, the maternal centrosome's ability to replicate is selectively destroyed during meiosis I (Sluder *et al.,* 1989, 1993). Similarly, *Spisula* oocytes, which can complete meiosis I and II after artificial activation with KCl, enter the mitotic cycles, but cannot replicate the remaining maternal centrosome, which still forms a maternal monaster during subsequent mitotic cell cycles (Kuriyama *et al.,* 1986). In the absence of a paternal centrosome, these eggs do not undergo cleavage. Importantly, in both starfish and *Spisula,* the male centrosome is present in the egg cytoplasm during the time when the maternal centrosome's ability to replicate is destroyed (Sluder *et al.,* 1989, 1993; Wu and Palazzo, 1999). The fact that the paternal centrosome's reproductive capacity is spared suggests the action of a selective mechanism.

Like many other systems, *Spisula* sperm centrioles cannot organize microtubules into astral arrays before fertilization (Fig. 2D; Wu and Palazzo, 1999). As a result of differentiation, sperm centrosomes typically lose many PCM components (see chapter by Manandhar *et al.,* this volume, for review), including proteins such as pericentrin (Doxsey *et al.,* 1994) and γ-tubulin (Stearns and Kirschner, 1994; Wu and Palazzo, 1999), rendering them incapable of organizing microtubules into astral arrays. However, sperm centrioles are able to recruit the maternal components needed for the assembly of a centrosome that is capable of organizing a sperm aster (Fig. 2E). Thus, restoration of the zygotic centrosome requires both maternal and paternal components (Holy and Schatten, 1991; Schatten, 1994).

The mechanisms involved in the restoration of the paternal centrosome's microtubule nucleation potential are poorly understood. Considering the results of Bobinnec *et al.* (1998) that suggest a physical linkage between centrioles and PCM (see Küntziger and Bornens, this volume), we propose the following model for paternal centrosome maturation. Following fertilization, maternal cytoplasm activates sperm centrioles or centriole associated proteins to initiate centromatrix assembly, which allows the subsequent binding of elements required for microtubule nucleation, thus resulting in sperm aster formation (Fig. 2E). This model predicts that activation of centromatrix assembly by paternal centrioles is regulated, and there is evidence to support this contention. First, the assembly and maturation of *Xenopus* sperm centrosomes in *Xenopus* lysates requires ATP (Stearns and Kirschner, 1994), as well as the action of Ran GTP binding proteins (Wilde and Zheng, 1999; also see Nakamura *et al.,* 1998, and Ohba *et al.,* 1999). Second, centrioles associated with isolated human sperm heads do not acquire microtubule nucleation potential in cytoplasmic extracts unless they are first primed by treatment with disulfide reducing agents (Navara *et al.,* 1995). Finally, preliminary experiments using isolated *Spisula* sperm heads indicate that activated M-phase oocyte extracts prepared shortly after oocyte activation can induce paternal centrosome maturation (Fig. 2E), but interphase extracts prepared from unactivated oocytes cannot (Fig. 2D; work in progress). However, both interphase and M-phase extracts support the recovery of microtubule nucleation potential by salt-stripped maternal *Spisula* centrosomes that already contain a centromatrix structure (Schnackenberg *et al.,* 2000). Previous studies revealed that the recovery of microtubule nucleation potential by salt-stripped maternal centrosomes is independent of energy, kinase activity, and cell cycle stage, indicating that this aspect of centrosome maturation is not under regulatory control (Schnackenberg and Palazzo, 1999). Taken together, these studies suggest that if a centromatrix is present, an aster will form, regardless of cell cycle stage and independently of regulatory molecules. Yet, preliminary results suggest that introduction of sperm centrioles into interphase extracts does not result in aster formation, possibly because they lack an associated centromatrix.

Based on the collective data, we suggest that sperm centrioles do not have an associated centromatrix (Fig. 2D). After entry into the egg cytoplasm, a regulatory event is required to trigger the initiation of centromatrix assembly by the centrioles and/or by centriole-associated proteins (Fig. 2E). Once centromatrix assembly is initiated, γ-TuRCs bind to the centromatrix, the PCM can complete assembly, and paternal centrosome maturation ensues.

A. Differential Regulation of Maternal vs Paternal Centrosome Maturation

As stated previously, recent studies using *Spisula* oocytes and embryos suggest that maternal and paternal centrosome assembly and maturation can be highly selective, even within the same cytoplasm. *Spisula* oocytes are fertilized at an unusually early stage of the meiotic cell cycle, during late prophase of meiosis I. While fertilization induces the assembly of the two maternal centrosomes necessary for the completion of meiosis I and II, it also results in the presence of three centrosomes, two maternal and one paternal (Wu and Palazzo, 1999).

It was previously thought that *Spisula* sperm centrosomes remain dormant and are not activated to function as MTOCs until meiosis is completed, when embryos enter the mitotic cycle (Kuriyama *et al.,* 1986). However, we recently showed that during meiosis I, sperm centrosomes undergo a unique transient maturation process (Wu and Palazzo, 1999). Upon entry into the egg, during prometaphase of meiosis I, *Spisula* sperm centrosomes recruit maternal γ-tubulin, nucleate microtubules, and form asters (Fig. 2E). At the same time, and in the same cytoplasm, two maternal centrosomes mature, resulting in the presence of three active centrosomes. This situation could lead to tripolar spindle assembly and disruption of chromosome segregation during meiosis I. However, by metaphase of meiosis I, sperm centrosomes lose γ-tubulin and the ability to nucleate microtubules (Fig. 2F). At the same time and in the same cell, both maternal centrosomes retain γ-tubulin and continue to function as MTOCs (Fig. 2C). By metaphase of meiosis II, the paternal centrosomes are once again reactivated, recruit γ-tubulin, again mature, and form sperm asters (Wu and Palazzo, 1999).

Based on these results, it was suggested that maternal and paternal centrosomes can be distinguished, and that the maturation of each can be selectively controlled (Wu and Palazzo, 1999). If centromatrix assembly is a prerequisite for the recruitment of γ-tubulin and the acquisition of MTOC function, the fact that γ-tubulin is suddenly and selectively dissociated from the sperm centrosome could be an indicator of selective centromatrix disassembly. Thus, regulation of centrosome structure and MTOC function could be at the level of either centromatrix polymer stability or centriole–centromatrix interactions.

VII. Summary

In the past, centrosome maturation has been described as the change in microtubule nucleation potential that occurs as cells pass through specific phases of the cell cycle. It is suggested that the idea of centrosome maturation be expanded to include gain of functions that are not necessarily related to microtubule nucleation. Some of these functions could be transient and dependent on the temporary association of molecules with the centrosome as cells progress through the cell cycle. Thus, the centrosome may best be viewed as a site for mediating macromolecular interactions, perhaps as a central processing station within the cell. The centromatrix, a relatively stable lattice of polymers within the centrosome's PCM, could serve as a scaffold for the transient binding of mediator molecules, as well as allow the dynamic exchange of centrosome constituents with a soluble cytoplasmic pool.

New evidence adds support to the idea that centrioles are crucial for the maintenance of PCM structure. However, significant evidence indicates that aspects of centrosome structure and function can be maintained in the absence of centrioles. In the case of paternal centrosome maturation, sperm centrioles may not contain an associated centromatrix. It is proposed that regulation of paternal centrioles or centriole associated proteins could mediate centriole-dependent centromatrix assembly following fertilization. Thus, regulation of centromatrix–centriole interactions could be involved in maintaining the integrity of the centrosome's PCM and play an important role in centrosome disassembly during cell differentiation and morphogenesis.

Acknowledgments

Supported in part by the National Institutes of Health (NIH) (GM43264), the American Cancer Society (JFRA-314) and the Robert Day Allen Fellowship of the Marine Biological Laboratory (to R.E. Palazzo).

References

Adams, I. R., and Kilmartin, J. V. (1999). Localization of core spindle pole body (SPB) components during SPB duplication in *Saccharomyces cerevisiae*. *J. Cell Biol.* **145,** 809–823.

Akashi, T., Yoon, Y., and Oakley, B. R. (1997). Characterization of γ-tubulin complexes in *Aspergillus nidulans* and detection of putative γ-tubulin interacting proteins. *Cell Motil. Cytoskeleton* **37,** 149–158.

Albertson, D. G., and Thomson, J. N. (1993) Segregation of holocentric chromosomes at meiosis in the nematode, *Caenorhabditis elegans*. *Chromosome Res.* **1,** 15–26.

Albrecht-Buehler, G. (1977). Phagokinetic tracks of 3T3 cells: parallels between the orientation of track segments and of cellular structures which contain actin or tubulin. *Cell* **12**(2), 333–339.

Alfa, C. E., Ducommun, B., Beach, D., and Hyams, J. S. (1990). Distinct nuclear and spindle pole body population of cyclin-cdc2 in fission yeast. *Nature* **347,** 680–682.

Allen, R. (1953). Fertilization and artificial activation in the eggs of the surf clam, *Spisula solidissima*. *Biol. Bull.* **105,** 213–239.

Bajer, A. S., and Mole-Bajer, J. (1982). Asters, poles and transport properties within spindle-like microtubule arrays. *Cold Spring Harbor Symp. Quant. Biol.* **46,** 263–283.

Balczon, R. (1996). The centrosome in animal cells and its functional homologs in plant and yeast cells. *Inter. Rev. Cytol.* **169,** 25–82.

Balczon, R., Bao, L., Zimmer, W. E., Brown, K., Zinkowski, R. P., and Brinkley, B. R. (1995). Dissociation of centrosome replication events from DNA synthesis and mitotic division in hydroxyurea arrested Chinese hamster ovary cells. *J. Cell Biol.* **130,** 105–115.

Bastmeyer, M., Steffen, W., and Fuge, H. (1986). Immunostaining of spindle components in tipulid spermatocytes using a serum against pericentriolar material. *Eur J. Cell Biol.* **42,** 305–310.

Berns, M. W., and Richardson, S. M. (1977). Continuation of mitosis after selective laser microbeam destruction of the centriolar region. *J. Cell Biol.* **75**(3), 977–982.

Bobinnec, Y., Khodjakov, A., Mir, L. M., Rieder, C. L., Edde, B., and Bornens, M. (1998). Centriole disassembly in vivo and its effect on centrosome structure and function in vertebrate cells. *J. Cell Biol.* **143**(6), 1575–1589.

Bonaccorsi, S., Giansanti, M. G., and Gatti, M. (1998). Spindle self-organization and cytokinesis during male meiosis in asterless mutants of *Drosophila melanogaster. J. Cell Biol.* **142,** 751–761.

Bornens, M., Paintrand, M., Berges, J., Marty, M-C., and Karsenti, E. (1987). Structural and chemical characterization of isolated centrosomes. *Cell Motil. Cytoskeleton* **8,** 238–249.

Boveri, T. (1901). *Zellen-Studien: Über die Natur der Centrosomen,* IV. Fisher, Jena, Germany.

Bridge, A. J., Morphew, M., Bartlett, R., and Hagan, I. M. (1998). The fission yeast SPB component Cut 12 links bipolar spindle formation to mitotic control. *Genes and Dev.* **12,** 927–942.

Brinkley, B. R. (1985). Microtubule organizing centers. *Annu. Rev. Cell Biol.* **1,** 145–172.

Brinkley, B. R., and Goepfert, T. M. (1998). Supernumerary centrosomes and cancer: Boveri's hypothesis resurrected. *Cell Motil. Cytoskeleton* **41,** 281–288.

Carroll, P. E., Okuda, M., Horn, H. F., Biddinger, P., Stambrook, P. J., Gleich, L. L., Li, Y. Q., Tarapore, P., and Fukasawa, K. (1999). Centrosome hyperamplification in human cancer: chromosome instability induced by p53 mutation and/or Mdm2 overexpression. *Oncogene* **18,** 1935–1944.

Cerrutti, L., and Simanis, V. (1999). Asymmetry of the spindle pole bodies and Spg1p GAP segregation during mitosis in fission yeast. *J. Cell Sci.* **112,** 2313–2321.

Cole, N. B., Sciaky, N., Marotta, A., Song, J., and Lippincott-Schwartz, J. (1996). Golgi dispersal during microtubule disruption: Regeneration of Golgi stacks at peripheral endoplasmic reticulum exit sites. *Mol. Biol. Cell.* **7,** 631–650.

Conklin, E. G. (1902). Karyokinesis and cytokinesis in the maturation, fertilization and cleavage of Crepidula and other Gasteropoda. I. Karyokinesis. II. Cytokinesis. *J. Acad. Nat. Sci., Philadelphia, ser. 2,* **12,** 1–121.

Crozet, N. (1990). Behavior of the sperm centriole during sheep oocyte fertilization. *Eur. J. Cell Biol.* **53**(2), 326–332.

de Saint Phalle, B., and Sullivan, W. (1998). Spindle assembly and mitosis without centrosomes in parthenogenetic *Sciara* embryos. *J. Cell Biol.* **141,** 1383–1391.

Detraves, C., Mazarguil, H., Lajoie-Mazenc, I., Julian, M., Raynaud-Messina, B., and Wright, M. (1997). Protein complexes containing gamma-tubulin are present in mammalian brain microtubule protein preparations. *Cell Motil. Cytoskeleton* **36**(2), 179–189.

Dictenberg, J. B., Zimmerman, W., Sparks, C. A., Young, A., Vidair, C., Zheng, Y., Carrington, W., Fay, F. S., and Doxsey, S. J. (1998). Pericentrin and gamma-tubulin form a protein complex and are organized into a novel lattice at the centrosome. *J. Cell Biol.* **141**(1), 163–174.

Dirksen, E. R. (1978). Assembly of the microtubular organelles, centrioles and cilia. In *Cell Reproduction.* Academic Press, New York, 711 pp.

do Carmo Avides, M., and Glover, D. M. (1999). Abnormal Spindle Protein, Asp, and the integrity of mitotic centrosomal microtubule organizing centers. *Science* **283,** 1733–1735.

Doxsey, S. J., Stein, P., Evans, L., Calarco, P. D., and Kirschner, M. (1994). Pericentrin, a highly conserved centrosome protein involved in microtubule organization. *Cell* **76,** 639–650.

Duesberg, P. (1999). Are centrosomes or aneuploidy the key to cancer? *Science* **284,** 2091–2092.

Ferreira, A., Palazzo, R. E., and Rebhun, L. I. (1993). Preferential dendritic localization of pericentriolar material in hippocampul pyramidal neurons in culture. *Cell Motil. Cytoskeleton* **25,** 336–344.

Fukasawa, K., Choi, T., Kuriyama, R., Rulong, S., and Vandevoude, G. F. (1996). Abnormal centrosome amplification in the absence of p53. *Science* **271,** 1744–1747.

Gotlieb, A. I., May, L. M., Subrahmanyan, L., and Kalnins, V. I. (1981). Distribution of microtubule organizing centers in migrating sheets of endothelial cells. *J. Cell Biol.* **91,** 589–594.

Gould, R. R., and Borisy, G. G. (1977). The pericentriolar material in Chinese hamster ovary cells nucleates microtubule formation. *J. Cell Biol.* **73,** 601–615.

Graf, R., Euteneuer, U., Ueda, M., and Schliwa, M. (1998). Isolation of nucleation-competent centrosomes from *Dictyostelium discoideum. Eur. J. Cell Biol.* **76,** 167–175.

Heald, R., Tournebize, R., Blank, T., Sandaltzopoulos, R., Becker, P., Hyman, A., and Karsenti, E. (1996). Self-organization of microtubules into bipolar spindles around artificial chromosomes in *Xenopus* egg extracts. *Nature* (*Lond*). **382,** 420–425.

Heald, R., Tournebize, R., Habermann, A., Karsenti, E., and Hyman, A. (1997). Spindle assembly in *Xenopus* egg extracts: Respective roles of centrosomes and microtubule self-organization. *J. Cell Biol.* **138,** 615–628.

Hinchcliffe, E. H., Li, C., Thompson, E. A., Maller, J. L., and Sluder, G. (1999). Cdk2-cyclin E activity is required for repeated centrosome reproduction in *Xenopus* egg extracts. *Science* **283,** 851–854.

Holy, J., and Schatten, G. (1991). Spindle pole centrosomes of sea urchin embryos are partially composed of material recruited from maternal stores. *Dev. Biol.* **147**(2), 343–353.

Horio, T., and Oakley, B. R. (1994). Human γ-tubulin functions in fission yeast. *J. Cell Biol.* **126,** 1465–1473.

Joshi, H. C., Palacios, M. J., McNamara, L., and Cleveland, D. W. (1992). γ-Tubulin is a centrosomal protein required for cell cycle-dependent microtubule nucleation. *Nature* **356,** 80–83.

Kalnins, V. I. (1992). *The Centrosome.* Academic Press, San Diego.

Kellogg, D. R., Moritz, M., and Alberts, B. M. (1994). The centrosome and cellular organization. *Annu. Rev. Biochem.* **63,** 639–674.

Keryer, G., Rios., R. M., Landmark, B. F., Skalhegg, B., Lohmann, S. M., and Bornens, M. (1993). A high-affinity binding protein for the regulatory subunit of cAMP-dependent protein kinase II in the centrosome of human cells. *Exp. Cell Res.* **204,** 230–240.

Khodjakov, A., and Rieder, C. L. (1999). The sudden recruitment of gamma-tubulin to the centrosome at the onset of mitosis and its dynamic exchange throughout the cell cycle, do not require microtubules. *J. Cell Biol.* **146,** 585–596.

Kilmartin, J. V., and Goh, P. Y. (1996). Spc110p: assembly properties and role in the connection of nuclear microtubules to the yeast spindle pole body. *EMBO J.* **15,** 4592–4602.

Kilmartin, J. V., Dyos, S. L., Kershaw, D., and Finch, J. T. (1993). A spacer protein in the *Saccharomyces cerevisiae* spindle pole body whose transcript is cell cycle-regulated. *J. Cell Biol.* **123,** 1175–1184.

Kirschner, M., and Mitchison, T. (1986). Beyond self-assembly: from microtubules to morphogenesis. *Cell* **45,** 329–342.

Klotz, C., Dabauvalle, M.-C., Paintrand, M., Weber, T., Bornens, M., and Karsenti, E. (1990). Parthenogenesis in Xenopus eggs requires centrosomal integrity. *J. Cell Biol.* **110,** 405–415.

Knop, M., and Schiebel, E. (1997). Spc98p and Spc97p of the yeast γ-tubulin complex mediate binding to the spindle pole body via their interaction with Spc110p. *EMBO J.* **16,** 6985–6995.

Knop, M., Pereira, G., Geissler, S., Grein, K., and Schiebel, E. (1997). The spindle pole body component Spc97p interacts with the γ-tubulin of *Saccharomyces cerevisiae* and functions in microtubule organization and spindle pole body duplication. *EMBO J.* **16,** 1550–1564.

Komesli, S., Tournier, F., Paintrand, M., Margolis, R. L., Job, D., and Bornens, M. (1989). Mass isolation of calf thymus centrosomes: Identification of a specific configuration. *J. Cell Biol.* **109,** 2869–2878.

Kuriyama, R., and Borisy, G. G. (1981a). Centriole cycle in chinese hamster ovary cells as determined by whole-mount electron microscopy. *J. Cell Biol.* **91,** 814–821.

Kuriyama, R., and Borisy, G. G. (1981b). Microtubule-nucleating activity of centrosomes in chinese hamster ovary cells is independent of the centriole cycle but coupled to the mitotic cycle. *J. Cell Biol.* **91,** 822–826.

Kuriyama, R., Borisy, G. G., and Masui, Y. (1986). Microtubule cycle in oocytes of the surf clam, *Spisula solidissima:* An immunofluorescence study. *Develop. Biol.* **114,** 151–160.

Lacey, K. R., Jackson, P. K., and Stearns, T. (1999). Cyclin-dependent kinase control of centrosome duplication. *Proc. Natl. Acad. Sci. USA* **96,** 2817–2822.

Lange, B. M. H., and Gull, K. (1995). A molecular marker for centriole maturation in the mammalian cell cycle. *J. Cell Biol.* **130,** 919–927.

Lee, C., Ferguson, M., and Chen, L. B. (1989). Construction of the endoplasmic reticulum. *J. Cell Biol.* **109,** 2045–2055.

Le Guen, P., and Crozet, N. (1989). Microtubule and centrosome distribution during sheep fertilization. *Eur. J. Cell Biol.* **48**(2), 239–249.

Lingle, W. L., Lutz, W. H., Ingle, J. N., Maihle, N. J., and Salisbury, J. L. (1998). Centrosome hypertrophy in human breast tumors: implications for genomic stability and cell polarity. *Proc. Natl. Acad. Sci. USA* **95,** 2950–2955.

Longo, F. J., and Anderson, E. (1970). An ultrastructural analysis of fertilization in the surf clam, *Spisula solidissima.* II. Development of the male pronucleus and the association of the maternally and paternally derived chromosomes. *J. Ultrastruct. Res.* **33,** 495–514.

Luykx, P. (1991). Behavior of egg and sperm centrioles in fertilized eggs of *Urechis caupo. Cytobios* **66,** 7–19.

Maro, B., Houliston, E., and Paintrand, M. (1988). Purification of meiotic spindles and cytoplasmic asters from mouse oocytes. *Dev. Biol.* **129**(2), 275–282.

Martin, O. C., Gunawardane, R. N., Iwamatsu, A., and Zheng, Y. (1998). Xgrip 109: A γ tubulin-associated protein with an essential role in γ tubulin ring complex (γTuRC) assembly and centrosome function. *J. Cell Biol.* **141,** 675–687.

Mazia, D. (1984). Centrosomes and mitotic poles. *Exp. Cell Res.* **153**(1), 1–15.

Mazia, D. (1987). The chromosome cycle and the centrosome cycle in the mitotic cycle. *Int. Rev. Cytol.* **100,** 49–92.

McIntosh, J. R. (1983). The centrosome as organizer of the cytoskeleton. *Mod. Cell Biol.* **2,** 115–142.

McIntosh, J. R. (1987). In *Biomechanics of Cell Division* (N. Akkis, Ed.). Plenum Press, p. 123.

McKim, K. S., and Hawley, R. S. (1995). Chromosomal control of meiotic cell division. *Science* **270,** 1595–1601.

McNiven, M. A., Wang, M., and Porter, K. R. (1984). Microtubule polarity and the direction of pigment transport reverse simultaneously in surgically severed melanophore arms. *Cell* **37,** 753–765.

Mitchison, T. J. (1989). Mitosis; basic concepts. *Curr. Opin. Cell Biol.* **1,** 67–74.

Mitchison, T. J., and Kirschner, M. (1984). Microtubule assembly nucleated by isolated centrosomes. *Nature (Lond.)* **312,** 232–237.

Mitchison, T. J., and Kirschner, M. (1986). Isolation of mammalian centrosomes. *Methods Enzymol.* **134,** 261–168.

Moritz, M., Braunfeld, M. B., Fung, J. C., Sedat, J. W., Alberts, B. M., and Agard, D. A. (1995a). Three-dimensional structural characterization of centrosomes from early *Drosophila* embryos. *J. Cell Biol.* **130,** 1149–1159.

Moritz, M., Braunfeld, M. B., Sedat, J. W., Alberts, B. M., and Agard, D. A. (1995b). Microtubule nucleation by γ-tubulin-containing rings in the centrosome. *Nature* **378,** 638–640.

Moritz, M., Zheng, Y., Alberts, B. M., and Oegema, K. (1998). Recruitment of the γ-tubulin ring complex to *Drosophila* salt-stripped centrosome scaffolds. *J. Cell Biol.* **142,** 775–786.

Moudjou, M., Bordes, N., Paintrand, M., and Bornens, M. (1996). γ-Tubulin in mammalian cells: the centrosomal and cytosolic forms. *J. Cell Sci.* **109,** 875–887.

Murphy, S. M., Urbani, L., and Stearns, T. (1998). The mammalian γ-tubulin complex contains homologues of the yeast spindle pole body components Spc97p and Spc98p. *J. Cell Biol.* **141,** 663–674.

Nakamura, M., Masuda, H., Horii, J., Kuma, Ki., Yokoyama, N., Ohba, T., Nishitani, H., Miyata, T., Tanaka, M., and Nishimoto, T. (1998). When overexpressed, a novel centrosomal protein, RanBPM, causes ectopic microtubule nucleation similar to gamma-tubulin. *J. Cell Biol.* **143,** 1041–1052.

Navara, C. S., Simerly, C., Zoran, S., and Schatten, G. (1995). The sperm centrosome during fertilization in mammals: implications for fertility and reproduction. *Reprod. Fertil. Dev.* **7**(4), 747–754.

Nemere, I., Kupfer, A., and Singer, S. J. (1985). Reorientation of the Golgi apparatus and the microtubule-organizing center inside macrophages subjected to a chemotactic gradient. *Cell Motil.* **5**(1), 17–29.

Nigg, E. A., Schafer, G., Hilz, H., and Eppenberger, H. M. (1985). Cyclic-AMP-dependent protein kinase type II is associated with the Golgi complex and with centrosomes. *Cell* **41**(3), 1039–1051.

Oakley, B. R., Oakley, C. E., Yoon, Y., and Jung, M. K. (1990). γ-Tubulin is a component of the spindle pole body that is essential for microtubule function in *Aspergillus nidulans. Cell* **61,** 1289–1301.

Oegema, K., Wiese, C., Martin, O. C., Milligan, R. A., Iwamatsu, A., Mitchison, T. J., and Zheng, Y. (1999). Characterization of two related *Drosophila* gamma-tubulin complexes that differ in their ability to nucleate microtubules. *J. Cell Biol.* **144,** 721–733.

Ohba, T., Nakamura, M., Nishitani, H., and Nishimoto, T. (1999). Self-organization of microtubule asters induced in *Xenopus* egg extracts by GTP-bound Ran. *Science* **284,** 1356–1358.

Ohta, K., Shiina, N., Okumura, E., Hisanaga, S.-I., Kishimoto, T., Endo, S., Gotoh, Y., Nishida, E., and Sakai, H. (1993). Microtubule nucleating activity of centrosomes in cell-free extracts from *Xenopus* eggs: Involvement of phosphorylation and accumulation of pericentriolar material. *J. Cell Sci.* **104,** 125–137.

Palazzo, R. E., and Vogel, J. V. (1999). Isolation of centrosomes from *Spisula solidissima* oocytes. In *Mitosis and Meiosis: Methods in Cell Biology,* Vol. 61. (C. L. Rieder, Ed.) Academic Press, New York, pp. 35–56.

Palazzo, R. E., Vaisberg, E., Cole, R. W., and Rieder, C. L. (1992). Centriole duplication in lysates of *Spisula solidissima* oocytes. *Science* **256,** 219–221.

Palazzo, R. E., Vaisberg, E. A., Weiss, D. G., Kuznetsov, S. A., and Steffen, W. (1999). Dynein is required for spindle assembly in cytoplasmic extracts of *Spisula solidissima* oocytes. *J. Cell Sci.* **112,** 1291–1302.

Paoletti, A., and Bornens, M. (1997). Organisation and functional regulation of the centrosome in animal cells. *Prog. Cell Cycle Res.* **3,** 285–299.

Pickett-Heaps, J. D. (1969). The evolution of the mitotic apparatus: an attempt at comparative ultrastructural cytology in dividing plant cells. *Cytobios.* **3,** 257–280.

Pihan, G. A., Purohit, A., Wallace, J., Knecht, H., Woda, B., Quesenberry, P., and Doxsey, S. J. (1998). Centrosome defects and genetic instability in malignant tumors. *Can. Res.* **58,** 3974–3985.

Rieder, C. L. (1991). Towards a molecular understanding of chromosome behavior. *Curr. Opin. Cell Biol.* **3,** 59–66.

Rieder, C. L., and Alexander, S. P. (1990). Kinetochores are transported poleward along a single astral microtubule during chromosome attachment to the spindle in newt lung cells. *J. Cell Biol.* **110,** 81–95.

Rieder, C. L., and Borisy, G. G. (1982). The centrosome cycle in PtK_2 cells: asymmetric distribution and structural changes in the pericentriolar material. *Biol. Cell.* **44,** 117–132.

Rodionov, V. I., and Borisy, G. G. (1997). Self-centering activity of cytoplasm. *Nature* (*Lond.*) **386,** 170–173.

Rothwell, W. F., Fogarty, P., Field, C. M., and Sullivan, W. (1998). Nuclear-fallout, a *Drosophila* protein that cycles from the cytoplasm to the centrosomes, regulates cortical microfilament organization. *Development* **125,** 1295–1303.

Rothwell, W. F., Zhang, C. X., Zelano, C., Hsieh, T. S., and Sullivan, W. (1999). The *Drosophila* centrosomal protein Nuf is required for recruiting Dah, a membrane associated protein, to furrows in the early embryo. *J. Cell Sci.* **112,** 2885–2893.

Saunders, R. D., Avides, M. C., Howard, T., Gonzalez, C., and Glover, D. M. (1997). The *Drosophila* gene abnormal spindle encodes a novel microtubule-associated protein that associates with the polar regions of the mitotic spindle. *J. Cell Biol.* **137**(4), 881–890.

Sawin, K. E., and Mitchison, T. J. (1991). Mitotic spindle assembly by two different pathways in vitro. *J. Cell. Biol.* **112,** 925–940.

Schatten, G. (1994). The centrosome and its mode of inheritance: The reduction of the centrosome during gametogenesis and its restoration during fertilization. *Dev. Biol.* **165,** 299–335.

Schatten, G., Simerly, C., and Schatten, H. (1985). Microtubule configurations during fertilization, mitosis, and early development in the mouse and the requirement for egg microtubule-mediated motility during mammalian fertilization. *Proc. Natl. Acad. Sci. USA* **82**(12), 4152–4156.

Schatten, G., Simerly, C., and Schatten, H. (1991). Maternal inheritance of centrosomes in mammals? Studies on parthenogenesis and polyspermy in mice. *Proc. Natl. Acad. Sci. USA* **88**(15), 6785–6789.

Schnackenberg, B. J., and Palazzo, R. E. (1999). Identification and function of the centrosome centromatrix. *Biol. Cell* **91,** 429–438.

Schnackenberg, B. J., Khodjakov, A., Rieder, C. L., and Palazzo, R. E. (1998). The disassembly and reassembly of functional centrosomes in vitro. *Proc. Natl. Acad. Sci. USA* **95,** 9295–9300.

Schnackenberg, B. J., Hull, D. R., Balczon, R. D., and Palazzo, R. E. (2000). Reconstitution of microtubule nucleation potential in centrosomes isolated from *Spisula solidissima* oocytes. *J. Cell Sci.* **113,** 943–953.

Schroer, T. A., and Sheetz, M. P. (1991). Functions of microtubule-based motors. *Annu. Rev. Physiol.* **53,** 629–652.

Sharp, G. A., Osborn, M, and Weber, K. (1981). Ultrastructure of multiple microtubule initiation sites in mouse reuroblastoma cells. *J. Cell Sci.* **47,** 1–24.

Sluder, G, and Rieder, C. L. (1985). Centriole number and the reproductive capacity of spindle poles. *J. Cell Biol.* **100,** 887–896.

Sluder, G., Miller, F. J., Lewis, K., Davison, E. D., and Rieder, C. L. (1989). Centrosome inheritance in starfish zygotes: selective loss of the maternal centrosome after fertilization. *Dev. Biol.* **131**(2), 567–579.

Sluder, G., Miller, F. J., and Lewis, K. (1993). Centrosome inheritance in starfish zygotes. II: selective suppression of the maternal centrosome during meiosis. *Dev. Biol.* **155**(1), 58–67.

Soltys, B. J., and Gupta, R. S. (1992). Interrelationships of endoplasmic reticulum, mitochondria, intermediate filaments, and microtubules—a quadruple fluorescence labeling study. *Biochem. Cell Biol.* **70,** 1174–1186.

Stearns, T., and Kirschner, M. W. (1994). In vitro reconstitution of centrosome assembly and function: The central role of γ-tubulin. *Cell* **76,** 623–637.

Stearns, T., Evans, L., and Kirschner, M. (1991). γ-Tubulin is a highly conserved component of the centrosome. *Cell* **65,** 825–836.

Steffen, W., Fuge, H., Dietz, R., Bastmeyer, M., and Muller, G. (1986). Aster-free spindle poles in insect spermatocytes: Evidence for chromosome-induced spindle formation. *J. Cell Biol.* **102,** 1679–1687.

Stoppin-Mellet, V., Peter, C., Buendia, B., Karsenti, E., and Lambert, A. M. (1999). Tobacco BY-2 cell-free extracts induce the recovery of microtubule nucleating activity of inactivated mammalian centrosomes. *Biochim. Biophys. Acta.* **1449**(1), 101–106.

Sunkel, C. E, Gomes, R., Sampaio, P., Perdigao, J., and Gonzalez, C. (1995). γ-Tubulin is required for the structure and function of the microtubule organizing centre in *Drosophila* neuroblasts. *EMBO J.* **14,** 28–36.

Takahashi, M., Shibata, H., Shimakawa, M., Miyamoto, M., Mukai, H., and Ono, Y. (1999). Characterization of a novel giant scaffolding protein, CG-NAP, that anchors multiple signaling enzymes to centrosome and the golgi apparatus. *J. Biol. Chem.* **274,** 17267–17274.

Telzer, B. R., and Rosenbaum, J. L. (1979). Cell cycle-dependent, in vitro assembly of microtubules onto pericentriolar material of HeLa cells. *J. Cell Biol.* **81**(3), 484–497.

Theurkauf, W. E., and Hawley, R. S. (1992). Meiotic spindle assembly in *Drosophila* females: behavior of nonexchange chromosomes and the effects of mutations in the nod kinesin-like protein. *J. Cell Biol.* **116,** 1167–1180.

Thompson-Coffe, C., Coffe, G., Schatten, H., Mazia, D., and Schatten, G. (1996). Cold-treated centrosome: Isolation of centrosomes from mitotic sea urchin eggs, production of an anticentrosomal antibody, and novel ultrastructural imaging. *Cell Motil. Cytoskeleton* **33,** 197–207.

Tilney, L. G., and Goddard, J. (1970). Nucleated sites for the assembly of cytoplasmic microtubules in the ectodermal cells of blastulae of *Arbacia punctulata*. *J. Cell Biol.* **46**(3), 564–575.

Verde, F., Berrez, J. M., Antony, C., and Karsenti, E. (1991). Taxol-induced microtubule asters in mitotic extracts of *Xenopus* eggs: requirement for phosphorylated factors and cytoplasmic dynein. *J. Cell Biol.* **112,** 1177–1187.

Vogel, J. M., Stearns, T., Rieder, C. L., and Palazzo, R. E. (1997). Centrosomes isolated from *Spisula solidissima* oocytes contain rings and an unusual stoichiometric ratio of α/β tubulin. *J. Cell Biol.* **137,** 193–202.

Weisenberg, R. C., and Rosenfeld, A. C. (1975). In vitro polymerization of microtubules into asters and spindles in homogenates of surf clam eggs. *J. Cell Biol.* **64**(1), 146–158.

Wheatley, D. N. (1982). *The Centriole: A Central Enigma of Cell Biology.* Elsevier, New York, 232 pp.

Wiese, C., and Zheng, Y. (1999). γ-Tubulin complexes and their interaction with microtubule-organizing centers. *Curr. Opin. Struct. Biol.* **9,** 250–259.

Wigge, P. A., Jensen, O. N., Holmes, S., Soues, S., Mann, M., and Kilmartin, J. V. (1998). Analysis of the *Saccharomyces* spindle pole by matrix-assisted laser desorption/ionization (MALDI) mass spectrometry. *J. Cell Biol.* **141,** 967–977.

Wilde, A., and Zheng, Y. (1999). Stimulation of microtubule aster formation and spindle assembly by the small GTPase Ran. *Science* **284**(5418), 1359–1362.

Wilson, E. B. (1925). *The Cell in Development and Heredity.* Macmillan, New York, 1232 pp.

Wu, X., and Palazzo, R. E. (1999). Differential regulation of maternal vs paternal centrosomes. *Proc. Natl. Acad. Sci. USA* **96,** 1397–1402.

Zhang, D., and Nicklas, R. B. (1995). The impact of chromosomes and centrosomes on spindle assembly as observed in living cells. *J. Cell. Biol.* **129,** 1287–1300.

Zheng, Y., Jung, M. K., and Oakley, B. R. (1991). γ-Tubulin is present in *Drosophila melanogaster* and *Homo sapiens* and is associated with the centrosome. *Cell* **65,** 817–823.

Zheng, Y., Wong, M. L., Alberts, B. M., and Mitchison, T. (1995). Nucleation of microtubule assembly by a γ-tubulin-containing ring complex. *Nature* **378,** 578–583.

Index

C

N

O

S

V

Y

Contents of Previous Volumes

Volume 42

Volume 43

Volume 44

Volume 45

Volume 46

Volume 47

Volume 48